Metabolic Engineering in Plants

Tariq Aftab • Khalid Rehman Hakeem
Editors

Metabolic Engineering in Plants

Springer

Editors
Tariq Aftab
Department of Botany
Aligarh Muslim University
Aligarh, Uttar Pradesh, India

Khalid Rehman Hakeem
Department of Biological Sciences
King Abdulaziz University, Faculty of Science
Jeddah, Saudi Arabia

ISBN 978-981-16-7261-3 ISBN 978-981-16-7262-0 (eBook)
https://doi.org/10.1007/978-981-16-7262-0

This Springer imprint is published by the registered company Springer Nature Singapore Pte Ltd.
The registered company address is: 152 Beach Road, #21-01/04 Gateway East, Singapore 189721, Singapore

Preface

Metabolic engineering is usually defined as the re-routing of one or more enzymatic reactions to generate new compounds, increase the production of existing compounds, or facilitate the degradation of compounds. Metabolic engineering is chiefly focused on metabolic intermediates or products, such as starch, vitamin E, amino acids, and enzymes. Metabolic engineering is different from genetic engineering because it investigates the properties of integrated metabolic pathways and genetic regulatory networks, not with individual genes and enzymes. Plants are the foundation of numerous compounds which are synthesized via assimilated complex biosynthetic routes. Plant metabolites are organic compounds synthesized via enzyme-mediated processes. These plant metabolites are important for both essential (proteins, lipids, and carbohydrates) as well as specific and specialized functions (anthocyanin, carotenoid, etc.). The primary metabolites are essential for the survival of the organisms and any alterations in their levels are likely to have severe major manifestations. However, secondary metabolites are considered to be the metabolites having specialized functions and playing a role in providing quality of life to the producer organism. Plants have evolved an incredible arrangement of metabolic pathways leading to molecules/compounds capable of responding promptly and effectively to stress situations imposed by biotic and abiotic factors, some of which supply the ever-growing needs of humankind for natural chemicals, such as pharmaceuticals, nutraceuticals, agrochemicals, food and chemical additives, biofuels, and biomass. Humans rely on plants for various needs since ancient time, and their population still seems enough for fulfilling our demands. But, in foreseeable future we will be forced to think about the accessibility of resources for the generations to come. For these reasons, we must look for alternative options of resources about food/food supplement, medicines, and other essential items. For achieving this task, we are choosing plant metabolic engineering through which we can generate higher amount of compounds/products in less time and space. The aim of metabolic engineering is enhancement of desired product through upregulation / downregulation of protein expression. With an estimated 100,000 unique compounds produced in the plant kingdom, elucidating these metabolic networks is an exciting endeavour. By using plant metabolic engineering approach, tremendous changes at metabolic level in different plants have been done.

The book comprises 19 chapters, review articles written by experts, highlighting a wide range of topics such as whole-plant and cell/organ culture systems, and environmental and genetic transformation-based modulation of biochemical pathways. Special focus is given to microRNA, RNAi, and CRISPR-Cas-mediated approaches for metabolic engineering applications using both model and non-model plant species. We are hopeful that this volume would furnish the need of all researchers who are working or have interest in this particular field.

We are highly grateful to all our contributors, for accepting our invitation, and for not only sharing their knowledge and research, but also for venerably integrating their expertise in dispersed information from diverse fields in composing the chapters and enduring editorial suggestions to finally produce this venture. We also thank Springer Nature team for their generous cooperation at every stage of the book production.

Lastly, thanks are also due to well-wishers, research students, and editors' family members for their moral support, blessings, and inspiration in the compilation of this book.

Aligarh, India — Tariq Aftab
Jeddah, Saudi Arabia — Khalid Rehman Hakeem

Contents

About the Editors

Tariq Aftab received his Ph.D. in the Department of Botany at Aligarh Muslim University, India, and is currently an Assistant Professor there. He is the recipient of a prestigious Leibniz-DAAD fellowship from Germany, Raman Fellowship from the Government of India, and Young Scientist Awards from the State Government of Uttar Pradesh (India) and Government of India. After completing his doctorate, he has worked as Research Fellow at National Bureau of Plant Genetic Resources, New Delhi, and as Post-doctorate Fellow at Jamia Hamdard, New Delhi, India. Dr. Aftab also worked as Visiting Scientist at the Leibniz Institute of Plant Genetics and Crop Plant Research (IPK), Gatersleben, Germany, and in the Department of Plant Biology, Michigan State University, USA. He is a member of various scientific associations from India and abroad.

He has edited 15 books with international publishers, including Elsevier Inc., Springer Nature and CRC Press (Taylor & Francis Group), co-authored several book chapters, and published over 75 research/review papers in peer-reviewed international journals. His research interests include physiological, proteomic, and molecular studies on medicinal and crop plants.

Khalid Rehman Hakeem Ph.D. is Professor at King Abdulaziz University, Jeddah, Saudi Arabia. After completing his doctorate (Botany; specialization in Plant Eco-physiology and Molecular Biology) from Jamia Hamdard, New Delhi, India, in 2011, he worked as a lecturer at the University of Kashmir, Srinagar, for a short period. Later, he joined Universiti Putra Malaysia, Selangor, Malaysia, and worked there as Post-doctorate Fellow in 2012 and Fellow Researcher (Associate Prof.) from 2013 to 2016. Dr. Hakeem has more than 10 years of teaching and research experience in plant eco-physiology, biotechnology and molecular biology, medicinal plant research, plant–microbe–soil interactions as well as in environmental studies. He is the recipient of several fellowships at both national and international levels; also, he has served as the visiting scientist at Jinan University, Guangzhou, China. Currently, he is involved with a number of international research projects with different government organizations.

So far, Dr. Hakeem has authored and edited more than 70 books with international publishers, including Springer Nature, Academic Press (Elsevier), and CRC Press. He also has to his credit more than 140 research publications in peer-reviewed international journals and 60 book chapters in edited volumes with international publishers.

At present, Dr. Hakeem serves as an editorial board member and reviewer of several high-impact international scientific journals from Elsevier, Springer Nature, Taylor & Francis, Cambridge, and John Wiley Publishers. He is included in the advisory board of Cambridge Scholars Publishing, UK. He is also a fellow of Plantae group of the American Society of Plant Biologists, member of the World Academy of Sciences, member of the International Society for Development and Sustainability, Japan, and member of Asian Federation of Biotechnology, Korea. Dr. Hakeem has been listed in Marquis Who's Who in the World since 2014–2019. Currently, Dr. Hakeem is engaged in studying the plant processes at eco-physiological as well as molecular levels.

Chapter 1
Metabolic Engineering: New Approaches in Pharmaceutical Production

Ahmed H. El-Desoky, Mohamed A. M. Atia, and Elsayed A. Omer

Abstract Natural products, natural products-derived, and natural products-inspired compounds represent a major sector of drugs and drug leads. Nature has the potential to produce compounds built to interact with biological systems. Mostly are produced as secondary metabolites by a living organism for different purposes such as defense and communication. Natural compounds usually contain multiple chiral centers within strictly specific molecular architecture that are very difficult to be obtained by combinatorial synthesis for economic supply of drugs. Meanwhile, natural compounds serving as drugs or precursors for synthesis are having several drawbacks with insufficient and fluctuating supply as the most serious problems. For these reasons metabolic engineering exploiting advanced biotechnology techniques such as sequencing, recombinant DNA, and protein engineering can add new incentives to produce secondary metabolites. Metabolic engineering is based on manipulating metabolic networks/pathways within prokaryotic organisms or cultured eukaryotes. Both organisms have common features, including a cytoplasmic membrane, DNA that codes for genetic information, and RNA that is subsequently translated into proteins. Successful study of the biosynthetic pathways responsible for the production of natural products together with advances in fermentation technologies and expression systems, variation in tool kits, and mutation tools allowed metabolic engineering to be a promising approach for the targeted production of a wide variety of natural compounds (taxol, camptothecin, vincristine, artemisinin, silymarin) with bacterial, fungal, and plant origins.

Keywords Metabolic engineering · Plant · Pathway · Secondary metabolites · Taxol · Camptothecin · Vincristine · Artemisinin · Silymarin

---

A. H. El-Desoky
Pharmacognosy Department, National Research Centre, Giza, Egypt

M. A. M. Atia
Genome Mapping Department, Agricultural Genetic Engineering Research Institute (AGERI), Agricultural Research Center (ARC), Giza, Egypt

E. A. Omer (✉)
Medicinal and Aromatic Plants Research Department, National Research Centre, Giza, Egypt

T. Aftab, K. R. Hakeem (eds.), *Metabolic Engineering in Plants*,
https://doi.org/10.1007/978-981-16-7262-0_1

Introduction

Metabolic engineering (ME) is defined as the directed development and improvement for the production of biochemical, biofuels, pharmaceuticals, and medicinal products by modifying and bioengineering a particular metabolic pathway(s). This method includes functional alteration of metabolic pathways toward better understanding and utilizing cellular/biological pathways. In addition, metabolic engineering is driven to advance different commercial applications by which we can enhance the strain's efficiency for producing valuable metabolites. The standard metabolic engineering method requires controlling the expression of a specific protein(s) either in overexpression or downregulation manner within a metabolic pathway, which promotes the biological cell to produce a new product (Yadav et al., 2018).

Directionality is the core of a successful metabolic engineering system, but scaling, stability, and productivity are essential factors. Unfortunately, many trials have failed during the last decade due to problems confronted in scaling up or other critical factors. Notably, meanwhile, the plant production systems are generally slower than microbial systems; plants are more suitable for commercial production due to their autotrophic and scale-up features, in addition to their low cost, especially when agricultural infrastructure is available. However, for plant production systems to be effective, using either fermentation or agriculture, stability and high yields of the bioengineered metabolites must be ensured (Nielsen, 2001).

Since Bailey's formal recognition of metabolic engineering in 1991, it demonstrated its powerful capability as a modern approach in engineering numerous microbial strains to produce hundreds of chemicals and bioproducts from raw substances (Nielsen, 2001; Bailey, 1991; Birchfield & McIntosh, 2020).This production is classically achieved by introducing targeted genetic changes through recombinant DNA technologies and analyzing all possible consequences of these changes at the cellular/biological levels. However, the number of bioproducts touching industrial-scale production is limited due to the high production costs, including raw materials, fermentation process, and purification. Thus, although immense efforts have been paid in the metabolic engineering field, it is still needed to enhance strain performance in order to create competitive bioprocesses compared to traditional chemical processes. Also, ME projects at their beginning stages frequently fail due to many critical practical issues occurring during industrial scaling production (Choi et al., 2019). Moreover, inaccurate trial-and-error cycles to enhance strain/plant performance usually resulted from a poor understanding of the metabolic and/or genetic mechanisms controlling a particular pathway in the host organisms, especially in the laboratory (Courdavault et al., 2021).

In a global sense, metabolic engineering is not considerably different from genetic or cellular engineering since they all intended to manipulate genes to generate a product of interest. However, metabolic engineering is focused explicitly on understanding the complex metabolic pathways/networks; meanwhile, genetic and cellular engineering focuses on particular enzymatic reactions inside the cell. Thus, a systemic approach is required for the straightforward and precise design of complex pathways through remodeling the optimal cell pathways, which can divert cell supplies to achieving more beneficial fitness (Choi et al., 2019; Courdavault et al., 2021; Liu & Nielsen, 2019). Accordingly, metabolic engineering represents a robust framework for dissecting differential gene expression at a genome-wide level, protein production, and the levels of a wide range of intracellular metabolic fluxes. Therefore, in order to increase the metabolic engineering approach productivity of certain products, many questions need to be carefully answered: (1) which host/organism must be chosen, (2) what commodity could be produced, and (3) which ME approach could be applied (Liu & Nielsen, 2019). Therefore, this chapter highlighted the different biological systems that can be used in the ME, examples of current commodities produced, and lastly, the various engineering strategies that can be employed to engineer and improve cellular production capabilities of specific metabolites that are in clinical use and are facing difficulties in commercial production hindering stable market supplies.

Strategies

Several approaches are adopted for secondary metabolites development or enhancement using bacteria, such as (1) overexpressing of a gene family or set of transcription factors to simultaneously stimulate/initiate diverse pathway-related endogenous genes to improve the synthesis of particular metabolites, (2) engineering of regulatory networks controlling a particular pathway through delicate tuning of many enzymes within the target pathway to assure an adequate accumulation of precursors and consequently boost metabolic flux through the targeted pathway, (3) gene deletion (knocking out) or/and insertion of essential enzyme genes regulating or controlling the pathway of the product of interest, (4) stimulation using specific substrates/elicitors, and (5) combining one or more of the above approaches to maximize the synthesis of the target metabolites. Recently, gene editing via CRISPR/Cas9-based activation and/or repression schemes can also be used in metabolic engineering development and improvement. Therefore, metabolic engineering represents potential cell factories to produce pharmaceuticals, amino acids, bioplastics, biofuels, new bioproducts, silk proteins, etc. (Liu & Nielsen, 2019; Zhu et al., 2020).

Many designed tools were implemented in metabolic engineering during the last decade for more precise pathway construction and optimized pathway development. These include selecting hosts with well-known and knowledgeable omics databases, computational pathway simulations resources/tools paired with custom enzyme designs, and guided evolution approaches that attempt to maximize the output of the desired product (Zhu et al., 2020). In addition, advancements in microbial engineering have enabled a wide range of plant metabolites to be manufactured (Fu et al., 2018).

Host Selection

Metabolic engineering is based on manipulating metabolic networks/pathways within prokaryotic organisms or cultured eukaryotes. Both organisms have common features, including a cytoplasmic membrane, DNA that codes for genetic information, and RNA that is subsequently translated into proteins. However, the distinguishing factors between the two kinds of organisms are revealed intracellular when comparing their genomes or organelles. For example, prokaryotes, including the Bacteria kingdom, include mostly stable chromosomal DNA with one inherent single loop and lacks membrane-bound organelles. In contrast, eukaryotes, such as Fungi, Plantae, and Animalia kingdoms, comprise DNA molecules organized in tightly bound chromosomes and contain membrane-bound organelles (e.g., chloroplasts, mitochondria, and the Golgi apparatus). The bacteria, fungi, plant, and animal cells are currently used in different metabolic engineering applications to manufacture products, and accurate analysis is required to decide the most fitting host for a particular target (Nabavi et al., 2020).

The selection of an organism that will be used in the production for an industrial biotechnology process is mainly driven by its potential to manufacture the product of interest efficiently. Genetically engineered cells are usually used to make two main kinds of products, proteins, and nonproteins (Zhu et al., 2020). Protein products comprise those used for human therapeutics, food processing, and industrial catalysts, and are generated by introducing their encoding genes into the host genome. Besides, nonprotein products, including metabolites such as amino acids, biofuels, antioxidants, and vitamins, are created by inserting genes coding for necessary enzymes that manipulate precursors into desired metabolites. Cultures of complex eukaryotes (yeast, plant cells, mammalian cells) are used to produce proteins, whereas prokaryotes and cultures of simple eukaryotes (yeast, filamentous fungi, and plants) are used more frequently in the production of nonprotein metabolites (Nabavi et al., 2020).

For long decades, bacteria, fungi, and plants have been acknowledged as excellent sources of natural compounds, especially those with multiple bioactive attributes and associated applications in the domains of nutrition and human health. Among the common valuable plant natural products (PNPs), scientists have identified those usually used to treat different pathologies such as cancers, infectious diseases, or cardiovascular diseases. Additionally, selected PNPs have been further processed into various active pharmaceutical constituents at an industrial scale. Examples of the most famous PNPs, and their associated applications, include tropane alkaloids from Solanaceae (mostly anticholinergics), wormwood sesquiterpene lactones (antimalarial drugs), monoterpene indole alkaloids of Apocynaceae (antihypertensive and anticancer drugs), poppy isoquinoline alkaloids (antitussives, antimicrobials, analgesics, vasodilators), yew taxane-type terpenoids (anticancer drugs), and mayapple lignans (anticancer and immunosuppressive drugs) (García-Granados et al., 2019).In the following pages, we will throw the light on the engineering of some vital metabolites that are used in the treatment of some vital diseases such as cancer, malaria, and hepatitis. The drug markets need these metabolites, and the demand for them is increasing day by day. These metabolites as arranged in the text are taxol, camptothecin, vincristine, artemisinin, and silymarin.

Examples of Engineered Metabolites

Taxol (TXL)

Paclitaxel, commercially known as taxol, is a terpenoidal pseudo-alkaloid that has potent antitumor activity (Sabzehzari & Naghavi, 2019). Paclitaxel is one of the strong plant-derived anticancer drugs that was first isolated from the park of Pacific yew. It was first isolated from the bark of the slow-growing coniferous tree *Taxus parvifolia* known as Pacific yew or western yew (Mansukhlal et al., 1971) in a very low yield 0.01–0.05 % (Cragg, 1988). Being only isolated from the bark, it is very hard to meet the increasing clinical consumption of taxol. One tree yields 18 kg of green bark, which becomes about 9 kg after drying (Connolly, 1988). 27,700 kg of dried bark yield about 4 kg of taxol (Cragg, 1988). Despite its clinical success, limited supply of paclitaxel for clinical trials is a very serious drawback. Meanwhile, combinatorial synthesis is very hard due to multiple chirality. Meanwhile, consumption of taxol is expected to show a yearly growth of 8.2% between 2020 and 2025 (Ning et al., 2020).

Biosynthetic Background

Biosynthesis of paclitaxel is divided into three stages (Fig. 1.1). The first is the MVA/MEA pathway-dependent synthesis of isopentenyl pyrophosphate (IPP) as terpene precursor which is the rate-limiting step (Soliman & Tang, 2015), the second is the synthesis of baccatin III as taxol's main carbon framework, and the third is the synthesis of nitrogenous side chain originated from the amino acid phenylalanine (Sabzehzari & Naghavi, 2019; Croteau et al., 2006).

Conventional Strategies to Increase TXL Yields

Due to its enormously increasing clinical importance, many attempts have been made to increase the production of paclitaxel including chemical synthesis that mainly depended on the exploitation of commercially abundant biosynthetic precursors as starting materials (Denis et al., 1988; Holton et al., 1994; Borah et al., 2007), searching for microbial sources that can be bioengineered to increase the production such as *Taxomyces andreanae* isolated from *Taxus chinensis* (Stierle

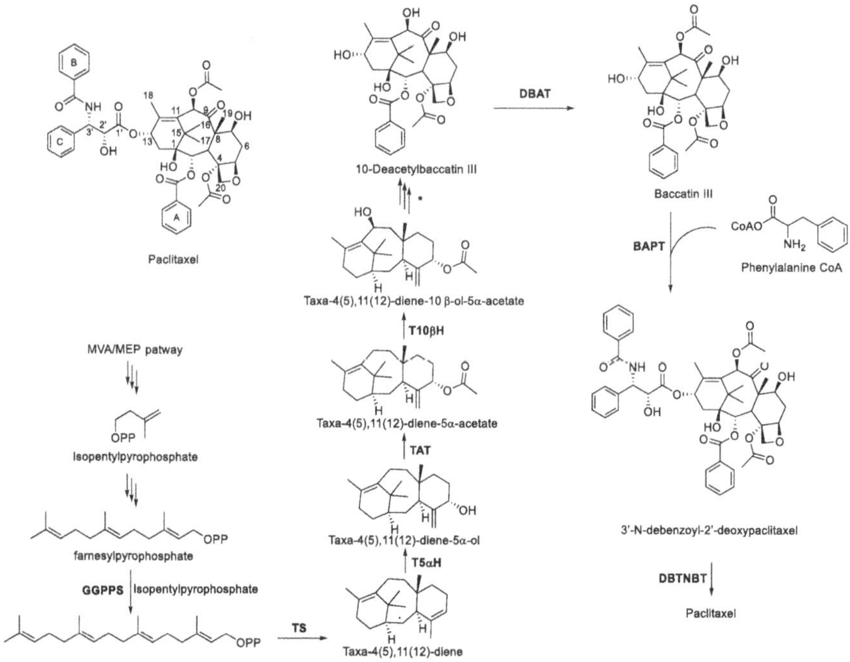

Fig. 1.1 Biosynthetic pathway for taxol [abbreviations: *MVA* mevalonic acid, *MEP* 2-C-methyl-D-erythritol-4-phosphate, *GGPPS* geranylgeranyl pyrophosphate synthase, *TS* taxa-4(5),11(12)-diene synthase (committed step), *T5αH* taxa-4(5),11(12)-diene-5α-hydroxylase, *TAT* taxa-4(5),11(12)-diene-5α-ol-O-acetyltransferase, *T10βH* taxane-10β-hydroxylase; *oxetane ring formation and branch migration enzymes taxane 2α-O-benzoyltransferase (*T2BT* or *DBBT* debenzoyltaxane-2/-α-O-benzoyltransferase) and C-13 hydroxylation; *DBAT* 10-deacetylbaccatin III-O-acetyltransferase, *BAPT* baccatin III 13-O-(3-amino-3-phenylpropanoyl)transferase, *DBTNBT* 3/-N-debenzoyl-2/-deoxytaxol-N-benzoyltransferase, after side chain hydroxylation by unknown enzyme; multiple arrows mean more than one step]

et al., 1993), and finding other plant species that can produce paclitaxel such as *T. chinensis*, *T. wallichiana*, *T. brevifolia*, *T. canadensis*, *T. floridana*, *T. cuspidata*, *T. baccata*, and *T. globosa* (Liu et al., 2016).

Biotechnological Strategies

On the other hand, biotechnology and metabolic engineering were extensively used to increase the production of paclitaxel.

Cell Culture

The first report of *Taxus* cell culture was in a patent application and an abstract (Christen et al., 1989, 1991). Back in 1992, Fett-Neto and coworkers managed to obtain an in vitro culture of the taxol-producing plant *T. cuspidata* (Fett-Neto et al., 1992). With the development of tissue culture technologies via optimization of culture conditions such as media composition procedure, specific metals, light, pH, temperature, osmotic pressure, gas contents, and elicitors exploitation (Howat et al., 2014), higher amounts of taxol could be produced from a variety of species such as *T. cuspidate*, *T. canadensis*, *T. baccata*, *T. globosa*, *T. yunnanensis*, and *T. chinensis*, with taxol production improved from ~1–3 mg/L to ~77.5–153 mg/L (Liu et al., 2016; Onrubia et al., 2013; Bringi et al., 1995).

Heterologous Expression Systems

Another biotechnological approach for steady and enhanced production of taxol is the use of heterologous expression systems. In these systems a part of genome encoding the biosynthetic pathway of taxol is transferred to another type of cells through a suitable transformation vector. New cells will start the expression of enzymes necessary for taxol production, whose speed and yield could be engineered by controlling the fermentation conditions. Some of the successful heterologous expression systems are *Saccharomyces cerevisiae* yeast model (Dejong et al., 2006), *Escherichia coli* bacteria model (Ajikumar et al., 2010), and *Arabidopsis thaliana* plant model (Besumbes et al., 2004). The highest taxol content from *E. coli* was 570 mg/L (Biggs et al., 2016). Taxadiene is the first diterpene intermediate involved in taxol biosynthesis through baccatin III. Its production exhibited 40-fold increase in *S. cerevisiae* via retarding the competition between steroid and taxol biosynthesis in favor to taxol. That was done by the expression of GGPPS, tHMGR1, and flux control by transcription factor UPC2-1 (Engels et al., 2008). Moreover, acetyl-CoA accumulation in *S. cerevisiae* via acetaldehyde dehydrogenase and acetyl-CoA synthetize overexpression was found to increase the terpenoid content over steroid content in transgenic yeasts which can help in increasing taxol production (Shiba et al., 2007). Recently, Abdallah and coworkers managed to engineer the first *B. subtilis* strain that can produce taxadiene. They expressed the plant-derived enzyme taxadiene synthetize and overexpressed the eight key biosynthetic enzymes of MEV pathway to enhance geranylgeranyl pyrophosphate (GGPP) flux that led to a ~82-fold increase in taxadiene content when compared to wild-type *B. subtilis*. Taxadiene content reached ~17.7 mg/L (Abdallah et al., 2019). In the transgenic plant heterologous expression model *Arabidopsis thaliana*, Besumbes and coworkers managed to produce taxadiene in a yield of 600 ng/g of dry weight by overexpressing TASY and GGPP synthase. However, growth retardation resulted from taxadiene accumulation (Besumbes et al., 2004). On the other hand, TASY overexpression in another model, *Physcomitrella patens* resulted in taxadiene production in a yield of 0.05% from fresh weight without growth retardation (Anterola et al., 2009).

IPP and GGPP are very important targets in taxol biosynthesis since they are the rate-limiting intermediates for this process. While trying to enhance IPP-mediated terpenoids biosynthesis, isopentenol utilization pathway (IUP) was transformed to *E. coli*, resulting in continuous phosphorylation of isopentenol (prenol or isoprenol) isomers and increased IPP production (Chatzivasileiou et al., 2019). IUP pathway optimization only needs the cofactor adenosine triphosphate, and it occurs within two reactions. Thus, IUP pathway is far much simpler than MEP or MVA pathways (Chatzivasileiou et al., 2019).

Overexpression Positive Effectors

Overexpression positive effectors can also play an important role in taxol biosynthesis. In 2017 Yu *et al.* suggested that manipulating the expression of some candidate genes in taxol biosynthetic pathway can affect taxol contents. Overexpression of the hydroxylation and acetylation enzymes (i.e., 2-debenzoyl-7, 13-diacetylbaccatin III-2-O-benzoyl-transferase (DBBT), 5-alpha-taxadienol-10-beta-hydroxylase (T10OH), taxadiene 5-alpha-hydroxylase (T5H), and taxadienol acetyltransferase (TAT)) increased taxol content. Later on, Wang and coworkers confirmed this approach when they revealed a ~ threefold increase in taxol content in *T. wallichiana* with overexpression of the genes WRKY, bHLH, MYB, and ERF (Wang et al., 2019). Taxadiene synthase (TS) is a pivotal enzyme in taxol biosynthesis (Köksal et al., 2011).

Knock-Outing Negative Effectors

Knock-outing negative effectors together with overexpression of positive effector genes can result in dramatic increase in taxol levels via blocking the competing downstream metabolic pathways. Sabzehzari and Naghavi (2018) used CRISPR/Cas9 and miRNA technology to achieve this strategy, in order to silence the genes involved in sterol biosynthesis that encode lanosterol/squalene synthase (Do et al., 2009) competing with terpene pathway for farnesyl pyrophosphate (FPP) (Sabzehzari & Naghavi, 2018).

Moreover, miR171 and miR164 were found to decrease paclitaxel biosynthesis in *T. baccata* (Hao et al., 2012). MiR171 and miR164 inhibit taxane 2α-obenzoyl-transferase and taxane 13α-hydroxylase, which are very important for the biosynthesis path of taxol (Ramírez-Estrada et al., 2016). Therefore, silencing MiR171 and miR164 using CRISPR/Cas9 or miRNA can promote paclitaxel production.

Camptothecin (CPT)

Camptothecin (CPT) is a modified monoterpene indole alkaloid (Fig. 1.2) possessing potent antitumor activities. FDA approved some camptothecin (CPT) derivatives, such as topotecan and irinotecan, which were developed to overcome poor water solubility and severe side effect, for clinical treatment of some human tumors such as ovarian cancer, colorectal cancer, and leukemia via binding DNA topoisomerase I and DNA leading to ternary structure inhibiting DNA relegation and DNA damage (Venditto & Simanek, 2010; Hsiang et al., 1985; Sirikantaramas et al., 2007). The main commercial source of CPT, which is the main starting material for the clinically used synthetic derivative, is some woody plant species such as *Camptotheca acuminata* and *Nothapodytes foetida* (Wall et al., 1966; Kai et al., 2015). The slow growth of such species, the low yield of CPT (average, 1 mg/g) (Lopez-Meyer et al., 1994), and environmental concerns of overharvesting woody plants cannot fulfill the increasing market demand that recently reached 2.2 billion US dollars in 2008 in the form of topotecan and irinotecan sales with high probability of increase (Kai et al., 2014). Meanwhile, combinatorial synthesis is yet not effective (Kai et al., 2015). Hence, improvement of CPT production could be achieved through the CPT production through biotechnology techniques such as introducing biosynthetic genes encoding enzymes responsible of its production to rapidly growing vector cells such as hairy roots which are rapidly growing, with genetic stability, easily elicited, and hormone free (Shi et al., 2020).

Biosynthetic Background

The biosynthetic pathway for CPT production is similar to several other terpenoid indole alkaloids (TIA), originating from strictosidine which is a condensed product of both the amino acid derivative tryptamine (nitrogen bearing moiety) and the monoterpene secologanin (terpenoid moiety). CPT biosynthesis (Fig. 1.3) was extensively studied over the last three decades (Kai et al., 2015). Knowledge of biosynthetic genes encoding enzymes for TIA and CPT enabled the production of transgenic cell suspension, hairy roots or seedlings with increased CPT yields, and

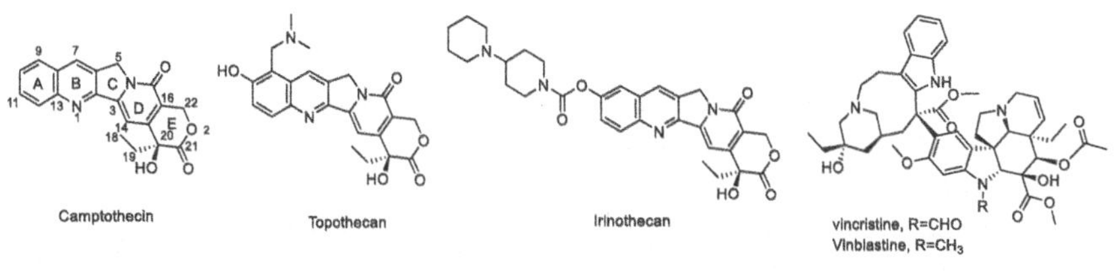

Fig. 1.2 Structures of some clinically important indole alkaloids

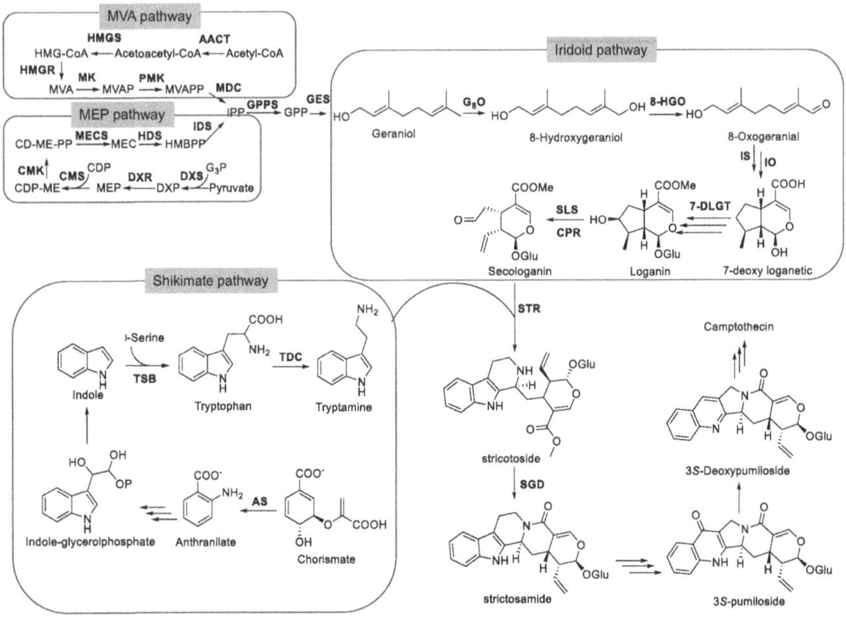

Fig. 1.3 Biosynthetic pathway for camptothecin [*CMS* 4-(cytidine 5-diphospho)-2-C-methylerythritol synthase, *CMK* 4-(cytidine 5-diphospho)-2-C-methylerythritolkinase, *MECS* 2-C-methylerythritol-2,4-cyclodiphosphate synthase, *HDS* hydroxymethylbutenyl 4-diphosphate synthase, *IDS* IPP/DMAPP synthase, *GPPS* geranyl pyrophosphate synthase, *CPR* NADPH-cytochrome P450 reductase, *GES* geraniol synthase, *G8O* geraniol-8-hydroxylase, *8-HGO* 8-hydroxy-geraniol oxidoreductase, *IS* iridoid synthase, *IO* iridoid oxidase (CYP76A26), *DLGT* 7-deoxyloganetic acid UDP-glucosyltransferase, *SLS* secologanin synthetase, *AACT* acetyl-CoA C-acetyltransferase, *HMGS* 3-hydroxy-3-methylglutaryl-CoA synthase, *HMGR* 3-hydroxy-3-methylglutaryl-CoA reductase, *MK* mevalonate kinase, *PMK* phosphomevalonate kinase, *MDC* mevalonate 5-diphosphate decarboxylase, *AS* anthranilic acid synthetase, *TSB* tryptophan synthase beta, *TDC* tryptophan decarboxylase, *STR* strictosidine synthase, *SGD* strictosidine beta-glucosidase; multiple arrows indicate multiple steps]

stable rapid production and that can overcome the difficulties of finding a stable transformation system for woody plants (Ni et al., 2011).

Hairy Root Culture

Interestingly in 2001, Saito and coworkers established a successful hairy root culture system of *Ophiorrhiza pumila* (Rubiaceae) transformed by *Agrobacterium rhizogenes* strain 15834, which carries the PGSGluc1 gene plasmid. The hairy roots grew in liquid culture media with 16-fold increase in the first 5 weeks. It produced CPT as the main alkaloid with 0.1% yield of dry weight. Meanwhile, CPT was found to accumulate in culture media and its production was increased by adding the polystyrene resin (Diaion HP-20) to the culture media to absorb CPT and facilitate its regeneration and later release from the resin, rendering this method more suitable for commercial production (Saito et al., 2001).

Moreover, the jasmonate-responsive APETALA2 domain transcription factor called ORCA3 isolated from *C. roseus* strongly upregulated the expression of several TIA biosynthetic genes (van der Fits & Memelink, 2000). *C. acuminata* transgenic hairy root lines with overexpressed ORCA3 produced CPT with 1.5-fold increase (1.12 mg/g dw) (Ni et al., 2011).

In 2020 Shi and coworkers managed to increase the production of CPT from the hairy roots of *Ophiorrhiza pumila* which originally produce CPT in higher amounts compared to the wild-type plantlets and cell suspension culture that did not accumulate CPT. Comparative transcriptome analysis revealed that hairy roots of *Ophiorrhiza pumila* contained genes encoding key enzymes involved in CPT biosynthesis, which were mainly identified as *OpG10H* and *OpSLS*. The knockout of these genes by CRISPR-Cas9 editing resulted in CPT-free hairy roots whereas individual overexpression of both genes resulted in two transgenic hairy root lines, G line for *OpG10H* and S line for *OpSLS*, showing increased CPT contents up to 2.40 mg/g and 3.28 mg/g, respectively. On the other hand, SG line with overexpression of both genes exhibited CPT contents up to 3.50 mg/g (Shi et al., 2020).

Endophytes

On the other hand, several reports stated the isolation of CPT from plant endophytic fungi such as *Entrophospora infrequens*, isolated from the inner bark of *Nothapodytes foetida* (Puri et al., 2005; Amna et al., 2006), a novel endophytic fungus from the bark of *C. acuminata* (Kusari et al., 2009), and *Fusarium solani* from *Apodytes dimidiata* (Shweta et al., 2010). Although the yield of CPT from endophytic fungi was very low, it provides incentives for stable production of CPT through the improvement of culture conditions or metabolic engineering of the fungi via identification and manipulation of its biosynthetic gene clusters (BGCs).

Vinca Alkaloids (Vincristine (VCT) and Vinblastine (VBL)

Other anticancer alkaloids produced through TIA biosynthetic pathway and have increased market demand and clinical use are *Vinca* alkaloids, vincristine (VCT) and vinblastine (VBL) produced by *Catharanthus roseus* (Jacobs et al., 2004).

Such alkaloids are produced in very low yield and difficult to meet commercial pharmaceutical production. Meanwhile, they have multiple chiral centers, rendering their synthesis very difficult (O'Connor & Maresh, 2006).

Biosynthetic Background

They are formed from strictosidine through a stereoselective Pictet-Spengler reaction between tryptamine and secologanin (Kutchan, 1993; Stöckigt & Zenk, 1977; Maresh et al., 2008), catalyzed by the enzyme STR1 that is highly substrate-specific, limiting precursor-directed biosynthesis in plant cell culture to produce natural product analogs (McCoy & O'Connor, 2006).

Several attempts have been made to produce bis-indole *Vinca* alkaloids through biochemical engineering and cell suspension cultures without any successful result. Therefore, other trials have been directed toward metabolic engineering and manipulation of TIA pathway (Zhao & Verpoorte, 2007).

In 2007 Peebles and coworkers exploited ethanol-inducible promoter system to increase the biomass of *C. roseus* hairy roots to enable the production of higher amounts of the target *Vinca* alkaloids. Ethanol concentration greatly affected ethanol-inducible promoter at very low concentration of 0.005% that resulted in a sixfold increase in chloramphenicol acetyltransferase (CAT) reporter activity after 24 h and 24-fold after 72 h with 0.5% ethanol. Being commercially available and biodegradable, ethanol induction can be promising tool to increase *Vinca* alkaloids production (Peebles et al., 2007).

Sharma and coworkers found that overexpression of two upstream TIA pathway genes tryptophan decarboxylase (CrTDC) and strictosidine synthase (CrSTR) enhanced terpenoid indole alkaloid (TIA) pathway activity and antineoplastic vinblastine biosynthesis in *Catharanthus roseus* in the transgenic whole plant produced via transformation with *Agrobacterium tumefaciens* LBA1119. The produced transgenic plant exhibited twofold increase alkaloid production, ninefold increase in vindoline and catharanthine, and fivefold increase in vinblastine production (Sharma et al., 2018).

Recently, McGehee and coworkers found that carbon-based nanomaterials (CBNs) such as multiwalled carbon nanotubes (MWCNT) and graphene enhanced biomass production of *C. roseus* cell culture under dark conditions. It also doubled the production of total alkaloids under both dark and light conditions. Further HPLC analysis revealed that vincristine and vinblastine production was dramatically increased in CBN-treated cell cultures (McGehee et al., 2019)

Artemisinin (ATN)

Another example of clinically relevant natural compound with increasing demand for pharmaceutical industry is the sesquiterpene lactone artemisinin.

Artemisinin-based combination therapies (ACTs) are the frontline treatments for malaria (Dalrymple, 2013). Moreover, both artemisinin and its semisynthetic congener artesunate proved to be cytotoxic against several cancer cell lines and *Schistosoma* (Efferth, 2006; Utzinger et al., 2007; Efferth et al., 2011). Artesunate exhibited strong activity against ganciclovir-resistant human *Cytomegalovirus* (Shapira et al., 2008). According to the published statistics of WHO, 229 million people are infected with malaria in 2019 and 409,000 people died of malaria in 2019 (World Health Organization, 2020). However, patients in endemic regions cannot benefit these therapies due to low yield of natural sources (White, 2008). Given the large number of cases, world demand for artemisinin is 130 tons per year (Artepal, 2010).

Biosynthetic Background

Artemisinin is a sesquiterpene lactone endoperoxide assembled from three isoprene units. The common sesquiterpene precursor is farnesyl pyrophosphate (FPP) that is generated via either the mevalonate-dependent or mevalonate-independent (MEP/DXP) pathway (Lange et al., 2000; Hunter, 2007). The committed step of artemisinin biosynthesis is the cyclization of FPP to amorpha-4,11-diene (amorphadiene) as shown in Fig. 1.4. This step is carried out by the enzyme amorphadiene synthase (ADS) (Bouwmeester et al., 1999). Amorphadiene is then oxidized by the cytochrome P450 monooxygenase, CYP71AV1, to artemisinic aldehyde (Ro et al., 2006; Teoh et al., 2009). Its reduction to dihydroartemisinic aldehyde is catalyzed by artemisinic aldehyde Δ11 (13) reductase, BDR2 (Zhang et al., 2008). Artemisinin is mainly produced in the leaves in glandular hairs. Artemisinin biosynthesis is mediated by three key enzymes exclusively expressed in apical cells of glandular trichomes. Although both apical trichome cells and

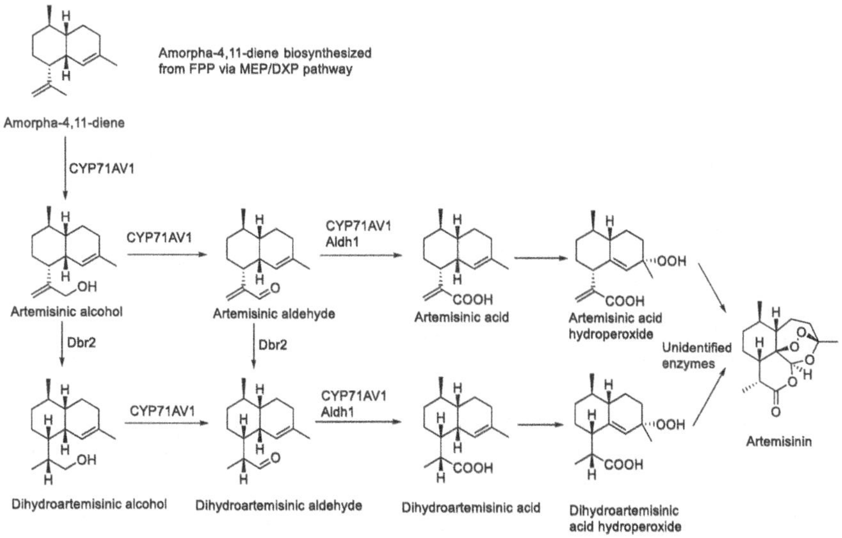

Fig. 1.4 Biosynthesis of artemisinin

subapical chloroplast-containing cells express farnesyl pyrophosphate (FPP) synthase, only glandular trichomes can producc artemisinin (Olsson et al., 2009). Artemisinin production is sensitive to variation in many factors such as genetic variation, cultivation processes, and harvesting time. Moreover, biosynthesis is affected by biotic and abiotic conditions (Omer et al., 2013, 2014, 2016; Ferreira et al., 1995; Davies et al., 2009, 2011). Artemisinin yield from *Artemisia annua* is ~0.5% of dry weight (Brisibe & Chukwurah, 2014), whereas the high-yield variety "CIM-Arogya," developed for marker-assisted breeding, has maximum yields of ~1% (Khanuja et al., 2008) which are very low yields.

All the aforementioned reasons render the development of transgenic production platforms for artemisinin, including microbes and plants essential to lower artemisinin prices and stabilize supply for millions of people who depend on the drug.

Heterologous Expression Systems

Engineering *E. coli* to produce a high level of artemisinin precursor was exploited. Instead of using the native MEP/DXP pathway, Martin and coworkers expressed the heterologous mevalonate pathway from *S. cerevisiae* in *E. coli* to improve isoprenoid flux (Martin et al., 2003). When coupled with the expression of ADS, amorphadiene (amorpha-4,11-diene) was produced, demonstrating the usefulness of this approach. However, a high level of expression of the mevalonate pathway resulted in accumulation of pathway intermediates and inhibition of cell growth (Pitera et al., 2007). Metabolite analysis revealed the accumulation of HMG-CoA, suggesting that the bottleneck was HMG-CoA reductase (HMGR). By modulating the expression level of this enzyme via addition of another copy of the truncated HMGR (tHMGR), the growth rate was restored. To further alleviate this rate-limiting step, bacterial HMGR genes, mvaA from *S. aureus* and mvaE from *Enterococcus faecalis*, were investigated. The strain containing mvaA from *S. aureus* produced the highest titer, compared to those containing the yeast or *E. faecalis* HMGR gene. In light of this success, another strain was constructed to replace the yeast HMG synthase (HMGS) gene with the *S. aureus* HMGS gene mvaS. The strain containing both mvaS and mvaA had an even higher amorphadiene titer. Under fermentation conditions where carbon and nitrogen were strictly controlled, this strain had a high titer with commercial potential (27.4 g/L) (Tsuruta et al., 2009).

Metabolic engineering of *S. cerevisiae* for artemisinic acid production was adopted as an approach to increase the production of an important artemisinin production in transgenic yeast model. As demonstrated in *E. coli*, the key to a high titer is to optimize the isoprenoid pathway flux. In yeast, this was accomplished by the upregulation of several key genes in the mevalonate pathway and the downregulation of genes for sterol biosynthesis (Ro et al., 2006). The HMG-CoA reductase in the mevalonate pathway is the principal target of complex regulation. Deletion of the N-terminal regulatory region of HMG-CoA reductase increases the carbon flux to isopentenyl diphosphate (Donald et al., 1997). Overexpression of tHMGR increased amorphadiene production by fivefold in yeast. When tHMGR overexpression was combined with reduction in the expression of squalene synthase using the methionine-repressible MET3 promoter, another twofold increase in titer was obtained. In yeast, the sterol biosynthesis competes for the common precursor FPP. To further downregulate the sterol pathway, the upc2-1 mutant allele was overexpressed. The transcriptional factor UPC2 is a key regulator of yeast steroid uptake (Vik & Rine, 2001). Expression of the mutant upc2-1 allele allows uptake of exogenous steroids, which inhibits endogenous sterol biosynthesis. After integration of another copy of the tHMGR gene, the final strain produced 150 mg L-1 amorphadiene, nearly 500-fold higher than previously reported levels.

Overexpression of Key Biosynthetic Genes

Efforts are underway to increase artemisinin level in *A. annua* through metabolic engineering. Similar to strategies used in yeast, attempts have been made to increase carbon flux through the isoprenoid pathway for artemisinin biosynthesis in *A. annua*. One example is the expression of HMGCoA reductase gene. The HMGR gene from *Catharanthus roseus* (L.) G. Don was integrated into *A. annua* using *Agrobacterium*-mediated transformation (Aquil et al., 2009).

Based on information gathered from *N. benthamiana*, Plant Research International and Dafra Pharma International NV are collaborating to engineer chicory (*Cichorium intybus*) to produce the artemisinin precursor dihydroartemisinic acid in the roots (Brisibe et al., 2008; Dafra Pharma, n.d.). Dihydroartemisinic acid, like artemisinic acid, can be chemically converted to artemisinin at low cost. As a member of the Asteraceae family, chicory already produces considerable amounts of sesquiterpene lactones through the isoprenoid pathway. Chicory is a well-established plant for industrial nonfood applications, and the entire chain of large-scale agricultural production is already in place. Experiments are underway to evaluate the expression of HMGR, FPS, and ADS in chicory (Ye & Bhatia, 2012).

Silymarin

Phenolic compounds are among the most ubiquitous plant secondary metabolites that have diverse structures and biological activities (Ammar et al., 2012; El-Desoky et al., 2018). They could be found in almost all plant genera. Phenolic compounds vary in their structure complexity from simple phenolic acids to highly complex tannins (Elkhateeb et al., 2019). Among phenolic compounds with high medicinal value is silymarin. Silymarin is a flavolignan mixture consisting of silybin (SB), isosilybin (ISB), silydianin (SD), silychristin (SC), and taxifolin (TXF) as shown in Fig. 1.5, extractable from milk thistle's (*Silybum marianum* L. Gaertn.) seeds and

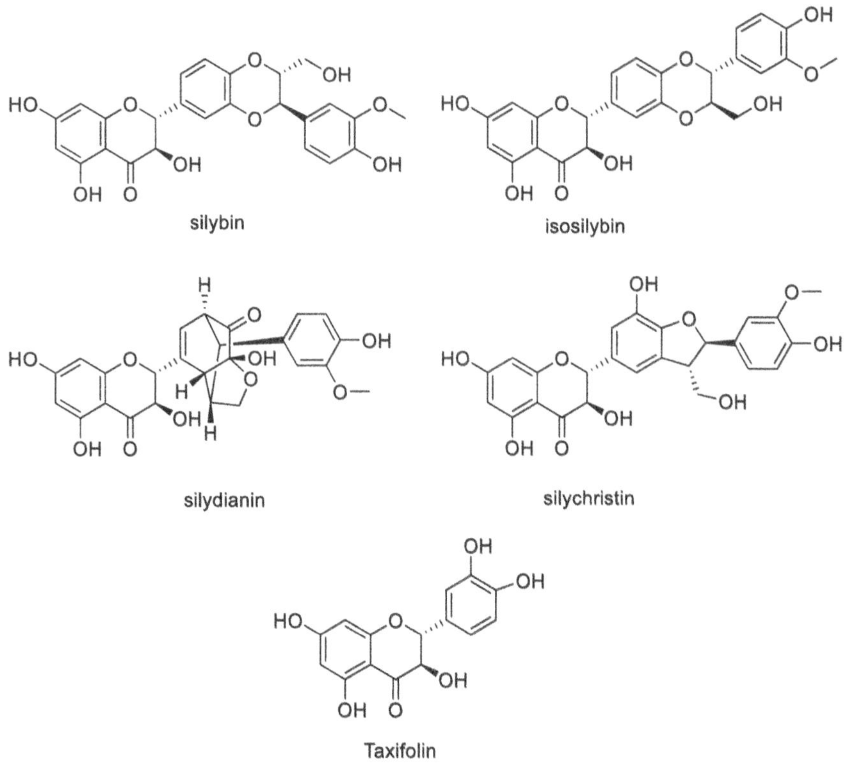

Fig. 1.5 The main compounds of silymarin (flavolignan mixture)

fruits (Khalili et al., 2009a, 2009b; Elyasi, 2021; AbouZid & Ahmed, 2013). Several research projects and studies have been conducted to increase the silymarin content of *Silybum marianum* seeds. A large part of these studies were concerned with environmental factors, whether edaphically or non-edaphically, and agricultural treatments such as irrigation, fertilization, planting dates, and harvesting in order to increase the content of silymarin in seeds (Elsayed et al., 1993; Omer et al., 1993). Silymarin is used worldwide as liver support treatment for several hepatic dysfunctions (Crocenzi & Roma, 2006). Being the main active component (60–70%), silybin reserves the main potential in silymarin hepatoprotective activities (Negi et al., 2007).

Chalcone synthase (CHS, EC 2.3.1.74) is a key enzyme for the naringenin chalcone biosynthesis catalyzing sequential addition of three acetate units from malonyl-CoA to p-coumaroyl-CoA (Spribille & Forkman, 1982). Then naringenin is produced by cyclization and hydroxylated to taxifolin and then silybin is produced by other steps. Therefore, CHS overexpression is a promising target to increase silybin production.

After detailed bioinformatics study, Rahnama and coworkers overexpressed chalcone synthase (chsA) of *Petunia hybrid* L. petals in the hairy root of *Silybum marianum*. ChsA gene was transformed to *S. marianum* hairy roots by *A. rhizogenes* AR15834 strain harboring pCamCHS vectors. That resulted in a sevenfold overexpression of chsA gene in transgenic hairy roots. And subsequent ten times increase in silybin content in transgenic hairy roots when compared to non-transgenic ones (Rahnama et al., 2013).

Some other factors were found to enhance the production of silymarin in the hairy root cultures of *S. marianum* such as the use of elicitors. Among these elicitors is the abiotic elicitor salicylic acid, which enhanced silymarin production (1.89 mg/g dry weight) in hairy roots of *S. marianum*. Salicylic acid (6 mg/50 ml culture) after 24 h increased silymarin content 2.42 times compared to control (0.78 mg/g dry weight). The content of silybin, isosilybin, silychristin, silydianin, and taxifolin were 0.703, 0.017, 0.289, 0.02, and 0.863 mg/g dry weight, respectively. Lipoxygenase activity was increased by ~1.57-fold (Khalili et al., 2009a, 2009b). Similarly, the same group studied the effect of silver ion $[Ag]^+$. Ag^+ was found to increase the activity of lipoxygenase that mediates signal transduction pathway for silymarin production. That resulted in a fourfold increase in silymarin production at 2 mM Ag^+ and after 96 h after elicitation, when compared to non-treated hairy roots (Khalili et al., 2010).

Another type of elicitors is biotic elicitors. Hasanloo and coworkers studied the effect of three fungal species, namely, *Fusarium proliferatum*, *Aspergillus niger*, and *Rhizoctonia solani*, as elicitors for hairy root cultures of *Silybum marianum* (L.) Gaertn. The maximum production for each of the fungal treatment was different: 10 mg *A. niger*/50 ml culture produced 0.18 mg/g dry weight of silymarin after 48 h while 20 mg *F. proliferatum*/50 ml culture produced 0.34 mg/g dry weight of silymarin. And 20 mg *R. solani*/50 ml produced 0.22 mg/g dry weight of silymarin after 72 h. The best profile owed to *F. proliferatum* treated hairy roots that contained taxifolin (0.068 mg/g dry weight), silydianin (0.11 mg/g dry weight), silybin (0.021 mg/g dry weight), and isosilybin (0.015 mg/g dry weight) which were 1.94-, 1.19-, 2.62-, and 1.25-fold greater than untreated hairy roots, respectively. The results indicated that the type of fungi, their concentrations, and their exposure time have significant effect on the stimulation of silymarin production (Hasanloo et al., 2013).

Conclusion

The abovementioned engineered metabolites highlight important aspects to be considered for such objective. Specifications of the host organism have a high impact on final commercial success. These specifications include metabolic capabilities, genetic systems, and scale-up potential. The limitations of the genetic system for plants or even microbes could present challenges. As the optimal host is chosen, the strategy for metabolic engineering may involve determining a specific biosynthetic pathway and enzymes to target. The full understanding of the biochemistry and gene regulation networks is critical for pathway construction. Optimization of pathways involves balancing the upstream and downstream steps as well as exclusion or decrease of challenging reactions.

Biotechnology and metabolic engineering are extensively used to increase the production of paclitaxel.

The production of CPT from the hairy roots of *Ophiorrhiza pumila* which originally produce CPT in higher amounts compared to the wild-type plantlets and cell suspension culture that did not accumulate CPT was managed to increase its production. Comparative transcriptome analysis revealed that hairy roots of *Ophiorrhiza pumila* contained genes encoding key enzymes involved in CPT biosynthesis, which were mainly identified as *OpG10H* and *OpSLS*. Knockout of these genes by CRISPR-Cas9 editing resulted in CPT-free hairy roots. Overexpression of two upstream TIA pathway genes tryptophan decarboxylase (CrTDC) and strictosidine synthase (CrSTR) enhanced terpenoid indole alkaloid (TIA) pathway activity and antineoplastic vinblastine biosynthesis in *Catharanthus roseus* in the transgenic whole plant produced via transformation with *Agrobacterium tumefaciens* LBA1119.

Engineering *E. coli* to produce a high level of artemisinin precursor was exploited. Instead of using the native MEP/DXP pathway, the heterologous mevalonate pathway from *S. cerevisiae* in *E. coli* to improve isoprenoid flux is used.

Efforts are underway to increase artemisinin level in *A. annua* through metabolic engineering. Similar to strategies used in yeast, attempts have been made to increase carbon flux through the isoprenoid pathway for artemisinin biosynthesis in *A. annua*. One example is the expression of HMGCoA reductase gene.

The studies with hairy roots indicated that the type of fungi, their concentrations, and their exposure time have significant effect on the stimulation of silymarin production.

Mechanisms of metabolite engineering are the most promising ways to contribute to the provision of many important entities in the field of pharmaceuticals and medicines, especially high-cost treatments, such as cancers, liver diseases, and viruses.

References

Abdallah, I. I., Pramastya, H., & Van Merkerk, R. (2019). Metabolic engineering of *Bacillus subtilis* toward taxadiene biosynthesis as the first committed step for Taxol production. *Frontiers in Microbiology, 10*, 218.

AbouZid, S., & Ahmed, O. M. (2013). Chapter 14 - Silymarin Flavonolignans: Structure–activity relationship and biosynthesis. *Studies in Natural Products Chemistry, 40*, 469–484.

Ajikumar, P. K., Xiao, W. H., Tyo, K. E. J., Wang, Y., Simeon, F., Leonard, E., Mucha, O., Phon, T. H., Pfeifer, B., & Stephanopoulos, G. (2010). Isoprenoid pathway optimization for taxol precursor overproduction in *Escherichia coli*. *Science, 330*, 70–74.

Ammar, N. M., El-Hawary, S. S. E. D., El-Anssary, A. A., Othman, N., Galal, M., & El-Desoky, A. H. (2012). Phytochemical and clinical studies of the bioactive extract of *Glycyrrhiza glabra* L. Family Leguminosae. *International Journal of Phytomedicine, 4*(3), 429.

Amna, T., Puri, S. C., Verma, V., Sharma, J. P., Khajuria, R. K., Musarrat, J., Spiteller, M., & Qazi, G. N. (2006). Bioreactor studies on the endophytic fungus *Entrophospora infrequens* for the production of an anticancer alkaloid camptothecin. *Canadian Journal of Microbiology, 52*(3), 189–196.

Anterola, A., Shanle, E., Perroud, P. F., & Quatrano, R. (2009). Production of taxa-4(5),11(12)-diene by transgenic *Physcomitrella patens. Transgenic Research, 18*, 655–660.

Aquil, S., Husaini, A. M., Abdin, M. Z., & Rather, G. M. (2009). Overexpression of the HMG-CoA reductase gene leads to enhanced artemisinin biosynthesis in transgenic *Artemisia annua* plants. *Planta Medica, 75*, 1453–1458.

Artepal, J. P. (2010). *Artemisinin market: Quantities and pricing*. Artemisinin Conference, Antananarivo, Madagascar.

Bailey, J. E. (1991). Toward a science of metabolic engineering. *Science, 252*(5013), 1668–1675.

Besumbes, O., Sauret-Güeto, S., Phillips, M. A., Imperial, S., Rodríguez-Concepción, M., & Boronat, A. (2004). Metabolic engineering of isoprenoid biosynthesis in Arabidopsis for the production of taxadiene, the first committed precursor of Taxol. *Biotechnology and Bioengineering, 88*, 168–175.

Biggs, B. W., Lim, C. G., Sagliani, K., Shankar, S., Gregory, S., De Marjan, M., et al. (2016). Overcoming heterologous protein interdependency to optimize P450-mediated Taxol precursor synthesis in *Escherichia coli. PNAS, 113*, 3209–3214.

Birchfield, A. S., & McIntosh, C. A. (2020). Metabolic engineering and synthetic biology of plant natural products – a mini review. *Current Plant Biology*, 100163.

Borah, J. C., Boruwa, J., & Barua, N. C. (2007). Synthesis of the C-13 side-chain of taxol. *Current Organic Synthesis, 4*, 175–199.

Bouwmeester, H. J., Wallaart, T. E., Janssen, M. H., van Loo, B., Jansen, B. J. M., Posthumus, M. A., Schmidt, C. O., De Kraker, J. W., König, W. A., & Franssen, M. C. R. (1999). Amorpha-4,11-diene synthase catalyses the first probable step in artemisinin biosynthesis. *Phytochemistry, 1999*(52), 843–854.

Bringi, V., Kadkade, P.G., Prince, C.L., Schubmehl, B.F., Kane, E.J., Roach, B., (1995). Enhanced production of taxol and taxanes by cell cultures of *Taxus* species. U.S. Patent No. 5,407,816.

Brisibe, E. A., & Chukwurah, P. N. (2014). Production of artemisinin in planta and in microbial systems need not be mutually exclusive. In *Artemisia annua—Pharmacology and biotechnology* (pp. 269–292). Springer.

Brisibe, E. A., Uyoh, E. A., Brisibe, F., Magalhaes, P. M., & Ferreira, J. F. S. (2008). Building a golden triangle for the production and use of artemisinin derivatives against falciparum malaria in Africa. *African Journal of Biotechnology, 7*, 4884–4896.

Chatzivasileiou, A. O., Ward, V., McBride Edgar, S., & Stephanopoulos, G. (2019). Two-step pathway for isoprenoid synthesis. *Proceedings of the National Academy of Sciences of the United States of America, 116*(2), 506–511.

Choi, K. R., Jang, W. D., Yang, D., Cho, J. S., Park, D., & Lee, S. Y. (2019). Systems metabolic engineering strategies: integrating systems and synthetic biology with metabolic engineering. *Trends in Biotechnology, 37*(8), 817–837.

Christen, A. A., Bland, J., & Gibson, D. (1989). Cell cultures as a means to produce taxol. In *Proceedings of the American Association for Cancer Research* (p. 566).

Christen, A.A., Gibson, D.M., Bland, J., (1991). Production of taxol or taxol-like compounds in cell culture. *U.S. Patent* No. 5,019,504.

Connolly, Patrick (1988–1989) . *Personal communications*. Yew Wood Industries, Portland, OR. (April 1988, December 1988, January 1989).

Courdavault, V., O'Connor, S. E., Jensen, M. K., & Papon, N. (2021). Metabolic engineering for plant natural products biosynthesis: new procedures, concrete achievements and remaining limits. *Natural Product Reports*.

Cragg, G. (1988). *Personal communications*. National Cancer Institute. (March, April, July 1988).

Crocenzi, F. A., & Roma, M. G. (2006). Silymarin as a new hepatoprotective agent in experimental cholestasis: new possibilities for an ancient medication. *Current Medicinal Chemistry, 13*, 1055–1074.

Croteau, R., Ketchum, R. E. B., Long, R. M., Kaspera, R., & Wildung, M. R. (2006). Taxol biosynthesis and molecular genetics. *Phytochemistry Reviews, 5*(1), 75–97.

Dafra Pharma. (n.d.) Chicory Project (http://www.dafra.be/content/chicory-project).

Dalrymple, D. G. (2013). *Artemisia annua, Artemisinin, ACTs and malaria control in Africa: Tradition, science and public policy*. American Journal of Agricultural Economics.

Davies, M., Atkinson, C., Burns, C., Arroo, R., & Woolley, J. (2011). Increases in leaf artemisinin concentration in *Artemisia annua* in response to the application of phosphorus and boron. *Industrial Crops and Products, 34*, 1465–1473.

Davies, M. J., Atkinson, C. J., Burns, C., Woolley, J. G., Hipps, N. A., Arroo, R. R., Dungey, N., Robinson, T., Brown, P., & Flockart, I. (2009). Enhancement of artemisinin concentration and yield in response to optimization of nitrogen and potassium supply to *Artemisia annua*. *Annals of Botany, 104*, 315–323.

Dejong, J. M., Liu, Y., Bollon, A. P., Long, R. M., Jennewein, S., Williams, D., & Croteau, R. B. (2006). Genetic engineering of taxol biosynthetic genes in *Saccharomyces cerevisiae*. *Biotechnology and Bioengineering, 93*(2), 212–224.

Denis, J. N., Greene, A. E., Guenard, D., Gueritte-Voegelein, F., Mangatal, L., & Potier, P. (1988). Highly efficient, practical approach to natural Taxol. *Journal of the American Chemical Society, 110*, 5917–5919.

Do, R., Kiss, R. S., Gaudet, D., & Engert, J. C. (2009). Squalene synthase: A critical enzyme in the cholesterol biosynthesis pathway. *Clinical Genetics, 75*, 19–29.

Donald, K. A., Hampton, R. Y., & Fritz, I. B. (1997). Effects of overproduction of the catalytic domain of 3-hydroxy-3-methylglutaryl coenzyme A reductase on squalene synthesis in *Saccharomyces cerevisiae*. *Applied and Environmental Microbiology, 63*, 3341–3344.

Efferth, T. (2006). Molecular pharmacology and pharmaco-genomics of artemisinin and its derivatives in cancer cells. *Current Drug Targets, 7*, 407–421.

Efferth, T., Herrmann, F., Tahrani, A., & Wink, M. (2011). Cytotoxic activity of secondary metabolites derived from *Artemisia annua* L. towards cancer cells in comparison to its designated active constituent artemisinin. *Phytomedicine, 18*, 959–969.

El-Desoky, A. H., Abdel-Rahman, R. F., Ahmed, O. K., El-Beltagi, H. S., & Hattori, M. (2018). Anti-inflammatory and antioxidant activities of naringin isolated from *Carissa carandas* L.: *In vitro* and *in vivo* evidence. *Phytomedicine, 42*, 126–134.

Elkhateeb, A., El-Shabrawy, M., Abdel-Rahman, R. F., Marzouk, M. M., El-Desoky, A. H., Abdel-Hameed, E. S. S., & Hussein, S. R. (2019). LC-MS-based metabolomic profiling of *Lepidium coronopus* water extract, anti-inflammatory and analgesic activities, and chemosystematic significance. *Medicinal Chemistry Research, 28*(4), 505–514.

Elsayed, A. A., Razin, A. M., Omer, E. A., & Ahmed, S. S. (1993). *Use of micronutrients in increasing yield of milk thistle (Silybum marianum, Gaertn) in a desert area under irrigation* (pp. 693–696). Kluwer Academic Publishers, Dordrecht.

Elyasi, S. (2021). Chapter 43 – *Silybum marianum*, antioxidant activity, and cancer patients. In *Cancer (Oxidative stress and dietary antioxidants)* (2nd ed., pp. 483–493).

Engels, B., Dahm, P., & Jennewein, S. (2008). Metabolic engineering of taxadiene biosynthesis in yeast as a first step towards Taxol (Paclitaxel) production. *Metabolic Engineering, 10*, 201–206.

Ferreira, J., Simon, J., & Janick, J. (1995). Relationship of artemisinin content of tissue-cultured, greenhouse-grown, and field-grown plants of *Artemisia annua*. *Planta Medica, 61*, 351–355.

Fett-Neto, A. G., DiCosmo, F., Reynolds, W. F., & Sakata, K. (1992). Cell culture of *Taxus* as a source of the antineoplastic drug taxol and related taxanes. *Biotechnology (N Y), 10*(12), 1572–1575.

Fu, R., Martin, C., & Zhang, Y. (2018). Next-generation plant metabolic engineering, inspired by an ancient Chinese irrigation system. *Molecular Plant, 11*(1), 47–57.

García-Granados, R., Lerma-Escalera, J. A., & Morones-Ramírez, J. R. (2019). Metabolic engineering and synthetic biology: Synergies, future, and challenges. *Frontiers in Bioengineering and Biotechnology, 7*, 36.

Hao, D. C., Yang, L., Xiao, P. G., & Liu, M. (2012). Identification of Taxus microRNAs and their targets with high-throughput sequencing and degradome analysis. *Physiologia Plantarum, 146*, 388–403.

Hasanloo, T., Ahmadi, M., Khayyam Nekoei, S. M., & Salehi Jouzani, G. H. (2013). Improvement of silymarin production in hairy root cultures of *Silybum marianum* (L.) Gaertn using fungal elicitors. *Romanian Biotechnology Letters, 18*(3), 8221–8232.

Holton, R. A., Somoza, C., Kim, H. B., Liang, F., Biediger, R. J., Boatman, P. D., et al. (1994). First total synthesis of Taxol. 1. Functionalization of the B ring. *Physiological Chemistry and Physics and Medical NMR, 34*(2), 103–117.

Howat, S., Park, B., Oh, I. S., Jin, Y. W., Lee, E. K., & Loake, G. J. (2014). Paclitaxel: Biosynthesis, production and future prospects. *New Biotechnology, 31*, 242–245.

Hsiang, Y. H., Hertzberg, R., Hecht, S., & Liu, L. F. (1985). Camptothecin induces protein linked DNA breaks via mammalian DNA topoisomerase I. *The Journal of Biological Chemistry, 260*, 14873–14878.

Hunter, W. N. (2007). The non-mevalonate pathway of isoprenoid precursor biosynthesis. *The Journal of Biological Chemistry, 282*, 21573–21577.

Kai, G., Wu, C., Geng, L., Zhang, L., Cui, L., & Ni, X. (2015). Biosynthesis and biotechnological production of anti-cancer drug camptothecin. *Phytochemistry Reviews, 14*, 525–539.

Kai, G. Y., Teng, X. J., Cui, L. J., Li, S. S., Hao, X. L., Shi, M., & Yan, B. (2014). Effect of three plant hormone elicitors on the camptothecin accumulation and gene transcript profiling in *Camptotheca acuminata* seedlings. *International Journal of Science, 3*(1), 86–95.

Khalili, M., Hasanloo, T., & Kazemi Tabar, S. K. (2010). Ag {+} enhanced silymarin production in hairy root cultures of *'Silybum marianum'* (L.) Gaertn. *Plant Omics, 3*(4), 109–114.

Khalili, M., Hasanloo, T., KazemiTabar, S. K., & Rahnama, H. (2009a). Influence of exogenous salicylic acid on flavonolignans and lipoxygenase activity in the hairy root cultures of *Silybum marianum. Cell Biology International, 33*, 988–994.

Khalili, M., Hasanloo, T., Tabar, S. K. K., & Rahnama, H. (2009b). Influence of exogenous salicylic acid on flavonolignans and lipoxygenase activity in the hairy root cultures of *Silybum marianum. Cell Biology International, 33*(9), 988–994.

Khanuja, S.P.S., Paul, S., Shasany, A.K., Gupta, A.K., Darokar, M.P.,Gupta, M.M., Verma, R.K., Ram, G., Kumar, A., & Lal, R.K.(2008). High Artemisinin Yielding Artemisia Plant Named 'CIMArogya': U.S. Patent 7, 375,260.

Köksal, M., Jin, Y., Coates, R. M., Croteau, R., & Christianson, D. W. (2011). Taxadiene synthase structure and evolution of modular architecture in terpene biosynthesis. *Nature, 469*(7328), 116–120.

Kusari, S., Zühlke, S., & Spiteller, M. (2009). An endophytic fungus from *Camptotheca acuminata* that produces camptothecin and analogues. *Journal of Natural Products, 72*(1), 2–7.

Kutchan, T. M. (1993). Strictosidine: from alkaloid to enzyme to gene. *Phytochemistry, 32*, 493–506.

Lange, B. M., Rujan, T., Martin, W., & Croteau, R. (2000). Isoprenoid biosynthesis: The evolution of two ancient and distinct pathways across genomes. *Proceedings of the National Academy of Sciences of the United States of America, 97*, 13172–13177.

Liu, W. C., Gong, T., & Zhu, P. (2016). Advances in exploring alternative Taxol sources. *Journal of RSC Advances Issue, 54*, 48272–49288.

Liu, Y., & Nielsen, J. (2019). Recent trends in metabolic engineering of microbial chemical factories. *Current Opinion in Biotechnology, 60*, 188–197.

Lopez-Meyer, M., Nessler, C. L., & McKnight, T. D. (1994). Sites of accumulation of the antitumor alkaloid camptothecin in *Camptotheca acuminata. Planta Medica, 60*(7), 558–560.

Mansukhlal, C. W., Harold, L. T., Monroe, E. W., Philip, C., & Andrew, T. M. (1971). Plant anticancer agents. VI. The isolation and structure of Taxol, a novel antileukemic and anticancer agent from *Taxus brevifolia*. *Journal of the American Chemical Society, 93*, 2325–2327.

Maresh, J., Giddings, L. A., Friedrich, A., Loris, E. A., Panjikar, S., Stockigt, J., Peters, B., & O'Connor, S. E. (2008). Strictosidine synthase: Mechanism of a Pictet-Spengler catalyzing enzyme. *Journal of the American Chemical Society, 130*, 710–723.

Martin, V. J., Pitera, D. J., Withers, S. T., Newman, J. D., & Keasling, J. D. (2003). Engineering a mevalonate pathway in *Escherichia coli* for production of terpenoids. *Nature Biotechnology, 21*, 796–802.

McCoy, E., & O'Connor, S. E. (2006). Directed biosynthesis of alkaloid analogs in the medicinal plant *Catharanthus roseus*. *Journal of the American Chemical Society, 128*, 14276–14277.

McGehee, D. L., Alimohammadi, M., & Khodakovskaya, M. V. (2019). Carbon-based nanomaterials as stimulators of production of pharmaceutically active alkaloids in cell culture of *Catharanthus roseus*. *Nanotechnology, 30*(27), 275102.

Nabavi, S. M., Šamec, D., Tomczyk, M., Milella, L., Russo, D., Habtemariam, S., Suntar, I., Rastrelli, L., Daglia, M., Xiao, J., & Giampieri, F. (2020). Flavonoid biosynthetic pathways in plants: Versatile targets for metabolic engineering. *Biotechnology Advances, 38*, 107316.

Negi, A. S., Kumar, J. K., Luqman, S., Shanker, K., Gupta, M. M., & Kbanuja, S. P. S. (2007). Recent advance in plant hepatoprotective: A chemical and biological profile of some important leads. *Medicinal Research Reviews, 28*, 746–772.

Ni, X. L., Wen, S. Q., Wang, W., Wang, X. Y., Xu, H., & Kai, G. Y. (2011). Enhancement of camptothecin production in *Camptotheca acuminata* hairy roots by overexpressing ORCA3 gene. *Journal of Applied Pharmaceutical Science, 1*(8), 85–88.

Nielsen, J. (2001). Metabolic engineering. *Applied Microbiology and Biotechnology, 55*(3), 263–283.

Ning, L., You, C., Zhang, Y., Li, X., & Wang, F. (2020). Synthesis and biological evaluation ofsurface-modified nanocellulose hydrogel loaded with paclitaxel. *Life Science, 241*, 117137.

O'Connor, S. E., & Maresh, J. J. (2006). Chemistry and biology of monoterpene indole alkaloid biosynthesis. *Natural Product Reports, 23*(4), 532–547.

Olsson, M. E., Olofsson, L. M., Lindahl, A.-L., Lundgren, A., Brodelius, M., & Brodelius, P. E. (2009). Localization of enzymes of artemisinin biosynthesis to the apical cells of glandular secretory trichomes of *Artemisia annua* L. *Phytochemistry, 70*, 1123–1128.

Omer, E. A., Abou Hussein, E. A., Hendawy, S. F., El-din, E., Azza, A., & El-Gendy, A. G. (2014). Introduction of three different *Artemisia* species into the Egyptian cultivation. *Journal of Essential Oil-Bearing Plants, 2*(2), 11–20.

Omer, E. A., Abou, H. E. A., Hendawy, S. F., Ezz El-din, A. A., & El-Gendy, A. G. (2013). Effect of soil type and seasonal variation on growth, yield, essential oil and artemisinin content of *Artemisia annua* L. *International Research Journal of Horticulture, 1*(1), 15–27.

Omer, E. A., Hendawy, S. F., Hussein, E. A. A., Gendy, A. G. E., & Ezz El-din, A. A. (2016). Introduction of *Artemisia* species into the Egyptian cultivation. *Journal of Essential Oil-Bearing Plants, 19*(4), 793–805.

Omer, E. A., Refaat, A. M., Ahmed, S. S., Kamel, A., & Hammouda, F. M. (1993). Effect of spacing and fertilization on the yield and active constituents of *Silybum marianum*. *Journal of Herbs Spices & Medicinal Plants, 1*(4), 17–23.

Onrubia, M., Cusido, R. M., Ramirez, K., Hernandez-Vazquez, L., Moyano, E., Bonfill, M., & Palazon, J. (2013). Bioprocessing of plant *in vitro* systems for the mass production of pharmaceutically important metabolites: paclitaxel and its derivatives. *Current Medicinal Chemistry, 20*, 880–891.

Peebles, C. A., Gibson, S. I., Shanks, J. V., & San, K. Y. (2007). Characterization of an ethanol-inducible promoter system in *Catharanthus roseus* hairy roots. *Biotechnology Progress, 23*(5), 1258–1260.

Pitera, D. J., Paddon, C. J., Newman, J. D., & Keasling, J. D. (2007). Balancing a heterologous mevalonate pathway for improved isoprenoid production in *Escherichia coli. Metabolic Engineering, 9*, 193–207.

Puri, S. C., Verma, V., Amna, T., Qazi, G. N., & Spiteller, M. (2005). An endophytic fungus from *Nothapodytes foetida* that produces camptothecin. *Journal of Natural Products, 68*(12), 1717–1719.

Rahnama, H., Razi, Z., Dadgar, M. N., & Hasanloo, T. (2013). Enhanced production of flavonolignans in hairy root cultures of *Silybum marianum* by over-expression of chalcone synthase gene. *Journal of Plant Biochemistry and Biotechnology, 22*(1), 138–143.

Ramírez-Estrada, K., Altabella, T., Onrubia, M., Moyano, E., Notredame, C., Osuna, L., VandenBossche, R., Goossens, A., Cusido, R. M., & Palazon, J. (2016). Transcript profiling of jasmonate-elicited Taxus cells reveals a β-phenylalanine-CoA ligase. *Plant Biotechnology Journal, 14*(1), 85–96.

Ro, D. K., Paradise, E. M., Ouellet, M., Fisher, K. J., Newman, K. L., Ndungu, J. M., Ho, K. A., Eachus, R. A., Ham, T. S., Kirby, J., Chang, M. C. Y., Withers, S. T., Shiba, Y., Sarpong, R., & Keasling, J. D. (2006). Production of the antimalarial drug precursor artemisinic acid in engineered yeast. *Nature, 440*, 940–943.

Sabzehzari, M., & Naghavi, M. R. (2018). Phyto-miRNA: A molecule with beneficial abilities for plant biotechnology. *Gene, 683*, 28–23.

Sabzehzari, M., & Naghavi, M. R. (2019). Phyto-miRNAs-based regulation of metabolites biosynthesis in medicinal plants. *Gene, 682*, 13–24.

Saito, K., Sudo, H., Yamazaki, M., Koseki-Nakamura, M., Kitajima, M., Takayama, H., & Aimi, N. (2001). Feasible production of camptothecin by hairy root culture of *Ophiorrhiza pumila. Plant Cell Reports, 20*(3), 267–271.

Shapira, M. Y., Resnick, I. B., Chou, S., Neumann, A. U., Lurain, N. S., Stamminger, T., Caplan, O., Saleh, N., Efferth, T., & Marschall, M. (2008). Artesunate as a potent antiviral agent in a patient with late drug-resistant cytomegalovirus infection after hematopoietic stem cell transplantation. *Clinical Infectious Diseases, 46*, 1455–1457.

Sharma, A., Verma, P., Mathur, A., & Mathur, A. K. (2018). Overexpression of tryptophan decarboxylase and strictosidine synthase enhanced terpenoid indole alkaloid pathway activity and antineoplastic vinblastine biosynthesis in *Catharanthus roseus. Protoplasma, 255*(5), 1281–1294.

Shi, M., Gong, H., Cui, L., Wang, Q., Wang, C., Wang, Y., & Kai, G. (2020). Targeted metabolic engineering of committed steps improves anti-cancer drug camptothecin production in *Ophiorrhiza pumila* hairy roots. *Industrial Crops and Products, 148*, 112277.

Shiba, Y., Paradise, E. M., Kirby, J., Ro, D. K., & Keasling, J. D. (2007). Engineering of the pyruvate dehydrogenase bypass in *Saccharomyces cerevisiae* for high-level production of isoprenoids. *Metabolic Engineering, 9*(2), 160–168.

Shweta, S., Zuehlke, S., Ramesha, B. T., Priti, V., Mohana Kumar, P., Ravikanth, G., Spiteller, M., Vasudeva, R., & Uma Shaanker, R. (2010). Endophytic fungal strains of *Fusarium solani*, from *Apodytes dimidiata* E. Mey. ex Arn (Icacinaceae) produce camptothecin, 10-hydroxycamptothecin and 9-methoxycamptothecin. *Phytochemistry, 71*(1), 117–122.

Sirikantaramas, S., Asano, T., Sudo, H., Yamazaki, M., & Saito, K. (2007). Camptothecin: Therapeutic potential and biotechnology. *Current Pharmaceutical Biotechnology, 8*, 196–202.

Soliman, S., & Tang, Y. (2015). Natural and engineered production of taxadiene with taxadiene synthase. *Biotechnology and Bioengineering, 112*(2), 229–235.

Spribille, R., & Forkman, G. (1982). Genetic control of chalcone synthase activity in flowers of *Antirrhinum majus. Phyotochemistry, 21*, 2231–2234.

Stierle, A., Strobel, G., & Stierle, D. (1993). Taxol and taxane production by *Taxomyces andreanae*, an endophytic fungus of pacific yew. *Science, 260*(9), 214–216.

Stöckigt, J., & Zenk, M. H. (1977). Strictosidine (isovincoside): the key intermediate in the biosynthesis of monoterpenoid indole alkaloids. *Journal of the Chemical Society, Chemical Communications, 18*, 646–648.

Teoh, K. H., Polichuk, D. R., Reed, D. W., & Covello, P. S. (2009). Molecular cloning of an aldehyde dehydrogenase implicated in artemisinin biosynthesis in *Artemisia annua*. *Botany, 87*, 635–642.

Tsuruta, H., Paddon, C. J., Eng, D., Lenihan, J. R., Horning, T., Anthony, L. C., Regentin, R., Keasling, J. D., Renninger, N. S., & Newman, J. D. (2009). High-level production of amorpha-4,11-diene, a precursor of the antimalarial agent artemisinin, in *Escherichia coli*. *PLoS One, 4* (2), e4489. https://doi.org/10.1371/journal.pone.0004489

Utzinger, J., Xiao, S.-H., Tanner, M., & Keiser, J. (2007). Artemisinins for *schistosomiasis* and beyond. *Current Opinion in Investigational Drugs, 8*, 105–116.

van der Fits, L., & Memelink, J. (2000). ORCA3, a jasmonate responsive transcriptional regulator of plant primary and secondary metabolism. *Science, 289*, 295–297.

van Der Heijden, R., Jacobs, D. I., Snoeijer, W., Hallard, D., & Verpoorte, R. (2004). The Catharanthus alkaloids: Pharmacognosy and biotechnology. *Current Medicinal Chemistry, 11*(5), 607–628.

Venditto, V. J., & Simanek, E. E. (2010). Cancer therapies utilizing the camptothecins: A review of the in vivo literature. *Molecular Pharmaceutics, 7*, 307–349.

Vik, A., & Rine, J. (2001). Upc2p and Ecm22p, dual regulators of sterol biosynthesis in *Saccharomyces cerevisiae*. *Molecular and Cellular Biology, 21*, 6395–6405.

Wall, M. E., Wani, M. C., Cook, C. E., Palmer, K. H., McPhail, A. T., & Sim, G. A. (1966). Plant antitumor agents. I. The isolation and structure of camptothecin, a novel alkaloidal leukemia and tumor inhibitor from *Camptotheca acuminata*. *Journal of the American Chemical Society, 88*, 3888–3890.

Wang, T., Chen, Y., Zhuang, W., Zhang, F., Shu, X., Wang, Z., & Yang, Q. (2019). Transcriptome sequencing reveals regulatory mechanisms of Taxol synthesis in *Taxus wallichiana* var. Mairei. *International Journal of Genomics, 2019*, 1596895.

White, N. J. (2008). Qinghaosu (Artemisinin): The price of success. *Science, 18*, 330–334.

World Health Organization. (2020, November 30). *World malaria report 2020*. Featured Publication.

Yadav, D., Tanveer, A., Malviya, N., & Yadav, S. (2018). Overview and principles of bioengineering: the drivers of omics technologies. In *Omics technologies and bio-engineering* (pp. 3–23). Academic Press.

Ye, M. V., & Bhatia, S. K. (2012). Metabolic engineering for the production of clinically important molecules: Omega-3 fatty acids, artemisinin, and taxol. *Biotechnology Journal, 7*, 20–33.

Zhang, Y., Teoh, K. H., Reed, D. W., Maes, L., Goossens, A., Olson, D. J. H., Ross, A. R. S., & Covello, P. S. (2008). The molecular cloning of artemisinic aldehyde Delta 11(13) reductase and its role in glandular trichome-dependent biosynthesis of artemisinin in *Artemisia annua*. *Journal of Biological Chemistry, 283*, 21501–21508.

Zhao, J., & Verpoorte, R. (2007). Manipulating indole alkaloid production by *Catharanthus roseus* cell cultures in bioreactors: From biochemical processing to metabolic engineering. *Phytochemistry Reviews, 6*(2), 435–457.

Zhu, Q., Wang, B., Tan, J., Liu, T., Li, L., & Liu, Y. G. (2020). Plant synthetic metabolic engineering for enhancing crop nutritional quality. *Plant Communications, 1*(1), 100017.

Chapter 2
Ameliorating Abiotic Stress Tolerance in Crop Plants by Metabolic Engineering

Sakshi Narula, Smita Chaudhry, and Gagan Preet Singh Sidhu

Abstract Environmental adversities like heat, cold, drought, salinity, ultraviolet radiation and flooding induces abiotic distress in plants and are the pioneer limiting factors for plant growth, development and productivity. Anthropogenic activities have fuelled changes in global climatic conditions and these changes have incremented multiple abiotic stresses in crop plants. Researchers are making unprecedented efforts to intercept heavy crop losses and in turn to generate more food and feed to meet the demands of the ever-increasing human population. Highlighting the techniques involved to combat abiotic stresses, their role in regulating plant growth and development under unfavourable climatic factors holds substantial importance. This chapter reviews the role of osmoprotectants, polyamines, flavonoids and phytohormones in plant growth and development under abiotic stress conditions and their metabolic engineering for producing abiotic stress-tolerant transgenic plants. This strategy can prove a vital tool to minimise heavy crop losses and alleviate the problem of increasing food demand of human populations.

Keywords Abiotic stress · Climate change · Crop plants · Metabolic engineering · Tolerance

Abbreviations

ABA	Abscisic acid
ADC	Arginine decarboxylase
APX	Ascorbate peroxidase
As	Arsenic
ATP	Adenosine triphosphate

S. Narula · S. Chaudhry
Institute of Environmental Studies, Kurukshetra University, Kurukshetra, Haryana, India

G. P. S. Sidhu (✉)
Department of Environmental Sciences, School of Vocational Studies and Applied Sciences, Gautam Buddha University, Greater Noida, Uttar Pradesh, India

T. Aftab, K. R. Hakeem (eds.), *Metabolic Engineering in Plants*,
https://doi.org/10.1007/978-981-16-7262-0_2

BR	Brassinosteroid
Cad	Cadaverine
CAT	Catalase
CHS	Chalcone synthase
CK	Cytokinin
DAO	Diamine oxidase
dcSAM	Decarboxylated S-adenosylmethionine
ET	Ethylene
GA	Gibberellic acid
GB	Glycine-betaine
Glc-6-P	Glucose-6-phosphate
GR	Glutathione reductase
H_2O_2	Hydrogen peroxide
IAA	Indole-3-acetic acid
JA	Jasmonic acid
LAT	L-type amino acid transporter
MaCHS	*Morus atropurpurea* Roxb chalcone synthase
MDA	Malondialdehyde
NADPH	Nicotinamide adenine dinucleotide phosphate hydrogen
NOX	NADH oxidase
O^2	Superoxide ion
OAT	Ornithine aminotransferase
ODC	Ornithine dicarboxylic acid
P5C	Pyrroline-5-carboxylate
P5CR	Pyrroline-5-carboxylate reductase
P5CS	Pyrroline-5-carboxylate synthetase
PA	Polyamine
POD	Peroxidase
Pro	Proline
ProDH	Proline dehydrogenase
PS II	Photosystem II
PUT	Polyamine uptake transporter
Put	Putrescine
ROS	Reactive oxygen species
RuBisCO	Ribulose-1,5-bisphosphate carboxylase/oxygenase
SA	Salicylic acid
SAM	S-adenosylmethionine
SAMDC	SAM decarboxylase
SOD	Superoxide dismutase
Spd	Spermidine
SPDS	Spd synthase
Spm	Spermine
SPMS	Spm synthase
Tre	Trehalose

Tre6P	Trehalose 6-phosphate
UDP-Glc	Uridine diphospho-glucose

Introduction

Plants being sessile in nature have to consecutively endure altering environmental circumstances including predominant abiotic stresses. Increased use of fossil fuels, rising population and deforestation are causing climate change at a faster pace. In the past few decades, climate change is a topic of concern in reinforcing the intensity of abiotic stresses (Songstad et al., 2017; Khan et al., 2019). The exacerbated abiotic stresses due to climate change include heat, cold, drought, water logging, salinity, heavy metal toxicity and nutrient deficiency. Among these stresses, drought, salt and temperature stress affect the most and limit the crop productivity. These stresses also concern food security worldwide.

The environmental challenges causing abiotic stresses affect plant growth adversely by altering morphological and physiological processes and induce alterations in the cellular pathways (Egamberdieva et al., 2017). These stresses impede photosynthesis by affecting stomatal conductance and diminish RuBisCO (ribulose-1,5-bisphosphate carboxylase/oxygenase) activity (Kohli et al., 2017). In addition, these stresses negatively influence photosystem I, II (PS I, II) and chlorophyll biosynthesis (Sharma et al., 2016). Abiotic stresses are known to cause overproduction of ROS (reactive oxygen species). These stresses prompt multifaceted responses to avoid injuries in plants and upsurge their survivability in adverse circumstances. In response to abiotic stresses, plants modify cellular and molecular processes that ultimately mount their growth and development (Sharma et al. 2019a, b).

Plants have evolved highly coordinated and interactive processes to cope with environmental adversities. To protect cellular machinery from adverse environmental severities, plants accumulate osmolytes, flavonoids, polyamines and various other pigments and solutes (Fig. 2.1). Furthermore, plants also produce phytohormones to alleviate abiotic stress effects through signalling and play an important role in various physiological and biochemical mechanisms. The perception of stresses triggers the activation of signal transduction cascades that interact with the baseline pathways transduced by phytohormones, compatible solutes and different metabolites produced by plants (Fig. 2.1).

Osmoprotectants are compatible solutes that augment the cell potential to sustain water in cell without obstructing the normal metabolism (Sharma et al. 2019a, b). These regulate osmotic adjustment and protect the plants from oxidative damage. Polyamines are small polycationic aliphatic amines. They play substantial roles in growth and tolerance to wide range of abiotic stresses in plants (Khare et al., 2018). Flavonoids are the biosynthate compounds and are important stress mitigator. They play a significant role by inducing resistance against abiotic stresses in plants (Shah

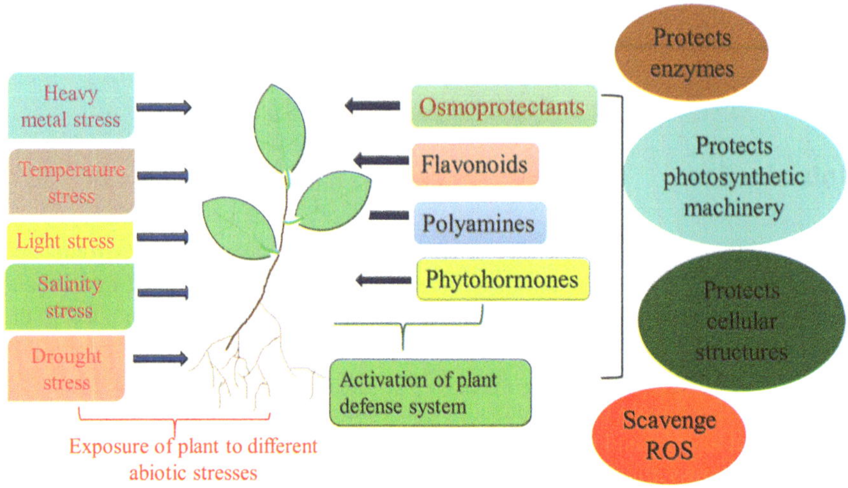

Fig. 2.1 Schematic representation of biosynthates providing tolerance to plant on exposure to different abiotic stresses

& Smith, 2020). Phytohormones are the chemical messengers required and produced by plants in little quantities, which communicate cellular activities in higher plants (Vob et al., 2014). Their role is critical in alleviating abiotic stresses and promote plant tolerance under the exposure of adverse environmental conditions.

Altered environmental conditions fuelled by climate change pose harsh impacts on plant productivity with greater intensities. These stress spells further pose a threat to food security of rising global population. To cope up with increasing food demand and to increase productivity, innovative approaches should be formulated. There is a necessity to devise strategies to deal with environmental stresses and to prevent heavy crop losses, so that food demands of ever-increasing population can be met. Numerous strategies, such as exogenous application of compatible solutes and micronutrient in form of amendments, plant breeding, tissue culture methods, genetic engineering and metabolic engineering, are being applied to enhance agricultural productivity in hostile environmental circumstances (Helaly, 2017). Metabolic engineering of new cultivars via genes and metabolic pathways for enhancing abiotic stress tolerance in plants is gaining attention.

Metabolic engineering is "the improvement of cellular activities by manipulation of enzymatic, transport and regulatory functions of cell with use of recombinant DNA technology" (Bailey, 1991). Metabolic engineering of osmolytes, flavonoids, polyamines and phytohormones can be applied as one strategy to produce high-yielding crops with enhanced levels of antioxidants and increased abiotic stress tolerance. It can be an attractive technique by introducing genes for improvement in metabolic pathways, stress tolerance and to increase crop productivity. This chapter presents an overview about biosynthates produced in plants, their role in growth, development, abiotic stress response and their metabolic engineering for inducing enhanced stress tolerance and increased productivity in plants.

Abiotic Distress in Crop Plants

Abiotic stress in plants is the outcome of interactions between plants and their physical environment, like water, light, temperature fluctuations, salinity, heavy metals and mineral imbalances, which causes unfavourable effects on plants. Abiotic stresses, such as drought, salinity, heavy metal and temperature extremes, may cause massive losses in yield and productivity of agricultural systems (Yadav et al., 2020, 2021). Photosynthesis is a major process affected by abiotic stresses in plants as these stresses disturb the photosynthetic apparatus (Brestic et al., 2016). Also, these stresses lead to the generation and acclimatisation of ROS in plants due to interruption of cellular homeostasis (Zang et al., 2017).

Drought is one of the major stresses that affects crop growth and productivity all around the world (Kour et al., 2019, 2020). Initial symptoms of plants exposed to drought stress are leaf wilting, decrease in plant height and interruption in establishment of buds and flowers (Bhatt & Rao, 2005). For example, in *Lathyrus sativus* shoot length was reduced under water-deficit conditions (Gheidary et al., 2017). Similarly, Bhargavi et al. (2017) reported a decline in number of leaves in *Andrographis paniculate* on exposure to drought stress. Reduction in leaf area is also a stress-avoiding strategy adopted by plants as observed in *Petroselinum crispum* and *Stevia rebaudiana* (Najla et al., 2012; Srivastava & Srivastava, 2014). Drought stress poses a negative impact on the metabolism of photosynthetic pigments, decreases RuBisCO function and further induces disruption of photosynthetic pigments (Foyer & Noctor, 2000). In *Glycine max*, chlorophyll synthesis and chlorophyll ratio (a/b proportion) are altered by drought stress (Chowdhury et al., 2017). It results in increased production of ROS and reduced generation of photosynthetic electron transport constituents (Basu et al., 2016). Disrupted photosynthetic apparatus causes reduction in photosynthetic rate, stomatal conductance, transpiration rate and photochemical efficiency of PSII (Campos et al., 2019; Ye et al., 2016). In *Vigna mungo,* photosynthetic activity has been reported to be reduced due to alterations in chlorophyll pigments and photosynthetic apparatus (Gurumurthy et al., 2019). It causes disparity in production of ROS and their scavenging, which leads to oxidative stress (Hussain et al., 2018). Reduced scavenging and overaccumulation of ROS results in protein oxidation and peroxidation of lipid membranes, and it may damage genetic material (Apel & Hirt, 2004). Generation of various osmoprotectants, such as proline and glycine-betaine, gets enhanced and assists in maintaining leaf turgor pressure. Plant's enzymatic antioxidants also work in conjugation with non-enzymatic antioxidants to counteract oxidative stress (Hussain et al., 2018). Numerous studies have reported enhanced activities of antioxidant enzymes such as catalase (CAT), superoxide dismutase (SOD), peroxidase (POD) and ascorbate peroxidase (APX) under water-deficit conditions (Cao et al., 2017). Prathyusha and Chaitanya (2019) have reported enhancement of non-enzymatic antioxidants and antioxidant enzyme activities in *Coleus (Plectranthus)* in water stressed conditions. Activities of SOD, POD and APX were also increased in *Vicia faba* under drought conditions (Abid et al., 2017). In *Vigna mungo,* increased activities of SOD and POD reduced the production and accumulation of ROS that enabled the plants to cope the oxidative outburst (Gurumurthy et al., 2019).

Salinity stress is becoming a huge challenge leading to limited plant performance. It causes substantial reduction in plant height, root length and dry weight (Abbas et al., 2010; Mahjoor et al., 2016). Salt stress causes reduced fresh and dry weights of *Gossypium* seedlings (Balal et al., 2017; Tabatabaie & Nazari, 2007) and declined seed germination percentage in *Triticum aestivum* (Le Gall et al., 2015). It is reported to reduce dry weight, plant height, fruit weight and relative water content in *Solanum lycopersicum* and *Piper nigrum* (Kaya et al., 2007; Saini et al., 2015; Shahid et al., 2020). Elevated concentration of salts causes alterations in gas exchange, photosynthesis, stomatal conductance and transpiration negatively (Muradoglu et al., 2015). In *Piper nigrum* and *Cucumis melo*, chlorophyll pigment is reported to decline significantly under salt stress (Wu et al., 2014). Salinity stress hinders water movement from soil to plants by reduction in water conductivity of roots and affecting relative water content at cellular level (Saha et al., 2015). As a consequence of salt stress, ROS generation increased and induced injury in plants. Plants tackle ROS through enzymatic and non-enzymatic mechanisms. ROS scavenging enzymes include SOD, CAT, POD, APX and glutathione reductase (GR) (Xia et al., 2014). In *Triticum aestivum,* SOD, APX, CAT and GR activities have been reported to increase under stress (Sairam & Tyagi, 2004). Salt stress enhances APX, POD and SOD activity and alterations in isoenzymes activities in *Solanum tuberosum* and *Broussonetia papyrifera* (Hu et al., 2016; Talaat & Shawky, 2011).

Temperature is a major factor which affects plant growth and development. Plant growth is affected adversely by heat stress, and it leads to a wide range of morphological, physiological and biochemical responses (Zandalinas et al., 2018). At early growth stages of plant, heat stress arrests seed germination and delays seedling emergence (Deng et al., 2015). Heat stress impacts photosynthetic system in plants. Reduced photosynthetic activity is attributed to the inhibition of PS II, which causes decreased chlorophyll synthesis in plants (Chen et al., 2017; Guo et al., 2016). Decreased chlorophyll content was observed in *Triticum aestivum* (Bouchemal et al., 2017) and reduced PS II activity in *Solanum lycopersicum* (Faik et al., 2016). Fahad et al. (2016) reported decline in photosynthesis, stomatal conductance and transpiration rate in *Oryza sativa* on exposure to heat stress. Heat stress causes elevated production of ROS. ROS reduces the synthesis of nicotinamide adenine dinucleotide phosphate hydrogen (NADPH) and adenosine triphosphate (ATP) and ultimately results in decreased photosynthetic rate (Schrader et al., 2004). ROS can damage organic constituents in the cell and causes oxidative stress. In *Brassica campestris*, heat stress upsurged the formation of hydrogen peroxide (H_2O_2), malondialdehyde (MDA) and superoxide ion (O^{2-}) and decreased the SOD, POD and CAT activities (Sun et al., 2016). Zou et al. (2017) reported elevated H_2O_2,

MDA and O^{2-} content on exposure to heat stress, whereas SOD, POD and CAT activities first increased and then decreased with increasing heat stress.

Light is a fundamental feature for the process of photosynthesis. Light intensities above the light saturation point of photosynthesis are harmful to plants and are termed as high light stress (Lichtenthaler & Burkart, 1999). Fluctuations in solar radiation impact various physiological and biochemical processes in plants. High intensities of solar light cause increment in production of ROS, which may result in photoinhibition and may lead to reduction in primary productivity of plants (Gururani et al., 2015). Under extreme low light conditions, plant growth is adversely affected because of insufficient energy. Low light has adverse impact on plants like reduction in photosynthetic efficiency and enhanced oxidative stress (Wang et al., 2013). Reduction in photosynthesis by low light is attributed to decline in stomatal conductance and upsurge intercellular CO_2 concentration in leaves and ultimately leads to reduced photosynthetic efficiency (Liu et al., 2014).

Heavy metal stress is a serious risk to plant growth and development (Bali et al., 2020; Bali & Sidhu, 2021). Heavy metal contamination is increasing due to urbanisation and industrialisation. Heavy metals adversely affect transportation and accumulation of essential elements and cause alterations in morphological parameters (Seneviratne et al., 2017). Heavy metal excess induces oxidative stress in plants by generation of ROS (Sidhu et al., 2017a, 2017b, 2020). ROS acclimatisation alters the cellular, physiological and biochemical mechanisms in plants (Sidhu et al., 2018; Bali & Sidhu, 2021). Upadhyay et al. (2016) reported reduction in root length and shoot length in *Oryza sativa* on exposure to arsenic (As). Under heavy metal exposure, metabolic pathways of plants such as photosynthesis, respiration, gaseous exchange and nutrient absorption get altered (Arif et al., 2019). Sidhu et al. (2017a, 2017b) found that lead (Pb) contamination caused drop in concentration of leaf pigment in *Coronopus didymus,* attributed to decline in chlorophyll synthesis. Under copper (Cu) stress, a decline in gas exchange and chlorophyll content in *Brassica napus* plants was observed (Zaheer et al., 2015).

Mediators of Plant Responses Towards Abiotic Stresses

Plants have evolved various defence mechanisms to counteract environmental adversities. These adaptive defence mechanisms embrace physiological, biochemical, cellular and molecular methods to deal with hostile environmental circumstances. In response to these abiotic stresses, plants produce many biosynthates in the form of osmolytes, flavonoids, polyamines and phytohormones. These biosynthetic species aids in alleviating abiotic stress effects, signalling pathways and also scavenging of ROS leading to reduction in oxidative stress.

Osmoprotectants

Exposure to different abiotic stresses brings modifications in plant cellular functions. Among these modifications, a myriad of highly soluble, low-molecular-weight, neutral and non-toxic organic compounds are generated, and these compounds are known as osmolytes or osmoprotectants (Per et al., 2017; Riaz et al., 2019). Osmoprotectants have protective role against damage to cellular machinery in response to stressed conditions. Osmoprotectants stabilise the cell turgor pressure and counteract high levels of ROS, regaining the redox balance to tackle environmental adversities (Cortleven et al., 2019). On the basis of chemical properties, osmoprotectants are categorised into amino acids, betaines and sugars (Slama et al., 2015) (Fig. 2.2). Osmoprotectants facilitate osmotic adjustment under inadequate water supply and regulate internal osmotic potential and also macromolecule structures. They improve stress tolerance in plants by diverse roles: protection of biological membranes, upregulation of photosynthesis, stabilisation of proteins and other structures, scavenging of ROS, maintenance of cellular redox balance, protection of key antioxidant enzymes, stabilisation of thylakoid membranes and activation of defence-related genes (Hasanuzzaman et al., 2014; Suprasanna et al., 2016; Wani et al., 2018).

Among the amino acids group, proline (Pro) is a proteinogenic amino acid, which plays an important role in both metabolism and in defence system as osmoprotectant (Kaur & Asthir, 2015) (Fig. 2.3). It plays a key role in regulation of enzyme activities and maintenance of protein integrity. Under abiotic stress, plants accumulate Pro in cytoplasm and chloroplast (Ben Rejeb et al., 2015). It has antioxidant properties as it can scavenge singlet oxygen as well as ROS (Suprasanna et al., 2016). Its synthesis in plants can occur via two pathways: glutamate pathway and ornithine pathway. Glutamate pathway is the major source of Pro accumulation. The major enzymes in

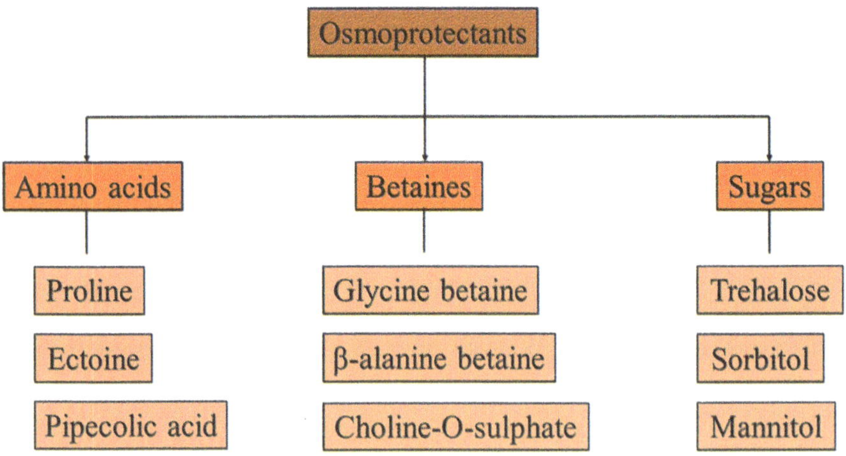

Fig. 2.2 Categorisation of osmoprotectants

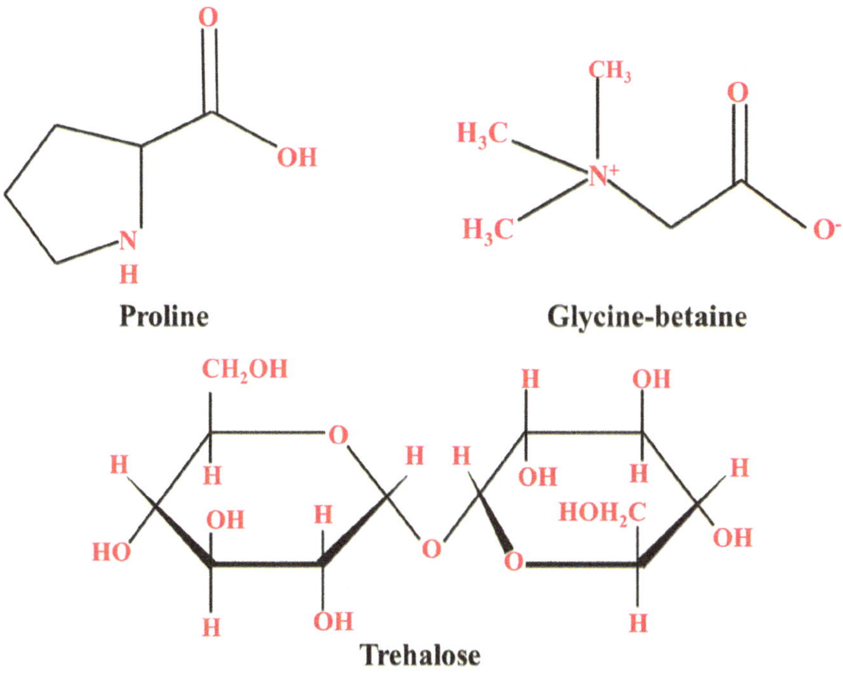

Fig. 2.3 Chemical structure of major osmoprotectants

glutamate pathway are pyrroline-5-carboxylate synthetase (P5C) and pyrroline-5-carboxylate reductase (P5CR) (Sekhar et al., 2007). Ornithine pathway is activated in cytoplasm or chloroplasts, and Pro is produced from glutamic acid through an intermediate pyrroline-5-carboxylate (P5C) under stress situations (Dar et al., 2016).

Glycine betaine (GB) is a quaternary ammonium compound and is N-methylated derivative of glycine and is synthesised in chloroplasts (Fariduddin et al., 2013) (Fig. 2.3). It protects plants against harsh environmental conditions. Usually, its synthesis occurs via two substrates: choline and glycine. Synthesis from choline is a two-step process. Its synthesis in chloroplast occurs from serine through ethanolamine, choline and betaine aldehyde (Zulfiqar et al., 2020). In another pathway, direct N-methylation of glycine occurs.

Trehalose (Tre) is a non-reducing disaccharide sugar and a major osmoprotectant as a reserve carbohydrate and plays a vital role in plant metabolic processes (Fig. 2.3). It is composed of two glucose residues joined by a very stable α-α linkage (Fernandez et al., 2010). It is synthesised via an intermediate trehalose-6-phosphate (Tre6P) (Figueroa & Lunn, 2016). OtsA–OtsB pathway is followed for its biosynthesis in plants which is a two-reaction process depending on two key molecules uridine diphospho-glucose (UDP-Glc) and glucose-6-phosphate (Glc-6-P) (Kosar et al., 2019).

In crop plants, osmoprotectants are found in trace amount, and there is growing interest to develop stress-tolerant and high-yielding cultivars via manipulation of osmoprotectants predominantly proline, glycine betaine and trehalose. Metabolic engineering of proline in stress-tolerant plants is mediated through overexpression and accumulation of P5CS, pyrroline-5-carboxylate reductase (P5CR) and ornithine aminotransferase (OAT) or via domination of proline dehydrogenase (ProDH) (Kaur & Asthir, 2015). For manipulation of glycine betaine concentration and accumulation in plants, genes identified are *ApGSMT2g, ApDMT2g* and *BADH* (Song et al., 2018; Tian et al., 2017). Transgenic lines with higher concentrations of trehalose can be developed by metabolic engineering via genes *otsA, otsB, OsTPS1* and *TPP* (Garg et al., 2002; Li et al., 2011; Nuccio et al., 2015).

Polyamines

Polyamines (PAs) are small polycationic aliphatic amines and are profusely present in plants. They have a great role in plant growth, development of plants and stress responses in plants (Sequeramutiozabal et al., 2016). PAs exist in free, soluble conjugated, insoluble bound forms (Gholami et al., 2013) and are synthesised from the decarboxylation of amino acid's ornithine, arginine, methionine and lysine (Falahi et al., 2018). The common PAs in free state in higher plants include spermidine (Spd, *N-(3-Aminopropyl)-1,4-diaminobutane*), spermine (Spm, *N, N′-Bis(3-aminopropyl)-1,4-diaminobutane*) and putrescine (Put, *1,4-diaminobutane*) as these are involved in molecular signalling pathways in interaction between plants and microbes (El Ghachtouli et al., 1996; Mustafavi et al., 2018) (Fig. 2.4). These three PAs are the most commonly found in higher plants and differ in number of amino acid groups. Cadaverine (Cad, *1,5-diaminopentane*) is a diamine and is a less common PA commonly found in plants belonging to families Gramineae and Leguminosae (Saha et al., 2015). Polyamines Put and Spd are supposed to be crucial for the survival of cells, while Spm is not considered as

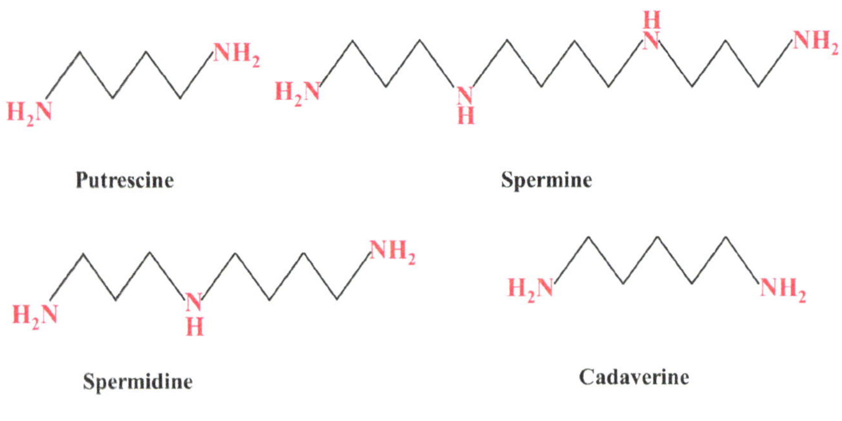

Fig. 2.4 Chemical structure of most common polyamines

important, owing to its silencing of genes in PA biosynthetic pathway. Put is the central product of common PA biosynthetic pathway formed by arginine by the sequential action of enzymes arginine decarboxylase (ADC), N- carbamoyl putrescine amidohydrolase and agmatine deiminase (Pegg, 2016). Enzyme ornithine decarboxylase directly converts ornithine to Put (Docimo et al., 2012; Pegg, 2016). For triamine Spd, Put acts as a precursor. In Spd, the third amino group is donated from decarboxylated S-adenosylmethionine (dcSAM) formed by decarboxylation of S-adenosylmethionine (SAM), catalysed by Spd synthase (SPDS). For the formation of tetramine Spm, another amino group is added to Spd by dcSAM by methionine, and this reaction is catalysed by Spm synthase (SPMS) (Vuosku et al., 2018).

PA's homeostasis is achieved by regulating the activities of biosynthetic enzymes, predominantly SAM decarboxylase (SAMDC). SAMDC is further processed to form the active enzymes as earlier it is synthesised as proenzyme. When Put concentration becomes high in cells, SAMDC activity is promoted, and it leads to the synthesis of Spd and Spm. Further, when the product of SAMDC-dcSAM is accumulated, it inactivates SAMDC by irretrievable alteration at the cysteine residue in one of its subunits.

PA's transportation is arbitrated by PA uptake transporters (PUTs), which belong to L-type amino acid transporter (LAT) family, located in the endoplasmic reticulum, Golgi membranes and plasma membrane. Long-distance transportation of PAs is attributed to xylem, and storage forms of PAs occur as conjugates with phenolic acids such as cinnamic acid, coumaric acid or sinapic acid (Wuddineh et al., 2018). PAs may get attached to either glutamine or lysine residue of protein. PAs have the capability to bind large macromolecules, such as DNA, proteins, membrane phospholipids and pectic polysaccharides, and are also recognised to have a great role in protein phosphorylation, post-transcriptional modifications and also conformational transition of DNA (Khare et al., 2018). This attributes to prevention in damage induced by stresses to macromolecules. PAs are known to regulate fundamental processes such as cell proliferation, enzyme activities, transportation of ions and membrane stabilisation as these are present in all compartments of plant cell. PAs also regulate plant growth and developmental processes. All these properties make them vital for the usual functioning of cells.

PAs impart a dynamic role in alleviating plant stress by scavenging ROS, neutralisation of acids and stabilisation of cell membranes (Pál et al., 2015; Romero et al., 2018). PA metabolism and their expression levels of pathway genes are linked positively with enhanced tolerance of abiotic stresses in plants (Shi & Chan, 2014). Spd is reported to be involved in enhancing antioxidant enzymes production in *Lycopersicon esculentum* (tomato) seedlings under high temperature conditions (Sang et al., 2017). According to the study of Saha and Giri (2017), Spd is found to be vital for salt stress tolerance in rice. Furthermore, PAs trigger defence response in plants against abiotic stresses, by regulation of ion channel homeostasis, stomatal responses to maintain water status, by activation of antioxidant metabolism to cope ROS (Shi & Chan, 2014). Exogenous application of polyamines enhances plant biomass and promotes stress tolerance (Table 2.1). Apart from exogenous

Table 2.1 Effects of exogenous application of polyamines on different plant species

Polyamine	Plant	Effect on plant	Reference
Putrescine	*Malus sylvestris*	– Increased fresh weight – Reduced electrolyte leakage in salt stress conditions	Kuznetsov and Shevyakova (2007)
Putrescine	*Scenedesmus obliquus*	– Increased cell growth – Improved status of photosynthetic apparatus	Demetriou et al. (2007)
Putrescine	*Glycine max*	– Elevated plant growth and biomass – Increased Put, SAMDC and DAO levels – Enhanced antioxidant enzymes activity	Zhang et al. (2014)
Putrescine	*Saccharum officinarum*	– Induces somatic embryo development	Reis et al. (2016)
Putrescine	*Oryza sativa*	– Increased Spm and Spd content – Improvement in growth and callus embryogenic traits	Tan et al. (2017)
Putrescine	*Thymus vulgaris*	– Foliar application triggers physiological processes – Increased tolerance to drought stress	Mohammadi et al. (2018)
Spermidine	*Cucumis sativus*	– Increased plant growth – Increased PA levels and antioxidant enzymes	Parvin et al. (2014)
Spermidine	*Zoysia japonica*	– Decreased ROS and MDA level and increased antioxidant enzyme activity – Enhanced activity of SAMDC and DAO and reduced activity of ADC	Li et al. (2016)
Spermidine	*Solanum lycopersicum*	– Enhanced assimilation of excess NH_4^+ – Increased activity of antioxidant enzymes	Zhang et al. (2013)
Spermidine	*Dendranthema morifolium*	– Significantly affects endogenous PA levels and phytohormones – Acceleration in the process of flower bud differentiation	Xu (2015)
Spermine	*Triticum aestivum*	– Increased endogenous Spm, Spd, ABA and IAA content in grains – Increased grain filling rate and grain weight	Liu et al. (2013)
Spermine	*C. sativus*	– Induced antioxidant system in chloroplast – Improved efficiency of photosystem-II	Shu et al. (2013)
Spermine	*Zea mays*	– Increased activity of antioxidant enzymes – Alleviation of negative impacts of drought stress by foliar application	Talaat et al. (2015)
Spermine	*Rosa damascene*	– Improved characteristics of growth – Enhanced relative water content, chlorophyll content and stomatal conductance	Hassan et al. (2018)

ABA abscisic acid, *ADC* arginine decarboxylase, *DAO* diamine oxidase, *IAA* indole-3-acetic acid, *MDA* malondialdehyde, *PA* polyamine, *PS II* photosystem II, *Put* putrescine; *SAMDC* SAM decarboxylase, *Spd* spermidine, *Spm* spermine, *Spd* spermidine application of PAs, a strategy of producing transgenic lines with altered PA levels and overexpressing PA biosynthetic genes can be implemented to reduce stress sensitivity in crop plants. The genes identified for metabolic engineering and augmenting stress tolerance in plants are *ADC*, *ODC*, *SAMDC*, *SPDS* and *SPMS* (Marco et al., 2015).

Flavonoids

Flavonoids are secondary metabolites and can be synthesised by plants, fungi and microorganisms. These are fundamental compounds for vital processes of life and include terpenoids, phenylpropanoids, alkaloids, etc. These secondary metabolites play important roles in plant's life to combat abiotic stress conditions and changing environmental challenges. Flavonoids can be known as "specialised metabolites" because species-specific flavonoids are synthesised by plants (Harborne & Williams, 2000). Flavonoids are comprised of polyphenolic compounds containing a 15-carbon skeleton (C6-C3-C6), which is made up of a heterocyclic benzopyran ring, an aromatic ring and a phenyl ring in their structures. A benzopyran ring has asymmetric carbon, whereas a phenyl ring has attached hydroxyl groups. In most of the flavonoids, linking chain further forms a heterocyclic pyran (Corradini et al., 2011). Flavonoids are called as chalcones in the absence of the third ring (aromatic ring) and mostly serve as the precursors of various flavonoids.

A simple structural unit is 2-phenylchroman (flavan) and structural diversity arises from the degree of unsaturation and oxidation of linking ring. Different flavonoids are (2-phenylchromenyliums) anthocyanins; (2-phenylochromones) flavanones, flavanols, flavones and di-OH-flavonols; (2-phenylchromanes) flavans, flavan-3-ols and flavan-3,4-diols (proanthocyanidins); chalcones/dihydrochalcones, etc. (Khoo et al., 2017). Flavonoid are classified into subgroups, such as flavanones, flavanols, flavones and flavonols. Other classes of flavonoids include isoflavones, biflavonoids, aurones and chalcones (Fig. 2.5).

There are two distinct biosynthetic pathways for the synthesis of flavonoid compounds in plants. These include the acetate pathway, which serves as a building block for polymeric 2- carbon units, and the shikimic acid pathway, which generates the phenylpropanoids skeleton (Croteau et al., 2000). Benzopyrene is synthesised from three malonyl-CoA molecules which is generated via the transformation of glucose, whereas the phenyl ring is synthesised from 4-coumaroyl-CoA, which is produced from phenylalanine through the shikimate pathway. Further condensation of both rings occurs to generate chalcone, and the action of chalcone synthase (CHS) characterises the beginning of specific flavonoid compound. This compound subsequently undergoes isomerase catalysed cyclisation to produce flavanone. All known flavonoid compounds have this same biosynthetic pathway and also the same basic skeleton.

A strong link has been found between the content of flavonoids and stress tolerance capabilities of plants. It is demonstrated by studies that oxidative stress

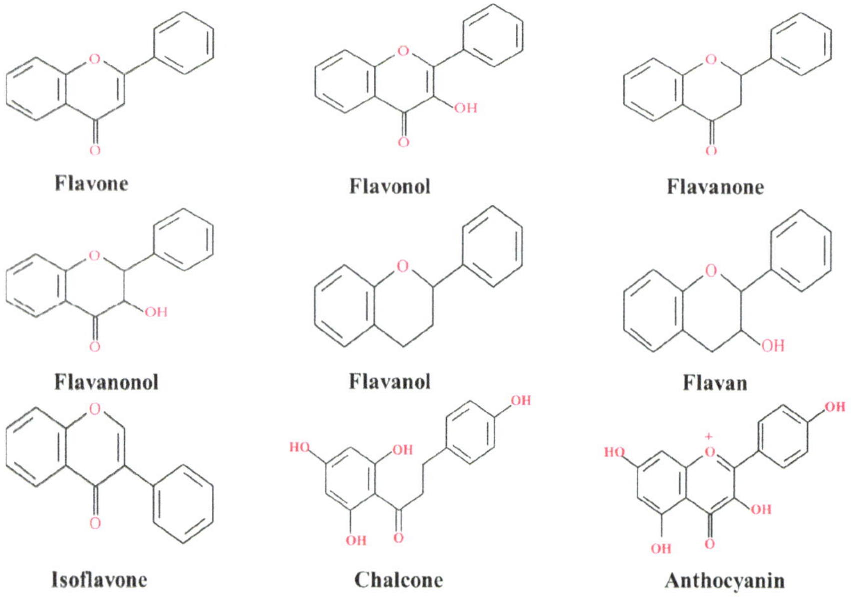

Fig. 2.5 Chemical structure of common flavonoids in plants

leads to stimulation of flavonoid synthesis, which is triggered more by ROS. In addition to scavenging of ROS by directly donating the hydrogen atom, flavonoids upregulate antioxidant enzyme system for the alleviation of oxidative stress (Halliwell & Gutteridge, 2015). Flavonoids conquer ROS formation by uncoupling of proteins triggered by superoxide. They constrain enzymes which produce ROS such as NADH oxidase (NOX), glutathione-S-transferase, mitochondrial succinoxidase and microsomal monooxygenase, and they also have the potential to scavenge free radicals directly by donating the hydrogen atoms (Procházková et al., 2011). These compounds also protect the plants against the damage caused by UV radiation (Agati & Tattini, 2010). Flavonoids are known to have the ability to chelate free metal ions and are also capable to prevent peroxidation of lipids and oxidative damage of membrane lipids (Mishra et al., 2013a, 2013b; Kumar et al., 2013).

For metabolic engineering of flavonoids, a critical enzyme *Morus atropurpurea* Roxb chalcone synthase (MaCHS) has been identified (Wang et al., 2017). This novel enzyme has been found to modulate metabolite profile in transgenic *Nicotiana tabacum* and also provide resistance against abiotic stresses. A number of genes are identified for flavonoid synthesis from several species, such as *Camellia sinensis* Kuntze (Wang et al., 2018), *Citrus unshiu* (Wellmann et al., 2002), *Allium cepa* (Park et al., 2017) and *Mangifera indica* (Bajpai et al., 2018).

Phytohormones

Phytohormones are produced in lower concentrations, are able to regulate various abiotic stresses in plants and play an important role in the growth of plants. Phytohormones include auxins, abscisic acid, gibberellic acid, cytokinin, ethylene, brassinosteroids, salicylic acid and jasmonic acid (Fig. 2.6). Phytohormones coordinate signal transduction pathways during abiotic stresses and act as chemical messengers in cellular activities in plants (Vob et al., 2014; Kazan, 2015). These plant hormones have protective effects in plants especially in abiotic stress acclimatisation by affecting plant nutrition, growth and particularly signalling pathways. These act either at the site of synthesis or elsewhere in plants after transportation.

Auxin (indole-3-acetic acid; IAA) is an important phytohormone in plant growth and development. It also regulates plant growth in conditions of abiotic stress tolerance (Kazan, 2013). Auxins stimulate the transcription of many genes called as primary auxin response genes leading to abiotic stress tolerance in plants. Abscisic acid (ABA) plays a significant role in the adaptation of plants to various

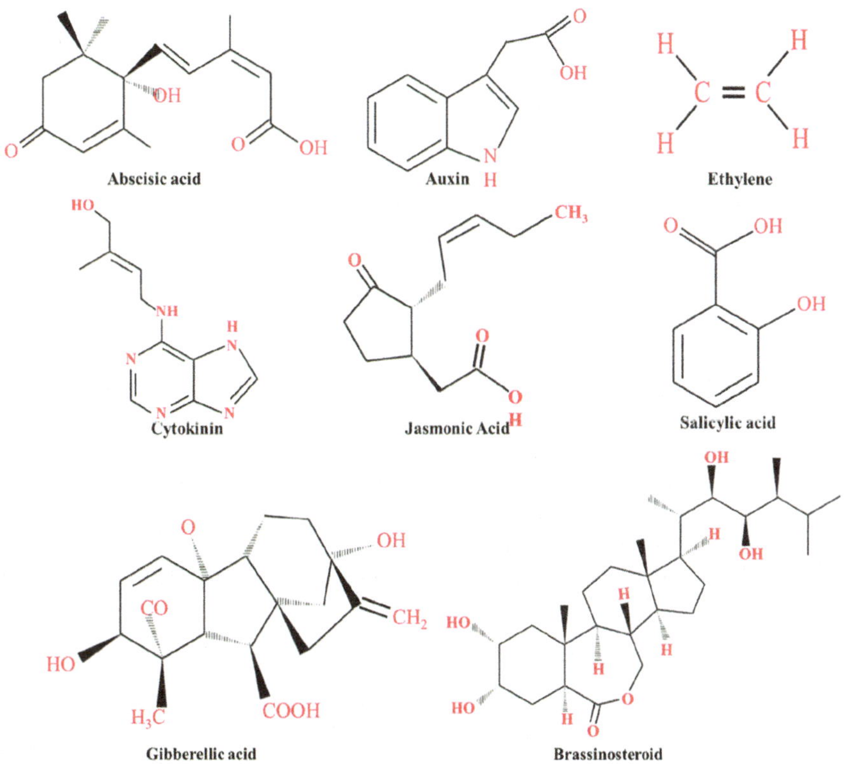

Fig. 2.6 Chemical structure of major phytohormones in plants

stresses by initiating signalling pathways (Bücker-Neto et al., 2017). It is synthesised in the plastidial 2-C-methyl-D-erythritol 4-phosphate pathway. It is used in numerous developmental processes like seed germination, seed dormancy and stomata closure. The higher amount of ABA activates signalling cascade in guard cells, which leads to the outflow of K^+ ions from guard cells, resulting in diminished turgor pressure and stomata closure (Lim et al., 2015). ABA also induces the feedback inhibition of photosynthesis by accumulation of carbohydrates and reduction in the concentration of photosynthetic enzymes (Vishwakarma et al., 2017). It is considered to be involved in modulation of various key physiological processes in extreme environmental conditions.

Brassinosteroids (BRs) are steroidal plant hormone regulating plant growth and development by producing an array of physiological variations. They are involved in numerous activities, such as seed germination, cell growth, reproductive growth and production of flowers and fruits. Numerous studies indicate BR-induced stress tolerance in plants to various environmental stresses, including high temperature, cold stress, drought, salinity and heavy metals (Bajguz, 2010; Wang et al., 2014). Cytokinin (CK) is a multibranched phytohormone, important for plant development in addition to stress management. Furthermore, it plays an imperative role in assimilation, transportation and monitoring reactions to changes in levels of both macro- and micronutrients (Tiwari et al., 2020). Abiotic stresses lead to change in concentration of endogenous cytokinin in plants and indicate their role in abiotic stress tolerance (Kang et al., 2012). Ethylene (ET) is a gaseous hormone, which regulates plant growth and development. Often, jasmonic acid (JA) and ET act conjointly for plant defence against stresses. JA and ET signalling hubs, such as JAZ proteins, *CTR1*, *MYC2* and several members of the *AP2/ERF* transcription factor gene family, are found to have defensive roles in abiotic stress conditions (Kazan, 2015). Gibberellic acid (GA) is a growth hormone responsible for abiotic stress regulation in plants (Rastegari et al., 2020). Several researches have specified linking of GA with stress tolerance in plants by altering the production and distribution of GA. Growth and stress tolerance in plants can be regulated by promoting existence and escape from various environmental stresses. Also, it influences seed germination, leaf expansion, stem elongation and fruit development. Salicylic acid (SA) is a phenolic compound found to play critical role in regulation of plant growth and development. Its function is found to be evident in seed germination, flowering, fruiting, ion uptake transport, photosynthesis, stomatal conductance and transpiration (Khan et al., 2003). During stressed situations, many genes, which respond to SA, are related to genes encoding heat shock proteins, antioxidants and genes responsible for production of secondary metabolites (Rasool et al., 2018).

Metabolic Engineering of Biosynthates for Promoting Plant Tolerance Towards Abiotic Stress

Metabolic engineering imparts a critical role in the production of plant cultivars resistant towards changing climatic conditions (Kour et al., 2019; Yadav et al., 2019). Metabolic engineering can be a significant technique for enhancing abiotic stress tolerance in plants leading to sustainable agriculture. Also, in recent times with the advancement in genome technologies, genes and metabolic pathway responsible for developing abiotic stress-tolerant varieties can be identified. For the development of plants with improved abiotic stress tolerance, profound knowledge and understanding of mechanisms for enduring different abiotic stresses is required. Metabolic engineering is a dynamic but complex process that requires the understanding of core pathways, genes and transcription factors. Currently, changes in physiological, cellular, biochemical and molecular processes during stressed conditions are known. Additionally, several genes accounting for abiotic stress tolerance in plants have been identified and are being studied for their expression pattern under stressed conditions. Owing to the knowledge of plant cellular functions and genomic structures, the ability of plants to increase the concentration of compounds, which help in enhancing tolerance of crop plants to abiotic stresses, to make them better adapted to harsh physical circumstances and also to get higher yields and biomass has been facilitated.

Plants naturally produce osmoprotectants as an adaptive mechanism on interaction with various kinds of abiotic stresses, but their concentration is less and may offer only partial protection. There are variations in the level of osmoprotectants even between the cultivars of same species. There can be an opportunity to intensify the tolerance towards incompatible environmental conditions by utilising the cultivars with high osmoprotectant production (Li et al. 2019a, 2019b). Furthermore, progress has been made in the identification and characterisation of genes, which are involved in the biosynthesis of osmoprotectants (Alzahrani, 2021). A listing of few genes and transgenic plants with overexpression of osmoprotectants concentration in plants is mentioned in Table 2.2.

Polyamines are an assemblage of polycationic amines affecting processes like plant growth, metabolism and development. Plant cultivars with improved stress tolerance can be formed with altered PA levels, overexpressing PA biosynthetic genes (Pathak et al., 2014). The extent of amendment of endogenous levels of one or more specific PAs in transgenic varieties may vary, but all these transgenic varieties show communal expansion in tolerance towards abiotic stresses. Genes acquired for producing transgenic lines overexpressed with PAs include *ODC*, *ADC*, *SAMDC*, *SPDS*, *SAMS*, *MYB* and *SPMS* (Marco et al., 2015). Transgenic varieties of plants with genes overexpressing polyamines and their impacts are summarised in Table 2.3.

Flavonoid synthesis in plants plays an imperative role in abiotic stress alleviation. Enforced manipulation and metabolic engineering of gene expressions of flavonoids can be a useful technique to increase stress tolerance in plants. Important

Table 2.2 Transgenic plants, their genes expressing osmoprotectants and their impacts on plants

Transgenic plant	Gene	Source plant	Impacts on transgenic plant	Reference
Lotus tenuis	*pRD29A; oatADC*	*A. thaliana*	– Improved tolerance to salinity stress – Improved vegetative, physiological and biochemical characteristics	Espasandin et al. (2018)
Medicago truncatula	*Oat ADC*	*Avena sativa*	– Improved tolerance to drought stress – Improved physiological status – Enhanced polyamines concentration under drought stress	Duque et al. (2016)
Arabidopsis thaliana	*IbRAP2–12*	*Ipomoea batatas*	– Enhanced tolerance to salinity and drought stress – Upregulation of genes associated with abscisic acid and jasmonic acid signalling – Increased ROS scavenging and maintained plant-water relation in stress conditions	Li et al. (2019b)
A. thaliana	*TaMIPS2*	*T. aestivum*	– Improved tolerance to drought, heat, salinity and cold stress – Improvement in morphological, physiological and biochemical attributes	Khurana et al. (2017)
Triticum aestivum	*BADH*	*Atriplex hortensis*	– Improvement in salt tolerance – Enhanced protection of thylakoid membrane and enhanced photosynthesis under stressed conditions	Tian et al. (2017)
Zea mays	*BADH*	*Atriplex micrantha*	– Enhanced tolerance to salinity stress – Increased fresh weight, plant height and grain yield – Lower malondialdehyde content and higher chlorophyll content	Di et al. (2015)
Cajanus cajan	*P5CSF129A*	*Vigna aconitifolia*	– Enhanced proline content in transgenic plant – Enhanced tolerance to salinity stress	Surekha et al. (2014)
A. thaliana	*GsMIPS2*	*Glycine soja*	– Overexpression in osmoprotectant concentration and ultimately enhanced tolerance to salt stress	Nisa et al. (2016)
Gossypium hirsutum	*ApGSMT2 g; ApDMT2 g*	*Aphanothece halophytica*	– Increased tolerance to salinity stress – Enhancement in yield	(Song et al. (2018)
Lotus tenuis	*pRD29A; oat ADC*	*A. sativa*	– Enhanced tolerance to water-deficit conditions	Espasandin et al. (2014)

(continued)

Table 2.2 (continued)

Transgenic plant	Gene	Source plant	Impacts on transgenic plant	Reference
Solanum lycopersicum	*codA*	*A. sativa*	– Enhanced glycine betaine accumulation in transgenic varieties – Increased tolerance to salinity stress	Wei et al. (2017)
A. thaliana	*SHMT3*	*Oryza sativa*	– Enhanced tolerance to salinity stress	Mishra et al. (2019)
Arachis hypogea	*AtHDG11*	*A. thaliana*	– Enhanced tolerance to salinity and drought stress	Banavath et al. (2018)
Solanum tuberosum	*codA*	*Arthrobacter globiformis*	– Enhanced performance in water-deficit circumstances	You et al. (2019)
A. thaliana	*AhERF019*	*Arachis hypogea*	– Improved tolerance to drought, heat and salinity stress	Wan et al. (2014)

ROS reactive oxygen species

transcription factors responsible for biosynthesis of flavonoids include *MYB*, *bHLH* and *WD40* (Qiu et al., 2014). CHS is a critical enzyme, and in recent study, a novel *Morus atropurpurea* Roxb chalcone synthase (MaCHS) has been identified (Wang et al., 2017). Genes responsible for the synthesis of flavonoid compounds include *GaCHS1*, *GaCHS2*, *Bra013652*, *Bra019350*, *Bra027457*, *Bra003021*, *Bra038445*, *Bra035004*, *CsMYB6A* and *CsUGT72AM1* (Wani et al., 2017; Yahyaa et al., 2017; Zhang et al., 2017; He et al., 2018).

Phytohormones have a functional role in biological processes and in regulation of cell signalling pathways under abiotic stress condition. But they are produced in low concentration in plants. In the current scenario, phytohormone production in plants is a potential method to produce abiotic stress-tolerant crops (Verma et al., 2015, 2016; Yadav et al., 2015a, 2015b). Table 2.4 summarises genes identified for phytohormones and their impacts in transgenic species.

Conclusions

Climate change has aggravated abiotic stresses, which limit crop productivity by affecting plant growth and developmental processes. In this era, sustainable production approaches with fewer inputs in agricultural fields are desirable. With the advancement in genomic technology, much research has been performed towards understanding plant abiotic stress response. Metabolic engineering for introgression of stress-tolerant genes or genes responsible for enhanced accumulation of biosynthates in plants are gaining attention. However, genes involved in biosynthesis of osmoprotectants, polyamines, flavonoids and phytohormones were identified earlier. But their biosynthetic pathways have been demonstrated recently. Stress-tolerant plants accumulate osmoprotectants under abiotic stress conditions. To

Table 2.3 Transgenic plants with overexpression of polyamines genes for increased stress tolerance

Transgenic plant	Gene	Source plant	Effects on endogenous PA and response to abiotic stresses in transgenic lines	Reference
Solanum lycopersicum	*SAMS*	*S. lycopersicum*	– Enhanced concentration of Spd and Spm – Improved yield and tolerance to alkali stress – Higher photosynthetic capacity and lower oxidative stress	Gong et al. (2014)
Nicotiana tabacum	*MYB*	*Poncirus trifoliata*	– Enhanced levels of PAs – Enhanced tolerance to drought stress – Reduced concentration of ROS and MDA	Sun et al. (2014)
N. tabacum	*MYB*	*Gossypium barbadense*	– Increased transcript levels of ADC and SAMDC – Improved survival and reduced water loss under drought stress – Increased concentration of antioxidant enzymes	Chen et al. (2015)
S. lycopersicum	*ODC*	*Mus musculus*	– Enhanced levels of PAs in transgenic plants – Tomato plants with enhanced fruit quality	Pandey et al. (2015)
Medicago truncatula	*ADC*	*Avena sativa*	– Elevated levels of Put and Spd – Increased efficiency of photosynthetic apparatus during drought stress – Higher seed yield	Duque et al. (2016)
Citrus sinensis	*PAO*	*Clonorchis sinensis*	– Decrease in Spm and Spd content – Better germination of transgenic seeds under salt stress	Wang and Liu (2016)
A. thaliana	*PAO*	*Gossypium hirsutum*	– Declined Spm and increased Put content – Disagreement on salt resistance during germination	Cheng et al. (2017)
Eremochloa ophiuroides	*SAMDC*	*Cynodon dactylon*	– Higher level of polyamine oxidase activity – Improved tolerance to cold stress through NO signalling	Luo et al. (2017)
N. tabacum	*MYB*	*Pyrus betulaefolia*	– Modulation of polyamine synthesis by regulation of ADC expression – Positive role in drought tolerance	Li et al. (2017)

(continued)

Table 2.3 (continued)

Transgenic plant	Gene	Source plant	Effects on endogenous PA and response to abiotic stresses in transgenic lines	Reference
Lotus tenuis	*ADC*	*Avena sativa*	– Improved tolerance to salt stress – Increase in root growth in response to stress – Increased osmotic adjustment via proline production	Espasandin et al. (2018)

ADC arginine decarboxylase, *MDA* malondialdehyde, *PA* polyamine, *Put* putrescine, *ROS* reactive oxygen species, *SAMDC* SAM decarboxylase, *Spd* spermidine, *Spm* spermine

enhance stress tolerance in plant cultivars, enhanced accumulator genes encoding osmoprotectants can be engineered. PAs are known as important compounds to regulate plant growth and development. The concentration of PAs can be modified by altering the expression of ADC, ornithine dicarboxylic acid (ODC) and SAMDC. Phytohormone engineering represents an imperative podium for the production of high-yielding and abiotic stress-tolerant crops. Nevertheless, comprehensive work is still required at genetic level to understand and recognise the complexity concealed in stress signal-transduction pathways.

Most of the studies on metabolic engineering are based on either lab or pot scale experiments. In future, there is need to evaluate the response of transgenic plant varieties under natural field conditions, so that their economic and practical value can be assessed. Under natural field conditions, plants face interactive effect of these abiotic stresses, so there is a need of introgression of multiple genes to enhance plant tolerance towards different abiotic stresses and also to obtain optimum yields to feed the ever-increasing population.

Metabolic engineering approaches can be very useful for the biosynthesis and accumulation of compounds reducing oxidative stress in plants. These compounds alleviate stress symptoms in plants as well as enhance the yield of plants. This chapter provides information on the manipulation of osmoprotectants, polyamines, flavonoids and phytohormones to confer abiotic stress tolerance in plants. Metabolic engineering can be proven as an encouraging field, but significant measures are still needed to reach its pinnacle. It is necessary to investigate genes and metabolic pathways carefully to develop stress-resistant cultivars. The greatest challenge to be addressed is the development of stable engineered stress-tolerant cultivars required to fulfil the demand of agricultural production. Furthermore, commercialisation of these advancements is likely to face challenging regulatory hurdles and may also face consumer acceptance issues. Therefore, progress in metabolic engineering of osmoprotectants, polyamines, flavonoids, phytohormones and other metabolites needs to be accompanied with progress in societal acceptance.

Table 2.4 Genes depicting phytohormones and their impacts on transgenic lines

Phytohormone	Gene	Function of gene	Transgenic plant	Impact on transgenic species	Reference
Abscisic acid	*NCED*	Imperative part in rate-limiting step of biosynthesis of ABA for feedback control	*Petunia hybrida*	– Enhanced levels of endogenous ABA – Increased drought tolerance – Decreased stomatal conductance	Estrada-Melo et al. (2015)
Abscisic acid	*MsZEP*	Vital role in ABA biosynthesis	*Nicotiana tabacum*	– Improved salt and drought tolerance	Zhang et al. (2015)
Abscisic acid	*OsPYL3; OsPYL9*	Receptor of ABA	*Oryza sativa*	– Increase in resistance to drought and cold stress	Tian et al. (2015)
Abscisic acid	*AtLOS5*	Key regulator of ABA biosynthesis	*Zea mays*	– Enhanced salinity stress tolerance which attributes to increased Na^+ efflux and H^+ influx	Zhang et al. (2016)
Abscisic acid	*MLP43*	Positive regulator of ABA response	*Arabidopsis thaliana*	– Improved drought tolerance on overexpression of *MLP43*	Wang et al. (2016)
Abscisic acid	*GhNAC2*	ABA-induced leaf senescence	*Gossypium hirsutum*	– Improved vegetative growth and productivity	Gunapati et al. (2016)
Auxin	*OsIAA6*	Gene of IAA family	*Oryza sativa*	– Enhanced drought tolerance via auxin biosynthesis regulation	Jung et al. (2015)
Auxin	*YUCCA6*	Important gene for IAA synthesis	*Populus alba*× *Populus glandulosa*	– Increased tolerance to drought and oxidative stress	Ke et al. (2015)
Cytokinin	*AtCKX1*	Cytokinin dehydrogenase	*Hordeum vulgare*	– Improved drought tolerance via improved dehydration avoidance	Pospíšilová et al. (2016)
Ethylene	*ACC synthase*	Catalysation of rate limiting step in ethylene biosynthesis	*Z. mays*	– Enhanced drought tolerance – Reduced ethylene levels	Habben et al. (2014)

Ethylene	*ZmARGOS*	Negative regulator of ethylene signal transduction	*Z. mays*	– Increased drought tolerance	Shi et al. (2015)
Brassinosteroid	*BdBR11*	BR receptor gene	*Brachypodium distachyon*	– Enhanced drought tolerance.	Feng et al. (2015)
Brassinosteroid	*AtDWF4*	Encrypts C-22 hydroxylase; Catalyses rate determining step in BR synthesis	*Brassica napus*	– Increased root length during drought stress and high temperature – Increased seed yield and productivity	Sahni et al. (2016)

References

Abbas, W., Ashraf, M., & Akram, N. A. (2010). Alleviation of salt-induced adverse effects in eggplant (*Solanum melongena* L.) by glycinebetaine and sugarbeet extracts. *Scientia Horticulturae, 125*(3), 188–195.

Abid, G., M'hamdi, M., Mingeot, D., Aouida, M., Aroua, I., Muhovski, Y., Sassi, K., Souissi, F., Mannai, K., & Jebara, M. (2017). Effect of drought stress on chlorophyll fluorescence, antioxidant enzyme activities and gene expression patterns in faba bean (*Vicia faba* L.). *Archives of Agronomy and Soil Science, 63*(4), 536–552.

Agati, G., & Tattini, M. (2010). Multiple functional roles of flavonoids in photoprotection. *New Phytologist, 186*(4), 786–793.

Alzahrani, F. O. (2021). Metabolic engineering of osmoprotectants to elucidate the mechanism (s) of salt stress tolerance in crop plants. *Planta, 253*(1), 1–17.

Apel, K., & Hirt, H. (2004). Reactive oxygen species: Metabolism, oxidative stress, and signal transduction. *Annual Review of Plant Biology, 55*, 373–399.

Arif, M. S., Yasmeen, T., Shahzad, S. M., Riaz, M., Rizwan, M., Iqbal, S., Asif, M., Soliman, M. H., & Ali, S. (2019). Lead toxicity induced phytotoxic effects on mung bean can be relegated by lead tolerant *Bacillus subtilis* (PbRB3). *Chemosphere, 234*, 70–80.

Bailey, J. E. (1991). Toward a science of metabolic engineering. *Science, 252*(5013), 1668–1675.

Bajguz, A. (2010). An enhancing effect of exogenous brassinolide on the growth and antioxidant activity in *Chlorella vulgaris* cultures under heavy metals stress. *Environmental and Experimental Botany, 68*(2), 175–179.

Bajpai, A., Khan, K., Muthukumar, M., Rajan, S., & Singh, N. K. (2018). Molecular analysis of anthocyanin biosynthesis pathway genes and their differential expression in mango peel. *Genome, 61*(3), 157–166.

Balal, R. M., Shahid, M. A., Javaid, M. M., Iqbal, Z., Liu, G. D., Zotarelli, L., & Khan, N. (2017). Chitosan alleviates phytotoxicity caused by boron through augmented polyamine metabolism and antioxidant activities and reduced boron concentration in *Cucumis sativus* L. *Acta Physiologiae Plantarum, 39*(1), 1–15.

Bali, A. S., & Sidhu, G. P. S. (2021). Arsenic acquisition, toxicity and tolerance in plants-From physiology to remediation: A review. *Chemosphere*, 131050.

Bali, A. S., Sidhu, G. P. S., & Kumar, V. (2020). Root exudates ameliorate cadmium tolerance in plants: A review. *Environmental Chemistry Letters, 18*(4), 1243–1275.

Banavath, J. N., Chakradhar, T., Pandit, V., Konduru, S., Guduru, K. K., Akila, C. S., Podha, S., & Puli, C. O. (2018). Stress inducible overexpression of AtHDG11 leads to improved drought and salt stress tolerance in peanut (*Arachis hypogaea* L.). *Frontiers in Chemistry, 6*, 34.

Basu, S., Ramegowda, V., Kumar, A., & Pereira, A. (2016). Plant adaptation to drought stress. *F1000Research*, 5.

Ben Rejeb, K., Lefebvre-De Vos, D., Le Disquet, I., Leprince, A. S., Bordenave, M., Maldiney, R., Jdey, A., Abdelly, C., & Savouré, A. (2015). Hydrogen peroxide produced by NADPH oxidases increases proline accumulation during salt or mannitol stress in *Arabidopsis thaliana*. *New Phytologist, 208*(4), 1138–1148.

Bhargavi, B., Kalpana, K., & Reddy, J. K. (2017). Influence of water stress on morphological and physiological changes in *Andrographis paniculata*. *International Journal of Pure & Applied Bioscience, 5*, 1550–1556.

Bhatt, R. M., & Rao, N. S. (2005). Influence of pod load on response of okra to water stress. *Indian Journal of Plant Physiology, 10*(1), 54.

Bouchemal, K., Bouldjadj, R., Belbekri, M. N., Ykhlef, N., & Djekoun, A. (2017). Differences in antioxidant enzyme activities and oxidative markers in ten wheat (*Triticum durum* Desf.) genotypes in response to drought, heat and paraquat stress. *Archives of Agronomy and Soil Science, 63*(5), 710–722.

Brestic, M., Zivcak, M., Kunderlikova, K., & Allakhverdiev, S. I. (2016). High temperature specifically affects the photoprotective responses of chlorophyll b-deficient wheat mutant lines. *Photosynthesis Research, 130*(1), 251–266.

Bücker-Neto, L., Paiva, A. L. S., Machado, R. D., Arenhart, R. A., & Margis-Pinheiro, M. (2017). Interactions between plant hormones and heavy metals responses. *Genetics and Molecular Biology, 40*(1), 373–386.

Campos, C. N., Ávila, R. G., de Souza, K. R. D., Azevedo, L. M., & Alves, J. D. (2019). Melatonin reduces oxidative stress and promotes drought tolerance in young *Coffea arabica* L. plants. *Agricultural Water Management, 211*, 37–47.

Cao, Y., Luo, Q., Tian, Y., & Meng, F. (2017). Physiological and proteomic analyses of the drought stress response in *Amygdalus Mira (Koehne) Yü et Lu* roots. *BMC Plant Biology, 17*(1), 1–16.

Chen, T., Li, W., Hu, X., Guo, J., Liu, A., & Zhang, B. (2015). A cotton MYB transcription factor, GbMYB5, is positively involved in plant adaptive response to drought stress. *Plant and Cell Physiology, 56*(5), 917–929.

Chen, Y., Wang, H., Hu, W., Wang, S., Snider, J. L., & Zhou, Z. (2017). Co-occurring elevated temperature and waterlogging stresses disrupt cellulose synthesis by altering the expression and activity of carbohydrate balance-associated enzymes during fiber development in cotton. *Environmental and Experimental Botany, 135*, 106–117.

Cheng, X. Q., Zhu, X. F., Tian, W. G., Cheng, W. H., Sun, J., Jin, S. X., & Zhu, H. G. (2017). Genome-wide identification and expression analysis of polyamine oxidase genes in upland cotton (*Gossypium hirsutum* L.). *Plant Cell, Tissue and Organ Culture (PCTOC), 129*(2), 237–249.

Chowdhury, J. A., Karim, M. A., Khaliq, Q. A., & Ahmed, A. U. (2017). Effect of drought stress on bio-chemical change and cell membrane stability of soybean genotypes. *Bangladesh Journal of Agricultural Research, 42*(3), 475–485.

Corradini, E., Foglia, P., Giansanti, P., Gubbiotti, R., Samperi, R., & Lagana, A. (2011). Flavonoids: Chemical properties and analytical methodologies of identification and quantitation in foods and plants. *Natural Product Research, 25*(5), 469–495.

Cortleven, A., Leuendorf, J. E., Frank, M., Pezzetta, D., Bolt, S., & Schmülling, T. (2019). Cytokinin action in response to abiotic and biotic stresses in plants. *Plant, Cell & Environment, 42*(3), 998–1018.

Croteau, R., Kutchan, T. M., & Lewis, N. G. (2000). Natural products (secondary metabolites). *Biochemistry and Molecular Biology of Plants, 24*, 1250–1319.

Dar, M. I., Naikoo, M. I., Rehman, F., Naushin, F., & Khan, F. A. (2016). Proline accumulation in plants: Roles in stress tolerance and plant development. In *Osmolytes and plants acclimation to changing environment: Emerging omics technologies* (pp. 155–166). Springer.

Demetriou, G., Neonaki, C., Navakoudis, E., & Kotzabasis, K. (2007). Salt stress impact on the molecular structure and function of the photosynthetic apparatus—The protective role of polyamines. *Biochimica et Biophysica Acta (BBA)-Bioenergetics, 1767*(4), 272–280.

Deng, B., Yang, K., Zhang, Y., & Li, Z. (2015). The effects of temperature on the germination behavior of white, yellow, red and purple maize plant seeds. *Acta Physiologiae Plantarum, 37*(8), 1–11.

Di, H., Tian, Y., Zu, H., Meng, X., Zeng, X., & Wang, Z. (2015). Enhanced salinity tolerance in transgenic maize plants expressing a *BADH* gene from *Atriplex micrantha*. *Euphytica, 206*(3), 775–783.

Docimo, T., Reichelt, M., Schneider, B., Kai, M., Kunert, G., Gershenzon, J., & D'Auria, J. C. (2012). The first step in the biosynthesis of cocaine in *Erythroxylum coca*: The characterization of arginine and ornithine decarboxylases. *Plant Molecular Biology, 78*(6), 599–615.

Duque, A. S., López-Gómez, M., Kráčmarová, J., Gomes, C. N., Araújo, S. S., Lluch, C., & Fevereiro, P. (2016). Genetic engineering of polyamine metabolism changes *Medicago truncatula* responses to water deficit. *Plant Cell, Tissue and Organ Culture (PCTOC), 127*(3), 681–690.

Egamberdieva, D., Wirth, S. J., Alqarawi, A. A., Abd Allah, E. F., & Hashem, A. (2017). Phytohormones and beneficial microbes: Essential components for plants to balance stress and fitness. *Frontiers in Microbiology, 8*, 2104.

El Ghachtouli, N., Martin-Tanguy, J., Paynot, M., & Gianinazzi, S. (1996). First-report of the inhibition of arbuscular mycorrhizal infection of *Pisum sativum* by specific and irreversible inhibition of polyamine biosynthesis or by gibberellic acid treatment. *FEBS Letters, 385*(3), 189–192.

Espasandin, F. D., Calzadilla, P. I., Maiale, S. J., Ruiz, O. A., & Sansberro, P. A. (2018). Overexpression of the arginine decarboxylase gene improves tolerance to salt stress in *Lotus tenuis* plants. *Journal of Plant Growth Regulation, 37*(1), 156–165.

Espasandin, F. D., Maiale, S. J., Calzadilla, P., Ruiz, O. A., & Sansberro, P. A. (2014). Transcriptional regulation of 9-cis-epoxycarotenoid dioxygenase (NCED) gene by putrescine accumulation positively modulates ABA synthesis and drought tolerance in *Lotus tenuis* plants. *Plant Physiology and Biochemistry, 76*, 29–35.

Estrada-Melo, A. C., Reid, M. S., & Jiang, C. Z. (2015). Overexpression of an ABA biosynthesis gene using a stress-inducible promoter enhances drought resistance in petunia. *Horticulture Research, 2*(1), 1–9.

Fahad, S., Hussain, S., Saud, S., Hassan, S., Tanveer, M., Ihsan, M. Z., Shah, A. N., Ullah, A., Khan, F., Ullah, S., & Alharby, H. (2016). A combined application of biochar and phosphorus alleviates heat-induced adversities on physiological, agronomical and quality attributes of rice. *Plant Physiology and Biochemistry, 103*, 191–198.

Faik, A., Popova, A. V., & Velitchkova, M. (2016). Effects of long-term action of high temperature and high light on the activity and energy interaction of both photosystems in tomato plants. *Photosynthetica, 54*(4), 611–619.

Falahi, H., Sharifi, M., Chashmi, N. A., & Maivan, H. Z. (2018). Water stress alleviation by polyamines and phenolic compounds in *Scrophularia striata* is mediated by NO and H_2O_2. *Plant Physiology and Biochemistry, 130*, 139–147.

Fariduddin, Q., Varshney, P., Yusuf, M., Ali, A., & Ahmad, A. (2013). Dissecting the role of glycine betaine in plants under abiotic stress. *Plant Stress, 7*(1), 8–18.

Feng, Y., Yin, Y., & Fei, S. (2015). Down-regulation of BdBRI1, a putative brassinosteroid receptor gene produces a dwarf phenotype with enhanced drought tolerance in *Brachypodium distachyon*. *Plant Science, 234*, 163–173.

Fernandez, O., Béthencourt, L., Quero, A., Sangwan, R. S., & Clément, C. (2010). Trehalose and plant stress responses: Friend or foe? *Trends in Plant Science, 15*(7), 409–417.

Figueroa, C. M., & Lunn, J. E. (2016). A tale of two sugars: Trehalose 6-phosphate and sucrose. *Plant Physiology, 172*(1), 7–27.

Foyer, C. H., & Noctor, G. (2000). Tansley review No. 112: Oxygen processing in photosynthesis: Regulation and signalling. *New Phytologist, 146*(3), 359–388.

Garg, A. K., Kim, J. K., Owens, T. G., Ranwala, A. P., Do Choi, Y., Kochian, L. V., & Wu, R. J. (2002). Trehalose accumulation in rice plants confers high tolerance levels to different abiotic stresses. *Proceedings of the National Academy of Sciences, 99*(25), 15898–15903.

Gheidary, S., Akhzari, D., & Pessarakli, M. (2017). Effects of salinity, drought, and priming treatments on seed germination and growth parameters of *Lathyrus sativus* L. *Journal of Plant Nutrition, 40*(10), 1507–1514.

Gholami, M., Fakhari, A. R., & Ghanati, F. (2013). Selective regulation of nicotine and polyamines biosynthesis in tobacco cells by enantiomers of ornithine. *Chirality, 25*(1), 22–27.

Gong, B., Li, X., VandenLangenberg, K. M., Wen, D., Sun, S., Wei, M., Li, Y., Yang, F., Shi, Q., & Wang, X. (2014). Overexpression of S-adenosyl-l-methionine synthetase increased tomato tolerance to alkali stress through polyamine metabolism. *Plant Biotechnology Journal, 12*(6), 694–708.

Gunapati, S., Naresh, R., Ranjan, S., Nigam, D., Hans, A., Verma, P. C., Gadre, R., Pathre, U. V., Sane, A. P., & Sane, V. A. (2016). Expression of GhNAC2 from *G. herbaceum*, improves root growth and imparts tolerance to drought in transgenic cotton and Arabidopsis. *Scientific Reports, 6*(1), 1–14.

Guo, F., Yang, S., Feng, Y., Zhang, J. L., Meng, J. J., Li, X. G., & Wan, S. B. (2016). Effects of heat and high irradiance stress on energy dissipation of photosystem II in low irradiance-adapted peanut leaves. *Russian Journal of Plant Physiology, 63*(1), 62–69.

Gurumurthy, S., Sarkar, B., Vanaja, M., Lakshmi, J., Yadav, S. K., & Maheswari, M. (2019). Morpho-physiological and biochemical changes in black gram (*Vigna mungo* L. Hepper) genotypes under drought stress at flowering stage. *Acta Physiologiae Plantarum, 41*(3), 1–14.

Gururani, M. A., Venkatesh, J., & Tran, L. S. P. (2015). Regulation of photosynthesis during abiotic stress-induced photoinhibition. *Molecular Plant, 8*(9), 1304–1320.

Habben, J. E., Bao, X., Bate, N. J., DeBruin, J. L., Dolan, D., Hasegawa, D., Helentjaris, T. G., Lafitte, R. H., Lova, N., Mo, H., Reimann, K., & Schussler, J. R. (2014). Transgenic alteration of ethylene biosynthesis increases grain yield in maize under field drought-stress conditions. *Plant Biotechnology Journal, 12*(6), 685–693.

Halliwell, B., & Gutteridge, J. M. (2015). *Free radicals in biology and medicine*. Oxford University Press.

Harborne, J. B., & Williams, C. A. (2000). Advances in flavonoid research since 1992. *Phytochemistry, 55*(6), 481–504.

Hasanuzzaman, M., Alam, M., Rahman, A., Hasanuzzaman, M., Nahar, K., & Fujita, M. (2014). Exogenous proline and glycine betaine-mediated upregulation of antioxidant defense and glyoxalase systems provides better protection against salt-induced oxidative stress in two rice (*Oryza sativa* L.) varieties. *BioMed Research International, 2014*.

Hassan, F. A. S., Ali, E. F., & Alamer, K. H. (2018). Exogenous application of polyamines alleviates water stress-induced oxidative stress of *Rosa damascena* Miller var. *trigintipetala* Dieck. *South African Journal of Botany, 116*, 96–102.

He, X., Zhao, X., Gao, L., Shi, X., Dai, X., Liu, Y., Xia, T., & Wang, Y. (2018). Isolation and characterization of key genes that promote flavonoid accumulation in purple-leaf tea (*Camellia sinensis* L.). *Scientific Reports, 8*(1), 1–13.

Helaly, A. A. (2017). Strategies for improvement of horticultural crops against abiotic stresses. *Journal of Hortic, 4*, e107.

Hu, Y., Xia, S., Su, Y., Wang, H., Luo, W., Su, S., & Xiao, L. (2016). Brassinolide increases potato root growth in vitro in a dose-dependent way and alleviates salinity stress. *BioMed Research International, 2016*.

Hussain, H. A., Hussain, S., Khaliq, A., Ashraf, U., Anjum, S. A., Men, S., & Wang, L. (2018). Chilling and drought stresses in crop plants: Implications, cross talk, and potential management opportunities. *Frontiers in Plant Science, 9*, 393.

Jung, H., Lee, D. K., Do Choi, Y., & Kim, J. K. (2015). OsIAA6, a member of the rice Aux/IAA gene family, is involved in drought tolerance and tiller outgrowth. *Plant Science, 236*, 304–312.

Kang, N. Y., Cho, C., Kim, N. Y., & Kim, J. (2012). Cytokinin receptor-dependent and receptor-independent pathways in the dehydration response of *Arabidopsis thaliana*. *Journal of Plant Physiology, 169*(14), 1382–1391.

Kaur, G., & Asthir, B. J. B. P. (2015). Proline: A key player in plant abiotic stress tolerance. *Biologia Plantarum, 59*(4), 609–619.

Kaya, C., Tuna, A. L., Ashraf, M., & Altunlu, H. (2007). Improved salt tolerance of melon (*Cucumis melo* L.) by the addition of proline and potassium nitrate. *Environmental and Experimental Botany, 60*(3), 397–403.

Kazan, K. (2013). Auxin and the integration of environmental signals into plant root development. *Annals of Botany, 112*(9), 1655–1665.

Kazan, K. (2015). Diverse roles of jasmonates and ethylene in abiotic stress tolerance. *Trends in Plant Science, 20*(4), 219–229.

Ke, Q., Wang, Z., Ji, C. Y., Jeong, J. C., Lee, H. S., Li, H., Xu, B., Deng, X., & Kwak, S. S. (2015). Transgenic poplar expressing *Arabidopsis* YUCCA6 exhibits auxin-overproduction phenotypes and increased tolerance to abiotic stress. *Plant Physiology and Biochemistry, 94*, 19–27.

Khan, M. I. R., Reddy, P. S., Ferrante, A., Khan, N. A., & (Eds.). (2019). *Plant signaling molecules: Role and regulation under stressful environments*. Woodhead Publishing.

Khan, W., Prithiviraj, B., & Smith, D. L. (2003). Photosynthetic responses of corn and soybean to foliar application of salicylates. *Journal of Plant Physiology, 160*(5), 485–492.

Khare, T., Srivastav, A., Shaikh, S., & Kumar, V. (2018). Polyamines and their metabolic engineering for plant salinity stress tolerance. In *Salinity responses and tolerance in plants* (Vol. 1, pp. 339–358). Springer.

Khoo, H. E., Azlan, A., Tang, S. T., & Lim, S. M. (2017). Anthocyanidins and anthocyanins: Colored pigments as food, pharmaceutical ingredients, and the potential health benefits. *Food & Nutrition Research, 61*(1), 1361779.

Khurana, N., Sharma, N., & Khurana, P. (2017). Overexpression of a heat stress inducible, wheat myo-inositol-1-phosphate synthase 2 (TaMIPS2) confers tolerance to various abiotic stresses in *Arabidopsis thaliana*. *Agri Gene, 6*, 24–30.

Kohli, S. K., Handa, N., Sharma, A., Kumar, V., Kaur, P., & Bhardwaj, R. (2017). Synergistic effect of 24-epibrassinolide and salicylic acid on photosynthetic efficiency and gene expression in *Brassica juncea* L. under Pb stress. *Turkish Journal of Biology, 41*(6), 943–953.

Kosar, F., Akram, N. A., Sadiq, M., Al-Qurainy, F., & Ashraf, M. (2019). Trehalose: A key organic osmolyte effectively involved in plant abiotic stress tolerance. *Journal of Plant Growth Regulation, 38*(2), 606–618.

Kour, D., Rana, K. L., Kaur, T., Sheikh, I., Yadav, A. N., Kumar, V., Dhaliwal, H. S., & Saxena, A. K. (2020). Microbe-mediated alleviation of drought stress and acquisition of phosphorus in great millet (*Sorghum bicolour* L.) by drought-adaptive and phosphorus-solubilizing microbes. *Biocatalysis and Agricultural. Biotechnology, 23*, 101501.

Kour, D., Rana, K. L., Yadav, N., Yadav, A. N., Singh, J., Rastegari, A. A., & Saxena, A. K. (2019). *Agriculturally and industrially important fungi: Current developments and potential biotechnological applications. Recent advancement in white biotechnology through fungi* (pp. 1–64). Springer.

Kumar, S., Chashoo, G., Saxena, A. K., & Pandey, A. K. (2013). *Parthenium hysterophorus*: A probable source of anticancer, antioxidant and anti-HIV agents. *BioMed Research International, 2013*.

Kuznetsov, V. V., & Shevyakova, N. I. (2007). Polyamines and stress tolerance of plants. *Plant Stress, 1*(1), 50–71.

Le Gall, H., Philippe, F., Domon, J. M., Gillet, F., Pelloux, J., & Rayon, C. (2015). Cell wall metabolism in response to abiotic stress. *Plants, 4*(1), 112–166.

Li, H., Mo, Y., Cui, Q., Yang, X., Guo, Y., Wei, C., Yang, J., Zhang, Y., Ma, J. X., & Zhang, X. (2019a). Transcriptomic and physiological analyses reveal drought adaptation strategies in drought-tolerant and-susceptible watermelon genotypes. *Plant Science, 278*, 32–43.

Li, H. W., Zang, B. S., Deng, X. W., & Wang, X. P. (2011). Overexpression of the trehalose-6-phosphate synthase gene OsTPS1 enhances abiotic stress tolerance in rice. *Planta, 234*(5), 1007–1018.

Li, K., Xing, C., Yao, Z., & Huang, X. (2017). Pbr MYB 21, a novel MYB protein of *Pyrus betulaefolia*, functions in drought tolerance and modulates polyamine levels by regulating arginine decarboxylase gene. *Plant Biotechnology Journal, 15*(9), 1186–1203.

Li, S., Jin, H., & Zhang, Q. (2016). The effect of exogenous spermidine concentration on polyamine metabolism and salt tolerance in Zoysiagrass (*Zoysia japonica* Steud) subjected to short-term salinity stress. *Frontiers in Plant Science, 7*, 1221.

Li, Y., Zhang, H., Zhang, Q., Liu, Q., Zhai, H., Zhao, N., & He, S. (2019b). An AP2/ERF gene, IbRAP2-12, from sweetpotato is involved in salt and drought tolerance in transgenic *Arabidopsis*. *Plant Science, 281*, 19–30.

Lichtenthaler, H. K., & Burkart, S. (1999). Photosynthesis and high light stress. *Bulg. J. Plant Physiol, 25*(3–4), 3–16.

Lim, C. W., Baek, W., Jung, J., Kim, J. H., & Lee, S. C. (2015). Function of ABA in stomatal defense against biotic and drought stresses. *International Journal of Molecular Sciences, 16*(7), 15251–15270.

Liu, Q. H., Xiu, W. U., Chen, B. C., & Jie, G. A. O. (2014). Effects of low light on agronomic and physiological characteristics of rice including grain yield and quality. *Rice Science, 21*(5), 243–251.

Liu, Y., Gu, D., Wu, W., Wen, X., & Liao, Y. (2013). The relationship between polyamines and hormones in the regulation of wheat grain filling. *PLoS One, 8*(10), e78196.

Luo, J., Liu, M., Zhang, C., Zhang, P., Chen, J., Guo, Z., & Lu, S. (2017). Transgenic centipedegrass (*Eremochloa ophiuroides* [Munro] Hack.) overexpressing S-adenosylmethionine decarboxylase (SAMDC) gene for improved cold tolerance through involvement of H_2O_2 and NO signaling. *Frontiers in Plant Science, 8*, 1655.

Mahjoor, F., Ghaemi, A. A., & Golabi, M. H. (2016). Interaction effects of water salinity and hydroponic growth medium on eggplant yield, water-use efficiency, and evapotranspiration. *International Soil and Water Conservation Research, 4*(2), 99–107.

Marco, F., Bitrián, M., Carrasco, P., Alcázar, R., & Tiburcio, A. F. (2015). Polyamine biosynthesis engineering as a tool to improve plant resistance to abiotic stress. In *Genetic manipulation in plants for mitigation of climate change* (pp. 103–116). Springer.

Mishra, A., Kumar, S., & Pandey, A. K. (2013a). Scientific validation of the medicinal efficacy of *Tinospora cordifolia*. *The Scientific World Journal, 2013*.

Mishra, A., Sharma, A. K., Kumar, S., Saxena, A. K., & Pandey, A. K. (2013b, 2013). *Bauhinia variegata* leaf extracts exhibit considerable antibacterial, antioxidant, and anticancer activities. *BioMed Research International*.

Mishra, P., Jain, A., Takabe, T., Tanaka, Y., Negi, M., Singh, N., Jain, N., Mishra, V., Maniraj, R., Krishnamurthy, S. L., Sreevathsa, R., Singh, N. K., & Rai, V. (2019). Heterologous expression of serine hydroxymethyltransferase-3 from rice confers tolerance to salinity stress in *E. coli* and *Arabidopsis*. *Frontiers in Plant Science, 10*, 217.

Mohammadi, H., Ghorbanpour, M., & Brestic, M. (2018). Exogenous putrescine changes redox regulations and essential oil constituents in field-grown *Thymus vulgaris* L. under well-watered and drought stress conditions. *Industrial Crops and Products, 122*, 119–132.

Muradoglu, F., Gundogdu, M., Ercisli, S., Encu, T., Balta, F., Jaafar, H. Z., & Zia-Ul-Haq, M. (2015). Cadmium toxicity affects chlorophyll a and b content, antioxidant enzyme activities and mineral nutrient accumulation in strawberry. *Biological Research, 48*, 1–7.

Mustafavi, S. H., Badi, H. N., Sękara, A., Mehrafarin, A., Janda, T., Ghorbanpour, M., & Rafiee, H. (2018). Polyamines and their possible mechanisms involved in plant physiological processes and elicitation of secondary metabolites. *Acta Physiologiae Plantarum, 40*(6), 1–19.

Najla, S., Sanoubar, R., & Murshed, R. (2012). Morphological and biochemical changes in two parsley varieties upon water stress. *Physiology and Molecular Biology of Plants, 18*(2), 133–139.

Nisa, Z. U., Chen, C., Yu, Y., Chen, C., Mallano, A. I., Xiang-bo, D., Xiao-li, S., & Yan-ming, Z. (2016). Constitutive overexpression of myo-inositol-1-phosphate synthase gene (GsMIPS2) from *Glycine soja* confers enhanced salt tolerance at various growth stages in *Arabidopsis*. *Journal of Northeast Agricultural University (English Edition), 23*(2), 28–44.

Nuccio, M. L., Wu, J., Mowers, R., Zhou, H. P., Meghji, M., Primavesi, L. F., Paul, M. J., Gao, Y., Haque, E., Basu, S. S., & Lagrimini, L. M. (2015). Expression of trehalose-6-phosphate phosphatase in maize ears improves yield in well-watered and drought conditions. *Nature Biotechnology, 33*(8), 862–869.

Pál, M., Szalai, G., & Janda, T. (2015). Speculation: Polyamines are important in abiotic stress signaling. *Plant Science, 237*, 16–23.

Pandey, R., Gupta, A., Chowdhary, A., Pal, R. K., & Rajam, M. V. (2015). Over-expression of mouse ornithine decarboxylase gene under the control of fruit-specific promoter enhances fruit quality in tomato. *Plant Molecular Biology, 87*(3), 249–260.

Park, S., Kim, D. H., Lee, J. Y., Ha, S. H., & Lim, S. H. (2017). Comparative analysis of two flavonol synthases from different-colored onions provides insight into flavonoid biosynthesis. *Journal of Agricultural and Food Chemistry, 65*(26), 5287–5298.

Parvin, S., Lee, O. R., Sathiyaraj, G., Khorolragchaa, A., Kim, Y. J., & Yang, D. C. (2014). Spermidine alleviates the growth of saline-stressed ginseng seedlings through antioxidative defense system. *Gene, 537*(1), 70–78.

Pathak, M. R., Teixeira da Silva, J. A., & Wani, S. H. (2014). Polyamines in response to abiotic stress tolerance through transgenic approaches. *GM Crops & Food, 5*(2), 87–96.

Pegg, A. E. (2016). Functions of polyamines in mammals. *Journal of Biological Chemistry, 291*(29), 14904–14912.

Per, T. S., Khan, N. A., Reddy, P. S., Masood, A., Hasanuzzaman, M., Khan, M. I. R., & Anjum, N. A. (2017). Approaches in modulating proline metabolism in plants for salt and drought stress tolerance: Phytohormones, mineral nutrients and transgenics. *Plant Physiology and Biochemistry, 115*, 126–140.

Pospíšilová, H., Jiskrova, E., Vojta, P., Mrizova, K., Kokáš, F., Čudejková, M. M., Bergougnoux, V., Plihal, O., Klimesova, J., Novak, O., Dzurova, L., Frebort, I., & Galuszka, P. (2016). Transgenic barley overexpressing a cytokinin dehydrogenase gene shows greater tolerance to drought stress. *New Biotechnology, 33*(5), 692–705.

Prathyusha, I. V. S. N., & Chaitanya, K. V. (2019). Effect of water stress on the physiological and biochemical responses of two different *Coleus (Plectranthus)* species. *Biologia Futura, 70*, 312–322.

Procházková, D., Boušová, I., & Wilhelmová, N. (2011). Antioxidant and prooxidant properties of flavonoids. *Fitoterapia, 82*(4), 513–523.

Qiu, J., Sun, S., Luo, S., Zhang, J., Xiao, X., Zhang, L., Wang, F., & Liu, S. (2014). Arabidopsis AtPAP1 transcription factor induces anthocyanin production in transgenic *Taraxacum brevicorniculatum*. *Plant Cell Reports, 33*(4), 669–680.

Rasool, S., Urwat, U., Nazir, M., Zargar, S. M., & Zargar, M. Y. (2018). Cross talk between phytohormone signaling pathways under abiotic stress conditions and their metabolic engineering for conferring abiotic stress tolerance. In *Abiotic stress-Mediated Sensing and signaling in plants: An omics perspective* (pp. 329–350).

Rastegari, A. A., Yadav, A. N., & Yadav, N. (2020). *New and future developments in microbial biotechnology and bioengineering: Trends of microbial biotechnology for sustainable agriculture and biomedicine systems: Perspectives for human health*. Elsevier.

Reis, R. S., de Moura Vale, E., Heringer, A. S., Santa-Catarina, C., & Silveira, V. (2016). Putrescine induces somatic embryo development and proteomic changes in embryogenic callus of sugarcane. *Journal of Proteomics, 130*, 170–179.

Riaz, M., Arif, M. S., Ashraf, M. A., Mahmood, R., Yasmeen, T., Shakoor, M. B., Shahzad, S. M., Ali, M., Saleem, I., Arif, M., & Fahad, S. (2019). A comprehensive review on rice responses and tolerance to salt stress. In *Advances in rice research for abiotic stress tolerance* (pp. 133–158).

Romero, F. M., Maiale, S. J., Rossi, F. R., Marina, M., Ruíz, O. A., & Gárriz, A. (2018). Polyamine metabolism responses to biotic and abiotic stress. *Polyamines*, 37–49.

Saha, J., Brauer, E. K., Sengupta, A., Popescu, S. C., Gupta, K., & Gupta, B. (2015). Polyamines as redox homeostasis regulators during salt stress in plants. *Frontiers in Environmental Science, 3*, 21.

Saha, J., & Giri, K. (2017). Molecular phylogenomic study and the role of exogenous spermidine in the metabolic adjustment of endogenous polyamine in two rice cultivars under salt stress. *Gene, 609*, 88–103.

Sahni, S., Prasad, B. D., Liu, Q., Grbic, V., Sharpe, A., Singh, S. P., & Krishna, P. (2016). Overexpression of the brassinosteroid biosynthetic gene DWF4 in *Brassica napus* simultaneously increases seed yield and stress tolerance. *Scientific Reports, 6*(1), 1–14.

Saini, S., Sharma, I., & Pati, P. K. (2015). Versatile roles of brassinosteroid in plants in the context of its homoeostasis, signaling and crosstalks. *Frontiers in Plant Science, 6*, 950.

Sairam, R. K., & Tyagi, A. (2004). Physiology and molecular biology of salinity stress tolerance in plants. *Current Science*, 407–421.

Sang, Q., Shan, X., An, Y., Shu, S., Sun, J., & Guo, S. (2017). Proteomic analysis reveals the positive effect of exogenous spermidine in tomato seedlings' response to high-temperature stress. *Frontiers in Plant Science, 8*, 120.

Schrader, S. M., Wise, R. R., Wacholtz, W. F., Ort, D. R., & Sharkey, T. D. (2004). Thylakoid membrane responses to moderately high leaf temperature in Pima cotton. *Plant, Cell & Environment, 27*(6), 725–735.

Sekhar, P. N., Amrutha, R. N., Sangam, S., Verma, D. P. S., & Kishor, P. K. (2007). Biochemical characterization, homology modeling and docking studies of ornithine δ-aminotransferase—an important enzyme in proline biosynthesis of plants. *Journal of Molecular Graphics and Modelling, 26*(4), 709–719.

Seneviratne, M., Weerasundara, L., Ok, Y. S., Rinklebe, J., & Vithanage, M. (2017). Phytotoxicity attenuation in *Vigna radiata* under heavy metal stress at the presence of biochar and N fixing bacteria. *Journal of Environmental Management, 186*, 293–300.

Sequeramutiozabal, M. I., Erban, A., Kopka, J., Atanasov, K. E., Bastida, J., Fotopoulos, V., Alcazar, R., & Tiburcio, A. F. (2016). Global metabolic profiling of *Arabidopsis polyamine oxidase 4 (AtPAO4)* loss-of-function mutants exhibiting delayed dark-induced senescence. *Frontiers in Plant Science, 7*, 173.

Shah, A., & Smith, D. L. (2020). Flavonoids in agriculture: Chemistry and roles in, biotic and abiotic stress responses, and microbial associations. *Agronomy, 10*(8), 1209.

Shahid, M. A., Sarkhosh, A., Khan, N., Balal, R. M., Ali, S., Rossi, L., Gomez, C., Mattson, N., Nasim, W., & Garcia-Sanchez, F. (2020). Insights into the physiological and biochemical impacts of salt stress on plant growth and development. *Agronomy, 10*(7), 938.

Sharma, A., Shahzad, B., Kumar, V., Kohli, S. K., Sidhu, G. P. S., Bali, A. S., Handa, N., Kapoor, D., Bhardwaj, R., & Zheng, B. (2019b). Phytohormones regulate accumulation of osmolytes under abiotic stress. *Biomolecules, 9*(7), 285.

Sharma, A., Shahzad, B., Rehman, A., Bhardwaj, R., Landi, M., & Zheng, B. (2019a). Response of phenylpropanoid pathway and the role of polyphenols in plants under abiotic stress. *Molecules, 24*(13), 2452.

Sharma, A., Thakur, S., Kumar, V., Kanwar, M. K., Kesavan, A. K., Thukral, A. K., Bhardwaj, R., Alam, P., & Ahmad, P. (2016). Pre-sowing seed treatment with 24-epibrassinolide ameliorates pesticide stress in *Brassica juncea* L. through the modulation of stress markers. *Frontiers in Plant Science, 7*, 1569.

Shi, H., & Chan, Z. (2014). Improvement of plant abiotic stress tolerance through modulation of the polyamine pathway. *Journal of Integrative Plant Biology, 56*(2), 114–121.

Shi, J., Habben, J. E., Archibald, R. L., Drummond, B. J., Chamberlin, M. A., Williams, R. W., Lafitte, H. R., & Weers, B. P. (2015). Overexpression of ARGOS genes modifies plant sensitivity to ethylene, leading to improved drought tolerance in both Arabidopsis and maize. *Plant Physiology, 169*(1), 266–282.

Shu, S., Yuan, L. Y., Guo, S. R., Sun, J., & Yuan, Y. H. (2013). Effects of exogenous spermine on chlorophyll fluorescence, antioxidant system and ultrastructure of chloroplasts in *Cucumis sativus* L. under salt stress. *Plant Physiology and Biochemistry, 63*, 209–216.

Sidhu, G. P. S., Bali, A. S., Singh, H. P., Batish, D. R., & Kohli, R. K. (2018). Ethylenediamine disuccinic acid enhanced phytoextraction of nickel from contaminated soils using *Coronopus didymus* (L.) Sm. *Chemosphere, 205*, 234–243.

Sidhu, G. P. S., Bali, A. S., Singh, H. P., Batish, D. R., & Kohli, R. K. (2020). Insights into the tolerance and phytoremediation potential of *Coronopus didymus* L. (Sm) grown under zinc stress. *Chemosphere, 244*, 125350.

Sidhu, G. P. S., Singh, H. P., Batish, D. R., & Kohli, R. K. (2017a). Appraising the role of environment friendly chelants in alleviating lead by *Coronopus didymus* from Pb-contaminated soils. *Chemosphere, 182*, 129–136.

Sidhu, G. P. S., Singh, H. P., Batish, D. R., & Kohli, R. K. (2017b). Tolerance and hyperaccumulation of cadmium by a wild, unpalatable herb *Coronopus didymus* (L.) Sm. (Brassicaceae). *Ecotoxicology and Environmental Safety, 135*, 209–215.

Slama, I., Abdelly, C., Bouchereau, A., Flowers, T., & Savouré, A. (2015). Diversity, distribution and roles of osmoprotective compounds accumulated in halophytes under abiotic stress. *Annals of Botany, 115*(3), 433–447.

Song, J., Zhang, R., Yue, D., Chen, X., Guo, Z., Cheng, C., Hu, M., Zhang, J., & Zhang, K. (2018). Co-expression of *ApGSMT2g* and *ApDMT2g* in cotton enhances salt tolerance and increases seed cotton yield in saline fields. *Plant Science, 274*, 369–382.

Songstad, D. D., Petolino, J. F., Voytas, D. F., & Reichert, N. A. (2017). Genome editing of plants. *Critical Reviews in Plant Science, 36*, 1–23.

Srivastava, S., & Srivastava, M. (2014). Morphological changes and antioxidant activity of *Stevia rebaudiana* under water stress. *American Journal of Plant Sciences, 5*(22), 3417.

Sun, P., Zhu, X., Huang, X., & Liu, J. H. (2014). Overexpression of a stress-responsive MYB transcription factor of *Poncirus trifoliata* confers enhanced dehydration tolerance and increases polyamine biosynthesis. *Plant Physiology and Biochemistry, 78*, 71–79.

Sun, Y., Liu, X., Zhai, H., Gao, H., Yao, Y., & Du, Y. (2016). Responses of photosystem II photochemistry and the alternative oxidase pathway to heat stress in grape leaves. *Acta Physiologiae Plantarum, 38*(10), 1–8.

Suprasanna, P., Nikalje, G. C., & Rai, A. N. (2016). Osmolyte accumulation and implications in plant abiotic stress tolerance. In *Osmolytes and plants acclimation to changing environment: Emerging omics technologies* (pp. 1–12). Springer.

Surekha, C. H., Kumari, K. N., Aruna, L. V., Suneetha, G., Arundhati, A., & Kishor, P. K. (2014). Expression of the *Vigna aconitifolia P5CSF129A* gene in transgenic pigeonpea enhances proline accumulation and salt tolerance. *Plant Cell, Tissue and Organ Culture (PCTOC), 116*(1), 27–36.

Tabatabaie, S. J., & Nazari, J. (2007). Influence of nutrient concentrations and NaCl salinity on the growth, photosynthesis, and essential oil content of peppermint and lemon verbena. *Turkish Journal of Agriculture and Forestry, 31*(4), 245–253.

Talaat, N. B., & Shawky, B. T. (2011). Influence of arbuscular mycorrhizae on yield, nutrients, organic solutes, and antioxidant enzymes of two wheat cultivars under salt stress. *Journal of Plant Nutrition and Soil Science, 174*(2), 283–291.

Talaat, N. B., Shawky, B. T., & Ibrahim, A. S. (2015). Alleviation of drought-induced oxidative stress in maize (*Zea mays* L.) plants by dual application of 24-epibrassinolide and spermine. *Environmental and Experimental Botany, 113*, 47–58.

Tan, Y., Hu, W., Xu, X., Zhou, J., Wang, C., Liu, X., & Cheng, G. (2017). *Polyamine plays a role in subculture growth of in vitro callus of indica rice*. Acta Biologica Cracoviensia s. Botanica.

Tian, F., Wang, W., Liang, C., Wang, X., Wang, G., & Wang, W. (2017). Overaccumulation of glycine betaine makes the function of the thylakoid membrane better in wheat under salt stress. *The Crop Journal, 5*(1), 73–82.

Tian, X., Wang, Z., Li, X., Lv, T., Liu, H., Wang, L., Niu, H., & Bu, Q. (2015). Characterization and functional analysis of pyrabactin resistance-like abscisic acid receptor family in rice. *Rice, 8*(1), 1–13.

Tiwari, P., Bajpai, M., Singh, L. K., Mishra, S., & Yadav, A. N. (2020). Phytohormones producing fungal communities: Metabolic engineering for abiotic stress tolerance in crops. In *Agriculturally important fungi for sustainable agriculture* (pp. 171–197). Springer.

Upadhyay, A. K., Singh, N. K., Singh, R., & Rai, U. N. (2016). Amelioration of arsenic toxicity in rice: Comparative effect of inoculation of *Chlorella vulgaris* and *Nannochloropsis* sp. on growth, biochemical changes and arsenic uptake. *Ecotoxicology and Environmental Safety, 124*, 68–73.

Verma, P., Yadav, A. N., Khannam, K. S., Kumar, S., Saxena, A. K., & Suman, A. (2016). Molecular diversity and multifarious plant growth promoting attributes of Bacilli associated with wheat (*Triticum aestivum* L.) rhizosphere from six diverse agro-ecological zones of India. *Journal of Basic Microbiology, 56*(1), 44–58.

Verma, P., Yadav, A. N., Khannam, K. S., Panjiar, N., Kumar, S., Saxena, A. K., & Suman, A. (2015). Assessment of genetic diversity and plant growth promoting attributes of psychrotolerant bacteria allied with wheat (*Triticum aestivum*) from the northern hills zone of India. *Annals of Microbiology, 65*(4), 1885–1899.

Vishwakarma, K., Upadhyay, N., Kumar, N., Yadav, G., Singh, J., Mishra, R. K., Kumar, V., Verma, R., Upadhyay, R. G., Pandey, M., & Sharma, S. (2017). Abscisic acid signaling and abiotic stress tolerance in plants: A review on current knowledge and future prospects. *Frontiers in Plant Science, 8*, 161.

Vob, U., Bishopp, A., Farcot, E., & Bennett, M. J. (2014). Modelling hormonal response and development. *Trends in Plant Science, 19*(5), 311–319.

Vuosku, J., Karppinen, K., Muilu-Mäkelä, R., Kusano, T., Sagor, G. H. M., Avia, K., Alakärppä, E., Kestilä, J., Suokas, M., Nickolov, K., Hamberg, L., Savolainen, O., Haggman, H., & Sarjala, T. (2018). Scots pine aminopropyltransferases shed new light on evolution of the polyamine biosynthesis pathway in seed plants. *Annals of Botany, 121*(6), 1243–1256.

Wan, L., Wu, Y., Huang, J., Dai, X., Lei, Y., Yan, L., Jiang, H., Zhang, J., Varshney, R. K., & Liao, B. (2014). Identification of ERF genes in peanuts and functional analysis of *AhERF008* and *AhERF019* in abiotic stress response. *Functional & Integrative Genomics, 14*(3), 467–477.

Wang, C., Zhi, S., Liu, C., Xu, F., Zhao, A., Wang, X., Tang, X., Li, Z., Huang, P., & Yu, M. (2017). Isolation and characterization of a novel chalcone synthase gene family from mulberry. *Plant Physiology and Biochemistry, 115*, 107–118.

Wang, L., Deng, F., Ren, W. J., & Yang, W. Y. (2013). Effects of shading on starch pasting characteristics of indica hybrid rice (*Oryza sativa* L.). *PLoS One, 8*(7), e68220.

Wang, P., Zhang, L., Jiang, X., Dai, X., Xu, L., Li, T., Xing, D., Li, Y., Li, M., Gao, L., & Xia, T. (2018). Evolutionary and functional characterization of leucoanthocyanidin reductases from *Camellia sinensis*. *Planta, 247*(1), 139–154.

Wang, W., & Liu, J. H. (2016). CsPAO4 of *Citrus sinensis* functions in polyamine terminal catabolism and inhibits plant growth under salt stress. *Scientific Reports, 6*(1), 1–15.

Wang, X., Shu, C., Li, H., Hu, X., & Wang, Y. (2014). Effects of 0.01% brassinolide solution application on yield of rice and its resistance to autumn low-temperature damage. *Acta Agriculturae Jiangxi, 26*(5), 36–38.

Wang, Y., Yang, L., Chen, X., Ye, T., Zhong, B., Liu, R., Wu, Y., & Chan, Z. (2016). Major latex protein-like protein 43 (MLP43) functions as a positive regulator during abscisic acid responses and confers drought tolerance in *Arabidopsis thaliana*. *Journal of Experimental Botany, 67*(1), 421–434.

Wani, M. A., Jan, N., Qazi, H. A., Andrabi, K. I., & John, R. (2018). Cold stress induces biochemical changes, fatty acid profile, antioxidant system and gene expression in *Capsella bursa pastoris* L. *Acta Physiologiae Plantarum, 40*(9), 1–14.

Wani, T. A., Pandith, S. A., Gupta, A. P., Chandra, S., Sharma, N., & Lattoo, S. K. (2017). Molecular and functional characterization of two isoforms of chalcone synthase and their expression analysis in relation to flavonoid constituents in *Grewia asiatica* L. *PLoS One, 12*(6), e0179155.

Wei, D., Zhang, W., Wang, C., Meng, Q., Li, G., Chen, T. H., & Yang, X. (2017). Genetic engineering of the biosynthesis of glycinebetaine leads to alleviate salt-induced potassium efflux and enhances salt tolerance in tomato plants. *Plant Science, 257*, 74–83.

Wellmann, F., Lukačin, R., Moriguchi, T., Britsch, L., Schiltz, E., & Matern, U. (2002). Functional expression and mutational analysis of flavonol synthase from *Citrus unshiu*. *European Journal of Biochemistry, 269*(16), 4134–4142.

Wu, X., He, J., Chen, J., Yang, S., & Zha, D. (2014). Alleviation of exogenous 6-benzyladenine on two genotypes of eggplant (*Solanum melongena* Mill.) growth under salt stress. *Protoplasma, 251*(1), 169–176.

Wuddineh, W., Minocha, R., & Minocha, S. C. (2018). Polyamines in the context of metabolic networks. *Polyamines*, 1–23.

Xia, X. J., Gao, C. J., Song, L. X., Zhou, Y. H., Shi, K. A. I., & Yu, J. Q. (2014). Role of H_2O_2 dynamics in brassinosteroid-induced stomatal closure and opening in *Solanum lycopersicum*. *Plant, Cell & Environment, 37*(9), 2036–2050.

Xu, L. (2015). The effect of polyamineon flower bud differentiation and bud germination of chrysanthemum. *Shandong Agricultural University, 2*, 31–36.

Yadav, A. N., Kour, D., Rana, K. L., Yadav, N., Singh, B., Chauhan, V. S., Rastegari, A. A., Hesham, A. E. L., & Gupta, V. K. (2019). Metabolic engineering to synthetic biology of secondary metabolites production. In *New and future developments in microbial biotechnology and bioengineering* (pp. 279–320). Elsevier.

Yadav, A. N., Rastegari, A. A., Yadav, N., & (Eds.). (2021). *Microbiomes of extreme environments: Biodiversity and biotechnological applications*. CRC Press.

Yadav, A. N., Sachan, S. G., Verma, P., & Saxena, A. K. (2015a). Prospecting cold deserts of north western Himalayas for microbial diversity and plant growth promoting attributes. *Journal of Bioscience and Bioengineering, 119*(6), 683–693.

Yadav, A. N., Sachan, S. G., Verma, P., Tyagi, S. P., Kaushik, R., & Saxena, A. K. (2015b). Culturable diversity and functional annotation of psychrotrophic bacteria from cold desert of Leh Ladakh (India). *World Journal of Microbiology and Biotechnology, 31*(1), 95–108.

Yadav, A. N., Singh, J., Rastegari, A. A., & Yadav, N. (Eds.). (2020). *Plant microbiomes for sustainable agriculture* (Vol. 25). Springer.

Yahyaa, M., Ali, S., Davidovich-Rikanati, R., Ibdah, M., Shachtier, A., Eyal, Y., Lewinsohn, E., & Ibdah, M. (2017). Characterization of three chalcone synthase-like genes from apple (*Malus* x *domestica* Borkh.). *Phytochemistry, 140*, 125–133.

Ye, J., Wang, S., Deng, X., Yin, L., Xiong, B., & Wang, X. (2016). Melatonin increased maize (*Zea mays* L.) seedling drought tolerance by alleviating drought-induced photosynthetic inhibition and oxidative damage. *Acta Physiologiae Plantarum, 38*(2), 48.

You, L., Song, Q., Wu, Y., Li, S., Jiang, C., Chang, L., Yang, X., & Zhang, J. (2019). Accumulation of glycine betaine in transplastomic potato plants expressing choline oxidase confers improved drought tolerance. *Planta, 249*(6), 1963–1975.

Zaheer, I. E., Ali, S., Rizwan, M., Farid, M., Shakoor, M. B., Gill, R. A., Najeeb, U., Iqbal, N., & Ahmad, R. (2015). Citric acid assisted phytoremediation of copper by *Brassica napus* L. *Ecotoxicology and Environmental Safety, 120*, 310–317.

Zandalinas, S. I., Mittler, R., Balfagón, D., Arbona, V., & Gómez-Cadenas, A. (2018). Plant adaptations to the combination of drought and high temperatures. *Physiologia Plantarum, 162*(1), 2–12.

Zang, X., Geng, X., Wang, F., Liu, Z., Zhang, L., Zhao, Y., Tian, X., Ni, Z., Yao, Y., Xin, M., Hu, Z., Sun, Q., & Peng, H. (2017). Overexpression of wheat ferritin gene *TaFER-5B* enhances tolerance to heat stress and other abiotic stresses associated with the ROS scavenging. *BMC Plant Biology, 17*(1), 1–13.

Zhang, G. W., Xu, S. C., Hu, Q. Z., Mao, W. H., & Gong, Y. M. (2014). Putrescine plays a positive role in salt-tolerance mechanisms by reducing oxidative damage in roots of vegetable soybean. *Journal of Integrative Agriculture, 13*(2), 349–357.

Zhang, J., Yu, H., Zhang, Y., Wang, Y., Li, M., Zhang, J., Duan, L., Zhang, M., & Li, Z. (2016). Increased abscisic acid levels in transgenic maize overexpressing *AtLOS5* mediated root ion fluxes and leaf water status under salt stress. *Journal of Experimental Botany, 67*(5), 1339–1355.

Zhang, L., Xu, B., Wu, T., Yang, Y., Fan, L., Wen, M., & Sui, J. (2017). Transcriptomic profiling of two Pak Choi varieties with contrasting anthocyanin contents provides an insight into structural and regulatory genes in anthocyanin biosynthetic pathway. *BMC Genomics, 18*(1), 1–11.

Zhang, Y., Hu, X. H., Shi, Y., Zou, Z. R., Yan, F., Zhao, Y. Y., Zhang, H., & Zhao, J. Z. (2013). Beneficial role of exogenous spermidine on nitrogen metabolism in tomato seedlings exposed to saline-alkaline stress. *Journal of the American Society for Horticultural Science, 138*(1), 38–49.

Zhang, Z., Wang, Y., Chang, L., Zhang, T., An, J., Liu, Y., Cao, Y., Zhao, X., Hu, T., & Yang, P. (2015). *MsZEP*, a novel zeaxanthin epoxidase gene from alfalfa (*Medicago sativa*), confers drought and salt tolerance in transgenic tobacco. Plant Cell Reports, 35(2), 439–453.

Zou, M., Yuan, L., Zhu, S., Liu, S., Ge, J., & Wang, C. (2017). Effects of heat stress on photosynthetic characteristics and chloroplast ultrastructure of a heat-sensitive and heat-tolerant cultivar of wucai (*Brassica campestris* L.). *Acta Physiologiae Plantarum, 39*(1), 1–10.

Zulfiqar, F., Akram, N. A., & Ashraf, M. (2020). Osmoprotection in plants under abiotic stresses: New insights into a classical phenomenon. *Planta, 251*(1), 1–17.

Chapter 3
Enhancing Photosynthetic Efficiency of Crop Through Metabolic Engineering

Misbah Naz, Muhammad Ammar Raza, Muhammad Tariq, Shanshan Qi, Zhicong Dai, and Daolin Du

Abstract Several current findings have shown that enhancing the photosynthetic process through genetic engineering could provide a technique to boost crop yield. Photosynthesis is the primary predictor of crop output and crop efficacy in capturing light, and converting it into biomass during the growing season, which is the main indicator of yield attributes, whether its biomass or grain. Boosting crop photosynthetic performance by metabolic changes in a changing environment is another area where information is lacking. In the present chapter, we would discuss present and prospective ways for boosting photosynthetic efficiency under different climate conditions, as well as their implications on photosynthesis activity. Our objective is to analyze the existing projects being made to better photosynthesis effectiveness. This paper investigates the impact of modifying the Calvin-Benson (CB) cycle, photorespiration, and electron transport on biomass and seeds yield. It highlights some surprise findings where harmful impacts were seen. In the preceding part, we discussed future possibilities such as integrating polygenic modulation of photosynthetic carbon absorption to boost yield potential and features that address yield variability.

M. Naz · Z. Dai (✉) · D. Du
Institute of Environment and Ecology, School of The Environment and Safety Engineering, Jiangsu University, Zhenjiang, Jiangsu Province, People's Republic of China
e-mail: zcdai@ujs.edu.cn

M. A. Raza
School of Food Science and Biotechnology, Key Laboratory of Fruits and Vegetables Postharvest and Processing Technology Research of Zhejiang Province, Zhejiang Gongshang University, Hangzhou, China

M. Tariq
Department of Pharmacology, Lahore Pharmacy Collage, Lahore, Pakistan

S. Qi
School of Agricultural Engineering, Jiangsu University, Zhenjiang, Jiangsu Province, People's Republic of China

Jiangsu Collaborative Innovation Center of Technology and Material of Water Treatment, Suzhou University of Science and Technology, Suzhou, Jiangsu Province, People's Republic of China

T. Aftab, K. R. Hakeem (eds.), *Metabolic Engineering in Plants*,
https://doi.org/10.1007/978-981-16-7262-0_3

Keywords Photosynthetic efficiency · Metabolic pathways · Enzymes

Introduction

Several new studies have found that genetically enhancing the photosynthetic process could provide a way to increase yield potential. Present article's aim is to strengthen the Calvin-Benson cycle and electron transport (Long et al., 2015; Hall & Richards, 2013). It also addressed how modifying the regulating mechanism could provide other approaches to improve photosynthetic efficiency. Thus, we discuss some obtained updates in photosynthetic genetic manipulation, demonstrating that yields may be enhanced by more than 40% in the greenhouse and the field (Murchie et al., 2009). Those outcomes additionally demonstrate the possibility of increasing output by boosting photosynthesis. Inside the final section, we argue that it is critical to stack and enhance photosynthetic traits as well as yield gap features in order to supply powerful germplasm for various crops in the world (Dahal et al., 2019). Photosynthesis is the most essential component in determining crop production. The effectiveness with which crops catch the light and convert it to biomass during the growth cycle is the most important factor in determining the final output, whether it is biomass or grain (Murchie et al., 2009). Potential yield is the highest yield that can be achieved from a specific crop. It is defined as the amount yield that could be produced when the most adaptable crop varieties are planted under optimal conditions without biotic or abiotic stress (Hatfield et al., 2011). Production potential is influenced by elements such as light availability, light capture, energy conversion, and plant structure. Energy conversion, which is based on photosynthetic efficiency, is the only one of the four components whose production is less than the potential maximum for our key crops, rice, wheat, and maize (Singh et al., 2013; Zhu et al., 2010). However, because photorespiration can result in the loss of up to 50% of fixed carbon under some conditions, the effectiveness of this energy conversion into harvestable biomass has not been thoroughly examined. Our purpose is to outline current work to improve the efficiency of photosynthesis. The above chapter looks into the effects of changing the Calvin-Benson (CB) cycle, photorespiration, and electron transport on biomass and seed yield, as well as some unexpected adverse outcomes (Ullah et al., 2019). In the final section, we looked at future possibilities, such as integrating several gene alterations of photosynthetic carbon absorption to boost yield potential and features that focus on yield gaps (Simkin et al., 2019).

Following the positive effects of increased photosynthetic tissue photorespiration on *Arabidopsis* and tobacco plant growth. Simkin et al. (2017a) investigated if overexpression of GCS H protein to boost photorespiration and the activity of two CB cycle enzymes (SBPase and FBPA) may have a cumulative effect on photosynthesis efficiency and yield (López-Calcagno et al., 2019). The current study looked at plants that produce SBPase, FBPA, and GCS H proteins alone or in combination Under both moderate and intense light conditions, our research shows that controlling photorespiration and the CB cycle at the same time can have a favorable synergistic effect on biomass output (Simkin et al., 2020). Surprisingly, the only alteration of the photorespiration pathway increased biomass yield, but no increase in seed yield, was observed in these plants (Hagemann & Bauwe, 2016).

While it has demonstrated the ability to increase yield by single-gene and multigene manipulation of various photosynthetic processes, it is also difficult to offer all our rising population need under all conditions and with these alone (Tkemaladze & Makhashvili, 2016). In the future, what alternative methods will be required to attain the increase in yield necessary to sustain a rising population (Ray et al., 2012). It may be essential to stack many other features for photosynthesis in accordance to the aims outlined in this chapter. Shortening the process by integrating additional biosynthetic pathways, for example, will accelerate the relaxation rate of NPQ and reduce the photorespiration losses (Kimura et al., 2020). Photosynthesis has been the topic of this review. Photosynthesis enhances the source's capability; however, given the sinking status of plants with enhanced source capacity, it is likely to be necessary. Two recent publications have focused on source/sink balance, emphasizing the possibilities of integrating source and sink capability enhancements (Parry et al., 2011).

Boosting photosynthesis is one tool for enhancing yield potential; however, it is also required to close the yield gap in order to give flexibility, which will necessitate increases in water use efficiency (WUE), nitrogen use efficiency (NUE), and responsiveness (Fischer et al., 2009). In these areas, some advancement has been made. A recent study, for example, demonstrated how variations in the quantity of PsbS protein might trigger changes in the redox status of QA (Głowacka et al., 2018). WUE should be increased. Furthermore, there have been studies of efforts to boost NUE by overexpressing glutamine synthetase, which will result in greater biomass and grain yields in a variety of plants such as tobacco, wheat, and rice (James et al., 2018; Kant et al., 2011). Several comprehensive reviews of nitrogen costs, uptake, and migration in plant photosynthesis have been published. Better tactics and technology, including new breeding techniques, as well as genome-editing technologies for endogenous gene modification CRISPR/Cas9, would be required to accomplish the ambitious goals required to sustain the world's expanding population (Simkin et al., 2019) and synthetic biology to produce designed promoters and proteins. Given the complexities of the process, modeling's involvement in identifying novel goals is hugely vital (Rodrigues & Kluskens, 2011). To fully exploit the potential of these prospects, new techniques for inserting many transgenes into plants rapidly, effectively, and inexpensively would be required. The production of novel crop plant promoters is currently limited if such possibilities are to be realized fully; policies limiting the use of genetic modification and genome-editing technologies must be reconsidered (Tylecote, 2019).

Photosynthetic Efficiency of Crop and WUE

WUE can be directly increased by improving photosynthetic efficiency and modifying the composition of plant products (Araus et al., 2002). Plants require higher stomatal conductance to allow for a higher rate of CO_2 fixation per unit leaf area in addition to increasing the utilization of soil and water during transpiration Limiting excessive transpiration rates through increased stomatal sensitivity, particularly during the high transpiration period around noon, is highly beneficial to WUE (El-Sharkawy, 2004). Nevertheless, the growing season must be extended to allow further carbon dioxide absorption to compensate for the reduction of carbon dioxide from the leaves (Gaastra, 1959). The tuber yam yield is determined by the time the cultivation, photosynthesis, and harvest indicator last. In the yam physiology literature, these three variables have varied definitions and formulations (Yang & Zhang, 2010).

The growth duration of the crops in yam lasts from the day of germination to the overall senescence of the leaves. Therefore, the preemergence time (from sowing time to emergence) is not calculated, because it depends on the stereotyped maturity and inherent dormancy (Buchanan-Wollaston et al., 2003). In general, there was a significant negative correlation between the tuber yield and the germination time of *P. sylvestris* and *D. sylvestris*. The death of eight bamboo shoots was slow at first, but all died soon after the end of the rainy season each year. As the bud dies, the photosynthetic product translocates into the tuber (Martín-Gómez et al., 2017). C_4 and CAM photosynthesis is a novel combination of complex anatomy and biochemistry, which can increase the efficiency of photosynthesis in aquatic environments with warmth, drought, and salinity and CO_2 deficiency. Despite their complexity, C_4 and CAM have each evolved independently in land plants many times. These origins are promoted by the presence of enablers and evolutionarily stable intermediates in specific plant lineages. When Miocene acidification and the opening of biomes provided new opportunities, the C_4 and CAM lineages began to diversify long after their initial origins. In this diversification process, the different integration of these types of photosynthesis in organisms has led to the diversification of new ecological strategies (Cavender-Bares, 2019). Obviously, reducing the PR rate by increasing the CO_2 concentration at Rubisco, and rerouting photorespiration metabolism by introducing a synthetic bypass and a new pathway that turns PR into a carbon-positive process, has excellent potential to increase production without increasing the need for arable land (Weber & Bar-Even, 2019). It is also conceivable to produce larger yields by combining photosynthetic carbon metabolism engineering with the most recent approaches targeted at adjusting to varying light circumstances faster by minimizing the effect of excitation energy induced by the light system's overprotection (Murchie & Niyogi, 2011). One of the critical issues that synthetic biology faces in the twenty-first century is the opportunity to sustain meeting the requirements of a growing population without negatively impacting the environment.

The Efficiency of Photosynthesis Genetic Engineering

Photosynthesis is a vital natural mechanism that plants use and is required for the average plant growth (Barber, 2009). Photosynthesis uses light to produce chemicals, which include organic molecules (primarily carbohydrates), that are obtained from carbon dioxide and hydrogen sources (such as water) (Murchie & Niyogi, 2011). However, photosynthetic efficiency is relatively low, and only a small portion of available light is utilized to make carbohydrates (Long et al., 2006). Therefore, scientists and farmers are interested in improving the efficiency of photosynthesis through various genetic engineering techniques, thereby increasing plant growth. If these methods prove to be successful and safe, they can increase the agricultural yields and yields of some of the most important plants in the world, including many food crops (Morison et al., 2008). See also agricultural science (plants), biotechnology, carbohydrates, crops, food, genetic engineering, genetically engineered plants, genetically modified crops, genetically modified organisms (GMO), genetics, photosynthesis, plants, plant growth, and plant physiology. By using genetic methods in tobacco (*Nicotiana*), the researchers modified a particularly inefficient operation called photorespiration, a process that occurs in many plants during photosynthesis (Abdelhamid et al., 2005). The results of this genetic manipulation are impressive. Compared with unmodified wild-type tobacco plants, transgenic tobacco plants have increased their growth by more than 40% and increased their biomass by about 25%. Genetically modified plants also grow faster (Simó et al., 2014).

Scientists genetically engineer tobacco plants to use shortcuts during photosynthesis to make their efficiency 40% higher than ordinary plants. Scientists are focusing on photosynthesis, because it provides one of the few options to increase crop yields dramatically (Fig. 3.1).

A schematic diagram of the various steps involved in the transformation of plastic bodies in the biolistic projectile delivery system.

(A) Initially, a *Agrobacterium tumefaciens* vector for transforming the plastid genome is constructed, which contains a selection marker and an expression cassette carrying the gene of interest, located between the left and right flanking regions amplified from the target chloroplast genome. These flanking regions mediate the specific integration of transgene sites into plastids through homologous recombination and are the primary requirement for the chloroplast transformation process.
(B) Next is the transformation of 4- to 6-week-old sterile leaves by bombarding helium-driven microcarriers (gold tungsten particles coated with plastid transformation vectors). After bombardment, the leaves were left in the dark for 48 hours and then cut into leaf discs.
(C) The leaf discs are then placed on a nutrient medium containing regeneration options for transplanted plants. The emergence of new shoots occurs within 2–3 months.

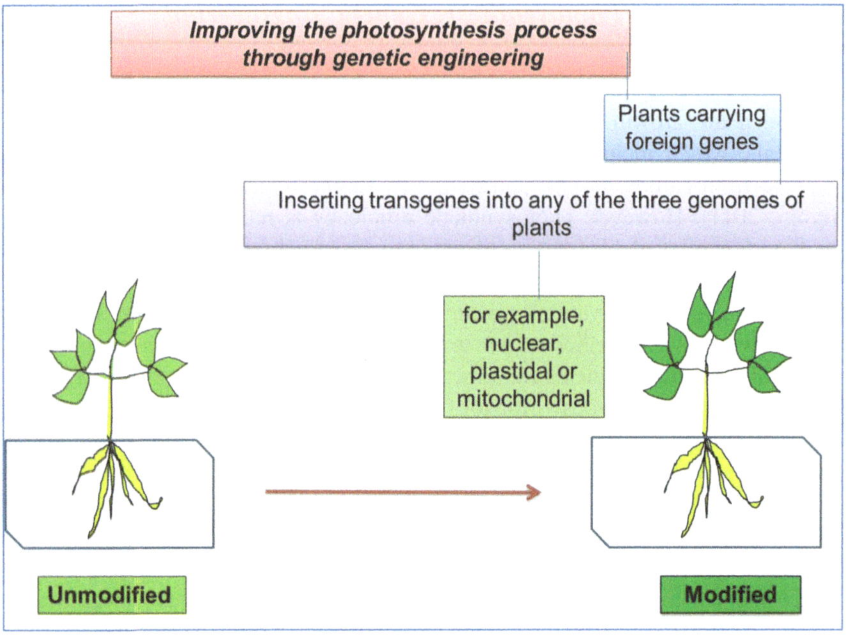

Fig. 3.1 Competence of photosynthesis genetic engineering

(D) The first few copies are transformed, and the cell contains both transformed and untransformed plastomer copies. This state is called heterogeneity. In order to obtain homogeneity (to obtain the cell state of a homogeneous population of transformed plastids), several rounds of regeneration were carried out. The shoots obtained by PCR screening have transgene integration at selected sites.

(E) Then, transfer the selected shoots to MS medium for rooting. When sufficient root mass is obtained, the transformation system is transferred to peat moss to reach maturity and seed collection (Ahmad & Mukhtar, 2017).

To feed the world's rising population, global wheat yields would climb from 3.3 tonnes to 5 tonnes per hectare by 2050. To achieve this goal, it is necessary to build on existing breeding practices and supplement with new technologies, based on the latest results of wheat genome sequencing, and accumulate knowledge on the genetic determinants of agricultural traits that determine crop yield and quality (Murtaza et al., 2016). This study focuses mainly on the methods and approaches available for obtaining gene functions that result in a distinct wheat phenotype (Borrill et al., 2019). The previous study provided a historical perspective on the development of wheat transformation technology. It summarized how the technology gains function through gene overexpression to obtain phenotype, through an expression of antisense RNA (RNA interference or *RNAi*), to obtain a loss-of-function phenotype, and the recent use of site-specific nucleases (such as *CRISPR/Cas9*), to perform gene structure and expression operations for gene structure editing

(Yang et al., 2020). This chapter summarized the latest successful experience in the application of wheat gene manipulation technology in increasing yield, improving wheat nutrition and health, and enhancing crop resistance to various biotic and abiotic stresses (Mittler & Blumwald, 2010). By 2050, due to population and prosperity growth, the demand for wheat is expected to grow at an annual rate of 1.6%. As a result, the global average wheat output per hectare will rise from 3.3 tonnes to almost 5 tonnes per hectare (Borrill et al., 2019). Bread wheat has a very complex hexaploid genome.

As a result, the future of crop breeding is heavily reliant on an understanding of functional genomics. The significant genes, structure, role, and function in the growth process of wheat plants are to be determined and, ultimately, bigger grain yields and better quality achieved (Bevan et al., 2017). Plant scientists can change the structure and function of selected important genes by "genetic manipulation" based on their knowledge of functional genomics. Genetic transformation is a valuable method for establishing scientific proofs of important gene roles and activities. The contributors of this review are not authorized to address the use of genetically modified wheat in global breeding methods, because it falls outside the scope of this chapter (Sumner et al., 2003). Furthermore, the word "genetic manipulation" is extensive and encompasses various molecular processes that yield products that are not classified as genetically modified "GM." RNA interference and *CRISPR/Cas9* are examples of cutting-edge transgenic technology. In an increasing number of nations (including the United States and Canada), the level of regulation applied to their goods is the same as that applied to traditional breeding technologies. Such "final product" instead of "process-based" regulation creates a more favorable environment for the development of molecular-based breeding technology, which can and should change the future of wheat breeding worldwide (Hsu et al., 2014). Yet, if all sophisticated approaches are used without any relationship to wheat breeders who currently use old methods, they will remain "laboratory tools"; as a result, we see the potential for real development and excellent prospects through effective collaboration between plant molecular geneticists and wheat breeders (Morris et al., 2006). The incorporation of genetic traits into traditional wheat breeding programs allows for the application of new methodologies and the analysis of genetically modified wheat plants in breeding to be turned into field applications. The public's awareness and concern about the massive economic disparity between industrialized and developing countries grows (Menz et al., 2020). The less developed countries in their overall economy are more dependent on agriculture. However, they are less able to develop or cooperate in the development of contemporary technological initiatives involving wheat and most other crops (Glover, 1984). As a result, experts in affluent nations must take the initiative and undertake responsibility for freely sharing and distributing their research findings and genetically modified wheat germplasm with breeders in underdeveloped countries (Chiffoleau & Desclaux, 2006). That type of "research donation" can improve the lives and communities in need while also ensuring the future safety and sustainability of wheat production, an important worldwide commodity (Reynolds et al., 2011).

Engineering Photosynthesis: Progress and Prospects

Photosynthesis is the basis of all primary productivity on the planet. Crop breeding continues to improve in terms of yield in order to keep up with population expansion (Sharwood, 2017). These advancements, however, do not result from improvements in the photosynthetic process itself but rather from changes in the way carbon is apportioned in plants; more and more evidence suggests that the rate at which agricultural yields are increasing through standard plant breeding methods is slowing and that they may reach the "yield limit" shortly (Long et al., 2015). A further improvement in yield could be achieved through targeted manipulation of plant metabolism (Snyder & Tegeder, 2021). Optimizing photosynthesis is one such strategy, and simulations suggest that it could have a significant and revolutionary influence on crop output. In this section, we describe the most recent breakthroughs in alternate methods of controlling and increasing photosynthesis, as well as their potential uses in crop improvement (Zhu et al., 2010).

Engineering Rubisco

Rubisco's known inefficiencies, such as slow CO_2 fixation rate and low CO_2 selectivity to O_2, make it a critical engineering objective for increasing crop photosynthesis (Weber & Bar-Even, 2019). Although the reaction process is well understood, engineering attempts have yet to generate the "Super Rubisco" holy grail since efforts to change one component of its catalytic biochemistry usually come at a cost. A single amino acid mutation works as a catalytic "switch," transforming Rubisco from distinct *Flaveri*a species from a "C_3-type" enzyme to a "C_4-type" enzyme, and vice versa (Sharwood et al., 2016). This indicates how targeted alterations predicted from the natural diversity of enzyme sequences and catalysis may be used to manipulate Rubisco's performance. It has resulted in a new surge of data on Rubisco's natural variation, intending to identify amino acid modifications that can increase its catalytic action in plants for guiding the improvement of the dynamics of the crop Rubisco under various environmental conditions; a thorough understanding of how Rubisco and other photosynthetic properties coevolve is required (Whitney et al., 2011). The Rubisco screen, which is not limited to plants, has also proven to be extremely useful Rubisco from diatoms, and tentacle plant microalgae, for example, have undergone different selection pressures, resulting in kinetic properties that differ from the typical trade-off between catalytic rate and CO_2 affinity, opening up a potential new field of investigation for improving Rubisco efficiency (Hanson, 2016).

Calvin-Benson Cycle Optimization

Optimizing the Irvine-Benson cycle, the other enzymes in the Calvin-Benson cycle have long been recognized as viable targets for speeding plant carbon fixation (Hagemann & Bauwe, 2016). Previous accomplishments in overexpression of SBP*ase* (sedoheptulose-bisphosphatase) to boost tobacco and rice growth have recently been reported (Raines, 2011). These efforts are now being expanded to incorporate the use of other Calvin-Benson enzymes in combination in order to avoid the formation of new bottlenecks in various stages of the cycle. Plant biomass has increased due to the combined modification of genes in the Calvin-Benson cycle and photorespiration pathway (Janssen et al., 2014). Other approaches would be to create a synthetic "photorespiration bypass" pathway in the chloroplast that directly releases carbon dioxide near Rubisco, so eliminating the carbon dioxide and energy expenditures of photorespiration. The advantages and disadvantages of various bypass techniques are discussed, except for overexpression of natural enzymes (Erb & Zarzycki, 2016) (Fig. 3.1).

Furthermore, numerous studies have indicated that the expression of exogenous membrane transporters in model and crop species (soybean and rice) has an effect. The membrane transporter IctB, which is intensively investigated but whose function is unknown, is one of the cyanobacteria-derived genes (Koester et al., 2021). The addition of IctB did not always promote crop development; for example, in one case, the photosynthetic rate was altered but biomass did not rise (Long et al., 2016). On the contrary, IctB expression promotes the growth of crop species like wheat in the greenhouse and soybeans in the field. Notably, the free air carbon dioxide enrichment (FACE) experiment is used to undertake long-term field experiments under future environmental scenarios in order to validate computer and greenhouse-based projections. One such procedure, for example, has been proven in soybeans to mitigate some of the detrimental effects of future climate on yields (Cammarano et al., 2016). The above inconsistencies highlight the frequent differences in crop yield projections between greenhouse trials and field trials, as well as the necessity of FACE field research in screening the adaptability of natural and synthetic crops for future climates (Köhler et al., 2017).

Photosynthesis and Crop Yield

Photosynthesis is the critical determinant of crop output, and crop effectiveness in absorbing light and turning it into biomass during the growing season is a critical factor in determining the final yield (whether biomass or grain) (Table 3.1) (Long et al., 2006). Potential yield is the maximum yield that can be obtained from a specific crop. It is defined as the maximum yield that can be obtained when the most adaptable crop varieties are planted under optimal conditions without biotic or abiotic stress (Bennett et al., 2012). The availability of light, the capture of light,

Table 3.1 Current and possible approaches for improving photosynthetic efficiency

Near-term opportunities	Longer-term opportunities	Midterm opportunities	References
1. Near-term opportunities include improving the display of leaves in crop canopies to avoid light saturation of individual leaves and further investigation of a photorespiratory bypass that has already improved the productivity of model species	Longer-term opportunities include engineering into plants carboxylases that are better adapted to current and forthcoming CO_2 concentrations and the use of modeling to guide molecular optimization of resource investment among the components of the photosynthetic apparatus, to maximize carbon gain without increasing crop inputs These modifications have the potential to more than quadruple the yield potential of our primary crops when taken together	Further increases in yield potential will rely in large part on improved photosynthesis Here, we examine inefficiencies in photosynthetic energy transduction in crops from light interception to carbohydrate synthesis Moreover, how classical breeding, systems biology, and synthetic biology provide new opportunities to develop more productive germplasm	Ustin and Middleton (2021), Prior et al. (2020)
2. Improving the efficiency of the primary CO_2-fixing enzyme Rubisco	Optimizing elements of the Calvin-Benson cycle	Introducing the carboxysome-based carbon concentrating mechanism (CCM) from cyanobacteria	Whitney et al. (2011)
3. Introducing an algal pyrenoid CCM	Improving the photochemical response of photosynthesis to rapid changes in light conditions	To attempt to convert C_3 crops such as rice to the more efficient C_4-type photosynthesis	Häusler et al. (2002)

the conversion of energy, and the plant structure are all elements that influence the yield potential (Blankenship et al., 2011). Energy conversion, which is based on photosynthetic efficiency, is the only one of the four components whose production is less than the potential maximum for our key crops, rice, wheat, and maize (Lobell et al., 2009). Nevertheless, because photorespiration can result in the loss of up to 50% of fixed carbon under certain conditions, this energy conversion's efficiency into harvestable biomass has not been properly investigated (Murchie et al., 2009). Whereas the ability to boost yield by single-gene and multigene manipulation of the various photosynthetic processes has been established, these genes alone cannot considerably increase yield in all crop species under all situations and satisfy the needs of our expanding population yield (Paleari et al., 2021). The world environment is changing. What alternative ways will be required in the future to achieve the increase in yield required to sustain a growing population? In addition to the goals outlined in this study, it may be necessary to stack a range of photosynthetic features

(Wang et al., 2021). This will involve, for example, shortening the process by introducing new biosynthetic pathways, which could accelerate the relaxation rate of NPQ and reduce photorespiration losses (Van Esse et al., 2020). This chapter has concentrated on photosynthesis. Photosynthesis improves the capacity of the source, but given the sinking status of plants with enhanced source capacity, it is likely to be needed; two recent publications have focused on source/sink balance, emphasizing the possibilities of integrating source and sink capacity enhancements (Peng et al., 2008; Sonnewald et al., 2020). Improved photosynthesis is one way to enhance yield potential, but closing the yield gap is also critical for flexibility, which will demand increased water use efficiency (WUE), nitrogen use efficiency (NUE), and responsiveness (Fischer et al., 2009). In these areas, some progress has also been made. A recent study, for example, demonstrated how variations in the quantity of PsbS protein might trigger changes in the redox status of QA (Głowacka et al., 2018). It has been proposed that altering the oxidation state of the plastoquinone pool can regulate stomatal motility (Suzuki et al., 2012), and the PsbS transgenic experiment showed that there is a linear relationship between stomatal conductivity and QA redox value, and response to light and phosgene reduces stomatal opening, increasing WUE. Furthermore, there have been indications of efforts to boost NUE through glutamine synthetase overexpression, resulting in improved biomass and grain yields in a variety of plants, including tobacco, wheat, and rice (Makino, 2011). Several evaluations of nitrogen costs, nitrogen uptake, and photosynthesis migration in plants have been published (Niu et al., 2020).

Current and Possible Approaches for Improving Photosynthetic Efficiency

In terms of the following aspects, photosynthesis is a highly inefficient process; the conversion from light energy to biomass has been included in the literature (Long et al., 2006). This is in the past decade alone; these processes have limited the efficiency of photosynthesis which have been understood to be able to gradually change our abilities to a certain extent manipulate light energy to absorb into carbon (Murchie & Niyogi, 2011). Therefore, it is expected that the grain yield potential of our major countries would improve in the future crops, which will be substantially dependent on improving photosynthetic efficiency (Fig. 3.2). The studies in this issue provide fresh insights into the nature of present photosynthesis restrictions and propose new targets that can be exploited for crop enhancement, as well as information regarding the influence of environmental changes on land and our photosynthetic production ocean. In the past 50 years, the increase in crop yields achieved through conventional plant breeding has been stable at around 1% per year. Following a significant improvement in cereal crops (De Souza et al., 2017). The achievements made during the Green Revolution to increase genetic yield potential appear to have plateaued. The yield of crops must increase by 100% to meet the expected

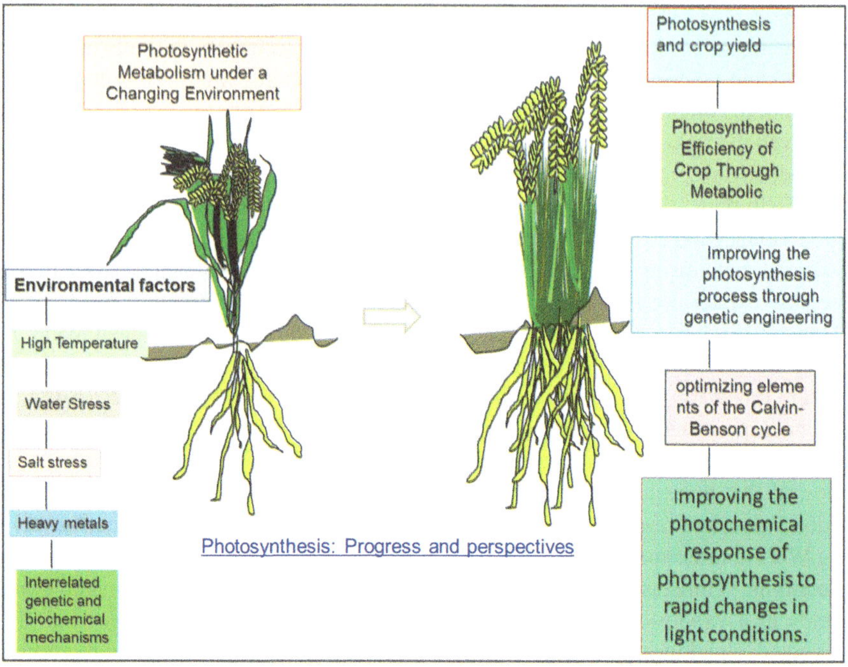

Fig. 3.2 Potential approaches for improving photosynthetic efficiency under changing environment

future food needs of the global population. It is necessary to improve the efficiency of photosynthesis while making more effective use of natural resources (Long et al., 2015). Comprehensive approach from multiple natural science disciplines improve photosynthesis prospects (Dann & Leister, 2017). Comprehensively discuss the introduction of new genetic diversity in crops once the authors discuss the different possible sources of nature change and believe that a novel approach is needed. According to the authors, the synthetic bacterium platform could provide options for improving eukaryotic photosynthesis (Schneeberger & Weigel, 2011).

Plant Productivity Depends on Many Genetic

Plant survival and crop enhancement in nature rely on genetic diversity. Plant breeders can generate new and improved varieties with excellent characteristics, thanks to the diversity of plant genetic resources (Hoisington et al., 1999). Such features comprise attributes desired by farmers (high yield potential, large seeds, and so on) as well as traits desired by breeders (pest and disease resistance, photosensitivity, etc.). Natural genetic variety in crop species has been employed to suit the needs of livelihood food since the beginning of agriculture. Then, the emphasis turned to produce surplus food to feed the world's rising population (Bhandari et al., 2017). In order to supply humans with a balanced diet, the current focus is on the quantity and quality of primary food crops. The breeding of climate-adaptive cultivars has become increasingly significant as a result of climate change (Porter & Semenov, 2005). The presence of genetic diversity in the form of wild species, related species, breeding populations, mutant lines, and other forms of genetic diversity can be exploited to find the optimum alleles and aid plant breeders in developing climate-adaptive cultivars (Yadav et al., 2021). Novel features, such as resistance to potential new pests and diseases, high heat and cold, and tolerance to various air and soil contaminants, are required when breeding varieties that adapt to climate change (Olesen et al., 2011). It is vital to have diverse genes in planted and plantable crops in the form of germplasm resources to keep shifting breeding goals. Breeders can choose superior genotypes directly from new varieties or the parents of hybridization projects because of the genetic variety inside and among crop plants (Bhandari et al., 2017). To achieve heterosis and the inheritance of transgressed sergeants, the two parents' inheritance is sufficient. Breeders' development is aided by genetics. It also encourages new lines for unusual purposes, such as sorghum, corn, and other biofuel kinds (Carter Jr et al., 2004). Adaptability to a variety of habitats is also critical, with a focus on variations in climatic circumstances. In several crops, there are some germplasm lines with ideal genes (Wassmann et al., 2009). Plant breeding must contend with the difficulty of feeding an ever-increasing population on ever-dwindling arable land. Modern plant breeding has had some success in this area.

Conversely, in many crops, the restricted genetic base of cultivars has resulted in genetic fragility (Ramsden & Hails, 2019). As a result, a paradigm shift in plant breeding is required, with a focus on varied genetic resources. Genetic diversity is currently acknowledged as a specialized area that can help to food and nutrition security (Toledo & Burlingame, 2006). A deeper understanding of genetic variety will aid in determining what and where conserved. Crop genetic diversity is the foundation for the long-term evolution of new kinds As a result, different statistical approaches must be used to describe various genetic resources and exploit them in breeding operations (Ortiz et al., 2007).

Photosynthetic Metabolism as a Basis for Crop Production and Yield Improvement Under Stressful Conditions

It has been suggested that the duration of water scarcity and high temperatures be extended, as well as the availability of nutrients be reduced, as these are the key variables that impede plant development (Seleiman et al., 2021). The uptake of photosynthetic CO_2 is the foundation of animal and human food crop production (Morales et al., 2020). As a result, it has been chosen as the primary focus of crop phenotyping/breeding research; understanding the mechanisms involved in photosynthetic CO_2 assimilation's response and adaptability to a variety of changing environmental variables (such as nutrients, water availability, and temperature change) is critical in this setting (Murchie et al., 2009). Also, build new tactics and technologies to help future plants grow. The purpose of this review is to look at the effects of changing environmental circumstances on photosynthetic device performance and plant growth from a variety of angles (involving breeding, gas exchange, genomics, etc.) (DaMatta et al., 2018). As previously said, one of the most challenging tasks of the twenty-first century is producing enough food to meet the requirements of a growing population while ensuring environmental sustainability (Goodland, 1995). Despite the fact that weather patterns related to global climate change are changing on a daily basis, this target must still be attained. Crop biomass production and yield are primarily determined by the rate of CO_2 uptake throughout the growing season (Dhillon & von Wuehlisch, 2013). According to mounting research, reaching quantum growth in cereal crop yield potential necessitates significant improvements in photosynthetic capacity or efficiency (Parry et al., 2011). According to Reynolds et al., the enhancement of photosynthetic capacity and efficiency (radiation use efficiency, RUE) could be a target method for the development of novel germplasm that is more appropriate for stress growth circumstances, together with traits that are optimally allocated to grain yield (Reynolds et al., 2016).

Crop yields must double by 2050 to fulfill the global population's predicted output requirement, but plant breeders will have tremendous difficulty accomplishing this target. Conventional plant breeding relies on phenotypic selection and subsequent offspring testing, which is usually followed by reselection, which may be a lengthy process that frequently necessitates the use of time-consuming and expensive phenotypes (Bradshaw, 2017). Addressing the physiological and molecular mechanisms underlying photosynthesis and plant development in response to changing environmental conditions could create new stress-resilient techniques and tools (Hasanuzzaman et al., 2013). There is strong evidence that a considerable increase in photosynthetic capacity or efficiency will be necessary to attain crop potential quantum growth. There is also evidence that the growth in wheat yield potential in the past has been linked to an increase in photosynthesis. As a result, crop breeding programs should strive to continuously improve plant CO_2 assimilation efficiency (via improved CO_2 stomata and/or mesophyll conductivity, higher photosynthetic electron transport and CO_2 fixation capabilities, and optimized sugar transport and use) while also improving water (WUE) and nutrient use efficiency to prepare for future environmental coexistence (Slafer et al., 2021).

Further research on plants with delayed senescence (to stay green) is also encouraged, although this condition usually keeps the plant's photosynthetic activity for a longer period of time, extending the grain-filling time (Gepstein & Glick, 2013). Absolutely no impact of increasing CO_2 on plant net photosynthesis and production, climate change projections include a 2.6 °C and 4.8 °C increase in average temperature in 2065 and 2100, respectively, as well as increased climate variability and change. Dry spells and heat waves are common. It is widely acknowledged that high temperatures and dryness have a negative impact on photosynthesis and plant productivity. Since only increased atmospheric CO_2

concentrations are included in the simulated climate change scenario, plant response in terms of photosynthesis, growth, and yield will normally rise (Morales et al., 2020). Yet, if plants are also subjected to extreme temperatures or a lack of water, growth and production may suffer in the future as atmospheric CO_2 concentrations rise. Rice and wheat yields have been reported to have declined dramatically as a result of a 1.5–2.0 °C increase in canopy temperature under increasing CO_2 levels (500 ppm) (Singh et al., 2007).

Photosynthetic Metabolism in a Changing Environment

Environmental stress can trigger a variety of physiological reactions in plants. Changes in photosynthetic rate, assimilation, and translocation, changes in water absorption and evapotranspiration, the impacts of antioxidant response and programmed cell death modification on nutrient absorption and translocation, and changes in gene expression and enzyme activity are among the most significant physiological changes (Lawlor & Cornic, 2002). These are the most important processes that are often affected under environmental stress conditions.

In general, the inhibitory impact of stress on photosynthesis could be attributable to CO_2 diffusion factor limitations and/or (ii) metabolic factors. Stomatal closure is the major occurrence under pressure settings, according to a lot of research (Niinemets & Keenan, 2014). Stomata closure causes a decrease in CO_2 assimilation by lowering CO_2 concentrations under the stomata and in the chloroplast (Ci and Cc, respectively). Water stress, on the other hand, might cause metabolic limitation. The influence on Calvin cycle enzyme activity (Rubisco, SBPase, etc.) and the lower availability of ATP and NADPH are the most generally deemed among them (Ennahli & Earl, 2005).

Environmental Factors

Plant development can be limited by a number of environmental conditions that affect photosynthetic equipment (Zlatev & Lidon, 2012) (Table 3.2). Light, temperature, water, nutrients, and carbon dioxide are among them. Because these factors cannot alter photosynthesis on their own, the stress response we see in the whole plant is usually an integral element of a number of metabolic consequences, of which photosynthesis is only one. It's also self-evident that in a wide range of ecosystems, conditions like excessive temperature and water stress are frequently relevant. When this happens, the interpretation of the factory's response can be a very complex issue (Chaves et al., 2009).

Plants will be exposed to abiotic challenges, such as heavy metals, salt, drought, nutritional insufficiency, light intensity, pesticide contamination, and severe temperatures, due to their fixative qualities. These forces placed significant constraints, restricting crop productivity and food security globally, because abiotic stress has a

Table 3.2 Environmental factors' effects on photosynthetic efficiency

Environmental factors	Crop	Effect	References
Drought	Wheat	Drought stress effects on soluble protein, sugar alcohols, Chl content, RWC, gs, and Pn	Saeidi et al. (2017), Siddique et al. (1999)
Drought	Rice (Zhenshan97B)	Limitation to photosynthesis occurred during drought stress	Ji et al. (2012)
Drought	Maize	Leaf rolling reduces photosynthetic loss in maize under severe drought	Saglam et al. (2014)
Drought	Maize and grain sorghum	Elevated CO_2 increases water use efficiency by sustaining photosynthesis of water-limited maize and sorghum	Allen Jr et al. (2011)
Drought	Sweet sorghum	Before anthesis growth is the predominant sink for photosynthesis	Massacci et al. (1996)
Drought	*Hibiscus rosa-sinensis*	Drought stress influences leaf water content, photosynthesis, and water use efficiency of *Hibiscus rosa-sinensis* at three potassium concentrations	Egilla et al. (2005)
Drought	Barley and wheat	Effect of water stress on photosynthesis and transpiration of flag leaves and spikes of barley and wheat	Johnson et al. (1974)
Drought	A plethora plant	Drought-induced responses of photosynthesis and antioxidant metabolism in higher plants	Reddy et al. (2004)
Drought selenium spraying	Barley	Effect of drought stress and selenium spraying on photosynthesis and antioxidant activity of spring barley	Ghotbi-Ravandi et al. (2014)
Drought	Bean (*Phaseolus vulgaris*)	Differential adaptation of two varieties of common bean to abiotic stress: I Effects of drought on yield and photosynthesis	Lizana et al. (2006)

detrimental effect on plant chlorophyll production, photosystem performance, electron transport mechanism, gas exchange parameters, etc. (Garg & Chandel, 2011). It mainly reduces the photosynthesis efficiency of plants. Understanding the photochemical characteristics of plants under various abiotic pressures can aid in the development of effective strategies to combat these stresses attempt to present an overview of the various mechanisms that influence these abiotic influences on plants' photosynthetic capability in this chapter (Bhattacharjee & Saha, 2014). The impacts of various abiotic pressures on plant photosynthesis are discussed in this chapter, which includes cell membranes, cell division and elongation, photosynthetic pigment production, and electron transport chains (Muhammad et al., 2021). Understanding the negative impacts of diverse abiotic stresses is critical for improved stress management, since a thorough understanding of plant responses is crucial for remedial treatments and management (Nouri et al., 2015).

The Effect of Abiotic Stress on Plants

Since abiotic stress has a negative impact on plant chlorophyll biosynthesis, photosystem performance, electron transport mechanism, gas exchange parameters, etc. (Table 3.2), it mainly reduces the photosynthesis efficiency of plants (Jahan et al., 2021).

High Temperature

Plant susceptibility to high temperatures varies depending on the severity, length, and stage of development of the stress; not all plant species can cope with heat stress to the same extent (Jagadish et al., 2021). They must all control their metabolism by reprogramming biological processes to adapt to stress in order to avoid irreparable damage (Ahuja et al., 2010). As a result, researchers are actively looking to improve crop heat tolerance using molecular breeding and genetic engineering, despite the fact that these technologies are still quite expensive and time-consuming. It must also contribute to our understanding of how plants respond to heat stress. Despite the fact that heat stress is not a biological stress, some study has focused on biochemical reactions, including hormones and major and minor metabolites such as antioxidants (Morales et al., 2020). Changes in gene expression can lead to the production of stress-related proteins, such as the activation of heat shock proteins (proteins involved in signal transduction during heat stress), which is thought to be a useful adaptation strategy (Warburton et al., 2017).

Water Stress

When faced with a water shortage, plants adjust their water relationships by lowering their leaf water potential, turgor pressure, relative water content (RWC), and transpiration rate (E) (Rodríguez et al., 2012). Because cell expansion is closely linked to the expansion pressure that drives cell expansion, cell development and leaf expansion are the most drought-sensitive processes. According to research on wheat, barley, and rice, drought is impeding crop development (Otieno et al., 2005). Reactive oxygen species (ROS), including peroxides (H_2O_2), superoxide (O_2^-), and singlet oxygen ($1O_2$), can tear membranes and inhibit photosynthesis when exposed to water stress. Due to the fact that the electron transport of mitochondria is predominantly controlled by the alternative oxidase route, which maintains electron transport while limiting the formation of ROS, plant respiration remains rather stable during water stress, unlike photosynthesis (Talaat, 2019). Plant growth regulators – aminobutyric acid and free amino acids and sugars – all contribute to ROS scavenging (Torabi et al., 2021). Scavenging enzyme activity is raised in other research; for example, durum wheat exhibits higher peroxidase activity in water-deficient conditions; drought has become one of the most significant obstacles to agricultural development, particularly in developing countries; the impacts of various levels of water stress on photosynthesis, growth, winter wheat yield, water usage efficiency (WUE), and irrigation water productivity (IWP) will be studied in order to develop water-saving agricultural scientific irrigation solutions (Farooq et al., 2009).

Drought is the most crucial nonbiological factor limiting growth, negatively impacting growth and agricultural yield. Stress causes aberrant physiological processes in one or more biological and environmental elements (Bita & Gerats, 2013). This abnormal metabolism can damage or reduce plant growth. According to the estimates of different scholars, the output is limited by environmental pressure, and there are only 10 types (Fageria & Baligar, 2005). In general, 100% of the global arable land is free of pressure, which is the main factor causing the difference between the two yields and potential performance, as well as environmental pressures. Drought and stress are the most common. Environmental pressure has caused almost 25% of agricultural land to be used for agricultural production (Evans & Fischer, 1999). The world is limited; the drought risk of successfully producing crops on a global scale can occur in the following situations: The combination of physical and environmental factors can cause plants to be stressed, thereby reducing productivity.

Under drought conditions, the water potential falls, the cell concentration of ABA increases, and the cell metabolism is regulated (Seleiman et al., 2021). Proline, glycine, and betaine increases could be one of the vital chemical processes. Drought is causing stress (Morison et al., 2008). Accumulation to lyse cells under stress maintains cell volume to prevent water loss; this is called osmotic adaptation. Free radicals induced by drought stress lead to lipid peroxidation and membrane deterioration. Drought stress causes an imbalance between the antioxidant defense and the number of antioxidants. Reactive oxygen species (ROS) cause oxidation pressure, 10 to −15 physiological potential Such as leaf growth, stomatal conductance, decreased photosynthetic rate and nitrogen (Fathi & Tari, 2016). According to the article, in research on the physiological phenomena of drought stress in plants, the factory's water condition is frequently assessed, and the organization's water potential is characterized by lowering the water level (Molina, 2021). In plants, cell development and protein synthesis can slow due to low water. Transpiration is slowed, although praline buildup and abscisic acid pressure rise (Rauf, 2008). Drought stress, affects stomatal photosynthesis and because of the confinement and transfer of carbon dioxide in chloroplasts, the cell's water potential decreases. Moreover, drought stress affects root and shoot growth, which may result in diminished plant growth.

The current crop yield trajectory is not enough to feed the world's population by 2051. Due to changes in diseases, insect pests and pathogens, precipitation, heat waves, and other factors caused climatic pressures to limit production; it is necessary to achieve more significant and more stable crop yields in extreme weather (Ray et al., 2013). Here, we consider the potential of plant science to solve agricultural challenges after the Green Revolution (GR) and explore emerging strategies to enhance sustainable crop production and resilience in climate change. Accelerated crop improvement must use natural evolutionary traits and mechanical understanding-driven transformation projects to produce the flexible production systems needed to ensure future harvests (Wise, 2013).

Plant Hormones and Interrelated Genetic and Biochemical Mechanisms

Plants' responses to water stress involve complex genes and processes that are governed by interconnected genetic and biochemical pathways (Henriet et al., 2021). Long-term water stress can cause downregulation of the photosynthetic metabolism process, leading to alterations in leaf biochemistry due to a lack of carbon substrates. Several enzymes are thus downregulated or inactivated at low CO_2 concentrations (Farooq et al., 2009). The reduction in the maximal carboxylation rate (*Vcmax*) implies that Rubisco activity is reduced, as demonstrated by *Medicago truncatula* plants. Other enzymes, such as nitrate reductase, similarly limit their activity. However, when exposed to water stress, the synthesis of antioxidant enzymes and stress-related proteins increases (Morales et al., 2020).

Other enzymes, such as nitrate reductase, limit their activity as well. Water stress, on the other hand, increases the synthesis of antioxidant enzymes and stress-related proteins (Khan et al., 2020). In this case, the probable function of genes in photosynthesis, as well as the correlation between the transcriptome profile and the plant hormone regulatory network, serves to connect numerous valuable data sets (Kumudini & Patil, 2019). More information on the cellular response to environmental stress and the transcriptional regulation of photosynthetic metabolism rate would be beneficial. Furthermore, a more detailed examination of its consequences for photosynthetic improvement, plant productivity, and stress tolerance should be done. Generally, the findings suggest that improved photosynthetic mechanisms and metabolic signaling, as well as improved plant stress resistance, necessitate a more thorough grasp of interdisciplinary tactics based on many study domains (Godde, 2018). Even though TF regulation of a specific gene is fundamentally dependent on the genome size of a given species, TF governs practically all cell functions in living creatures, hence strictly controlling gene expression; as a result, the primary goal of researchers is to identify the molecular mechanism of gene expression and to comprehend the role of individual genes in influencing the control of specific traits via the regulation of functional variables (Rochaix, 2011).

Salt Stress Significantly Affects Photosynthesis

Much salt or saline soil will alter biochemical and physiological processes, especially photosynthesis, causing plant development to be slowed and productivity to be reduced (Chaudhry & Sidhu, 2021). Crop productivity is reduced by approximately 50% as a result of salt stress (Majeed & Muhammad, 2019). Furthermore, osmotic stress caused by salinity inhibits photosynthesis by limiting subcellular organelle structures and metabolic processes (Arif et al., 2020). The cell membrane is stressed, and large concentrations of ions in the chloroplast (such as sodium (Na^+) and chloride ions (Cl^-)) cause substantial damage to the thylakoid's membrane. Furthermore, high concentrations of inorganic salts can inhibit electron transport by inactivating photophosphorylation in phosphorylation-like membranes (Solis et al., 2021). In conclusion, the initial investigations have shown that Chl content, photosynthetic pigments, membrane degradation, and metabolic alterations are the critical targets during salt stress. Membrane instability and pigment degradation, for example, would have a significant impact on plant growth, development, and physiological factors (Atia et al., 2018).

The Influence of Heavy Metals

Heavy metals, such as cadmium (Cd), copper (Cu), zinc (Zn), nickel (Ni), cobalt (Co), chromium (Cr), lead (Pb), and arsenic, are commonly responsible for soil pollution (As). Phosphate fertilizers, sewage sludge, industrial waste, wind-blown dust, incinerator emissions, traffic, volcanoes, and hard water practices are all factors to consider (Wittmann, 1981). Plant growth is severely hampered by heavy metal pollution, which causes Chl breakdown, DNA and protein damage, and enzyme inhibition (Fig. 3.2). Heavy metal contamination will wreak havoc on the environment when combined with other factors (Chmielowska-Bąk & Deckert, 2021). Heavy metals, for example, might create an excessive accumulation of ROS in plants, resulting in oxidative stress. Notably, metallothioneins (MTs) have been offered as a different weapon for plants to use to protect themselves against stress-induced oxidative damage (Wani et al., 2018). Also, Hasan et al. (2017) reported the role of MTs as ROS scavengers in abiotic stress tolerance, although the mechanism by which MTs mediate ROS homeostasis is still unclear.

Conclusions and Further Opportunities

Although it has been established that single-gene and multigene manipulation of the many processes of photosynthesis can improve yield, these genes alone cannot be greatly raised in all crop species under all situations and to satisfy the needs of our expanding population production. The world environment is changing and still there are questions in mind, what alternative strategies would be required in the future to attain the increase in yield required to maintain a growing population. Apart from the aims described in the chapter, numerous different features for photosynthesis may be required This will involve, for example, reducing the process by introducing new biosynthetic pathways, which will accelerate the relaxation rate of NPQ and reduce photorespiration losses. This review has concentrated on photosynthesis. Photosynthesis improves the capacity of the source, but the sink status of plants whose source capacity has increased must also be considered. Two recent publications have focused on source/sink balance, emphasizing the possibilities of integrating source and sink capacity enhancements. Improving photosynthesis is one technique for improving yield potential; however, it is also required to close the yield gap in order to give flexibility, which will necessitate increases in water use efficiency (WUE), nitrogen use efficiency (NUE), and responsiveness. The effects of stress on both living and nonliving entities. In these areas, some progress has also been made. A recent study, for example, demonstrated how variations in the quantity of PsbS protein might trigger changes in the redox status of QA. Changes in the oxidation state of the plastoquinone pool were proposed as a way of controlling stomach movement, and a linear relationship between the stomach conductivity and QA redox value was revealed with the psbS transgenic experiment as well as the reduced stomach opening in response to light and phosgene. WUE should be increased. Moreover, glutamine synthetase overexpression has been shown to improve NUE in a range of plants, including tobacco, wheat, and rice, resulting in increased biomass and grain yield (Andrews et al., 2004; Simkin et al., 2019). Numerous in-depth evaluations of the nitrogen costs of plant photosynthesis, nitrogen absorption, and nitrogen migration have been published. To fulfill the ambitious goals required to feed the world's expanding population, new breeding techniques will be required, including genome-editing technologies for endogenous gene modification *CRISPR/Cas9* and synthetic biology to create tailored promoters and proteins. Owing to the complexity of the process, modeling would play an important role in identifying unique goals. To fully realize the promise of these opportunities, new methods that can swiftly, efficiently, and cheaply introduce many transgenes into plants will be required, and the development of novel promoters for crop plants is currently limited. If these prospects are to be fully realized, regulations on genetic change and genome-editing technologies must be addressed. One of the most important tasks of today's biotechnology is to increase the productivity of crops to meet the growing demand for food and energy. Plant productivity depends on many genetic factors, including life cycle, harvest index, stress tolerance, and photosynthetic activity.

This chapter covers different suggestions and successful examples of improving plant photosynthesis and provides novel perspectives for future research.

References

Abdelhamid, A., Elshinnawy, M., Farrag, F., Abidi, S., Adom, K., Liu, R., Agrahar-Murugkar, D., Pal, P., Agte, V., & Tarwadi, K. (2005). Bibliography from web of science March 2005, Golden rice and (pro)-vitamin A. *Bioresource Technology, 60*(1), 91–93.

Ahmad, N., & Mukhtar, Z. (2017). Genetic manipulations in crops: Challenges and opportunities. *Genomics, 109*(5–6), 494–505.

Ahuja, I., de Vos, R. C., Bones, A. M., & Hall, R. D. (2010). Plant molecular stress responses face climate change. *Trends in Plant Science, 15*(12), 664–674.

Allen, L. H., Jr., Kakani, V. G., Vu, J. C., & Boote, K. J. (2011). Elevated CO_2 increases water use efficiency by sustaining photosynthesis of water-limited maize and sorghum. *Journal of Plant Physiology, 168*(16), 1909–1918.

Andrews, M., Lea, P. J., Raven, J., & Lindsey, K. (2004). Can genetic manipulation of plant nitrogen assimilation enzymes result in increased crop yield and greater N-use efficiency? An assessment. *Annals of Applied Biology, 145*(1), 25–40.

Araus, J., Slafer, G., Reynolds, M., & Royo, C. (2002). Plant breeding and drought in C3 cereals: What should we breed for? *Annals of Botany, 89*(7), 925–940.

Arif, Y., Singh, P., Siddiqui, H., Bajguz, A., & Hayat, S. (2020). Salinity induced physiological and biochemical changes in plants: An omic approach towards salt stress tolerance. *Plant Physiology and Biochemistry, 156*, 64–77.

Atia, A., Barhoumi, Z., Debez, A., Hkiri, S., Abdelly, C., Smaoui, A., Haouari, C. C., & Gouia, H. (2018). Plant hormones: Potent targets for engineering salinity tolerance in plants. In *Salinity responses and tolerance in plants* (Vol. 1, pp. 159–184). Springer.

Barber, J. (2009). Photosynthetic energy conversion: Natural and artificial. *Chemical Society Reviews, 38*(1), 185–196.

Bennett, A. J., Bending, G. D., Chandler, D., Hilton, S., & Mills, P. (2012). Meeting the demand for crop production: The challenge of yield decline in crops grown in short rotations. *Biological Reviews, 87*(1), 52–71.

Bevan, M. W., Uauy, C., Wulff, B. B., Zhou, J., Krasileva, K., & Clark, M. D. (2017). Genomic innovation for crop improvement. *Nature, 543*(7645), 346–354.

Bhandari, H., Bhanu, A., Srivastava, K., Singh, M., & Shreya, H. A. (2017). Assessment of genetic diversity in crop plants – An overview. *Advances in Plants & Agriculture Research, 7*(3), 00255.

Bhattacharjee, S., & Saha, A. K. (2014). Plant water-stress response mechanisms. In *Approaches to plant stress and their management* (pp. 149–172). Springer.

Bita, C., & Gerats, T. (2013). Plant tolerance to high temperature in a changing environment: Scientific fundamentals and production of heat stress-tolerant crops. *Frontiers in Plant Science, 4*, 273.

Blankenship, R. E., Tiede, D. M., Barber, J., Brudvig, G. W., Fleming, G., Ghirardi, M., Gunner, M., Junge, W., Kramer, D. M., & Melis, A. (2011). Comparing photosynthetic and photovoltaic efficiencies and recognizing the potential for improvement. *Science, 332*(6031), 805–809.

Borrill, P., Harrington, S. A., & Uauy, C. (2019). Applying the latest advances in genomics and phenomics for trait discovery in polyploid wheat. *The Plant Journal, 97*(1), 56–72.

Bradshaw, J. E. (2017). Plant breeding: Past, present and future. *Euphytica, 213*(3), 60.

Buchanan-Wollaston, V., Earl, S., Harrison, E., Mathas, E., Navabpour, S., Page, T., & Pink, D. (2003). The molecular analysis of leaf senescence – A genomics approach. *Plant Biotechnology Journal, 1*(1), 3–22.

Cammarano, D., Rötter, R. P., Asseng, S., Ewert, F., Wallach, D., Martre, P., Hatfield, J. L., Jones, J. W., Rosenzweig, C., & Ruane, A. C. (2016). Uncertainty of wheat water use: simulated patterns and sensitivity to temperature and CO_2. *Field Crops Research, 198*, 80–92.

Carter, T. E., Jr., Nelson, R. L., Sneller, C. H., & Cui, Z. (2004). Genetic diversity in soybean. *Soybeans: Improvement, production, and uses, 16*, 303–416.

Cavender-Bares, J. (2019). Diversification, adaptation, and community assembly of the American oaks (*Quercus*), a model clade for integrating ecology and evolution. *New Phytologist, 221*(2), 669–692.

Chaudhry, S., & Sidhu, G. P. S. (2021). Climate change regulated abiotic stress mechanisms in plants: A comprehensive review. *Plant Cell Reports*, 1–31.

Chaves, M. M., Flexas, J., & Pinheiro, C. (2009). Photosynthesis under drought and salt stress: Regulation mechanisms from whole plant to cell. *Annals of Botany, 103*(4), 551–560.

Chiffoleau, Y., & Desclaux, D. (2006). Participatory plant breeding: The best way to breed for sustainable agriculture? *International Journal of Agricultural Sustainability, 4*(2), 119–130.

Chmielowska-Bąk, J., & Deckert, J. (2021). Plant recovery after metal stress—A review. *Plants, 10*(3), 450.

Dahal, K., Li, X.-Q., Tai, H., Creelman, A., & Bizimungu, B. (2019). Improving potato stress tolerance and tuber yield under a climate change scenario – A current overview. *Frontiers in Plant Science, 10*, 563.

DaMatta, F. B. M., Avila, R. T., Cardoso, A. A., Martins, S. C., & Ramalho, J. C. (2018). Physiological and agronomic performance of the coffee crop in the context of climate change and global warming: A review. *Journal of Agricultural and Food Chemistry, 66*(21), 5264–5274.

Dann, M., & Leister, D. (2017). Enhancing (crop) plant photosynthesis by introducing novel genetic diversity. *Philosophical Transactions of the Royal Society B: Biological Sciences, 372*(1730), 20160380.

De Souza, A. P., Massenburg, L. N., Jaiswal, D., Cheng, S., Shekar, R., & Long, S. P. (2017). Rooting for cassava: Insights into photosynthesis and associated physiology as a route to improve yield potential. *New Phytologist, 213*(1), 50–65.

Dhillon, R., & von Wuehlisch, G. (2013). Mitigation of global warming through renewable biomass. *Biomass and Bioenergy, 48*, 75–89.

Egilla, J., Davies, F., & Boutton, T. (2005). Drought stress influences leaf water content, photosynthesis, and water-use efficiency of *Hibiscus rosa-sinensis* at three potassium concentrations. *Photosynthetica, 43*(1), 135–140.

El-Sharkawy, M. A. (2004). Cassava biology and physiology. *Plant Molecular Biology, 56*(4), 481–501.

Ennahli, S., & Earl, H. J. (2005). Physiological limitations to photosynthetic carbon assimilation in cotton under water stress. *Crop Science, 45*(6), 2374–2382.

Erb, T. J., & Zarzycki, J. (2016). Biochemical and synthetic biology approaches to improve photosynthetic CO_2-fixation. *Current Opinion in Chemical Biology, 34*, 72–79.

Evans, L., & Fischer, R. (1999). Yield potential: Its definition, measurement, and significance. *Crop Science, 39*(6), 1544–1551.

Fageria, N. K., & Baligar, V. (2005). Enhancing nitrogen use efficiency in crop plants. *Advances in Agronomy, 88*, 97–185.

Farooq, M., Wahid, A., Kobayashi, N., Fujita, D., & Basra, S. (2009). Plant drought stress: Effects, mechanisms and management. *Sustainable Agriculture*, 153–188.

Fathi, A., & Tari, D. B. (2016). Effect of drought stress and its mechanism in plants. *International Journal of Life Sciences, 10*(1), 1–6.

Fischer, R., Byerlee, D. R., & Edmeades, G. O. (2009) *Can technology deliver on the yield challenge to 2050*?

Gaastra, P. (1959). *Photosynthesis of crop plants as influenced by light, carbon dioxide, temperature, and stomatal diffusion resistance*. Veenman.

Garg, N., & Chandel, S. (2011). Arbuscular mycorrhizal networks: Process and functions. *Sustainable Agriculture, 2*, 907–930.

Gepstein, S., & Glick, B. R. (2013). Strategies to ameliorate abiotic stress-induced plant senescence. *Plant Molecular Biology, 82*(6), 623–633.

Ghotbi-Ravandi, A., Shahbazi, M., Shariati, M., & Mulo, P. (2014). Effects of mild and severe drought stress on photosynthetic efficiency in tolerant and susceptible barley (Hordeum vulgare L.) genotypes. *Journal of Agronomy and Crop Science, 200*(6), 403–415.

Glover, D. J. (1984). Contract farming and smallholder outgrower schemes in less-developed countries. *World Development, 12*(11–12), 1143–1157.

Głowacka, K., Kromdijk, J., Kucera, K., Xie, J., Cavanagh, A. P., Leonelli, L., Leakey, A. D., Ort, D. R., Niyogi, K. K., & Long, S. P. (2018). Photosystem II subunit S overexpression increases the efficiency of water use in a field-grown crop. *Nature Communications, 9*(1), 1–9.

Godde, D. (2018). Adaptations of the photosynthetic apparatus to stress conditions. In *Plant responses to environmental stresses* (pp. 449–474). Routledge.

Goodland, R. (1995). The concept of environmental sustainability. *Annual Review of Ecology and Systematics, 26*(1), 1–24.
Hagemann, M., & Bauwe, H. (2016). Photorespiration and the potential to improve photosynthesis. *Current Opinion in Chemical Biology, 35*, 109–116.
Hall, A. J., & Richards, R. A. (2013). Prognosis for genetic improvement of yield potential and water-limited yield of major grain crops. *Field Crops Research, 143*, 18–33.
Hanson, D. T. (2016). Breaking the rules of Rubisco catalysis. *Journal of Experimental Botany, 67*(11), 3180.
Hasan, M., Cheng, Y., Kanwar, M. K., Chu, X.-Y., Ahammed, G. J., & Qi, Z.-Y. (2017). Responses of plant proteins to heavy metal stress—A review. *Frontiers in Plant Science, 8*, 1492.
Hasanuzzaman, M., Nahar, K., Alam, M., Roychowdhury, R., & Fujita, M. (2013). Physiological, biochemical, and molecular mechanisms of heat stress tolerance in plants. *International Journal of Molecular Sciences, 14*(5), 9643–9684.
Hatfield, J. L., Boote, K. J., Kimball, B. A., Ziska, L., Izaurralde, R. C., Ort, D., Thomson, A. M., & Wolfe, D. (2011). Climate impacts on agriculture: Implications for crop production. *Agronomy Journal, 103*, 351–370.
Häusler, R. E., Hirsch, H. J., Kreuzaler, F., & Peterhänsel, C. (2002). Overexpression of C4-cycle enzymes in transgenic C3 plants: A biotechnological approach to improve C3-photosynthesis. *Journal of Experimental Botany, 53*(369), 591–607.
Henriet, C., Balliau, T., Aimé, D., Le Signor, C., Kreplak, J., Zivy, M., Gallardo, K., & Vernoud, V. (2021). Proteomics of developing pea seeds reveals a complex antioxidant network underlying the response to sulfur deficiency and water stress. *Journal of Experimental Botany, 72*(7), 2611–2626.
Hoisington, D., Khairallah, M., Reeves, T., Ribaut, J.-M., Skovmand, B., Taba, S., & Warburton, M. (1999). Plant genetic resources: What can they contribute toward increased crop productivity? *Proceedings of the National Academy of Sciences, 96*(11), 5937–5943.
Hsu, P. D., Lander, E. S., & Zhang, F. (2014). Development and applications of CRISPR-Cas9 for genome engineering. *Cell, 157*(6), 1262–1278.
Jagadish, S. K., Way, D. A., & Sharkey, T. D. (2021). Plant heat stress: Concepts directing future research. *Plant, Cell & Environment.*
Jahan, M. S., Guo, S., Sun, J., Shu, S., Wang, Y., Abou El-Yazied, A., Alabdallah, N. M., Hikal, M., Mohamed, M. H., & Ibrahim, M. F. (2021). Melatonin-mediated photosynthetic performance of tomato seedlings under high-temperature stress. *Plant Physiology and Biochemistry, 167*, 309–320.
James, D., Borphukan, B., Fartyal, D., Ram, B., Singh, J., Manna, M., Sheri, V., Panditi, V., Yadav, R., & Achary, V. M. M. (2018). Concurrent overexpression of OsGS1; 1 and OsGS2 genes in transgenic rice (Oryza sativa L.): Impact on tolerance to abiotic stresses. *Frontiers in Plant Science, 9*, 786.
Janssen, P. J., Lambreva, M. D., Plumeré, N., Bartolucci, C., Antonacci, A., Buonasera, K., Frese, R. N., Scognamiglio, V., & Rea, G. (2014). Photosynthesis at the forefront of a sustainable life. *Frontiers in Chemistry, 2*, 36.
Ji, K., Wang, Y., Sun, W., Lou, Q., Mei, H., Shen, S., & Chen, H. (2012). Drought-responsive mechanisms in rice genotypes with contrasting drought tolerance during reproductive stage. *Journal of Plant Physiology, 169*(4), 336–344.
Johnson, R., Frey, N., & Moss, D. N. (1974). Effect of water stress on photosynthesis and transpiration of flag leaves and spikes of Barley and wheat 1. *Crop Science, 14*(5), 728–731.
Kant, S., Bi, Y.-M., & Rothstein, S. J. (2011). Understanding plant response to nitrogen limitation for the improvement of crop nitrogen use efficiency. *Journal of Experimental Botany, 62*(4), 1499–1509.
Khan, I., Awan, S. A., Ikram, R., Rizwan, M., Akhtar, N., Yasmin, H., Sayyed, R. Z., Ali, S., & Ilyas, N. (2020). Effects of 24-epibrassinolide on plant growth, antioxidants defense system, and endogenous hormones in two wheat varieties under drought stress. *Physiologia Plantarum.*

Kimura, H., Hashimoto-Sugimoto, M., Iba, K., Terashima, I., & Yamori, W. (2020). Improved stomatal opening enhances photosynthetic rate and biomass production in fluctuating light. *Journal of Experimental Botany, 71*(7), 2339–2350.

Koester, R. P., Pignon, C. P., Kesler, D. C., Willison, R. S., Kang, M., Shen, Y., Priest, H. D., Begemann, M. B., Cook, K. A., & Bannon, G. A. (2021). Transgenic insertion of the cyanobacterial membrane protein ictB increases grain yield in Zea mays through increased photosynthesis and carbohydrate production. *PLoS One, 16*(2), e0246359.

Köhler, I. H., Ruiz-Vera, U. M., VanLoocke, A., Thomey, M. L., Clemente, T., Long, S. P., Ort, D. R., & Bernacchi, C. J. (2017). Expression of cyanobacterial FBP/SBPase in soybean prevents yield depression under future climate conditions. *Journal of Experimental Botany, 68*(3), 715–726.

Kumudini, B. S., & Patil, S. V. (2019). Role of plant hormones in improving photosynthesis. *Photosynthesis, Productivity and Environmental Stress*, 215–240.

Lawlor, D. W., & Cornic, G. (2002). Photosynthetic carbon assimilation and associated metabolism in relation to water deficits in higher plants. *Plant, Cell & Environment, 25*(2), 275–294.

Lizana, C., Wentworth, M., Martinez, J. P., Villegas, D., Meneses, R., Murchie, E. H., Pastenes, C., Lercari, B., Vernieri, P., & Horton, P. (2006). Differential adaptation of two varieties of common bean to abiotic stress: I. Effects of drought on yield and photosynthesis. *Journal of Experimental Botany, 57*(3), 685–697.

Lobell, D. B., Cassman, K. G., & Field, C. B. (2009). Crop yield gaps: Their importance, magnitudes, and causes. *Annual Review of Environment and Resources, 34*, 179–204.

Long, B. M., Rae, B. D., Rolland, V., Förster, B., & Price, G. D. (2016). Cyanobacterial CO_2-concentrating mechanism components: Function and prospects for plant metabolic engineering. *Current Opinion in Plant Biology, 31*, 1–8.

Long, S. P., Marshall-Colon, A., & Zhu, X.-G. (2015). Meeting the global food demand of the future by engineering crop photosynthesis and yield potential. *Cell, 161*(1), 56–66.

Long, S. P., Zhu, X. G., Naidu, S. L., & Ort, D. R. (2006). Can improvement in photosynthesis increase crop yields? *Plant, Cell & Environment, 29*(3), 315–330.

López-Calcagno, P. E., Fisk, S., Brown, K. L., Bull, S. E., South, P. F., & Raines, C. A. (2019). Overexpressing the H-protein of the glycine cleavage system increases biomass yield in glasshouse and field-grown transgenic tobacco plants. *Plant Biotechnology Journal, 17*(1), 141–151.

Majeed A, Muhammad Z (2019). Salinity: A major agricultural problem—Causes, impacts on crop productivity and management strategies. In: Plant abiotic stress tolerance. Springer, pp. 83–99.

Makino, A. (2011). Photosynthesis, grain yield, and nitrogen utilization in rice and wheat. *Plant Physiology, 155*(1), 125–129.

Martín-Gómez, P., Aguilera, M., Pemán, J., Gil-Pelegrín, E., & Ferrio, J. P. (2017). Contrasting ecophysiological strategies related to drought: The case of a mixed stand of Scots pine (Pinus sylvestris) and a submediterranean oak (*Quercus subpyrenaica*). *Tree Physiology, 37*(11), 1478–1492.

Massacci, A., Battistelli, A., & Loreto, F. (1996). Effect of drought stress on photosynthetic characteristics, growth and sugar accumulation of field-grown sweet sorghum. *Functional Plant Biology, 23*(3), 331–340.

Menz, J., Modrzejewski, D., Hartung, F., Wilhelm, R., & Sprink, T. (2020). Genome edited crops touch the market: A view on the global development and regulatory environment. *Frontiers in Plant Science*, 11.

Mittler, R., & Blumwald, E. (2010). Genetic engineering for modern agriculture: Challenges and perspectives. *Annual Review of Plant Biology, 61*, 443–462.

Molina, L. T. (2021). Introductory lecture: Air quality in megacities. *Faraday Discussions, 226*, 9–52.

Morales, F., Ancín, M., Fakhet, D., González-Torralba, J., Gámez, A. L., Seminario, A., Soba, D., Ben Mariem, S., Garriga, M., & Aranjuelo, I. (2020). Photosynthetic metabolism under stressful growth conditions as a bases for crop breeding and yield improvement. *Plants, 9*(1), 88.

Morison, J., Baker, N., Mullineaux, P., & Davies, W. (2008). Improving water use in crop production. *Philosophical Transactions of the Royal Society B: Biological Sciences, 363*(1491), 639–658.

Morris, M., Edmeades, G., & Pehu, E. (2006). The global need for plant breeding capacity: What roles for the public and private sectors? *HortScience, 41*(1), 30–39.

Muhammad, I., Shalmani, A., Ali, M., Yang, Q.-H., Ahmad, H., & Li, F. B. (2021). Mechanisms regulating the dynamics of photosynthesis under abiotic stresses. *Frontiers in Plant Science, 11*, 2310.

Murchie, E., Pinto, M., & Horton, P. (2009). Agriculture and the new challenges for photosynthesis research. *New Phytologist, 181*(3), 532–552.

Murchie, E. H., & Niyogi, K. K. (2011). Manipulation of photoprotection to improve plant photosynthesis. *Plant Physiology, 155*(1), 86–92.

Murtaza, B., Murtaza, G., Sabir, M., Amjad, M., & Imran, M. (2016). Nitrogen management in rice-wheat cropping system in salt-affected soils. In *Soil science: Agricultural and environmental prospectives* (pp. 67–89). Springer.

Niinemets, Ü., & Keenan, T. (2014). Photosynthetic responses to stress in Mediterranean evergreens: Mechanisms and models. *Environmental and Experimental Botany, 103*, 24–41.

Niu, J., Liu, C., Huang, M., Liu, K., & Yan, D. (2020). Effects of foliar fertilization: A review of current status and future perspectives. *Journal of Soil Science and Plant Nutrition*, 1–15.

Nouri, M.-Z., Moumeni, A., & Komatsu, S. (2015). Abiotic stresses: Insight into gene regulation and protein expression in photosynthetic pathways of plants. *International Journal of Molecular Sciences, 16*(9), 20392–20416.

Olesen, J. E., Trnka, M., Kersebaum, K. C., Skjelvåg, A. O., Seguin, B., Peltonen-Sainio, P., Rossi, F., Kozyra, J., & Micale, F. (2011). Impacts and adaptation of European crop production systems to climate change. *European Journal of Agronomy, 34*(2), 96–112.

Ortiz, R., Trethowan, R., Ferrara, G. O., Iwanaga, M., Dodds, J. H., Crouch, J. H., Crossa, J., & Braun, H.-J. (2007). High yield potential, shuttle breeding, genetic diversity, and a new international wheat improvement strategy. *Euphytica, 157*(3), 365–384.

Otieno, D. O., Schmidt, M., Adiku, S., & Tenhunen, J. (2005). Physiological and morphological responses to water stress in two Acacia species from contrasting habitats. *Tree Physiology, 25*(3), 361–371.

Paleari, L., Movedi, E., Zoli, M., Burato, A., Cecconi, I., Errahouly, J., Pecollo, E., Sorvillo, C., & Confalonieri, R. (2021). Sensitivity analysis using Morris: Just screening or an effective ranking method? *Ecological Modelling, 455*, 109648.

Parry, M. A., Reynolds, M., Salvucci, M. E., Raines, C., Andralojc, P. J., Zhu, X.-G., Price, G. D., Condon, A. G., & Furbank, R. T. (2011). Raising yield potential of wheat. II. Increasing photosynthetic capacity and efficiency. *Journal of Experimental Botany, 62*(2), 453–467.

Peng, S., Khush, G. S., Virk, P., Tang, Q., & Zou, Y. (2008). Progress in ideotype breeding to increase rice yield potential. *Field Crops Research, 108*(1), 32–38.

Porter, J. R., & Semenov, M. A. (2005). Crop responses to climatic variation. *Philosophical Transactions of the Royal Society B: Biological Sciences, 360*(1463), 2021–2035.

Prior, H., Baldrick, P., Beken, S., Booler, H., Bower, N., Brooker, P., Brown, P., Burlinson, B., Burns-Naas, L. A., & Casey, W. (2020). Opportunities for use of one species for longer-term toxicology testing during drug development: A cross-industry evaluation. *Regulatory Toxicology and Pharmacology, 113*, 104624.

Raines, C. A. (2011). Increasing photosynthetic carbon assimilation in C3 plants to improve crop yield: Current and future strategies. *Plant Physiology, 155*(1), 36–42.

Ramsden, S. J., & Hails, R. S. (2019). Conclusions on agricultural resilience. *Agricultural Resilience*, 362.

Rauf, S. (2008). Breeding sunflower (*Helianthus annuus* L.) for drought tolerance. *Communications in Biometry and Crop Science, 3*(1), 29–44.

Ray, D. K., Mueller, N. D., West, P. C., & Foley, J. A. (2013). Yield trends are insufficient to double global crop production by 2050. *PLoS One, 8*(6), e66428.

Ray, D. K., Ramankutty, N., Mueller, N. D., West, P. C., & Foley, J. A. (2012). Recent patterns of crop yield growth and stagnation. *Nature Communications, 3*(1), 1–7.

Reddy, A. R., Chaitanya, K. V., & Vivekanandan, M. (2004). Drought-induced responses of photosynthesis and antioxidant metabolism in higher plants. *Journal of Plant Physiology, 161*(11), 1189–1202.

Reynolds, M., Bonnett, D., Chapman, S. C., Furbank, R. T., Manès, Y., Mather, D. E., & Parry, M. A. (2011). Raising yield potential of wheat. I. Overview of a consortium approach and breeding strategies. *Journal of Experimental Botany, 62*(2), 439–452.

Reynolds, M. P., Quilligan, E., Aggarwal, P. K., Bansal, K. C., Cavalieri, A. J., Chapman, S. C., Chapotin, S. M., Datta, S. K., Duveiller, E., & Gill, K. S. (2016). An integrated approach to maintaining cereal productivity under climate change. *Global Food Security, 8*, 9–18.

Rochaix, J.-D. (2011). Assembly of the photosynthetic apparatus. *Plant Physiology, 155*(4), 1493–1500.

Rodrigues, L. R., & Kluskens, L. D. (2011). Synthetic biology & bioinformatics: Prospects in the cancer arena. *Computational Biology and Applied Bioinformatics, 8*, 159–186.

Rodríguez, P., Mellisho, C., Conejero, W., Cruz, Z., Ortuno, M., Galindo, A., & Torrecillas, A. (2012). Plant water relations of leaves of pomegranate trees under different irrigation conditions. *Environmental and Experimental Botany, 77*, 19–24.

Saeidi, M., Moradi, F., & Abdoli, M. (2017). Impact of drought stress on yield, photosynthesis rate, and sugar alcohols contents in wheat after anthesis in semiarid region of Iran. *Arid Land Research and Management, 31*(2), 204–218.

Saglam, A., Kadioglu, A., Demiralay, M., & Terzi, R. (2014). Leaf rolling reduces photosynthetic loss in maize under severe drought. *Acta Botanica Croatica, 73*(2), 315–323.

Schneeberger, K., & Weigel, D. (2011). Fast-forward genetics enabled by new sequencing technologies. *Trends in Plant Science, 16*(5), 282–288.

Seleiman, M. F., Al-Suhaibani, N., Ali, N., Akmal, M., Alotaibi, M., Refay, Y., Dindaroglu, T., Abdul-Wajid, H. H., & Battaglia, M. L. (2021). Drought stress impacts on plants and different approaches to alleviate its adverse effects. *Plants, 10*(2), 259.

Sharwood, R. E. (2017). Engineering chloroplasts to improve Rubisco catalysis: Prospects for translating improvements into food and fiber crops. *New Phytologist, 213*(2), 494–510.

Sharwood, R. E., Ghannoum, O., Kapralov, M. V., Gunn, L. H., & Whitney, S. M. (2016). Temperature responses of Rubisco from Paniceae grasses provide opportunities for improving C3 photosynthesis. *Nature Plants, 2*(12), 1–9.

Siddique, M., Hamid, A., & Islam, M. (1999). Drought stress effects on photosynthetic rate and leaf gas exchange of wheat. *Botanical Bulletin of Academia Sinica, 40*.

Simkin, A. J., Faralli, M., Ramamoorthy, S., & Lawson, T. (2020). Photosynthesis in non-foliar tissues: Implications for yield. *The Plant Journal, 101*(4), 1001–1015.

Simkin, A. J., López-Calcagno, P. E., & Raines, C. A. (2019). Feeding the world: Improving photosynthetic efficiency for sustainable crop production. *Journal of Experimental Botany, 70*(4), 1119–1140.

Simó, C., Ibáez, C., Valdés, A., Cifuentes, A., & García-Cañas, V. (2014). Metabolomics of genetically modified crops. *International Journal of Molecular Sciences, 15*(10), 18941–18966.

Singh, R. P., Prasad, P. V., & Reddy, K. R. (2013). Impacts of changing climate and climate variability on seed production and seed industry. *Advances in Agronomy, 118*, 49–110.

Singh, R. P., Prasad, P. V., Sunita, K., Giri, S., & Reddy, K. R. (2007). Influence of high temperature and breeding for heat tolerance in cotton: A review. *Advances in Agronomy, 93*, 313–385.

Slafer, G. A., Satorre, E. H., & Andrade, F. H. (2021). Increases in grain yieid in bread wheat from breeding and associated physiological changes. In *Genetic improvement of field crops* (pp. 1–68). CRC Press.

Snyder, R., & Tegeder, M. (2021). Targeting nitrogen metabolism and transport processes to improve plant nitrogen use efficiency. *Frontiers in Plant Science, 11*, 2330.

Solis, C. A., Yong, M. T., Venkataraman, G., Milham, P., Zhou, M., Shabala, L., Holford, P., Shabala, S., & Chen, Z. H. (2021). Sodium sequestration confers salinity tolerance in an ancestral wild rice. *Physiologia Plantarum*.

Sonnewald, U., Fernie, A. R., Gruissem, W., Schläpfer, P., Anjanappa, R. B., Chang, S. H., Ludewig, F., Rascher, U., Muller, O., & van Doorn, A. M. (2020). The Cassava Source–Sink project: Opportunities and challenges for crop improvement by metabolic engineering. *The Plant Journal, 103*(5), 1655–1665.

Sumner, L. W., Mendes, P., & Dixon, R. A. (2003). Plant metabolomics: Large-scale phytochemistry in the functional genomics era. *Phytochemistry, 62*(6), 817–836.

Suzuki, N., Koussevitzky, S., Mittler, R., & Miller, G. (2012). ROS and redox signalling in the response of plants to abiotic stress. *Plant, Cell & Environment, 35*(2), 259–270.

Talaat, N. B. (2019). Role of reactive oxygen species signaling in plant growth and development. *Reactive oxygen, nitrogen and sulfur species in plants: Production, metabolism, signaling and defense mechanisms* (pp. 225–266).

Tkemaladze, G. S., & Makhashvili, K. (2016). Climate changes and photosynthesis. *Annals of Agrarian Science, 14*(2), 119–126.

Toledo, Á., & Burlingame, B. (2006). Biodiversity and nutrition: A common path toward global food security and sustainable development. *Journal of Food Composition and Analysis, 19*(6–7), 477–483.

Torabi, S., Taheri, M. A., & Semsarha, F. (2021). Alleviative effects of Faradarmani consciousness field on Triticum aestivum L. under salinity stress. *F1000Research, 9*(1089), 1089.

Tylecote, A. (2019). Biotechnology as a new techno-economic paradigm that will help drive the world economy and mitigate climate change. *Research Policy, 48*(4), 858–868.

Ullah, H., Santiago-Arenas, R., Ferdous, Z., Attia, A., & Datta, A. (2019). Improving water use efficiency, nitrogen use efficiency, and radiation use efficiency in field crops under drought stress: A review. *Advances in Agronomy, 156*, 109–157.

Ustin, S. L., & Middleton, E. M. (2021). Current and near-term advances in Earth observation for ecological applications. *Ecological Processes, 10*(1), 1–57.

Van Esse, H. P., Reuber, T. L., & van der Does, D. (2020). Genetic modification to improve disease resistance in crops. *New Phytologist, 225*(1), 70–86.

Wang, Y., Cai, Y., Liu, G., Zhang, P., Li, B., Li, B., Jia, Q., Huang, Y., & Shu, T. (2021). Evaluation of sustainable crop production from an ecological perspective based emergy analysis: A case of China's provinces. *Journal of Cleaner Production*, 127912.

Wani, W., Masoodi, K. Z., Zaid, A., Wani, S. H., Shah, F., Meena, V. S., Wani, S. A., & Mosa, K. A. (2018). Engineering plants for heavy metal stress tolerance. *Rendiconti Lincei Scienze Fisiche e Naturali, 29*(3), 709–723.

Warburton, M. L., Rauf, S., Marek, L., Hussain, M., Ogunola, O., & de Jesus Sanchez Gonzalez, J. (2017). The use of crop wild relatives in maize and sunflower breeding. *Crop Science, 57*(3), 1227–1240.

Wassmann, R., Jagadish, S., Heuer, S., Ismail, A., Redona, E., Serraj, R., Singh, R., Howell, G., Pathak, H., & Sumfleth, K. (2009). Climate change affecting rice production: The physiological and agronomic basis for possible adaptation strategies. *Advances in Agronomy, 101*, 59–122.

Weber, A. P., & Bar-Even, A. (2019). Update: Improving the efficiency of photosynthetic carbon reactions. *Plant Physiology, 179*(3), 803–812.

Whitney, S. M., Houtz, R. L., & Alonso, H. (2011). Advancing our understanding and capacity to engineer nature's CO_2-sequestering enzyme, Rubisco. *Plant Physiology, 155*(1), 27–35.

Wise TA (2013) Can we feed the world in 2050. A scoping paper to assess the evidence Global Development and Environment Institute Working Paper (13–04).

Wittmann, G. (1981). Toxic metals. In *Metal pollution in the aquatic environment* (pp. 3–70). Springer.

Yadav R, Thankappan R, Kumar A (2021) Novel approaches for genome editing to develop climate smart crops. In Microbiomes and the global climate change. Springer, pp. 267–291.

Yang, J., & Zhang, J. (2010). Crop management techniques to enhance harvest index in rice. *Journal of Experimental Botany, 61*(12), 3177–3189.

Yang, W., Feng, H., Zhang, X., Zhang, J., Doonan, J. H., Batchelor, W. D., Xiong, L., & Yan, J. (2020). Crop phenomics and high-throughput phenotyping: Past decades, current challenges, and future perspectives. *Molecular Plant, 13*(2), 187–214.

Zhu, X.-G., Long, S. P., & Ort, D. R. (2010). Improving photosynthetic efficiency for greater yield. *Annual Review of Plant Biology, 61*, 235–261.

Zlatev, Z., & Lidon, F. C. (2012). An overview on drought induced changes in plant growth, water relationsand photosynthesis. *Emirates Journal of Food and Agriculture*, 57–72.

Chapter 4
Transcription Factor Mediated Plant Metabolite Production in Response to Environmental Stress Factors: Current Understanding and Future Aspects

Samrat Banerjee, Pinaki Roy, and Sujit Roy

Abstract The current agriculture is facing a continuous challenge of reaching up to 70% increase in crop productivity by 2050. An alarming expansion of human population, global climate changes, increasing soil salinity, and freshwater scarcity put the sustainable food production in a serious question. The inanimate life of plants makes them surrounded by a myriad of diverse biotic and abiotic stress conditions, which are mostly unavoidable. During the course of evolution, plants have evolved a robust and complicated mechanism of growth and defense trade off when responding to stress conditions. Secondary metabolite compounds like terpenoids, flavonols, flavones, and stilbenes are considered as the stress-inducible phytochemicals play crucial role in the development of plant immunity. It is now well established that transcription factors enable plants to counteract unfavorable conditions via the modulation of secondary metabolite genes, and they are now considered as potential genomic candidates for their wide applications in crop breeding strategies. Defensive molecular switches involve transcription factors which act as mediators of stress signals and regulate stress-responsive gene expression. To counteract the stress factors, different transcription factor families including AP2/ERF, WRKY, bHLH, bZIP, MYB, and NAC regulate the secondary metabolite biosynthesis genes. Classical breeding approaches for the generation of stress-resilient crops seem to be time-consuming and often the outcome is less effective. Recent advancement in the applications related to promoter engineering for the metabolite biosynthesis found to be novel approach for improved crop yield. The transcriptional and posttranscriptional modulation of TFs can facilitate molecular breeding and genetic moderation of plants for the increased production of secondary metabolites. Moreover, the synthetic promoters and transcription factors have been immensely powerful and effective as components for the regulation of targeted plant transgene expression. In this present chapter, we have highlighted on the function of TFs and the recent

S. Banerjee · P. Roy · S. Roy (✉)
Department of Botany, UGC Centre for Advanced Studies, The University of Burdwan, Burdwan, West Bengal, India

T. Aftab, K. R. Hakeem (eds.), *Metabolic Engineering in Plants*,
https://doi.org/10.1007/978-981-16-7262-0_4

advancement of promoter engineering in the generation of stress-tolerant plants in the context of enhanced metabolite production.

Keywords Gene expression · Plant metabolites · Transcription factors · Phytochemicals

Introduction

Plants have the ability to sense numerous stress stimuli and adapt different stress responses through activation of signaling networks. Plants are continuously facing several adverse environmental conditions like different types of biotic and abiotic stress factors, for example, bacteria, virus, fungi, heat, cold, salinity, drought, and ultraviolet-B (UV-B) radiation (Khan et al., 2018; Meraj et al., 2020). They respond to unfavorable environmental conditions through the expression of diverse group of stress-inducible phytochemicals that play crucial role in the development of plant immunity. Being sessile in nature, plants develop several adaptive mechanisms at the biochemical, physiological, and cellular levels through the production of antioxidants, phenylpropanoid compounds, phytohormones, cuticular wax, and osmolytes (Yamaguchi et al., 2007; Kitsios & Doonan, 2011; Liu et al., 2015a, 2015b; Peleg & Blumwald, 2011). Many receptors or sensors after sensing the stress signals induce many biological molecules (phytohormones, Ca^{2+}, reactive oxygen species (ROS), protein kinases), and they transferred the signal to transcription factors (TFs) (Meraj et al., 2020). TFs are involved in plant stress response through acting as mediators upon sensing the signals from stress factors, and subsequently they transfer the signal to the downstream components for the activation of stress-responsive metabolite compounds (Fig. 4.1).

The plant secondary metabolite compounds have multiple functions including growth, development, reproduction, and cell division (Kutchan, 2001). For more than a thousand years, mankind uses plant products as effective pharmaceutical compounds. Plant metabolites are used as foods, medicines, and raw materials of industry (Oksman-Caldentey & Saito, 2005). Secondary metabolite compounds like terpenoids, flavonols, flavones, stilbenes, and glucosinolates are considered as the stress-inducible phytochemicals. Based on the chemical composition, secondary metabolites are divided in two groups such as nitrogen-containing metabolites (alkaloids) and nitrogen-deficient molecules (phenolics and terpenoids). Alkaloids are derived from different amino acids, namely, tryptophan, tyrosine, phenylalanine, and lysine (Luca & Pierre, 2000). It was observed that terpenoid indole alkaloids, tropane alkaloids, and purine alkaloids protect the plants from herbivores and UV-B radiation and have been used to cure many terminal diseases (Patra et al., 2013). Terpenes are synthesized via methylerythritol phosphate pathway (chloroplast) or mevalonate pathway (cytosol). Terpenoid compounds play a critical role in the development of blast disease resistance in rice (Miyamoto et al., 2014). Besides this, terpenes act as phytohormones (gibberellins, brassinosteroids) (Singh &

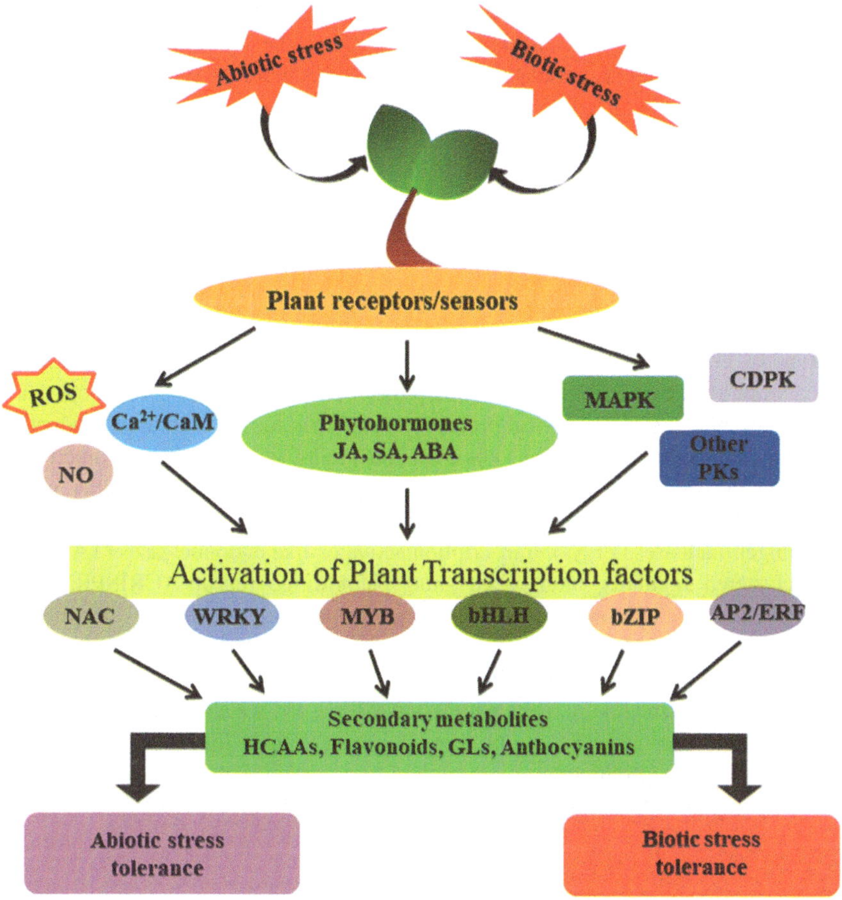

Fig. 4.1 Schematic representation showing signal transduction and regulation of secondary metabolite genes through the activation of different transcription factor family genes. The transcription factors modulate the production of plant secondary metabolite compounds under biotic or abiotic stress condition

Sharma, 2015) and give resistance against herbivores (Patra et al., 2013). Phenolics are aromatic compounds primarily synthesized from two pathways such as shikimic acid pathway and malonic acid pathway. It was observed that flavonoids act as antioxidant compounds in plants protecting them from UV-B stress (Takahashi & Ohnishi, 2004). Moreover, flavonoids and other phenolics are anticancerous, anti-inflammatory, antibacterial, and cardioprotective compound, widely used in medical science (Tungmunnithum et al., 2018).

The synthesis and accumulation of secondary metabolite compounds are spatially regulated by biotic and abiotic stress factors. It was observed that various transcription factor families (AP2/ERF, WRKY, bHLH, bZIP, MYB, NAC) regulate the biosynthetic enzymes of secondary metabolites. Transcription factors modulate the expression of the enzymes in the multistep biosynthesis pathway of metabolite compounds. Different transcription factor regulates the biotic and abiotic stress responses through the accumulation of secondary metabolite compounds as defense component (Table 4.1). Transcription factors not only interact with the promoters of the metabolite genes but they also alter the expression of metabolites via both posttranscriptionally and posttranslationally (Luo et al., 2012; Patra et al., 2013). For example, MdCOP1 interacts with the MdMYB1 to regulate the red apple color through accumulation of anthocyanin (Li et al., 2012). Moreover, GLYCOALKALOID METABOLISM 9 (GAME 9) is involved in the biosynthesis of alkaloid in Solanaceae (Cárdenas et al., 2016). Based on the above information, the modulation of TF genes can effectively change the metabolite production. Overexpression of activating TFs or knockout mutation of repressive TFs was found to be fruitful for efficient production of bioactive secondary metabolite compounds (Miyamoto et al., 2014; Wu et al., 2016). Classical breeding approaches for the increased production of metabolite compounds seems to be time-consuming and often the outcome is less effective. Recent advancement in genetic engineering shows new paths in the targeted modulation of gene expression. The site-directed endonucleases, namely, zinc finger nucleases (ZFNs), transcription activator-like endonucleases (TALENs), and clustered regularly interspaced short palindromic repeats (CRISPR), were emerged as new efficient tools for the manipulation of metabolite biosynthesis genes (Liu & Stewart Jr, 2016).

In this book chapter, we described the function of TFs in the regulation of abiotic stress-responsive metabolite gene expression. Moreover, we also explained how the synthetic promoter and TFs can be used as potential tools for the upregulation of desired metabolite production.

Transcription Factors in the Regulation of Secondary Metabolite Gene Expression Under Abiotic Stress

In plants, secondary metabolites such as alkaloids, terpenoids, and flavonoids have been found to be associated with various stress responses. Different families of TFs are induced by stress signals of environment transduced by different mediator compounds (MYC2, MAPK, ROS, and Ca^{2+}) (Meraj et al., 2020). However, among different families of TFs, only few groups of TF families were found to be directly involved in secondary metabolite biosynthesis. Previous studies revealed that six transcriptional families (AP2/ERF, WRKY, bHLH, bZIP, MYB, and NAC) are actively participating in the regulation of abiotic and biotic stress tolerance through accumulation of secondary metabolites.

Table 4.1 Transcription factor-mediated secondary metabolite production to combat against biotic and abiotic stress factors

Families	Transcription factor gene	Plant species	Secondary metabolite	References
AP2/ERF	JRE4 (GAME9)	*Solanum lycopersicum, Solanum tuberosum*	Steroidal glycoalkaloids (SGAs)	Nakayasu et al. (2017), Cárdenas et al. (2016)
	NtERF32	*Nicotiana tabaccum*	Nicotine	Sears et al. (2014)
	GbERF1	*Gossypium barbadense*	Lignin	Guo et al. (2016)
	ORA59	*Arabidopsis thaliana*	Hydroxycinnamic acid amides (HCAAs)	Li et al. (2018)
	TcERF12/ TcERF15	*Taxus chinensis*	Taxol	Zhang et al. (2015a, 2015b)
	VqERF114	*Vitis quinquangularis*	Resveratrol	Wang and Wang (2019)
	ORCA3	*Catharanthus roseus*	Terpenoid Alkaloid	Van der Fits and Memelink (2000)
	EREB58	*Zea mays*	Sesquiterpenes	Li et al. (2015)
	PnERF1	*Panax notoginseng*	Saponins	Deng et al. (2017)
	CrERF5	*Catharanthus roseus*	Bisindole alkaloids	Pan et al. (2019)
	Ii049	*Isatis indigotica*	Lignan/lignin biosynthesis	Ma et al. (2017)
WRKY	StWRKY1	*Solanum tuberosum*	Hydroxycinnamic acid amide (HCAAs)	Yogendra et al. (2015)
	StWRKY8	*Solanum tuberosum*	Benzylisoquinoline alkaloids (BIAs)	Yogendra et al. (2017)
	ZmWRKY79	*Zea mays*	Terpenoid phytoalexins	Fu et al. (2017)
	TcWRKY1	*Taxus chinensis*	Taxol	Li et al. (2013a, 2013b)
	WsWRKY1	*Withania somnifera*	Phytosterol	Singh et al. (2017)
	TaWRKY70	*Triticum aestivum*	Hydroxycinnamic acid amides (HCAAs)	Kage et al. (2017)
	SsWRKY18/ 40	*Salvia sclarea*	Diterpenoids	Alfieri et al. (2018)
	VviWRKY24/ 03/VvWRKY8	*Vitis vinifera*	Resveratrol	Vannozzi et al. (2018), Jiang et al. (2018)
	HvWRKY23	*Hordeum vulgare*	Flavonoid andhydroxycinnamic acid amides (HCAAs)	Karre et al. (2019)

(continued)

Table 4.1 (continued)

Families	Transcription factor gene	Plant species	Secondary metabolite	References
	CrWRKY1	*Catharanthus roseus*	Terpenoid indole alkaloids	Suttipanta et al. (2011)
bHLH	ILR3/bHLH104, bHLH04/05/06	*Arabidopsis thaliana*	Glucosinolates (GLs)	Samira et al. (2018), Frerigmann et al. (2014)
	VvbHLH1	*Arabidopsis thaliana*	Flavonoids	Wang et al. (2016a, 2016b, 2016c, 2016d)
	MdMYC2	*Malus domestica*	Anthocyanin	An et al. (2016)
	GL3	*Arabidopsis thaliana*	Anthocyanin	Gonzalez et al. (2008)
	NtAN1a and NtAN1b	*Nicotiana tabacum*	Flavonoids	Bai et al. (2011)
	NbbHLH1 and NbbHLH2	*Nicotiana benthamiana*	Nicotine alkaloid	Todd et al. (2010)
	CjbHLH1	*Coptis japonica*	Isoquinoline alkaloids	Yamada et al. (2015)
	DPF	*Oryza sativa*	Diterpenoid phytoalexins	Yamamura et al. (2015)
	TSAR1/TSAR2	*Medicago truncatula*	Saponins	Mertens et al. (2016)
	AH (Hoffman's Anthocyaninless)	*Solanum lycopersicum*	Anthocyanin	Qiu et al. (2016)
bZIP	MdHY5	*Malus domestica*	Anthocyanin	An et al. (2017)
	SlHY5	*Solanum lycopersicum*	Anthocyanin Monoterpenoids	Liu et al. (2018), Zhou et al. (2015)
	OsTGAP1	*Oryza sativa*	Diterpenoid phytoalexins	Miyamoto et al. (2014), Okada et al. (2009)
	OsbZIP79	*Oryza sativa*	Diterpenoid phytoalexins	Miyamoto et al. (2015)
MYB	AtMYB34/51/112	*Arabidopsis thaliana*	Indolicglucosinolate (IGs)	Frerigmann (Frerigmann, 2016)
	AtMYB75	*Arabidopsis thaliana*	Anthocyanin	Onkokesung et al. (2014)
	PtMYB115	*Populus tomentosa*	Proanthocyanidin	Wang et al. (2017)
	VvMYBC2-L1	*Vitis vinifera*	Proanthocyanidin	Huang et al. (2014)
	VvMYB14/ VviMYB14	*Vitis vinifera*	Resveratrol	Vannozzi et al. (2018), Jiang et al. (2018)

(continued)

Table 4.1 (continued)

Families	Transcription factor gene	Plant species	Secondary metabolite	References
	RrMYB5/ RrMYB10	*Rosa rugosa*	Proanthocyanidin	Shen et al. (2019)
	ROSEA1 and ROSEA2	*Antirrhinum majus*	Anthocyanin	Schwinn et al. (2006)
	CsMYBF1	*Citrus sinensis*	Phenylpropanoid	Liu et al. (2016)
	AtMYB11/12/ 111	*Arabidopsis thaliana*	Flavonoids	Misra et al. (2010), Pandey et al. (2014)
	MdMYB1	*Malus domestica*	Anthocyanin	Takos et al. (2006)
	MdMYB10	*Malus domestica*	Anthocyanin	Espley et al. (2007)
	CsMYB2/26	*Camellia sinensis*	Flavonoids	Wang et al. (2018)
	OsMYB30/55/ 110	*Oryza sativa*	Hydroxycinnamic acids (HCAAs)	Kishi-Kaboshi et al. (2018)
NAC	PtrNAC72	*Poncirus trifoliata*	Putrescine	Wu et al. (2016)
	PaNAC03	*Picea abies*	Flavonoid	Dalman et al. (2017)
	ANACO32	*Arabidopsis thaliana*	Anthocyanin	Mahmood et al. (2016)
	MfNACsa	*Medicago falcata*	Glutathione	Duan et al. (2017)
	ANAC078	*Arabidopsis thaliana*	Anthocyanin	Morishita et al. (2009)

WRKY Transcription Factors

The WRKY transcription factor (TF) family has been well recognized for its involvement in the regulation of various abiotic and biotic stress tolerances in plants. They are considered as one of the largest TF families in plants and are comprised of a WRKYGQK motif in the N-terminus and a C_2H_2 or C_2HC zinc finger motif at the C-terminus (Fig. 4.2) (Eulgem et al., 2000). The 60-amino acid-long WRKY domain interacts with the core cis-element, (T)TGAC(C) (also known as W-box), of targeted gene promoters (Rushton et al., 2010). In *Arabidopsis thaliana*, phylogenetic analysis of 70 members of WRKY proteins has placed them into three major groups (Group I-III) while the group II was further divided into five subgroups (Group IIa–IIe) (Eulgem et al., 2000). From previous works it was observed that WRKY TFs have been found to play pivotal roles in the regulation of abiotic stress tolerance against drought, heat, salt, cold, and wounding stress (Chen et al., 2012a, 2012b; Bakshi & Oelmüller, 2014; Tripathi et al., 2014; Schluttenhofer et al., 2014). Moreover, the WRKY TFs also regulate plant development such as male gametogenesis, size of seeds, seed coat color, trichome, and root hair formation (Johnson et al., 2002; Robatzek & Somssich, 2002; Luo et al., 2005; Guan et al., 2014).

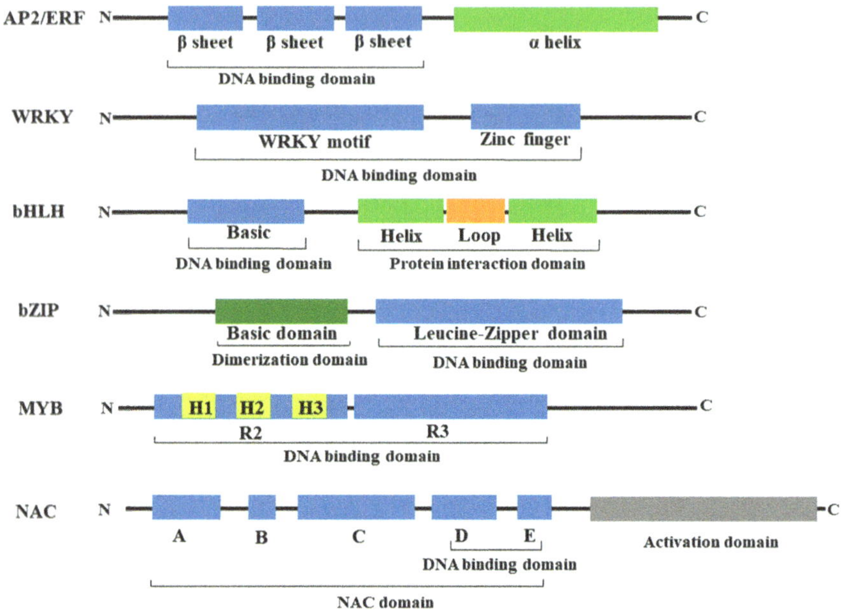

Fig. 4.2 Structure of different transcription factor family proteins with conserved domain organization

Extensive studies accumulate evidence about the involvement of WRKYs in the production of natural compounds (phenylpropanoids, terpenes, alkaloids) through the regulation of secondary metabolite biosynthesis genes (Suttipanta et al., 2011; Ma et al., 2009).

It was observed that hydroxy cinnamic acid amide (HCAA) is involved in the reinforcement of plant cell wall to resist *Phytophthora infestans*, causal organism of late blight of potato. In potato, StWRKY1 is directly interacting with the HCAA gene promoter region during the incidence of the disease (Yogendra et al., 2015). In wheat and barley, TaWRKY70 and HvWRKY23, respectively, were involved in the enhancement of the expression of HCAA and other flavonoid genes such as phenylalanine ammonia lyase (PAL) and chalcone synthase (CHS) (Kage et al., 2017; Karre et al., 2019). Moreover, StWRKY8 in potato positively modulates the activity of benzylisoquinoline alkaloids (BIAs) (Yogendra et al., 2017). Meanwhile, in maize, the WRKY TF ZmWRKY79 responds to various abiotic stresses via increasing the production of phytoalexins. Transient overexpression of ZmWRKY79 has been found to enhance the production of zealexins and kauralexins in maize protoplasts (Fu et al., 2017). Moreover, WRKYs affect the lignin production via regulating the phenylpropanoid biosynthesis pathway (Besseau et al., 2007). At WRKY12, a homolog of *Medicago truncatula* STP (secondary wall thickening in pith) positively affects the lignin biosynthesis through the interaction with the promoters of lignin biosynthesis genes (Wang et al., 2010a, 2010b). Regulation of the production of proanthocyanidins and pectinaceous seed mucilage is governed by

AtERKY44 while the expression of AtWRKY44 is further modulated by MYB-bHLH-WD40 TF complex (Johnson et al., 2002; Ishida et al., 2007). The accumulation of two alkaloids, catharanthine and serpentine, was found to be altered in hairy root cultures of *Catharanthus roseus*. Studies on RNAi lines revealed that CrWRKY1 is possibly involve in the regulation of tryptophan indole alkaloid (TIA) biosynthesis pathway of *Catharanthus roseus* (Suttipanta et al., 2011). Interestingly in *Papaver somniferum*, the production of benzylisoquinoline alkaloids (BIAs) is to be enhanced by the WRKY TFs (Mishra et al., 2013). WRKY1 has been found to play key roles in the upregulation of taxol biosynthesis in *Taxus chinensis* (Li et al., 2013a, 2013b). From previous studies it was observed that overexpression of OsWRKY13, OsWRKY53, and OsWRKY76 regulates the expression of phenylpropanoid and terpenoid biosynthesis genes (Qiu et al., 2008; Ishihama et al., 2011; Yokotani et al., 2013). Moreover, AtWRKY71 has been found to hasten flowering in *Arabidopsis* following salt stress (Yu et al., 2018) whereas AtWRKY53, a group of III WRKY protein, enhances the starch metabolism for modulation of stomatal opening during dehydration stress (Sun & Yu, 2015). Interestingly, TaWRKY10 of wheat when expressed in transgenic tobacco has been reported to participate in both drought and salt stress tolerance through the reduction in ROS accumulation and regulation of osmotic balance (Wang et al., 2013). From the above information it is confirmed that the WRKY TFs play crucial role in both biotic and abiotic stress responses through the regulation metabolite synthesis genes.

bHLH Transcription Factors

The basic helix-loop-helix (bHLH) TFs are considered as the second largest family of transcription factors in angiosperms (Le Hir et al., 2017). The conserved bHLH domain comprises of two regions, a basic amino acid residue containing N-terminal region which binds with the DNA (Fig. 4.2) and a helix-loop-helix at the C-terminus associated with the formation of homo- or heteromeric complexes (Feller et al., 2011). In plants, the bHLH TFs specifically recognize and bind to the E-box (5′-CANNTG-30-3′) and G-box (5′-CACGTG-3′) consensus DNA motifs present in the promoter elements of various secondary metabolite biosynthesis genes such as flavonoids, anthocyanin, glucosinolates, phytoalexins, and saponins (Babitha et al., 2013). Since their identification in maize, several researches on bHLH TFs have been proven that they play a crucial function in a myriad of biological processes such as flavonoid biosynthesis (Ohno et al., 2011), photosynthesis (Dong et al., 2014), and flowering (Ito et al., 2010). It was observed that the exogenous application of ABA results into alteration in the expression of AtbHLH68 in an organ-specific manner and plants overexpressing AtbHLH68 have been found to enhance drought tolerance (Le Hir et al., 2017). Overexpression lines of AtbHLH112 exhibit strong resistance against multiple stresses such as drought and salt (Liu et al., 2015a, 2015b). The bHLH TFs were found to be involved in the biosynthesis of flavonoids which confers resistance against multiple stresses (Wang et al., 2016a, 2016b, 2016c, 2016d). Similarly, another bHLH TF, TabHLH39, in wheat has been found to be induced in response to salt and cold stress (Zhai et al., 2016). The function of bHLH TFs in the development of abiotic stress resistance is conferred by their active participation in many metabolite biosynthesis. The NtMYC2 in tobacco directly involves in the regulation of the nicotine biosynthetic genes (Shoji & Hashimoto, 2011). In *Arabidopsis*, four bHLH TFs, namely, AtbHLH3, AtbHLH13, AtbHLH14, and AtbHLH17, have been found to interact with JA2 proteins resulting into jasmonic acid-mediated anthocyanin accumulation (Song et al., 2013). In *Petunia*, JAF13, a bHLH TF, was associated with flower color development through the production of anthocyanin (Quattrocchio et al., 1999). Moreover, anthocyanin biosynthesis is also coordinately regulated by R2R3-MYB, bHLH, and WD repeat (WDR) proteins which together form a MBW ternary complex (Ramsay & Glover, 2005). In apple, MdMYC2 is induced following wounding and jasmonic acid application while overexpression of MdMYC2 upregulates the biosynthesis of flavonol 3-hydroxylase (F3H) and CHS (An et al., 2016).

bZIP Transcription Factors

The basic leucine zipper (bZIP) family comprised of a highly conserved positively charged (basic) DNA-binding domain (Fig. 4.2) containing 16 amino acid residues and an adjacent dimerization domain known as leucine zipper (Zhang et al., 2017). Members of bZIP TF have been identified in various eukaryotes such as yeast (Liu et al., 2014), *Arabidopsis* (Jakoby, 2002), rice (Nijhawan et al., 2008), and maize (Wei et al., 2012). Besides their function in different plant developmental processes, bZIP family TFs are directly involved in different abiotic stress responses such as cold, high salinity, drought, and others (Banerjee & Roychoudhury, 2015). The light-responsive bZIP TF, HY5, has been well characterized and involved in the regulation of anthocyanin production under abiotic stresses. Beside their involvement in various pharmaceutically active compound production such as tanshinone, artemisinin, and catharanthine (Zhang et al., 2015a, 2015b, 2018; Roepke et al., 2010), bZIP TFs also take part in pigment accumulation. For example, in apple, MdHY5 in association with MdMYB10 is responsible for anthocyanin biosynthesis (An et al., 2017). In *Arabidopsis* bZIP TFs, ABF1, ABF2/AREB1, ABF3, ABF4, and ABI5 are directly involved in the abiotic stress responses (Khan et al., 2018). Interestingly, it has been reported in *Arabidopsis* that the expression of bZIP53 was enhanced following salt treatment in roots (Hartmann et al., 2015). Moreover, in rice, OsABF1 and OsbZIP40 were found to be responsible for the delay in the flowering time via suppression of early heading date 1 (EHD1) under drought stress (Zhang et al., 2016a, 2016b). Similarly, ABF3 also regulates the growth response in plants under drought stress (Wang et al., 2016a, b, c, d). OsbZIP23 and OsbZIP16 confirm salt and drought tolerance in rice seedlings, respectively (Todaka et al., 2015; Chen et al., 2012a, 2012b). The bZIP TF OsTGAP1 plays a crucial role in terpenoid phytoalexin biosynthesis in rice which provides resistance against blast disease (Miyamoto et al., 2014).

NAC Transcription Factors

NAC (NAM, ATAF1/ATAF2, CUC2) family TFs were first reported in *Petunia* NAM (Non apical meristem), and *Arabidopsis* ATAF1/ATAF2 and CUC2 (cup-shaped cotyledon 2) exhibit the conserved NAC domains (Souer et al., 1996; Aida et al., 1997). To date, about 117 genes in *Arabidopsis* and 151 genes in rice have been identified and characterized (Shao et al., 2015). This conserved family of TFs has been well known for their involvement in the regulation of biotic and abiotic stresses. The NAC proteins contain an N-terminal-conserved DNA-binding domain and a C-terminal highly variable transcription regulatory domain which might act as transcriptional activators or repressors to regulate the expression of downstream genes (Puranik et al., 2012). Flavonoids are directly involved in the development of abiotic stress tolerance in plants (Winkel-Shirley, 2002). In *Picea abies* the accumulation of flavonoids has been reported to be associated with increased tolerance against fungal pathogens (Danielsson et al., 2011). Meanwhile, in *Hevea brasiliensis* a NAC TF HbNAC1 was reported to be involved in latex biosynthesis via binding with CACG motif in the promoter of SRPP (small rubber particle protein) (Cao et al., 2017). Moreover, in *Medicago falcata*, the glyoxalase 1 activity is regulated by NAC TF resulting in the regulation of the glutathione level under various stresses (Duan et al., 2017). In rice, NAC TFs regulate the activity of a *PP2C* gene, *OsPP18*, in An ABA-independent pathways (You et al., 2014). Overexpression of three genes *ANAC019*, *ANAC055*, and *ANAC072* confers drought tolerance in *Arabidopsis* (Fujita et al., 2004). Besides this, OsNAC5 was also induced following exposure to drought, cold, methyl jasmonate, and ABA (Song et al., 2011; Takasaki et al., 2010). From previous studies it was also observed that ANAC078 directly affects the anthocyanin accumulation when exposed to high light stresses (Morishita et al., 2009). Camalexin, a major phytoalexin in *Arabidopsis*, has been reported to be induced by NAC TFs (Saga et al., 2012).

AP2/ERF Transcription Factor

The APETALA2/ethylene response factor (AP2/ERF) has been extensively studied in plants which possess a 60-amino acid-long DNA-binding domain (DBD) and three β-sheets adjacent to the α-helical region (Fig. 4.2) (Allen et al., 1998). The DBD directly interacts with GCC-box which has specific roles in ethylene-responsive transcription (Rashid et al., 2012). Genome-wide experimental studies were revealed that numerous AP2/ERF members are present in different plant species such as 170 in rice, 145 in *Arabidopsis*, and 178 in *Sorghum* (Joshi et al.,

2016). The AP2/ERF family has been further classified in four subgroups such as AP2, ERF, related to ABI3, and VP1 (RAV) and dehydration-responsive element-binding protein (DREB) of which ERF and DREB subfamilies have been studied extensively (Rashid et al., 2012). Overexpression of DREB1/CBF (C-repeat binding factor) showed increased expression of several drought-responsive genes improving drought tolerance in several plants (Datta et al., 2012; Iwaki et al., 2013). Transgenic rice overexpressing *DREB1A* gene has been found to be associated with higher accumulation of osmoprotectants such as proline and soluble sugars (Ito et al., 2006). Moreover, in leaves of barley and rice, the accumulation of DREB1 (HvDREB1 and OsDREB1F, respectively) was enhanced following drought, salt, and low temperature treatments (Xu et al., 2009; Wang et al., 2008). In *Salvia miltiorrhiza* ectopic expression of *Arabidopsis* DREB1A/CBF3 showed increased photosynthetic rate and antioxidant enzyme activities (Wei et al., 2016). Similarly, ERF6 in *Arabidopsis* positively regulates ROS signaling and biotic and abiotic stress tolerance (Sewelam et al., 2013).

From previous studies it was observed that in tobacco, jasmonic acid (JA)-inducible ORC1/ERF221 and ERF189 positively regulate the nicotine biosynthesis genes (Shoji et al., 2010). Interestingly, in *Artemisia annua* two JA-responsive AP2/ERF TFs, AaERF1 and AaERF2, regulate the expression of amorpha-4,11-diene synthase (ADS) and CYP sesquiterpene oxidase (CYP71AV1) (Yu et al., 2012). Another AP2/ERF TF from *Artemisia annua*, AaORA1, has been found to regulate the accumulation of artemisinin which provides resistance against necrotrophic pathogen (Lu et al., 2013). Moreover, octadecanoid-responsive *Catharanthus* AP2-domain protein 2 (ORCA2) and octadecanoid-responsive *Catharanthus* AP2-domain protein 3 (ORCA3) regulate the synthesis of terpenoid indole alkaloid (TIA) compounds (Menke et al., 1999). The ORCA3 directly binds with the JERE in the promoters of two TIA-responsive genes, namely, strictosidine synthase and tryptophan decarboxylase (Roepke et al., 2010). It was also observed that AP2/ERF TF family also regulates the expression of steroidal glycoalkaloids (SGAs). Another phyto-protective compound, saponin, was found to be regulated by AP2/ERF TFs (Avato et al., 2006). For example, the PnERF1 of *Panax notoginseng* directly interacts with the promoters of saponin biosynthesis genes, namely, 3-hydroxy-3-methylglutaryl-CoA reductase (HMGR), farnesyl pyrophosphate synthase gene (FPS), squalene epoxidase gene (SE), squalene synthase gene (SS), and dammarenediol synthase gene (DS) increasing the total saponin content (Deng et al., 2017). Moreover, the loss of function of GbERF1 in *Gossypium barbadense* results into the downregulation of ferulate-5 hydroxylase indicating their possible function in lignin biosynthesis (Guo et al., 2016). It was also observed that several anticancerous and antimicrobial compounds such as taxol, resveratrol, and HCAAs have been upregulated by ERF TFs (Young et al., 1992; Zhang et al., 2015a, 2015b; Montero et al., 2003).

MYB Transcription Factors

MYB (v-myb avian myeloblastosis viral oncogene homolog) transcription factor family represents one of the largest TF families in plants. This family of transcription factors is characterized by a MYB DNA-binding domain at the N-terminus (Fig. 4.2) (Zhou et al., 2016). The MYB TFs are further subdivided into four subgroups based on the number of adjacent repeats in MYB domain, namely, R1-MYB, R2R3-MYB, 3R-MYB, and 4R-MYB (Li et al., 2016). Each domain contains a 50–53-amino acid-long α-helical region with two adjacent α-helices which are connected through a helix-turn-helix that has a DNA-binding activity (Zhou et al., 2016). Among the four subfamilies, R2R3 members (Fig. 4.2) have been widely characterized and considered as the largest MYB subfamily.

Many secondary metabolite biosynthesis genes were ground to be regulated by MYB TFs including flavonoids, HCAAs, glucosinolates, and proanthocyanidins (Bednarek & Osbourn, 2009). In *Arabidopsis*, expression of two cytochrome-P450 (CYP79B2 and CYP79B3) related to indole glucosinolate biosynthesis pathway was found to be elevated by three MYB genes such as MYB34, MYB51, and MYB122. Triple mutants of *myb34/51/122*, however, exhibit higher susceptibility to fungal pathogen which is related to lower accumulation of indole glucosinolates (Frerigmann et al., 2014). Flavonoids are the most important secondary metabolite compounds conferring resistance against both biotic and abiotic stress. It has been reported that three MYB TFs in *Arabidopsis* AtMYB11, AtMYB12, and AtMYB111 positively regulate flavonoid biosynthesis through the interaction with MYB recognition elements (MREs) in the promoters of *CHS* and *FLS* genes (Misra et al., 2010; Pandey et al., 2015; Wang et al., 2016a, 2016b, 2016c, 2016d). In citrus, CsMYBF1 interacts with the promoters of chalcone synthase (*CHS*) gene enhancing flavonoid biosynthesis (Liu et al., 2016). Similarly, in *Scutellaria baicalensis*, SbMYB8 is responsible for flavonoid accumulation which enhances drought tolerance (Yuan et al., 2015). It has also been proposed in previous works that some MYB TFs act as negative regulators of stress-responsive genes. For example, overexpression of AtMYB75 in *Arabidopsis* results into the reduction of kaempferol-3,7-dirhamnoside level that develops resistance against insect herbivore (Onkokesung et al., 2014). Anthocyanin biosynthesis is dependent on light, sugar, and phytohormones. It has been reported that MYB TF, MYBL2, regulates the anthocyanin accumulation in *Arabidopsis* seedlings under high light and sucrose (Dubos et al., 2008). R2R3 MYB TFs in snapdragon, namely, ROSEA1, ROSEA2, and VENOSA, regulate anthocyanin accumulation in floral organs in association with bHLH TFs (Schwinn et al., 2006). In strawberry, FaMYB10 and FaMYB1 expression was found to be enhanced at the time of fruit ripening along with anthocyanin accumulation (Wang et al., 2010a, 2010b). It has been observed that three MYB TFs such as MYB30, MYB55, and MYB110 regulate the phenylpropanoid biosynthesis genes resulting into accumulation of HCAA compounds (Kishi-Kaboshi et al., 2018). Meanwhile, MYB96 promotes drought tolerance through generation of cuticular wax (Lee et al., 2014). Various other MYB TFs such as MtMYB3 and MtMYB61 in *Medicago truncatula* play important role in cold stress tolerance in plants (Zhang et al., 2016a, 2016b). Previous studies were revealed that MYB TFs stimulate the production of aliphatic and indolic glucosinolate compounds (Gigolashvili et al., 2008; Celenza et al., 2005). Overexpression of AtMYB12 in *Arabidopsis* results into enhancement of flavonoid, proline, and ABA biosynthesis under salt and drought stress (Wang et al., 2016a, 2016b, 2016c, 2016d). In *Rosa rugosa*, two MYB TFs, namely, RrMYB5 and RrMYB10, enhance the synthesis of proanthocyanidins in response to wound and oxidative stress (Shen et al., 2019).

Effective Manipulation of Secondary Metabolite Pathways by Transcription Factors

Genetic engineering provides new insights into the breeding approaches to improve crop tolerance against various abiotic stresses. The central dogma of abiotic stress research depends on how a plant senses and acclimatizes with certain abiotic stress (Mittler & Blumwald, 2010). Plants are the store house of numerous pharmaceutically important metabolite compounds which include both primary and secondary metabolites. Secondary metabolite compounds have multiple functions including growth, development, reproduction, cell division, and biotic and abiotic stress tolerance (Meraj et al., 2020). Plant metabolites are used as foods, medicines, and raw materials of industry for more than a thousand years (Oksman-Caldentey & Saito, 2005).

From previous works it was observed that plant TFs are the key regulators of the secondary metabolite biosynthesis genes (Meraj et al., 2020). Manipulations of TFs that are actively participating in metabolite biosynthetic pathways appear to be more effective for the production of useful metabolites (Capell & Christou, 2004). First of all, the TF of interest needs to be identified and characterized, i.e., whether they are activators or repressors (Iwase et al., 2009) to enhance the expression of a particular metabolite. The enhancement (gain of function) of activity of a TF can be regulated by fusion with an external activation domain or through ectopic expression (Iwase et al., 2009). Sometimes inhibition of branching pathways was found to be effective in the expression of the product of interest (Devic et al., 1999). Gene silencing technologies such as knockout mutant lines, RNA interference, and antisense technologies have been applied in this purpose. From previous observations it was analyzed that gain-of-function strategy was more effective than loss of functions (Moore & Purugganan, 2005; Hanada et al., 2008). For example, various MYB TFs such as PAP1, PAP2, MYB113, and MYB114 regulate the anthocyanin accumulation (Gonzalez et al., 2008). In this context to obtain the loss of function phenotype, multiple TF genes need to be knocked out simultaneously due to their functional redundancy (Gonzalez et al., 2008). Another approach is when a transcriptional activator converts into a dominant repressor simply by fusion with an ethylene-responsive element-binding factor-associated amphiphilic repression (EAR) domain, and this is known as CRES-T (Shikata & Ohme-Takagi, 2008). With this silencing machinery we can easily get a dominant negative phenotype. In recent times two current approaches are mostly applied for understanding the function of TFs such as generation of knockout mutant lines and overexpression of target genes. Though the RNAi approaches are found to be rapid and inexpensive, sometimes the inhibition of gene function is incomplete and also exhibits unpredictable off-target effects (Gaj et al., 2013). Meanwhile, the use of sequence-specific nucleases such as zinc finger nucleases (ZFNs), transcription activator like effector nucleases (TALENs), and meganucleases creates DNA double-strand breaks resulting into recombination events at TF gene loci (Lowder et al., 2015). Following the identification of clustered regularly interspaced short palindromic repeats (CRISPRs) in plants, CRISPR-Cas9 emerged as an effective tool for the analysis of gene expression and gene knockout (Gaj et al., 2013). Recent studies have been revealed that metabolite-responsive transcriptional factor (MRTF)-based biosensors have the ability to modulate target gene expression or regulate metabolic activity (Younger et al., 2017). These biosensors consist of an activator and repressor protein regulating transcriptional activity of promoter elements (Wan et al., 2019). Together these novel strategies based on TFs may have the potentiality to upregulate metabolite biosynthesis genes.

Synthetic Promoters and Synthetic TF-Mediated Metabolite Engineering

Promoter of a gene determines when the transcription of a gene will be initiated. A typical eukaryotic promoter (Fig. 4.3a) contains a core promoter element and a proximal promoter element (cis-motif) spanning a total region of about 1000 bps upstream of transcription initiation site (TSS) (Maston et al., 2006). The RNA Pol-II machinery specifically recognizes the core promoter which is located −40 to +40 bp of the TSS (Kadonaga, 2002). The core promoter further contains a TATA-box, BRE (TFIIB recognition element), Inr (initiator sequence), MTE (motif ten element), DPE (downstream promoter element), and DCE (downstream core element) (Srivastava et al., 2014). Meanwhile, the synthetic promoters and synthetic transcription factors are emerging as incredibly efficient and powerful tool for the regulation of targeted plant gene expression (Liu et al., 2013; Liu & Stewart Jr, 2016). They were found to be more advantageous over their native counterparts with regard to transgene specificity and expression pattern.

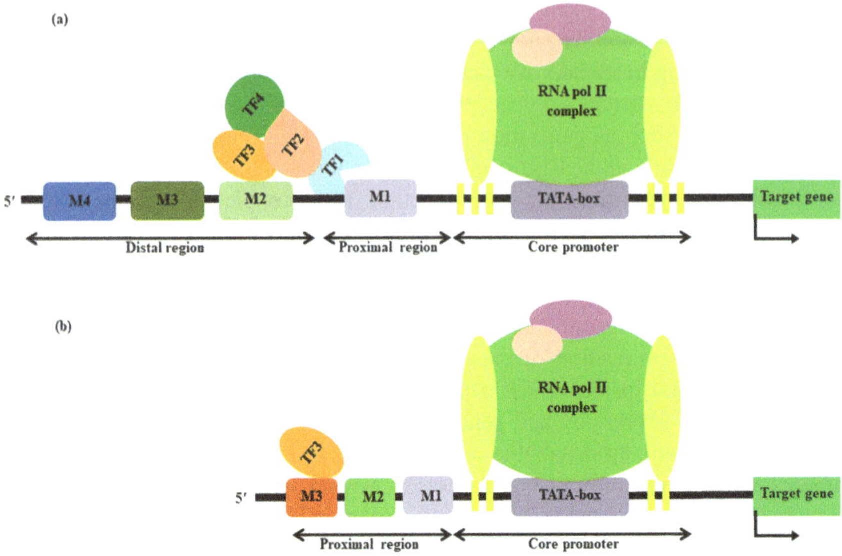

Fig. 4.3 Schematic representation showing the structure of natural promoter and synthetic promoter. (**a**) A natural promoter contains a core promoter region and a proximal and distal promoter region where multiple transcription factors with functional redundancy can bind. (**b**) A synthetic promoter is short and robust containing multiple cis-regulatory elements where only selected transcription factor can bind

Synthetic Promoters

A synthetic promoter as described in Fig. 4.3b is comprised of core promoter and synthetic cis-regulatory motif elements for the regulation of spatial and temporal transgene expression. The core promoter element is located ±50 bp from TSS and typically contains a TATA-box, CAAT-box, GA elements, and Inr region (Liu & Stewart Jr, 2016). It is not necessary that all the promoter sequences need to be presented in each core promoter. Generally, the TATA-box is an important element where preinitiation complex formation takes place and it is also recognized by RNA Pol-II and other TFs (Venter, 2007). Meanwhile, the cis-motif sequences are obtained from extant sequences which are either multiplied or recombined (Liu & Stewart Jr, 2016). The cauliflower mosaic virus 35 (CaMV35) minimal core promoters are commonly used as a core promoter element in synthetic promoter construction for efficient expression of transgene in plants (Ali & Kim, 2019). Another minimal promoter, ZmUbi1, of 126 bp length was identified from maize ubiquitin1 gene which implicated in the improvements of crop (Kumar et al., 2015). Various in silico approaches were found to be useful for the selection, copy number, and spacing of cis-regulatory elements. The use of computational tools for motif discovery and analysis has been already reported in *Arabidopsis* (Koschmann et al., 2012). Databases such as TRANSFAC (Matys et al., 2003), Plant CARE (Lescot et al., 2002), and PLACE (Higo et al., 1999) are helpful for the selection and identification of cis-regulatory motifs in the synthetic promoters. The engineering of core promoters and 5′ UTR has significant function in the strength and efficacy of transcription. Copy number of proximal promoter elements was found to be correlated with synthetic promoter strength as demonstrated in rice (Wu et al., 1998), *Arabidopsis* (Sahoo et al., 2014), and tobacco (Sawant et al., 2005). The multiple copies of cis-regulatory motif elements (Fig. 4.3b) should be properly spaced so their corresponding TFs successfully bind with them (Mehrotra & Mehrotra, 2010). From previous studies it was observed that there are multiple copies of cis-motifs in *Catharanthus roseus* regulate the expression of *CrWRKY1* gene (Yang et al., 2013). Recent advancement in plant biotechnology discovers different synthetic promoter elements such as constitutive, bidirectional, and inducible promoters for the enhancement of important metabolite synthesis in plants.

Synthetic Transcription Factors

Synthetic transcription factors (Fig. 4.4) are designed by the modification of DBDs with activation or repression domains. The activation domain (VP16, widely used)

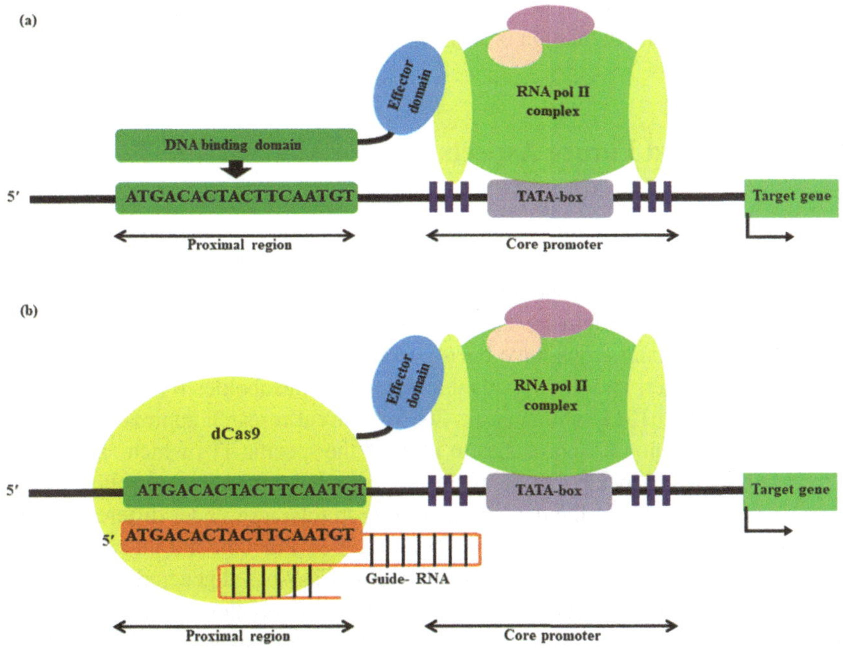

Fig. 4.4 Schematic model showing the structure of synthetic transcription factors. (**a**) ZF-TFs and TALE-TFs contain a genetically engineered DNA-binding domain. (**b**) dCas9-TFs are comprised of a catalytically inactive Cas9

recruits TFIID, TFIIH, and histone acetyl transferases facilitating formation of preinitiation complex on promoter elements (Hirai et al., 2010; Xiao et al., 1994). Synthetic DNA-binding domains such as C2H2-zinc finger (ZF) and transcription activation-like effectors (TALES) (Fig. 4.4a) can bind to any endogenous gene or transgene promoters. In *Arabidopsis*, maize, rice, and tobacco, various engineered ZF-TFs and TALE-TFs have been generated with synthetic DBDs which regulate the expression of several genes (Guan et al., 2002; Ordiz et al., 2002; Stege et al., 2002; Li et al., 2013a, 2013b). Each ZF protein contains a 30-amino acid-long ββα configuration which recognizes the DNA-binding sites (Guan et al., 2002). Meanwhile, the DBDs of TALEs contain tandem repeats of 34/35 amino acid and each repeat recognizes each nucleotide on DNA-binding sites of promoters (Gao et al., 2014). From previous studies it was observed that the binding sites located close to the TSS result into stronger expression of target gene (Boch & Bonas, 2010; Bogdanove et al., 2010). More recently in tobacco, a catalytically inactive Cas9, a catalytic subunit of clustered regularly interspaced short palindromic repeats (CRISPR)-Cas9, was reported to be used for targeted gene activation or repression (Piatek et al., 2015). With the help of guide RNAs, d-Cas9-TFs (Fig. 4.4b) can effectively target any genome loci which have significant sequence homology (Hilton et al., 2015). Moreover, coupled expression of synthetic TFs with their corresponding synthetic promoter elements results into optimal expression of desired gene (de Lange et al., 2013; Morbitzer et al., 2010). Therefore, it is cleared that synthetic TFs such as ZF-TFs, TALE-TFs, and dCas9-TFs can be used as effective tools for the regulation of any endogenous metabolite gene expression.

Conclusion and Future Aspects

Plants produce a large number of secondary metabolite components to combat against different abiotic and biotic stresses (Knight & Knight, 2001; Huot et al., 2014). Diverse group of metabolites do not only regulate plant growth and stress resistance but many of them are used as pharmaceutical compounds for more than a thousand years. Many crops have been genetically modified for the production of antibodies, vaccines, and highly valuable secondary metabolites (Gust et al., 2010; Madanala et al., 2015; Kashima et al., 2016). Current research approaches employ proteomics and genomic approaches to identify the specific TFs which regulate the expression of many metabolite biosynthesis genes (Fig. 4.5). Transcription factors play an important role in the development of crop tolerance under different abiotic stress conditions (Khan et al., 2018). However, their function and interaction with other proteins need to be further studied. CRISPR-Cas9-based gene-editing technologies (Fig. 4.4b) were found to be helpful for the elimination of the functionally redundant TFs. Moreover, reverse genetic approaches are emerging as a helpful tool for the identification of complex regulatory network because there are so many TFs which get activated under different stresses. Besides glycophytes, xerophytes and halophytes also produce different metabolite compounds but they are less explored.

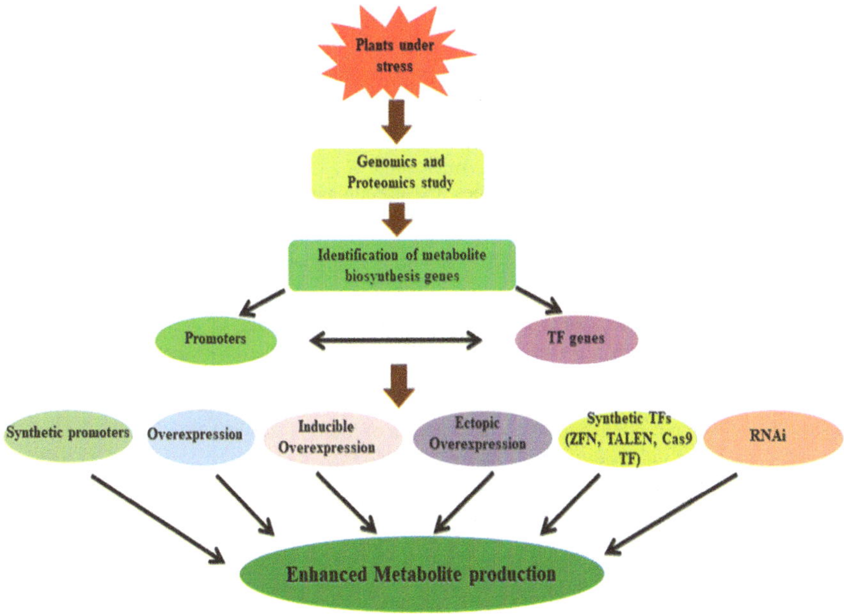

Fig. 4.5 Schematic representation showing different strategies can be used for the enhancement of plant secondary metabolite production

Thus the identification of metabolites in those plants may be fruitful for the effective production of different metabolites. However, due to the low concentration of secondary metabolite compounds in plants, extractions of the purified metabolites from plants were found to be expensive. Use of gene-editing strategies based on CRISPR-Cas9 and other similar approaches as described in Fig. 4.5 enables specific and targeted gene editing and increases the production of desired traits. Moreover, the combined approach involving both synthetic promoters and TFs revolutionizes the field of metabolite engineering. Synthetic approaches through engineering of metabolite pathways in microbes or plants were found to be efficient though limitations still exist.

Acknowledgments The authors gratefully acknowledge the Council of Scientific and Industrial Research, Govt. of India (Ref. No. 38(1587)/16/EMR-II, dated: 17/05/2016 to SR), UGC, Govt. of India (Start-Up research grant No. F.30-158/2015 (BSR), SERB, DST, Govt. of India (Ref. No. ECR/2016/000539 to SR) and DST & BT, Govt. of WB (Ref. No: ST/P/S & T/1G-5/2018 to SR) for providing financial supports for performing research related to the topic discussed in this review. SB and PR are thankful to CSIR, Govt. of India and UGC, Govt. of India for the research fellowship. We apologize to those authors whose work could not be cited due to space limitation.

References

Aida, M., Ishida, T., Fukaki, H., Fujisawa, H., & Tasaka, M. (1997). Genes involved in organ separation in *Arabidopsis*: An analysis of the cup-shaped cotyledon mutant. *Plant Cell, 9*, 841–857.

Alfieri, M., Vaccaro, M. C., Cappetta, E., Ambrosone, A., Tommasi, N. D., & Leone, A. (2018). Coactivation of MEP-biosynthetic genes and accumulation of abietane diterpenes in *Salvia sclarea* by heterologous expression of WRKY and MYC2 transcription factors. *Scientific Reports, 8*, 11009.

Ali, S., & Kim, W. C. (2019). A fruitful decade using synthetic promoters in the improvement of transgenic plants. *Frontiers in Plant Science, 10*, 1433. https://doi.org/10.3389/fpls.2019.01433

Allen, M. D., Yamasaki, K., Ohme-Takagi, M., Tateno, M., & Suzuki, M. A. (1998). Novel mode of DNA recognition by a β-sheet revealed by the solution structure of the GCC-box binding domain in complex with DNA. *The EMBO Journal, 17*, 5484–5496.

An, J. P., Li, H. H., Song, L. Q., Su, L., Liu, X., You, C. X., Wang, X. F., & Hao, Y. J. (2016). The molecular cloning and functional characterization of MdMYC2, a bHLH transcription factor in apple. *Plant Physiology and Biochemistry, 108*, 24–31.

An, J. P., Qu, F. J., Yao, J. F., Wang, X. N., You, C. X., Wang, X. F., & Hao, Y. J. (2017). The bZIP transcription factor MdHY5 regulates anthocyanin accumulation and nitrate assimilation in apple. *Horticulture Research, 4*, 17023.

Avato, P., Bucci, R., Tava, A., Vitali, C., Rosato, A., Bialy, Z., & Jurzysta, M. (2006). Antimicrobial activity of saponins from *Medicago* sp.: Structure-activity relationship. *Phytotherapy Research, 20*, 454–457.

Babitha, K. C., Ramu, S. V., Pruthvi, V., Mahesh, P., Nataraja, K. N., & Udayakumar, M. (2013). Co-expression of AtbHLH17 and AtWRKY28 confers resistance to abiotic stress in *Arabidopsis*. *Transgenic Research, 22*, 327–341.

Bai, Y., Pattanaik, S., Patra, B., Werkman, J. R., Xie, C. H., & Yuan, L. (2011). Flavonoid-related basic helix–loop–helix regulators, NtAn1a and NtAn1b, of tobacco have originated from two ancestors and are functionally active. *Planta, 234*, 363–375.

Bakshi, M., & Oelmüller, R. (2014). WRKY transcription factors: Jack of many trades in plants. *Plant Signaling & Behavior, 9*, e27700.

Banerjee, A., & Roychoudhury, A. (2015). Abscisic-acid-dependent basic leucine zipper (bZIP) transcription factors in plant abiotic stress. *Protoplasma, 254*, 3–16.

Bednarek, P., & Osbourn, A. (2009). Plant-microbe interactions: Chemical diversity in plant defense. *Science, 324*, 746–748.

Besseau, S., Hoffmann, L., Geoffroy, P., Lapierre, C., Pollet, B., & Legrand, M. (2007). Flavonoid accumulation in *Arabidopsis* repressed in lignin synthesis affects auxin transport and plant growth. *Plant Cell, 19*, 148–162.

Boch J, Bonas U (2010) *Xanthomonas* AvrBs3 family-type III effectors: Discovery and function. Annual Review of Phytopathology 48:419–436.

Bogdanove, A. J., Schornack, S., & Lahaye, T. (2010). TAL effectors: Finding plant genes for disease and defense. *Current Opinion in Plant Biology, 13*, 394–401.

Cao, Y., Zhai, J., Wang, Q., Yuan, H., & Huang, X. (2017). Function of *Hevea brasiliensis* NAC1 in dehydration-induced laticifer differentiation and latex biosynthesis. *Planta, 245*, 31–44.

Capell, T., & Christou, P. (2004). Progress in plant metabolic engineering. *Current Opinion in Biotechnology, 15*, 148–154.

Cárdenas, P. D., Sonawane, P. D., Pollier, J., Bossche, R. V., Dewangan, V., Weithorn, E., Tal, L., Meir, S., Rogachev, I., & Malitsky, S. (2016). GAME9 regulates the biosynthesis of steroidal alkaloids and upstream isoprenoids in the plant mevalonate pathway. *Nature Communications, 7*, 10654.

Celenza, J. L., Quiel, J. A., Smolen, G. A., Merrikh, H., Silvestro, A. R., Normanly, J., & Bender, J. (2005). The *Arabidopsis* ATR1 Myb transcription factor controls indolic glucosinolate homeostasis. *Plant Physiology, 137*, 253–262.

Chen, H., Chen, W., Zhou, J., He, H., Chen, L., Chen, H., & Deng, X. W. (2012a). Basic leucine zipper transcription factor OsbZIP16 positively regulates drought resistance in rice. *Plant Science, 193–194*, 8–17. https://doi.org/10.1016/j.plantsci.2012.05.003

Chen, L., Song, Y., Li, S., Zhang, L., Zou, C., & Yu, D. (2012b). The role of WRKY transcription factors in plant abiotic stresses. *Biochimica et Biophysica Acta, 1819*, 120–128.

Dalman, K., Wind, J. J., Nemesio-Gorriz, M., Hammerbacher, A., Lundén, K., Ezcurra, I., & Elfstrand, M. (2017). Overexpression of PaNAC03, a stress induced NAC gene family transcription factor in Norway spruce leads to reduced flavonol biosynthesis and aberrant embryo development. *BMC Plant Biology, 17*, 6.

Danielsson, M., Lundén, K., Elfstrand, M., Hu, J., Zhao, T., Arnerup, J., Ihrmark, K., Swedjemark, G., Borg-Karlson, A. K., & Stenlid, J. (2011). Chemical and transcriptional responses of Norway spruce genotypes with different susceptibility to *Heterobasidion* spp. infection. *BMC Plant Biology*, 11:154.

Datta, K., Baisakh, N., Ganguly, M., Krishnan, S., Yamaguchi Shinozaki, K., & Datta, S. K. (2012). Overexpression of *Arabidopsis* and rice stress genes inducible transcription factor confers drought and salinity tolerance to rice. *Plant Biotechnology Journal, 10*, 579–586. https://doi.org/10.1111/j.1467-7652.2012.00688.x

de Lange, O., Schreiber, T., Schandry, N., Radeck, J., Braun, K. H., Koszinowski, J., Heuer, H., Straub, A., & Lahaye, T. (2013). Breaking the DNA binding code of *Ralstonia solanacearum* TAL effectors provides new possibilities to generate plant resistance genes against bacterial wilt disease. *The New Phytologist, 199*, 773–786.

Deng, B., Huang, Z., Ge, F., Liu, D., Lu, R., & Chen, C. (2017). An AP2/ERF family transcription factor PnERF1 raised the biosynthesis of saponins in *Panax notoginseng*. *Journal of Plant Growth Regulation, 36*, 691–701.

Devic, M., Guilleminot, J., Debeaujon, I., Bechtold, N., Bensaude, E., Koornneef, M., Pelletier, G., & Delseny, M. (1999). The BANYULS gene encodes a DFR-like protein and is a marker of early seed coat development. *The Plant Journal, 19*, 387–398.

Dong, Y., Wang, C., Han, X., Tang, S., Liu, S., Xia, X., & Yin, W. (2014). A novel bHLH transcription factor PebHLH35 from *Populus euphratica* confers drought tolerance through regulating stomatal development, photosynthesis and growth in *Arabidopsis*. *Biochemical and Biophysical Research Communications, 450*, 453–458.

Duan, M., Zhang, R., Zhu, F., Zhang, Z., Gou, L., Wen, J., Dong, J., & Wang, T. (2017). A lipid-anchored NAC transcription factor is translocated into the nucleus and activates glyoxalase I expression during drought stress. *Plant Cell, 29*, 1748–1772.

Dubos, C., Gourrierec, J. L., Baudry, A., Huep, G., Lanet, E., Debeaujon, I., Routaboul, J. M., Alboresi, A., Weisshaar, B., & Lepiniec, L. (2008). MYBL2 is a new regulator of flavonoid biosynthesis in *Arabidopsis thaliana*. *The Plant Journal, 55*, 940–953.

Espley, R. V., Hellens, R. P., Putterill, J., Stevenson, D. E., Kutty-Amma, S., & Allan, A. C. (2007). Red colouration in apple fruit is due to the activity of the MYB transcription factor, MdMYB10. *The Plant Journal, 49*, 414–427.

Eulgem, T., Rushton, P. J., Robatzek, S., & Somssich, I. E. (2000). The WRKY superfamily of plant transcription factors. *Trends in Plant Science, 5*, 199–206.

Feller, A., Machemer, K., Braun, E. L., & Grotewold, E. (2011). Evolutionary and comparative analysis of MYB and bHLH plant transcription factors. *The Plant Journal, 66*, 94–116.

Frerigmann, H. (2016). Glucosinolate regulation in a complex relationship–MYC and MYB–no one can act without each other. *Advances in Botanical Research Elsevier, 80*, 57–97.

Frerigmann, H., Berger, B., & Gigolashvili, T. (2014). bHLH05 is an interaction partner of MYB51 and a novel regulator of glucosinolate biosynthesis in *Arabidopsis*. *Plant Physiology, 166*, 349–369.

Fu, J., Liu, Q., Wang, C., Liang, J., Liu, L., & Wang, Q. (2017). ZmWRKY79 positively regulates maize phytoalexin biosynthetic gene expression and is involved in stress response. *Journal of Experimental Botany, 69*, 497–510.

Fujita, M., Fujita, Y., Maruyama, K., Seki, M., Hiratsu, K., Ohme-Takagi, M., Tran, L. S., Yamaguchi-Shinozaki, K., & Shinozaki, K. (2004). A dehydration-induced NAC protein, RD26, is involved in a novel ABA-dependent stress-signaling pathway. *The Plant Journal, 39*, 863–876.

Gaj, T., Gersbach, C. A., & Barbas, C. F., III. (2013). ZFN, TALEN, and CRISPR/Cas-based methods for genome engineering. *Trends in Biotechnology, 31*, 397–405.

Gao, H., Wright, D. A., Li, T., Wang, Y., Horken, K., Weeks, D. P., Yang, B., & Spalding, M. H. (2014). TALE activation of endogenous genes in *Chlamydomonas reinhardtii*. *Algal Research, 5*, 52–60.

Gigolashvili, T., Engqvist, M., Yatusevich, R., Müller, C., & Flügge, U. I. (2008). HAG2/MYB76 and HAG3/MYB29 exert a specific and coordinated control on the regulation of aliphatic glucosinolate biosynthesis in *Arabidopsis thaliana*. *New Phytologist, 177*, 627–642. https://doi.org/10.1111/j.1469-8137.2007.02295.x

Gonzalez, A., Zhao, M., Leavitt, J. M., & Lloyd, A. M. (2008). Regulation of the anthocyanin biosynthetic pathway by the TTG1/bHLH/Myb transcriptional complex in *Arabidopsis* seedlings. *The Plant Journal, 53*, 814–827.

Guan, X., Stege, J., Kim, M., Dahmani, Z., Fan, N., Heifetz, P., Barbas, C. F., III, & Briggs, S. P. (2002). Heritable endogenous gene regulation in plants with designed polydactyl zinc finger transcription factors. *Proceedings of the National Academy of Sciences, 99*, 13296–13301.

Guan, Y., Meng, X., Khanna, R., LaMontagne, E., Liu, Y., & Zhang, S. (2014). Phosphorylation of a WRKY transcription factor by MAPKs is required for pollen development and function in *Arabidopsis*. *PLoS Genetics, 10*, e1004384.

Guo, W., Jin, L., Miao, Y., He, X., Hu, Q., Guo, K., Zhu, L., & Zhang, X. (2016). An ethylene response-related factor, GbERF1-like, from *Gossypium barbadense* improves resistance to *Verticillium dahliae* via activating lignin synthesis. *Plant Molecular Biology, 91*, 305–318.

Gust, A. A., Brunner, F., & Nürnberger, T. (2010). Biotechnological concepts for improving plant innate immunity. *Current Opinion in Biotechnology, 21*(2), 204–210. https://doi.org/10.1016/j.copbio.2010.02.004

Hanada, K., Zou, C., Lehti-Shiu, M. D., Shinozaki, K., & Shiu, S. H. (2008). Importance of lineage-specific expansion of plant tandem duplicates in the adaptive response to environmental stimuli. *Plant Physiology, 148*, 993–1003.

Hartmann, L., Pedrotti, L., Weiste, C., Fekete, A., Schierstaedt, J., Gottler, J., Kempa, S., Krischke, M., Dietrich, K., & Mueller, M. J. (2015). Crosstalk between two bZIP signaling pathways orchestrates salt-induced metabolic reprogramming in *Arabidopsis* roots. *Plant Cell, 27*, 2244–2260.

Higo, K., Ugawa, Y., Iwamoto, M., & Korenaga, T. (1999). Plant cis-acting regulatory DNA elements (PLACE) database. *Nucleic Acids Research, 27*(1), 297–300. https://doi.org/10.1093/nar/27.1.297

Hilton, I. B., D'Ippolito, A. M., Vockley, C. M., Thakore, P. I., Crawford, G. E., Reddy, T. E., & Gersbach, C. A. (2015). Epigenome editing by a CRISPRCas9-based acetyl transferase activates genes from promoters and enhancers. *Nature Biotechnology, 33*, 510–519.

Hirai, H., Tani, T., & Kikyo, N. (2010). Structure and functions of powerful trans activators: VP16, MyoO and FoxA. *The International Journal of Developmental Biology, 54*, 1589–1596.

Huang, Y. F., Vialet, S., Guiraud, J. L., Torregrosa, L., Bertrand, Y., Cheynier, V., This, P., & Terrier, N. (2014). A negative MYB regulator of proanthocyanidin accumulation, identified through expression quantitative locus mapping in the grape berry. *New Phytologist, 201*, 795–809.

Huot, B., Yao, J., Montgomery, B. L., & He, S. Y. (2014). Growth–defense tradeoffs in plants: A balancing act to optimize fitness. *Molecular Plant, 7*, 1267–1287.

Ishida T, Hattori S, Sano R, Inoue K, Shirano Y, Hayashi H, Shibata D, Sato S, Kato T, Tabata S (2007) *Arabidopsis* TRANSPARENT TESTA GLABRA2 is directly regulated by R2R3 MYB transcription factors and is involved in regulation of GLABRA2 transcription in epidermal differentiation. Plant Cell 19:2531–2543.

Ishihama, N., Yamada, R., Yoshioka, M., Katou, S., & Yoshioka, H. (2011). Phosphorylation of the *Nicotiana benthamiana* WRKY8 transcription factor by MAPK functions in the defense response. *Plant Cell, 23*, 1153–1170.

Ito, S., Song, Y. H., Josephson-Day, A. R., Miller, R. J., Breton, G., & Olmstead, R. G. (2010). Flowering bHLH transcriptional activators control expression of the photoperiodic flowering regulator CONSTANS in *Arabidopsis*. *Proceedings of the National Academy of Sciences, 109*, 3582–3587.

Ito, Y., Katsura, K., Maruyama, K., Taji, T., Kobayashi, M., & Seki, M. (2006). Functional analysis of rice DREB1/CBF-type transcription factors involved in cold-responsive gene expression in transgenic rice. *Plant & Cell Physiology, 47*, 141–153. https://doi.org/10.1093/pcp/pci230

Iwaki, T., Guo, L., Ryals, J. A., Yasuda, S., Shimazaki, T., & Kikuchi, A. (2013). Metabolic profiling of transgenic potato tubers expressing *Arabidopsis* dehydration response element-binding protein 1A (DREB1A). *Journal of Agricultural and Food Chemistry, 61*, 893–900. https://doi.org/10.1021/jf304071n

Iwase, A., Matsui, K., & Ohme-Takagi, M. (2009). Manipulation of plant metabolic pathways by transcription factors. *Plant Biotechnology, 26*, 29–38. https://doi.org/10.5511/plantbiotechnology.26.29

Jakoby, M. (2002). bZIP transcription factors in *Arabidopsis*. *Trends in Plant Science, 7*, 106–111.

Jiang, J., Xi, H., Dai, Z., Lecourieux, F., Yuan, L., Liu, X., Patra, B., Wei, Y., Li, S., & Wang, L. (2018). VvWRKY8 represses stilbene synthase genes through direct interaction with VvMYB14 to control resveratrol biosynthesis in grapevine. *Journal of Experimental Botany, 70*, 715–729.

Johnson, C. S., Kolevski, B., & Smyth, D. R. (2002). TRANSPARENT TESTA GLABRA2, a trichome and seed coat development gene of *Arabidopsis*, encodes a WRKY transcription factor. *Plant Cell, 14*, 1359–1375.

Joshi, R., Wani, S. H., & Singh, B. (2016). Transcription factors and plants response to drought stress: Current understanding and future directions. *Frontiers in Plant Science, 7*, 1029. https://doi.org/10.3389/fpls.2016.01029

Kadonaga, J. T. (2002). The DPE, a core promoter element for transcription by RNA polymerase II. *Experimental & Molecular Medicine, 34*, 259–264.

Kage, U., Yogendra, K. N., & Kushalappa, A. C. (2017). TaWRKY70 transcription factor in wheat QTL-2DL regulates downstream metabolite biosynthetic genes to resist *Fusarium graminearum* infection spread within spike. *Scientific Reports, 7*, 42596.

Karre, S., Kumar, A., Yogendra, K., Kage, U., Kushalappa, A., & Charron, J. B. (2019). HvWRKY23 regulates flavonoid glycoside and hydroxycinnamic acid amide biosynthetic genes in barley to combat *Fusarium* head blight. *Plant Molecular Biology*, 1–15.

Kashima, K., Yuki, Y., Mejima, M., Kurokawa, S., Suzuki, Y., & Minakawa, S. (2016). Good manufacturing practices production of a purification-free oral cholera vaccine expressed in transgenic rice plants. *Plant Cell Reports, 35*(3), 667–679. https://doi.org/10.1007/s00299-015-1911-9

Khan, S. A., Li, M. Z., Wang, S. M., & Yin, H. J. (2018). Revisiting the role of plant transcription factors in the battle against abiotic stress. *International Journal of Molecular Sciences, 19*(6), 1634. https://doi.org/10.3390/ijms19061634

Kishi-Kaboshi, M., Seo, S., Takahashi, A., & Hirochika, H. (2018). The MAMP-responsive MYB transcription factors MYB30, MYB55 and MYB110 activate the HCAA synthesis pathway and enhance immunity in Rice. *Plant & Cell Physiology, 59*, 903–915.

Kitsios, G., & Doonan, J. H. (2011). Cyclin dependent protein kinases and stress responses in plants. *Plant Signaling & Behavior, 6*, 204–209.

Knight, H., & Knight, M. R. (2001). Abiotic stress signalling pathways: Specificity and cross-talk. *Trends in Plant Science, 6*, 262–267.

Koschmann, J., Machens, F., Becker, M., Niemeyer, J., Schulze, J., & Bülow, L. (2012). Integration of bioinformatics and synthetic promoters leads to the discovery of novel elicitor-responsive

cis-regulatory sequences in *Arabidopsis*. *Plant Physiology, 160*(1), 178–191. https://doi.org/10.1104/pp.112.198259
Kumar, S., AlAbed, D., Whitteck, J. T., Chen, W., Bennett, S., Asberry, A., Wang, X., DeSloover, D., Rangasamy, M., & Wright, T. R. (2015). A combinatorial bidirectional and bicistronic approach for coordinated multi-gene expression in corn. *Plant Molecular Biology, 87*, 341–353.
Kutchan, T. M. (2001). Ecological arsenal and developmental dispatcher. The paradigm of secondary metabolism. *Plant Physiology, 125*(1), 58–60. https://doi.org/10.1104/pp.125.1.58
Le Hir, R., Castelain, M., Chakraborti, D., Moritz, T., Dinant, S., & Bellini, C. (2017). AtbHLH68 transcription factor contributes to the regulation of ABA homeostasis and drought stress tolerance in *Arabidopsis thaliana*. *Physiologia Plantarum, 160*, 312–327.
Lee, S. B., Kim, H., Kim, R. J., & Suh, M. C. (2014). Overexpression of *Arabidopsis* MYB96 confers drought resistance in *Camelina sativa* via cuticular wax accumulation. *Plant Cell Reports, 33*, 1535–1546.
Lescot, M., Déhais, P., Thijs, G., Marchal, K., Moreau, Y., & Van de Peer, Y. (2002). PlantCARE, a database of plant cis-acting regulatory elements and a portal to tools for in silico analysis of promoter sequences. *Nucleic Acids Research, 30*(1), 325–327. https://doi.org/10.1093/nar/30.1.325
Li, J., Zhang, K., Meng, Y., Hu, J., Ding, M., Bian, J., Yan, M., Han, J., & Zhou, M. (2018). Jasmonic acid/ethylene signaling coordinates hydroxycinnamic acid amides biosynthesis through ORA 59 transcription factor. *The Plant Journal, 95*, 444–457.
Li, S., Wang, H., Li, F., Chen, Z., Li, X., Zhu, L., Wang, G., Yu, J., Huang, D., & Lang, Z. (2015). The maize transcription factor EREB 58 mediates the jasmonate-induced production of sesquiterpene volatiles. *The Plant Journal, 84*, 296–308.
Li, S., Zhang, P., Zhang, M., Fu, C., & Yu, L. (2013a). Functional analysis of a WRKY transcription factor involved in transcriptional activation of the DBAT gene in *Taxus chinensis*. *Plant Biology, 15*, 19–26.
Li, T., Huang, S., Zhou, J., & Yang, B. (2013b). Designer TAL effectors induce disease susceptibility and resistance to *Xanthomonas oryzae* pv. *oryzae* in rice. *Molecular Plant, 6*, 781–789.
Li, X., Xue, C., Li, J., Qiao, X., Li, L., Yu, L., Huang, Y., & Wu, J. (2016). Genome-wide identification, evolution and functional divergence of MYB transcription factors in Chinese white pear (*Pyrus bretschneideri*). *Plant & Cell Physiology, 57*, 824–847.
Li, Y. Y., Mao, K., & Zhao, C. (2012). MdCOP1 ubiquitin E3 ligases interact with MdMYB1 to regulate light-induced anthocyanin biosynthesis and red fruit coloration in apple. *Plant Physiology, 160*(2), 1011–1022. https://doi.org/10.1104/pp.112.199703
Liu, C., Long, J., Zhu, K., Liu, L., Yang, W., Zhang, H., Li, L., Xu, Q., & Deng, X. (2016). Characterization of a citrus R2R3-MYB transcription factor that regulates the flavonol and hydroxycinnamic acid biosynthesis. *Scientific Reports, 6*, 25352.
Liu, C., Mao, B., Ou, S., Wang, W., Liu, L., Wu, Y., Chu, C., & Wang, X. (2014). OsbZIP71, a bZIP transcription factor, confers salinity and drought tolerance in rice. *Plant Molecular Biology, 84*(1–2), 19–36. https://doi.org/10.1007/s11103-013-0115-3
Liu, C. C., Chi, C., Jin, L. J., Zhu, J., Yu, J. Q., & Zhou, Y. H. (2018). The bZip transcription factor HY5 mediates CRY1a-induced anthocyanin biosynthesis in tomato. *Plant, Cell & Environment, 41*, 1762–1775.
Liu, J., Osbourn, A., & Ma, P. (2015a). MYB transcription factors as regulators of phenylpropanoid metabolism in plants. *Molecular Plant, 8*, 689–708.
Liu, W., & Stewart, C. N., Jr. (2016). Plant synthetic promoters and transcription factors. *Current Opinion in Biotechnology, 37*, 36–44. https://doi.org/10.1016/j.copbio.2015.10.001
Liu, W., Yuan, J. S., & Stewart, C. N., Jr. (2013). Advanced genetic tools for plant biotechnology. *Nature Reviews. Genetics, 14*(11), 781–793. https://doi.org/10.1038/nrg3583
Liu Y, Ji X, Nie X, Qu M, Zheng L, Tan Z, Zhao H, Huo L, Liu S, Zhang B (2015b) *Arabidopsis* AtbHLH112 regulates the expression of genes involved in abiotic stress tolerance by binding to their E-box and GCG-box motifs. The New Phytologist 207:692–709.

Lowder, L. G., Zhang, D., Baltes, N. J., Paul, J. W., III, Tang, X., Zheng, X., Voytas, D. F., Hsieh, T. F., Zhang, Y., & Qi, Y. (2015). A CRISPR/Cas9 Toolbox for multiplexed plant genome editing and transcriptional regulation. *Plant Physiology, 169*(2), 971–985. https://doi.org/10.1104/pp.15.00636

Lu, X., Zhang, L., Zhang, F., Jiang, W., Shen, Q., Zhang, L., Lv, Z., Wang, G., & Tang, K. (2013). AaORA, a trichome-specific AP2/ERF transcription factor of *Artemisia annua*, is a positive regulator in the artemisinin biosynthetic pathway and in disease resistance to *Botrytis cinerea*. *The New Phytologist, 198*, 1191–1202.

Luca, V. D., & Pierre, B. S. (2000). The cell and developmental biology of alkaloid biosynthesis. *Trends in Plant Science, 5*, 168–173.

Luo, M., Dennis, E. S., Berger, F., Peacock, W. J., & Chaudhury, A. (2005). MINISEED3 (MINI3), a WRKY family gene, and HAIKU2 (IKU2), a leucine-rich repeat (LRR) KINASE gene, are regulators of seed size in *Arabidopsis*. *Proceedings of the National Academy of Sciences of the United States of America, 102*, 17531–17536.

Luo, Q. J., Mittal, A., Jia, F., & Rock, C. D. (2012). An autoregulatory feedback loop involving PAP1 and TAS4 in response to sugars in *Arabidopsis*. *Plant Molecular Biology, 80*, 117–129.

Ma, D., Pu, G., Lei, C., Ma, L., Wang, H., Guo, Y., Chen, J., Du, Z., Wang, H., & Li, G. (2009). Isolation and characterization of AaWRKY1, an *Artemisia annua* transcription factor that regulates the amorpha-4,11-diene synthase gene, a key gene of artemisinin biosynthesis. *Plant & Cell Physiology, 50*, 2146–2161.

Ma, R., Xiao, Y., Lv, Z., Tan, H., Chen, R., Li, Q., Chen, J., Wang, Y., Yin, J., Zhang, L., & Chen, W. (2017). AP2/ERF transcription factor, Ii049, positively regulates lignan biosynthesis in *Isatis indigotica* through activating salicylic acid signaling and lignan/lignin pathway genes. *Frontiers in Plant Science, 8*, 1361.

Madanala, R., Gupta, V., Pandey, A. K., Srivastava, S., Pandey, V., & Singh, P. K. (2015). Tobacco chloroplasts as bioreactors for the production of recombinant superoxide dismutase in plants, an industrially useful enzyme. *Plant Molecular Biology, 33*(4), 1107–1115. https://doi.org/10.1007/s11105-014-0805-2

Mahmood, K., Xu, Z., El-Kereamy, A., Casaretto, J. A., & Rothstein, S. J. (2016). The *Arabidopsis* transcription factor ANAC032 represses anthocyanin biosynthesis in response to high sucrose and oxidative and abiotic stresses. *Frontiers in Plant Science, 7*, 1548.

Maston, G. A., Evans, S. K., & Green, M. R. (2006). Transcriptional regulatory elements in the human genome. *Annual Review of Genomics and Human Genetics, 7*, 29–59.

Matys, V., Fricke, E., Geffers, R., Gößling, E., Haubrock, M., & Hehl, R. (2003). TRANSFAC®: Transcriptional regulation, from patterns to profiles. *Nucleic Acids Research, 31*(1), 374–378. https://doi.org/10.1093/nar/gkg108

Mehrotra, R., & Mehrotra, S. (2010). Promoter activation by ACGT in response to salicylic and abscisic acids is differentially regulated by the spacing between two copies of the motif. *Journal of Plant Physiology, 167*, 1214–1218.

Menke, F. L. H., Champion, A., Kijne, J. W., & Memelink, J. (1999). A novel jasmonate- and elicitor-responsive element in the periwinkle secondary metabolite biosynthetic gene Str interacts with a jasmonate- and elicitor-inducible AP2-domain transcription factor, ORCA2. *EMBO, 18*, 4455–4463. https://doi.org/10.1093/emboj/18.16.4455

Meraj, T. A., Fu, J., & Raza, M. A. (2020). Transcriptional factors regulate plant stress responses through mediating secondary metabolism. *Genes, 11*(4), 346. https://doi.org/10.3390/genes11040346

Mertens, J., Pollier, J., Bossche, R. V., Lopez-Vidriero, I., Franco-Zorrilla, J. M., & Goossens, A. (2016). The bHLH transcription factors TSAR1 and TSAR2 regulate triterpene saponin biosynthesis in *Medicago truncatula*. *Plant Physiology, 170*, 194–210.

Mishra, S., Triptahi, V., Singh, S., Phukan, U. J., Gupta, M. M., Shanker, K., & Shukla, R. K. (2013). Wound induced transcriptional regulation of benzylisoquinoline pathway and characterization of wound inducible PsWRKY transcription factor from *Papaver somniferum*. *PLoS One, 8*, e52784.

Misra, P., Pandey, A., Tiwari, M., Chandrashekar, K., Sidhu, O. P., Asif, M. H., Chakrabarty, D., Singh, P. K., Trivedi, P. K., & Nath, P. (2010). Modulation of transcriptome and metabolome of tobacco by *Arabidopsis* transcription factor, AtMYB12, leads to insect resistance. *Plant Physiology, 152*, 2258–2268.

Mittler, R., & Blumwald, E. (2010). Genetic engineering for modern agriculture: Challenges and perspectives. *Annual Review of Plant Biology, 61*, 443–462. https://doi.org/10.1146/annurev-arplant-042809-12116

Miyamoto, K., Matsumoto, T., Okada, A., Komiyama, K., Chujo, T., Yoshikawa, H., Nojiri, H., Yamane, H., & Okada, K. (2014). Identification of target genes of the bZIP transcription factor OsTGAP1, whose overexpression causes elicitor-induced hyperaccumulation of diterpenoid phytoalexins in rice cells. *PLoS One, 9*, e105823.

Miyamoto, K., Nishizawa, Y., Minami, E., Nojiri, H., Yamane, H., & Okada, K. (2015). Overexpression of the bZIP transcription factor OsbZIP79 suppresses the production of diterpenoid phytoalexin in rice cells. *Journal of Plant Physiology, 173*, 19–27.

Montero, C., Cristescu, S., Jiménez, J., Orea, J., te Lintel, H. S., Harren, F., & Urena, A. G. (2003). Trans-resveratrol and grape disease resistance. A dynamical study by high-resolution laser-based techniques. *Plant Physiology, 131*, 129–138.

Moore, R. C., & Purugganan, M. D. (2005). The evolutionary dynamics of plant duplicate genes. *Current Opinion in Plant Biology, 8*, 122–128.

Morbitzer, R., Romer, P., Boch, J., & Lahaye, T. (2010). Regulation of selected genome loci using de novo-engineered transcription activator-like effector (TALE)-type transcription factors. *Proceedings of the National Academy of Sciences of the United States of America, 107*, 21617–21622.

Morishita T, Kojima Y, Maruta T, Nishizawa-Yokoi A, Yabuta Y, Shigeoka S (2009) *Arabidopsis* NAC transcription factor, ANAC078, regulates flavonoid biosynthesis under high-light. Plant & Cell Physiology 50:2210–2222.

Nakayasu, M., Umemoto, N., Ohyama, K., Fujimoto, Y., Lee, H. J., Watanabe, B., Muranaka, T., Saito, K., Sugimoto, Y., & Mizutani, M. (2017). A dioxygenase catalyzes steroid 16-hydroxylation in steroidal glycoalkaloid biosynthesis. *Plant Physiology, 175*, 120–133.

Nijhawan, A., Jain, M., Tyagi, A. K., & Khurana, J. P. (2008). Genomic survey and gene expression analysis of the basic leucine zipper transcription factor family in rice. *Plant Physiology, 146*(2), 333–350. https://doi.org/10.1104/pp.107.112821

Ohno, S., Hosokawa, M., Hoshino, A., Kitamura, Y., Morita, Y., Park, K. I., Nakashima, A., Deguchi, A., Tatsuzawa, F., & Doi, M. (2011). A bHLH transcription factor, DvIVS, is involved in regulation of anthocyanin synthesis in dahlia (*Dahlia variabilis*). *Journal of Experimental Botany, 62*, 5105–5116.

Okada, A., Okada, K., Miyamoto, K., Koga, J., Shibuya, N., Nojiri, H., & Yamane, H. (2009). OsTGAP1, a bZIP transcription factor, coordinately regulates the inductive production of diterpenoid phytoalexins in rice. *The Journal of Biological Chemistry, 284*, 26510–26518.

Oksman-Caldentey, K. M., & Saito, K. (2005). Integrating genomics and metabolomics for engineering plant metabolic pathways. *Current Opinion in Biotechnology, 16*, 174–179.

Onkokesung, N., Reichelt, M., van Doorn, A., Schuurink, R. C., van Loon, J. J., & Dicke, M. (2014). Modulation of flavonoid metabolites in *Arabidopsis thaliana* through overexpression of the MYB75 transcription factor: Role of kaempferol-3, 7-dirhamnoside in resistance to the specialist insect herbivore *Pieris brassicae*. *Journal of Experimental Botany, 65*, 2203–2217.

Ordiz, M. I., Barbas, C. F., III, & Beachy, R. N. (2002). Regulation of transgene expression in plants with polydactyl zinc finger transcription factors. *Proceedings of the National Academy of Sciences, 99*, 13290–13295.

Pan, Q., Wang, C., Xiong, Z., Wang, H., Fu, X., Shen, Q., Peng, B., Ma, Y., Sun, X., & Tang, K. (2019). CrERF5, an AP2/ERF transcription factor, positively regulates the biosynthesis of bisindole alkaloids and their precursors in *Catharanthus roseus*. *Frontiers in Plant Science, 10*, 931.

Pandey, A., Misra, P., Khan, M. P., Swarnkar, G., Tewari, M. C., Bhambhani, S., Trivedi, R., Chattopadhyay, N., & Trivedi, P. K. (2014). Co-expression of *Arabidopsis* transcription factor, AtMYB12, and soybean isoflavone synthase, GmIFS1, genes in tobacco leads to enhanced biosynthesis of isoflavones and flavonols resulting in osteoprotective activity. *Plant Biotechnology Journal, 12*, 69–80.

Pandey, A., Misra, P., & Trivedi, P. K. (2015). Constitutive expression of Arabidopsis MYB transcription factor, AtMYB11, in tobacco modulates flavonoid biosynthesis in favor of flavonol accumulation. *Plant Cell Reports, 34*, 1515–1528.

Patra, B., Schluttenhofer, C., Wu, Y., Pattanaik, S., & Yuan, L. (2013). Transcriptional regulation of secondary metabolite biosynthesis in plants. *Biochimica et Biophysica Acta, 1829*(11), 1236–1247. https://doi.org/10.1016/j.bbagrm.2013.09.006

Peleg, Z., & Blumwald, E. (2011). Hormone balance and abiotic stress tolerance in crop plants. *Current Opinion in Plant Biology, 14*, 290–295.

Piatek, A., Ali, Z., Baazim, H., Li, L., Abulfaraj, A., Al-Shareef, S., Aouida, M., & Mahfouz, M. M. (2015). RNA-guided transcriptional regulation in planta via synthetic dCas9-based transcription factors. *Plant Biotechnology Journal, 13*, 578–589.

Puranik, S., Sahu, P. P., Srivastava, P. S., & Prasad, M. (2012). NAC proteins: Regulation and role in stress tolerance. *Trends in Plant Science, 17*, 369–381.

Qiu, D., Xiao, J., Xie, W., Liu, H., Li, X., Xiong, L., & Wang, S. (2008). Rice gene network inferred from expression profiling of plants overexpressing OsWRKY13, a positive regulator of disease resistance. *Molecular Plant, 1*, 538–551.

Qiu, Z., Wang, X., Gao, J., Guo, Y., Huang, Z., & Du, Y. (2016). The tomato Hoffman's anthocyaninless gene encodes a bHLH transcription factor involved in anthocyanin biosynthesis that is developmentally regulated and induced by low temperatures. *PLoS One, 11*(3), e0151067.

Quattrocchio, F., Wing, J., van der Woude, K., Souer, E., de Vetten, N., Mol, J., & Koes, R. (1999). Molecular analysis of the anthocyanin2 gene of petunia and its role in the evolution of flower color. *Plant Cell, 11*, 1433–1444.

Ramsay, N. A., & Glover, B. J. (2005). MYB-bHLH-WD40 protein complex and the evolution of cellular diversity. *Trends in Plant Science, 10*, 63–70.

Rashid, M., Guangyuan, H., Guangxiao, Y., Hussain, J., & Xu, Y. (2012). AP2/ERF transcription factor in rice: Genome-wide canvas and syntenic relationships between monocots and eudicots. *Evolutionary Bioinformatics Online, 8*, 321–355. https://doi.org/10.4137/EBO.S9369

Robatzek, S., & Somssich, I. E. (2002). Targets of AtWRKY6 regulation during plant senescence and pathogen defense. *Genes & Development, 16*, 1139–1149.

Roepke, J., Salim, V., Wu, M., Thamm, A. M., Murata, J., Ploss, K., Boland, W., & De Luca, V. (2010). Vinca drug components accumulate exclusively in leaf exudates of Madagascar periwinkle. *Proceedings of the National Academy of Sciences of the United States of America, 107*, 15287–15292.

Rushton, P. J., Somssich, I. E., Ringler, P., & Shen, Q. J. (2010). WRKY transcription factors. *Trends in Plant Science, 15*, 247–258.

Saga, H., Ogawa, T., Kai, K., Suzuki, H., Ogata, Y., Sakurai, N., Shibata, D., & Ohta, D. (2012). Identification and characterization of ANAC042, a transcription factor family gene involved in the regulation of camalexin biosynthesis in *Arabidopsis*. *Molecular Plant-Microbe Interactions, 25*, 684–696.

Sahoo, D. K., Sarkar, S., Sumita, R., Maiti, I. B., & Dey, N. (2014). Comparative analysis of synthetic DNA promoters for high-level gene expression in plants. *Planta, 240*, 855–875.

Samira, R., Li, B., Kliebenstein, D., Li, C., Davis, E., Gillikin, J. W., & Long, T. A. (2018). The bHLH transcription factor ILR3 modulates multiple stress responses in *Arabidopsis*. *Plant Molecular Biology, 97*, 297–309.

Sawant, S. V., Kiran, K., Mehrotra, R., Chaturvedi, C. P., Ansari, S. A., Singh, P., Lodhi, N., & Tuli, R. (2005). A variety of synergistic and antagonistic interactions mediated by cis-acting

DNA motifs regulate gene expression in plant cells and modulate stability of the transcription complex formed on a basal promoter. *Journal of Experimental Botany, 56*, 2345–2353.

Schluttenhofer, C., Pattanaik, S., Patra, B., & Yuan, L. (2014). Analyses of *Catharanthus roseus* and *Arabidopsis thaliana* WRKY transcription factors reveal involvement in jasmonate signaling. *BMC Genomics, 15*, 50.

Schwinn, K., Venail, J., Shang, Y., Mackay, S., Alm, V., Butelli, E., Oyama, R., Bailey, P., Davies, K., & Martin, C. (2006). A small family of MYB-regulatory genes controls floral pigmentation intensity and patterning in the genus *Antirrhinum*. *The Plant Cell, 18*, 831–851.

Sears, M. T., Zhang, H., Rushton, P. J., Wu, M., Han, S., Spano, A. J., & Timko, M. P. (2014). NtERF32: A non-NIC2 locus AP2/ERF transcription factor required in jasmonate-inducible nicotine biosynthesis in tobacco. *Plant Molecular Biology, 84*, 49–66.

Sewelam, N., Kazan, K., Thomas-Hall, S. R., Kidd, B. N., Manners, J. M., & Schenk, P. M. (2013). Ethylene response factor 6 is a regulator of reactive oxygen species signaling in *Arabidopsis*. *PLoS One, 8*, e70289.

Shao, H., Wang, H., & Tang, X. (2015). NAC transcription factors in plant multiple abiotic stress responses: Progress and prospects. *Frontiers in Plant Science, 6*, 902.

Shen, Y., Sun, T., Pan, Q., Anupol, N., Chen, H., Shi, J., Liu, F., Deqiang, D., Wang, C., & Zhao, J. (2019). RrMYB5-and RrMYB10-regulated flavonoid biosynthesis plays a pivotal role in feedback loop responding to wounding and oxidation in *Rosa rugosa*. *Plant Biotechnology Journal, 17*, 2078–2095.

Shikata, M., & Ohme-Takagi, M. (2008). The utility of transcription factors for manipulation of floral traits. *Plant Biotechnology, 25*, 31–36.

Shoji, T., & Hashimoto, T. (2011). Tobacco MYC2 regulates jasmonate-inducible nicotine biosynthesis genes directly and by way of the NIC2-locus ERF genes. *Plant & Cell Physiology, 52*, 1117–1130.

Shoji, T., Kajikawa, M., & Hashimoto, T. (2010). Clustered transcription factor genes regulate nicotine biosynthesis in tobacco. *Plant Cell, 22*, 3390–3409.

Singh, A. K., Kumar, S. R., Dwivedi, V., Rai, A., Pal, S., Shasany, A. K., & Nagegowda, D. A. (2017). A WRKY transcription factor from *Withania somnifera* regulates triterpenoid withanolide accumulation and biotic stress tolerance through modulation of phytosterol and defense pathways. *New Phytologist, 215*, 1115–1131.

Singh, B., & Sharma, R. A. (2015). Plant terpenes: Defense responses, phylogenetic analysis, regulation and clinical applications. *3 Biotech, 5*(2), 129–151. https://doi.org/10.1007/s13205-014-0220-2

Song, S., Qi, T., Fan, M., Zhang, X., Gao, H., Huang, H., Wu, D., Guo, H., & Xie, D. (2013). The bHLH subgroup IIId factors negatively regulate jasmonate-mediated plant defense and development. *PLoS Genetics, 9*, e1003653.

Song, S. Y., Chen, Y., Chen, J., Dai, X. Y., & Zhang, W. H. (2011). Physiological mechanisms underlying OsNAC5-dependent tolerance of rice plants to abiotic stress. *Planta, 234*, 331–345.

Souer, E., Houwelingen, A. V., Kloos, D., Mol, J., & Koes, R. (1996). The no apical meristem gene of petunia is required for pattern formation in embryos and flowers and is expressed at meristem and primordia boundaries. *Cell, 85*, 159–170.

Srivastava, R., Rai, K. M., Srivastava, M., Kumar, V., Pandey, B., Singh, S. P., Bag, S. K., Singh, B. D., Tuli, R., & Sawant, S. V. (2014). Distinct role of core promoter architecture in regulation of light-mediated responses in plant genes. *Molecular Plant, 7*(4), 626–641. https://doi.org/10.1093/mp/sst146

Stege, J. T., Guan, X., Ho, T., Beachy, R. N., & Barbas, C. F., III. (2002). Controlling gene expression in plants using synthetic zinc finger transcription factors. *The Plant Journal, 32*, 1077–1086.

Sun, Y., & Yu, D. (2015). Activated expression of AtWRKY53 negatively regulates drought tolerance by mediating stomatal movement. *Plant Cell Reports, 34*, 1295–1306.

Suttipanta, N., Pattanaik, S., Kulshrestha, M., Patra, B., Singh, S. K., & Yuan, L. (2011). The transcription factor CrWRKY1 positively regulates the terpenoid indole alkaloid biosynthesis in *Catharanthus roseus*. *Plant Physiology, 157*, 2081–2093.

Takahashi, A., & Ohnishi, T. (2004). The significance of the study about the biological effects of solar ultraviolet radiation using the exposed facility on the international space station. *Biological Sciences in Space, 18*, 255–260.

Takasaki, H., Maruyama, K., Kidokoro, S., Ito, Y., Fujita, Y., Shinozaki, K., Yamaguchi-Shinozaki, K., & Nakashima, K. (2010). The abiotic stress-responsive NAC-type transcription factor OsNAC5 regulates stress-inducible genes and stress tolerance in rice. *Molecular Genetics and Genomics, 284*, 173–183.

Takos, A. M., Jaffe, F. W., Jacob, S. R., Bogs, J., Robinson, S. P., & Walker, A. R. (2006). Light-induced expression of a MYB gene regulates anthocyanin biosynthesis in red apples. *Plant Physiology, 142*, 1216–1232.

Todaka, D., Shinozaki, K., & Yamaguchi-Shinozaki, K. (2015). Recent advances in the dissection of drought-stress regulatory networks and strategies for development of drought-tolerant transgenic rice plants. *Frontiers in Plant Science, 6*, 84. https://doi.org/10.3389/fpls.2015.00084

Todd, A. T., Liu, E., Polvi, S. L., Pammett, R. T., & Page, J. E. (2010). A functional genomics screen identifies diverse transcription factors that regulate alkaloid biosynthesis in *Nicotiana benthamiana*. *The Plant Journal, 62*, 589–600.

Tripathi, P., Rabara, R., & Rushton, P. (2014). A systems biology perspective on the role of WRKY transcription factors in drought responses in plants. *Planta, 239*, 255–266.

Tungmunnithum, D., Thongboonyou, A., Pholboon, A., & Yangsabai, A. (2018). Flavonoids and other phenolic compounds from medicinal plants for pharmaceutical and medical aspects an overview. *Medicine, 5*(3), 93. https://doi.org/10.3390/medicines5030093

Van der Fits, L., & Memelink, J. (2000). ORCA3, a Jasmonate-responsive transcriptional regulator of plant primary and secondary metabolism. *Science, 289*, 295–297.

Vannozzi, A., Wong, D. C. J., Höll, J., Hmmam, I., Matus, J. T., Bogs, J., Ziegler, T., Dry, I., Barcaccia, G., & Lucchin, M. (2018). Combinatorial regulation of stilbene synthase genes by WRKY and MYB transcription factors in grapevine (*Vitis vinifera* L.). *Plant & Cell Physiology, 59*, 1043–1059.

Venter, M. (2007). Synthetic promoters: Genetic control through cis engineering. *Trends in Plant Science, 12*, 118–124.

Wan, X., Marsafari, M., & Xu, P. (2019). Engineering metabolite-responsive transcriptional factors to sense small molecules in eukaryotes: Current state and perspectives. *Microbial Cell Factories, 18*, 61. https://doi.org/10.1186/s12934-019-1111-3

Wang, C., Deng, P., Chen, L., Wang, X., Ma, H., Hu, W., Yao, N., Feng, Y., Chai, R., & Yang, G. (2013). A wheat WRKY transcription factor TaWRKY10 confers tolerance to multiple abiotic stresses in transgenic tobacco. *PLoS One, 8*, e65120.

Wang, F., Kong, W., Wong, G., Fu, L., Peng, R., Li, Z., & Yao, Q. (2016a). AtMYB12 regulates flavonoids accumulation and abiotic stress tolerance in transgenic *Arabidopsis thaliana*. *Molecular Genetics and Genomics, 291*, 1545–1559.

Wang, F., Zhu, H., Chen, D., Li, Z., Peng, R., & Yao, Q. (2016b). A grape bHLH transcription factor gene, VvbHLH1, increases the accumulation of flavonoids and enhances salt and drought tolerance in transgenic *Arabidopsis thaliana*. *Plant Cell, Tissue and Organ Culture, 125*, 387–398.

Wang, F., Zhu, H., Kong, W., Peng, R., Liu, Q., & Yao, Q. (2016c). The *Antirrhinum* AmDEL gene enhances flavonoids accumulation and salt and drought tolerance in transgenic *Arabidopsis*. *Planta, 244*, 59–73.

Wang, H., Avci, U., Nakashima, J., Hahn, M. G., Chen, F., & Dixon, R. A. (2010a). Mutation of WRKY transcription factors initiates pith secondary wall formation and increases stem biomass in dicotyledonous plants. *Proceedings of the National Academy of Sciences of the United States of America, 107*, 22338–22343.

Wang, K. L., Bolitho, K., Grafton, K., Kortstee, A., Karunairetnam, S., McGhie, T. K., Espley, R. V., Hellens, R. P., & Allan, A. C. (2010b). An R2R3 MYB transcription factor associated with regulation of the anthocyanin biosynthetic pathway in Rosaceae. *BMC Plant Biology, 10*, 50.

Wang, L., Ran, L., Hou, Y., Tian, Q., Li, C., Liu, R., Fan, D., & Luo, K. (2017). The transcription factor MYB115 contributes to the regulation of proanthocyanidin biosynthesis and enhances fungal resistance in poplar. *New Phytologist, 215*, 351–367.

Wang, L., & Wang, Y. (2019). Transcription factor VqERF114 regulates stilbene synthesis in Chinese wild *Vitis quinquangularis* by interacting with VqMYB35. *Plant Cell Reports, 38*, 1347–1360.

Wang, Q., Guan, Y., Wu, Y., Chen, H., Chen, F., & Chu, C. (2008). Overexpression of a rice OsDREB1F gene increases salt, drought, and low temperature tolerance in both *Arabidopsis* and rice. *Plant Molecular Biology, 67*, 589–602.

Wang, W. L., Wang, Y. X., Li, H., Liu, Z. W., Cui, X., & Zhuang, J. (2018). Two MYB transcription factors (CsMYB2 and CsMYB26) are involved in flavonoid biosynthesis in tea plant [*Camellia sinensis* (L.) O. Kuntze]. *BMC Plant Biol, 18*, 288.

Wang, Z., Su, G., Li, M., Ke, Q., Kim, S. Y., Li, H., Huang, J., Xu, B., Deng, X. P., & Kwak, S. S. (2016d). Overexpressing *Arabidopsis* ABF3 increases tolerance to multiple abiotic stresses and reduces leaf size in alfalfa. *Plant Physiology and Biochemistry, 109*, 199–208.

Wei, K., Chen, J., & Wang, Y. (2012). Genome-wide analysis of bZIP-encoding genes in maize. *DNA Research, 19*(6), 463–476. https://doi.org/10.1093/dnares/dss026

Wei, T., Deng, K., Liu, D., Gao, Y., Liu, Y., Yang, M., Zhang, L., Zheng, X., Wang, C., & Song, W. (2016). Ectopic expression of DREB transcription factor, AtDREB1A, confers tolerance to drought in transgenic *Salvia miltiorrhiza*. *Plant & Cell Physiology, 57*, 1593–1609.

Winkel-Shirley, B. (2002). Biosynthesis of flavonoids and effects of stress. *Current Opinion in Plant Biology, 5*(3), 218–223. https://doi.org/10.1016/s1369-5266(02)00256-x

Wu, C. Y., Suzuki, A., Washida, H., & Takaiwa, F. (1998). The GCN4 motif in a rice glutelin gene is essential for endosperm-specific gene expression and is activated by Opaque-2 in transgenic rice plants. *The Plant Journal, 14*, 673–683.

Wu, H., Fu, B., Sun, P., Xiao, C., & Liu, J. H. (2016). A NAC transcription factor represses putrescine biosynthesis and affects drought tolerance. *Plant Physiology, 172*, 1532–1547.

Xiao, H., Pearson, A., Coulombe, B., Truant, R., Zhang, S., Regier, J. L., Triezenberg, S. J., Reinberg, D., Flores, O., & Ingles, C. J. (1994). Binding of basal transcription factor TFIIH to the acidic activation domains of VP16 and p53. *Molecular and Cellular Biology, 14*, 7013–7024.

Xu, Z. S., Ni, Z. Y., Li, Z. Y., Li, L. C., Chen, M., Gao, D. Y., Yu, X. D., Liu, P., & Ma, Y. Z. (2009). Isolation and functional characterization of HvDREB1: A gene encoding a dehydration-responsive element binding protein in *Hordeum vulgare*. *Journal of Plant Research, 122*, 121–130.

Yamada, Y., Motomura, Y., & Sato, F. (2015). CjbHLH1 homologs regulate sanguinarine biosynthesis in *Eschscholzia californica* cells. *Plant & Cell Physiology, 56*, 1019–1030.

Yamaguchi, K., Takahashi, Y., Berberich, T., Imai, A., Takahashi, T., Michael, A. J., & Kusano, T. A. (2007). A protective role for the polyamine spermine against drought stress in *Arabidopsis*. *Biochemical and Biophysical Research Communications, 352*, 486–490.

Yamamura, C., Mizutani, E., Okada, K., Nakagawa, H., Fukushima, S., Tanaka, A., Maeda, S., Kamakura, T., Yamane, H., & Takatsuji, H. (2015). Diterpenoid phytoalexin factor, a bHLH transcription factor, plays a central role in the biosynthesis of diterpenoid phytoalexins in rice. *The Plant Journal, 84*, 1100–1113.

Yang, Z., Patra, B., Li, R., Pattanaik, S., & Yuan, L. (2013). Promoter analysis reveals cis-regulatory motifs associated with the expression of the WRKY transcription factor CrWRKY1 in *Catharanthus roseus*. *Planta, 238*, 1039–1049.

Yogendra, K. N., Dhokane, D., Kushalappa, A. C., Sarmiento, F., Rodriguez, E., & Mosquera, T. (2017). StWRKY8 transcription factor regulates benzylisoquinoline alkaloid pathway in potato conferring resistance to late blight. *Plant Science, 256*, 208–216.

Yogendra, K. N., Kumar, A., Sarkar, K., Li, Y., Pushpa, D., Mosa, K. A., Duggavath, R., & Kushalappa, A. C. (2015). Transcription factor StWRKY1 regulates phenylpropanoid metabolites conferring late blight resistance in potato. *Journal of Experimental Botany, 66*, 7377–7389.

Yokotani, N., Sato, Y., Tanabe, S., Chujo, T., Shimizu, T., Okada, K., Yamane, H., Shimono, M., Sugano, S., & Takatsuji, H. (2013). WRKY76 is a rice transcriptional repressor playing opposite roles in blast disease resistance and cold stress tolerance. *Journal of Experimental Botany, 64*, 5085–5097.

You, J., Zong, W., Hu, H., Li, X., Xiao, J., & Xiong, L. (2014). A STRESS-RESPONSIVE NAC1-regulated protein phosphatase gene rice protein phosphatase18 modulates drought and oxidative stress tolerance through abscisic acid-independent reactive oxygen species scavenging in rice. *Plant Physiology, 166*, 2100–2114. https://doi.org/10.1104/pp.114.251116

Young, D., Michelotti, E., Swindell, C., & Krauss, N. (1992). Antifungal properties of taxol and various analogues. *Experientia, 48*, 882–885.

Younger, A. K. D., Dalvie, N. C., Rottinghaus, A. G., & Leonard, J. N. (2017). Engineering modular biosensors to confer metabolite responsive regulation of transcription. *ACS Synthetic Biology, 6*(2), 311–325.

Yu, Y., Wang, L., Chen, J., Liu, Z., Park, C. M., & Xiang, F. (2018). WRKY71 acts antagonistically against salt-delayed flowering in *Arabidopsis thaliana*. *Plant & Cell Physiology, 59*, 414–422.

Yu, Z. X., Li, J. X., Yang, C. Q., Hu, W. L., Wang, L. J., & Chen, X. Y. (2012). The jasmonate responsive AP2/ERF transcription factors AaERF1 and AaERF2 positively regulate artemisinin biosynthesis in *Artemisia annua* L. *Molecular Plant, 5*, 353–365.

Yuan, Y., Qi, L., Yang, J., Wu, C., Liu, Y., & Huang, L. (2015). A *Scutellaria baicalensis* R2R3-MYB gene, SbMYB8, regulates flavonoid biosynthesis and improves drought stress tolerance in transgenic tobacco. *Plant Cell, Tissue and Organ Culture, 120*, 961–972.

Zhai, Y., Zhang, L., Xia, C., Fu, S., Zhao, G., Jia, J., & Kong, X. (2016). The wheat transcription factor, TabHLH39, improves tolerance to multiple abiotic stressors in transgenic plants. *Biochemical and Biophysical Research Communications, 473*, 1321–1327.

Zhang, C., Li, C., Liu, J., Lv, Y., Yu, C., Li, H., Zhao, T., & Liu, B. (2017). The OsABF1 transcription factor improves drought tolerance by activating the transcription of COR413-TM1 in rice. *Journal of Experimental Botany, 68*, 4695–4707.

Zhang, C., Liu, J., Zhao, T., Gomez, A., Li, C., Yu, C., Li, H., Lin, J., Yang, Y., & Liu, B. (2016a). A drought-inducible transcription factor delays reproductive timing in rice. *Plant Physiology, 171*, 334–343.

Zhang, F., Fu, X., Lv, Z., Lu, X., Shen, Q., Zhang, L., Zhu, M., Wang, G., Sun, X., & Liao, Z. (2015a). A basic leucine zipper transcription factor, AabZIP1, connects abscisic acid signaling with artemisinin biosynthesis in *Artemisia annua*. *Molecular Plant, 8*, 163–175.

Zhang, M., Li, S., Nie, L., Chen, Q., Xu, X., Yu, L., & Fu, C. (2015b). Two jasmonate-responsive factors, TcERF12 and TcERF15, respectively act as repressor and activator of tasy gene of taxol biosynthesis in *Taxus chinensis*. *Plant Molecular Biology, 89*, 463–473.

Zhang, Y., Xu, Z., Ji, A., Luo, H., & Song, J. (2018). Genomic survey of bZIP transcription factor genes related to tanshinone biosynthesis in *Salvia miltiorrhiza*. *Acta Pharmaceutica Sinica B, 8*, 295–305.

Zhang, Z., Hu, X., Zhang, Y., Miao, Z., Xie, C., Meng, X. Z., Deng, J., Wen, J., Mysore, K. S., & Frugier, F. (2016b). Opposing control by transcription factors MYB61 and MYB3 increases freezing tolerance by relieving c-repeat binding factor suppression. *Plant Physiology, 172*, 1306–1323.

Zhou, F., Sun, T. H., Zhao, L., Pan, X. W., & Lu, S. (2015). The bZIP transcription factor HY5 interacts with the promoter of the monoterpene synthase gene QH6 in modulating its rhythmic expression. *Frontiers in Plant Science, 6*, 304.

Zhou, H., Peng, Q., Zhao, J., Owiti, A., Ren, F., Liao, L., Wang, L., Deng, X., Jiang, Q., & Han, Y. (2016). Multiple R2R3-MYB transcription factors involved in the regulation of anthocyanin accumulation in peach flower. *Frontiers in Plant Science, 7*, 1557.

Chapter 5
Secondary Metabolite Engineering for Plant Immunity Against Various Pathogens

Devendra Singh, Himanshu Pandey, Vivek Kumar Dhiman, Nguyen Tran Hai Bang, Vinay Kumar Dhiman, Virendra Singh, and Devendra Pandey

Abstract In different natural conditions, plants are subjected to an ample of enemies, which include both biotic and abiotic stress related to climate change. Normally every ecosystem constitutes of different kinds of bacteria, fungi, mites, insects, viruses, nematodes, mammals and other plant-eating animals that cause damage to crop productivity and are responsible for low crop yield. In addition, plants produce some organic chemicals generally termed as secondary metabolites to protect themselves from invading pathogens and different abiotic stress conditions. This book chapter represents a broad picture about the various mechanisms that plants have evolved to safeguard themselves from the various pathogenic organisms and different abiotic stresses, as well as their peculiar responses to pathogen attack and genetic engineering principles to cope up with biotic stress conditions.

Keywords Plants · Biotic stress · Pathogens · Abiotic stress · Secondary metabolites

D. Singh (✉)
Motilal Nehru National Institute of Technology, Allahabad, Uttar Pradesh, India

H. Pandey · V. K. Dhiman
Dr. YSP UHF Nauni, Solan, Himachal Pradesh, India

V. K. Dhiman
Himachal Pradesh University, Shimla, Himachal Pradesh, India

N. T. H. Bang
Cuu Long Delta Rice Research Institute, Can Tho City, Vietnam

V. Singh
Maulana Azad Medical College, New Delhi, India

D. Pandey
Central Institute for Subtropical Horticulture, Lucknow, Uttar Pradesh, India

T. Aftab, K. R. Hakeem (eds.), *Metabolic Engineering in Plants*,
https://doi.org/10.1007/978-981-16-7262-0_5

Introduction

Plants are an important source of many naturally occurring bioactive molecules known as secondary metabolites that act as a defensive barrier against many pathogenic organisms and herbivores. These specialized (secondary) plant-derived metabolites exert various medicinal properties and are also used as an essential therapeutic agent in many herbal preparations. Most of the active compounds or specialized metabolites are produced with respect to sudden changes in climatic conditions. The major classes of these secondary metabolites include defence-related compounds such as terpenoids, flavonoids and phenylpropanoids like stilbenoids and other organic nitrogen-containing derivatives, namely, alkamides, glucosinolates and alkaloids. Sulphur is also present in many proactive metabolites in addition to nitrogen, such as found in many glucosinolates and indolic compounds. A particular plant produces only a small number of species-characteristic defence compounds, though the difference in intraspecific structure, composition or level of content can be significant (Großkinsky et al., 2012). Defensive metabolites can be retained as inactive forms or generated in response to an insect or microbial assault. A group of special metabolites or compounds referred to as phytoalexins are produced only upon pathogen infection. Their induced synthesis in response to the defensive action of plants makes them an integral part of the immune systems of plants. Another important group of defence bioactive compounds is phytoanticipins which are distinguished from phytoalexins on the basis of their constitutive expression and action in plants (Ahuja et al., 2012; Paxton, 1980).

Current modern agriculture requires variety of insecticides or pesticides. Therefore, it poses a major threat to the environment. However, these secondary metabolites can be used effectively to combat against many plant stresses like pest infestation (microbes or insects) and herbivores and to also promote plant fitness. Furthermore, most of these metabolites are used by mankind as fragrant, flavouring components and provide health benefits (Martin, 2013). Hence, this chapter deals with the engineering of metabolic processes for disease management and the role of plant chemical compounds or secondary metabolites in disease resistance. Figure 5.1 gives some information regarding the plant secondary metabolites and their functions.

Phytoanticipins and Phytoalexins

Phytoalexin is a term used in plant pathology to identify inducible plant chemical defence molecules. The fact that such plant-specific metabolites have a role in disease resistance was proposed nearly 80 years ago by Müller and Börger during the 1940s after observing the effects of unknown substances released by cells of potato tubers upon infection of *Phytophthora infestans* (Paxton, 1980). These are low-molecular antimicrobial metabolites synthesized as a reaction to biotic and

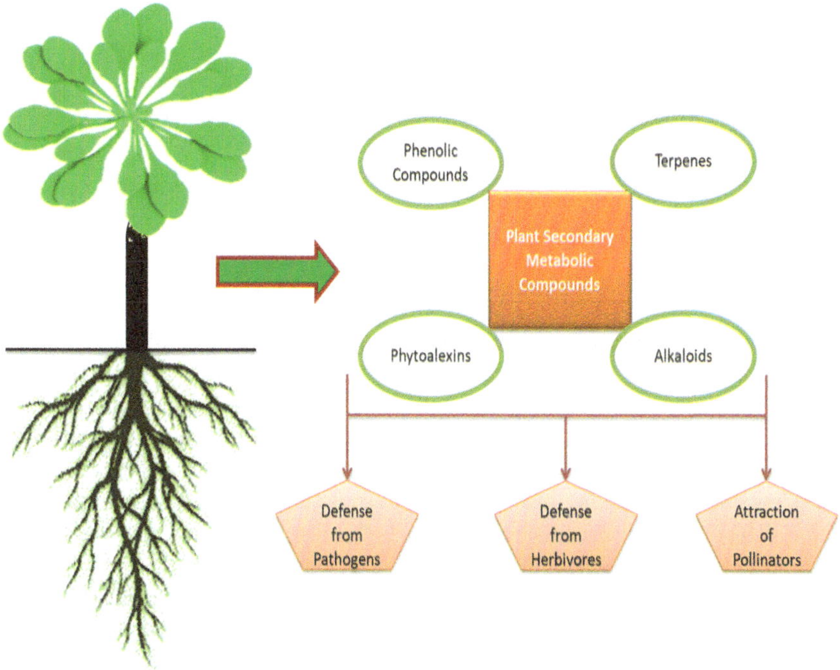

Fig. 5.1 A pictorial representation of plant secondary metabolites and their functions

abiotic stresses and accumulated in plants. The antibiotic role of phytoalexins, on the other hand, is challenging to study, and their active defensive function can only be presumed. Phytoalexins are microbicidal substances that require enzymes to synthesize complex precursors or molecules in response to elicitation and eventually their biosynthesis. Therefore, when the pathogen is discovered, the creation of phytoalexins needs transcriptional and translation work in the plant. Their mechanism of antimicrobial actions involves transport and secretion of bioactive metabolites at infection site or wounded parts of plants. Phytoalexins and phytoanticipins can be distinguished by the induction of their biosynthesis once the plant is confronted with a microbial elicitor (VanEtten et al., 2001).

Phytoanticipins are plant defence molecules constitutively expressed and have a pre-occurrence in plant cells or tissues at often high quantities. The distinction between phytoalexins and phytoanticipins, as VanEtten et al. (1994) pointed out, was based on establishing, simply, its synthesis and not its chemical structure. Since the de novo synthesis of the enzyme is not required in the final release of a chemical, they are therefore non- phytoalexins. A specialized metabolite might sometimes be regarded as either a phytoalexin or a phytoanticipin, even in an individual plant, depending on its environment. Therefore, it is not always feasible or useful to strictly classify the metabolite of the specific chemical defence as a phytoalexin or a phytoanticipin. Table 5.1 shows the different types of phytoanticipin found in plants.

Table 5.1 Phytoanticipin found in some plants

Plants	Phytoanticipin
Avena spp.	Avenacins, avenacosides
Zea mays, wheat (*Triticum aestivum*), rye (*Secale cereale*)	DIMBOA glucoside (benzoxazinone)
Sorghum (*Sorghum vulgare*)	Dhurrin
Tomato (*Lycopersicon esculentum*)	Tomatine
Cassava (*Manihot esculenta*), clover (Trifolium repens)	Laminarin
Onion bulbs	Catechol

Understanding metabolites as defence molecules that have antibacterial action and those that defend against herbivores is challenging. Therefore, to get a deeper insight into their mode of action, the studies need to be done by defining aspects of context. For example, herbivorous damage can promote the de novo biosynthesis of some chemical defence compounds, known as phytoalexins. The best-studied example is on small sap-sucking insects, aphids, feeding them on *Arabidopsis thaliana*-induced biosynthesis of camalexin (3-thiazol-2′-yl-indole), which inhibited aphid reproduction (Kuśnierczyk et al., 2008).

A substantial number of secondary metabolites possess pharmaceutical interest, health advantages and ambiguity used to them as describe plant defence metabolites (Martin, 2013). Bioactive natural products, therapeutic chemicals and phytonutrients are some of the terms used to scientifically refer and describe plant metabolites. Resveratrol is an important compound with dual characteristics, i.e. it provides an active role in plant protection against diseases and is beneficial for human consumption. It is a stilbene chemical that is present in grapes, cocoa, peanuts and mulberries and was the first transgenically expressed metabolite. Resveratrol is known for its antioxidant, anticancer and anti-inflammatory qualities, as well as its ability to enhance health and fight chronic illness like heart diseases, cancer or hyperglycemic conditions (Wood et al., 2004). Red wine contains high content of resveratrol and is sometimes blamed for the 'French paradox', which is the French population's low prevalence of coronary heart disease in spite of their fatty diet (Smoliga et al., 2011).

Metabolic Engineering and Infection Control

Agricultural crops are vulnerable to a variety of diseases and pests that can harm crops, reduce harvest quality or even wipe out entire harvests. Diseases and pests destroy over half of the world's entire yield each year. Furthermore, farmers are frequently confronted with many pests or diseases, as well as novel pesticide-resistant microorganisms. The major traits introduced into commercially cultivated crops are insect-resistant (Bt gene encoding endotoxin) and herbicide tolerance technologies (glyphosate). For crops like maize, soybeans and cotton, their introduction has been a huge success, accounting for nearly all of the 190.4 million hectares of transgenic crops planted globally in 2019. These traits need only the single gene transfer, making them easy to introduce; this may also indicate that research and development of transgene crops are in early stages. There is currently no commercially available transgenic crop that has been metabolically designed to produce a chemical defence molecule. The biosynthesis routes for plant defence chemicals sometimes include numerous enzymatic steps; introducing them by genetic engineering is more challenging than introducing single gene-based features. Evidently, identifying all the biosynthetic genes for synthesis of a metabolite is a time-consuming scientific process. The eventual transfer of these complex enzymatic pathways necessitates a balance between coordinating genes and enzymes. It is no surprise that the synthesis of resveratrol involves the transfer of stilbene synthase encoding gene; tobacco has been introduced to this gene which inferred resistance to an airborne pathogen *Botrytis cinerea* which is the most common example of the introduction of a plant chemical defence system to impart disease resistance to date (Hain et al., 1993). This section presents an overview of research and scientific approaches for genetically engineered metabolic resistance in plants against numerous insects or pathogens.

Engineering Pathogen Resistance

The insertion of stilbene synthase gene and its expression is a widely studied metabolic process to provide resistance against pests and plant diseases. It is a classic example of metabolic engineering applied for increasing plant defence to diseases (Delaunois et al., 2009). The core ethylene covalently bonded to a phenyl group in each arm characterize the Stilbenes (Chong et al., 2009). Stilbenoids in hydroxylated form have been discovered in plants; the most studied example of this category is resveratrol. These stilbenoids are obtained from phenylalanine metabolism involving the phenylpropanoid pathway. Stilbene synthase requires malonyl-coenzyme A and *p*-coumaroyl-coenzyme A metabolites as substrates for resveratrol production. Both metabolic precursors are commonly originated in each and every plant, and they are also utilized in the synthesis of different types of flavonoids by chalcone synthases. Evidently, transferring the stilbene biosynthetic gene to other species is sufficient to establish stable production of stilbene-related phytoalexins. For example, stilbene synthase enzyme in *Arachis hypogaea* (peanuts) has been used for transformation of the tobacco plant cells and induced resveratrol production in the non-host tobacco by using a fungal elicitor (Iwuala et al., 2020).

Earliest evidence of disease resistance by the use of resveratrol synthesis was done in genetically engineered tobacco plant. Hain et al. (1993) transferred the genomic segment of grapevine (*Vitis vinifera*), which is encoded for two sets of stilbene synthase genes. Its gene expression was increased and resistance to the infection improved upon inoculation with a disease-causing fungus *Botrytis cinerea* on tobacco leaves. The introduction of stilbenum synthase genes from similar grapevine DNA into *Lycopersicon esculentum* (tomato) also led to the increase in the synthesis of resveratrol upon fungal infections (Thomzik et al., 1997). However, these transgenic tomato plants were ineffective against *Botrytis cinerea* or *Alternaria solani*. Many phytopathogens develop tolerance against phytoalexins present in their host, and this tolerance can be considered as a virulence trait (VanEtten et al., 2001). Stilbene synthases from grapevine and peanut differed from pine (*Pinus sylvestris*) in that it requires malonyl-CoA and cinnamoyl-CoA precursors for pinosylvin synthesis. Antifungal experiments have revealed a substantially higher tolerance to pinosylvin by commonly producing fungi on conifers (Seppänen et al., 2004). Whereas *Phellinus tremulae*, known for serious damages in aspen trees (*Populus tremula*), was very much susceptible to pinosylvin. The pinosylvin synthase gene was transmitted into aspen, and its transcription analysis showed the activity of the pinosylvin synthase enzyme (Seppänen et al., 2004). But no significant stilbenes content was observed, and the amounts of cinnamoyl-CoA were considered to be restricting factors for the synthesis of pinosylvin in some plant species. The insertion of resveratrol synthase gene leads to the build-up of resveratrol glucosides in various plant species that are considered the storage form of molecule (Giorcelli et al., 2004; Pan et al., 2012; Schwekendiek et al., 2007). These findings show that metabolic engineering is feasible in providing disease resistance to crops, and their desired effects rely on the host plant metabolism and plant-pathogen interactions. Most of the pathogens are susceptible to phytoalexins or phytoanticipins, but natural tolerance could be developed over time.

Isoflavones

Isoflavones are synthesized from metabolism of phenylpropanoid found exclusively in Fabaceae plant species. Isoflavones are mostly present in leguminous plants, providing their function as phytoalexins and establishing symbiosis between legumes and nitrogen-fixing bacteria via functioning as root signalling molecules (Dakora & Phillips, 1996). This bacterial recognition leads to the formation of nodules and establishes symbiotic interactions with the host plant. Isoflavones also are known to be phytoestrogens and have claimed health advantages. Studies show that dietary isoflavones caused decreased risk of breast cancer (Pandey et al., 2014).

Isoflavone metabolism requires studies of their complex heterologous system; a key enzyme, isoflavone synthase, can be metabolically engineered to modify aromatic ring in the synthesis of isoflavones (Yu et al., 2000). Many synthetic enzymes are also required for core structure modifications in isoflavones. These modifications, substitutions or overall metabolic engineering processes could result in unexpected outcomes. One such enzyme is O-methyltransferases; a classic example is overexpression of 7-O-methyltransferase in *Medicago sativa* (legume), which resulted in 7-O-methyl daidzein production on the application of daidzein, whereas, when 7-O-methyltransferase-expressing plants were inoculated with fungus (*Phoma medicaginis*), no 7-O-methyl daidzein levels were detected, but a larger deposition of medicarpin and 4'-O-methyl daidzein were detected when compared to control

(He & Dixon, 2000). Under fungal infection, overexpressions of 7-O-methyltransferase resulted in a stronger induction of genes expressing isoflavonoid synthesis leading to resistance against a pathogenic mould named *P. medicaginis*. These findings on biosynthetic pathways lead to the concept of metabolic channelling of intermediary routes between biosynthetic enzymes organized in a higher complexity (He & Dixon, 2000). Such a structural-functional complex of multienzymes formed in metabolic pathways known as metabolons is challenging to study experimentally and scientifically (Jørgensen et al., 2005). However, their subsequent presence as intermediates deeply influences the design of metabolic engineering studies.

Hydroxycinnamic Acid Amides

Hydroxycinnamic acids or hydroxycinnamates are class C6-C3 skeletal aromatic acids derived from cinnamic acid. It is an extensively studied phytoalexin discovered in many plants belonging to several plants and horticultural crops such as strawberries, pineapple, coffee, rice, etc. (Jeandet et al., 2014; Okazaki et al., 2004). Metabolically engineered phyto-hydroxycinnamates are synthesized from agmatine coumaroyltransferase enzymes, which require *p*-coumaroyl-CoA and agmatine for production of these amine derivatives. These derivatives have antifungal properties and also essential for stilbenoid and flavonoid biosynthesis. Muroi et al. (2012) introduced the agmatine coumaroyltransferase gene from *Arabidopsis thaliana* to an ornamental plant wishbone flower (*Torenia hybrida*), leading to enhanced production of *p*-coumaroylagmatine and inferred resistance to *Botrytis cinerea*.

Terpenoids

Terpenoids or isoprenoids are a wide class of natural phytometabolites that have an important property attributed to their medicinal usage, flavour and fragrance-providing properties. These also have a diverse role in plant defence (Schmelz et al., 2014; Singh & Sharma, 2015; Zerbe et al., 2013). Many valuable terpenoids are a prominent target for metabolic engineering studies because of their very low availability in nature. For example, β-carotene from carrot (*Daucus carota*), Taxol (*Taxus brevifolia*), etc. are important anticancer drugs (De Luca et al., 2012). Terpenoids being an organic compound are derived from isoprene polymers or terpenes. Terpenes act as a substrate in a condensation process which is based on a number of isoprene units present in it. Biosynthesis of many terpenoids occurs in numerous cell components, including cytosol, peroxisomes, endoplasmic reticulum, plastids and mitochondria. Terpenoid production and accumulation are often confined on the tissue level in resin canals, Latinx and glandular hairs. Their synthesis and storage in plants occur in several unique structural tissues for like resin canals, glandular hairs and laticifers. The subclass of triterpenes is constituted by the cycling of 2,3-oxidosqualene and formed of six isoprene units; it also acts as a precursor in sterol biosynthesis.

Several dicots contain triterpenes as a natural antimicrobial agent but generally absent in monocots except for *Avena* spp. consisting of avenacins (Inagaki et al., 2011). The metabolic bioengineering of avenacin synthesis requires β-amyrin synthase. Its introduction from oat to rice leads to expression of β-amyrin synthase, inferring resistance to pathogens (Inagaki et al., 2011). A substantial inhibition of growth by monoterpene geranic acid against maize pathogens, namely, *Fusarium graminearum* and *Colletotrichum graminicola*, was obtained from scientific investigations (Yang et al., 2011). The geraniol synthase gene from the Aztec sweet herb (*Lippia dulcis*) was introduced into maize to metabolically produce geranic acid. It was observed that geraniol-derived compounds contained geranic acid glycoside, although no improved fungal resistance was seen. This glycosylation can detoxify the plant and store it, further demonstrating its connections with the metabolism of the host plant (Yang et al., 2011).

Camalexin

Camalexins are small indole alkaloids discovered in the *Arabidopsis thaliana* and cruciferous plants. This metabolite showed prominent antibacterial and antifungal properties (Beets et al., 2012). Camalexin biosynthesis originates and gets localized to the infection site. It is derived from tryptophan precursor indole-3-acetaldoxime from the biosynthetic pathway of indole glucosinolates (Glawischnig et al., 2004). *Nicotiana benthamiana* was genetically engineered for the synthesis of camalexin (Møldrup et al., 2013). The transient gene expression of three key genes encoding for P450 cytochrome enzymes is essential for camalexin synthesis. The insertion of additional genes, however, lowered the intermediate concentration levels and resulted in a more efficient synthesis of camalexin. These studies show that camalexin production in plant species safeguards them from pathogens.

Alkaloids

Alkaloids are a group of naturally synthesized organic and nitrogenous plant defence chemicals which has been of substantial interest due to their high usage in treatments of cancer, neurological disorders, analgesic substances and other medicinal usage (Desgagné-Penix, 2021; Ziegler & Facchini, 2008). The increased production of several essential alkaloid-derived metabolites could be achieved through a metabolic engineering of their suitable biosynthetic pathways (Glenn et al., 2013). The ecological and physiological role of alkaloids is far less researched, as opposed to their pharmaceutical activity, for example, alkaloids from California poppy (*Eschscholzia californica*) (Angelova et al., 2010). There are extremely limited studies on metabolic engineering of alkaloids to manage plant diseases. Caffeine, which comes from purine nucleotides, is a widely recognized alkaloid present in tea or coffee plants. Caffeine biosynthesis from xanthosine requires three N-methyltransferases (Ashihara et al., 2008). Tobacco was engineered for caffeine synthesis, which inferred resistance to caterpillars (*Spodoptera litura*). The transgenic tobacco line with caffeine synthesis also demonstrated improved resistance to the *Pseudomonas syringae* bacterial disease (Kim & Sano, 2008). However, it is still not clear whether the involvement of caffeine in plant disease resistance is explicit since, under microbial infection and natural conditions, the transgenic lines constitutively expressed defence-related genes.

Mode of Action

A number of phytochemicals are essential to provide plant immunity, and very little is understood about their mechanism. A wide variety of pathogen-induced metabolites have been examined with a potent microbicidal effect, indicating that such substances contribute to plant immunity (Bednarek, 2012). Saponins and steroidal glycoalkaloids are two plant secondary metabolites whose function has been studied in depth. These possess membrane lytic activity, which is detrimental to most of pathogens. Several members of this phytochemical class, including tomatine and avenacin A1, have been shown to break non-biological and biological lipid bilayers (Armah et al., 1999). The modular structure of these metabolites is responsible for this feature. The aglycone moiety can bind to membrane-resident sterol moieties in lipid bilayers, while the saponin moieties interact with one another through sugar chains, causing sterol reconfiguration in the target membrane and eventually creating porous holes in membranous structures (Armah et al., 1999). In the membrane lytic action of some alkaloids such as steroidal glycoalkaloids, saponins require the presence of a sugar chain at carbon 3 position in aglycone, but for sterols, a free 3-β-hydroxy group is required (Armah et al., 1999). The insensitivity of saponins to several oomycete infections, such as *Peronospora* species, has been conferred to their plasma membranes with low sterol content (Arneson & Durbin, 1968). The enzyme-mediated cleavage at C3-hydroxyl site corresponding to aglycone is also required for the virulence of many fungi dependent on steroidal glycoalkaloid or saponin generated in plants (Bouarab et al., 2002; Ökmen et al., 2013). Aside from these two metabolites, several other defensive metabolites are also capable of targeting potential plant pathogens by disintegrating their plasma membranes or endomembrane. Certain phenylpropanoids, such as phaseollin, glycinol and 3-deoxyanthocyanidines, also have a protective role in direct membrane disintegration or can indirectly impact on functions which are necessary for membrane integrity (Nielsen et al., 2004). The defensive plant secondary metabolites not only disrupt membrane proteins but also hinder the activity of essential proteins of a pathogen entering the plant cell. Several bioactive metabolites directly impact the function of certain proteins. Among these phytochemicals, isothiocyanates are produced during glucosinolate metabolism by enzyme myrosinase (Tierens et al., 2001; Wittstock & Burow, 2010). Isothiocyanate has a nucleophilic core, i.e. thiol group, which makes it a very reactive molecule. Therefore, it rapidly reacts within the selenocysteine, cysteine or amine residues, thus influencing action of certain enzymes, which may, in turn, impact the homeostasis of cell redox (Brown & Hampton, 2011). It is also reactive to thiol species present in and makes conjugation with glutathione which alters its levels of reduced and oxidized state, which in turn impacts the homeostasis of the cell (Brown & Hampton, 2011).

Studies on plant immunity inferred by phytometabolites or secondary metabolites in *Arabidopsis thaliana* have shown an active role in isothiocyanates and indole glucosinolates. Isothiocyanates act as antibiotics and reduced glutathione levels. Glutathione degradation influences reactive oxygen production, which in turn can have an adverse effect on programmed cell death and stomach closure. The products of isothiocyanate metabolism govern the entrance of fungal and oomycete infections in epidermal cells. They may also alter the deposition of callose and cell death. Benzoxazinone glucosides have a form of biological action akin to isothiocyanate. Aldehyde group of open ring tautomeric form of benzoxazinone glucosides aglycones operates as an electrophile for amino or thiol species of proteins and glutathiones (Perez & Niemeyer, 1989; Dixon et al., 2012).

Metabolic Engineering: Technical Challenges and Opportunities

In order to boost the production of a given chemical, substance or metabolite in plant cells, metabolic engineering of genetic and regulatory mechanisms needs to be extensively studied. The intricacy of plant metabolism is challenging and continues to be a technology under progress. It is a necessary first step to identify the genes involved in the pathway of interest, but a subsequent introduction into a heterologous system is a tedious task. There are just a few points to consider in order to achieve the coordinated expression of the biosynthesis enzymes introduced and their interaction with hosts. Some of the technical breakthroughs that allow more and more bioengineering of the metabolism of plants are covered briefly in this section.

Biosynthetic Gene Clusters and Gene Identity

An essential requisite for transforming a plant species with a desired exogenous gene conferring the effector molecule or plant defence compound is a thorough knowledge of the biological route and enzymes involved in its function (Glenn et al., 2013). In addition, the huge chemical differences in plant-specific metabolic systems have been discovered by high-throughput analytical approaches such as liquid or gas chromatography combined with mass spectrometry (Zhao et al., 2013).

The biosynthetic enzymes required for secondary metabolite synthesis predominantly belong to cytochrome P450s, glycosyltransferases, methyltransferases and acyltransferases. These enzymes are encoded by genes belonging to large gene families, and distinct substrate specificity is an inclusive feature of every distinct enzyme family. The difficulty is to determine the route of interest of those biosynthesis genes. Candidate gene identification was significantly facilitated by the increasing availability of sequences of plant genomes and by the use of transcriptome and co-expressional analyses data (Zerbe et al., 2013). With the convergence of genomic and proteomic tools, we can elucidate the path for the biosynthesis of these important metabolites, biochemical application of important herbal compounds (Zhao et al., 2013). Transient heterologous gene expression of synthetic enzymes in *Nicotiana benthamiana* has been experimentally designed and optimized; it is a simple and rapid tool which allows studies of metabolic pathways (Voinnet et al., 2003; Takos et al., 2011; Møldrup et al., 2013).

The fact that the biosynthetic genes coexist into the genome, their observation is essential in discovering the biosynthetic cluster of genes for creating a specific plant chemical defence (Chu et al., 2011; Takos & Rook, 2012). While the cluster of genes comprises of many non-homologous genes which encode for various enzymes required for the synthesis of specific effectors or metabolites. In the knowledge and availability of genome sequence along with genes of the biosynthetic pathway, the identification of candidate genes can be achieved with genetic colocalization. For example, Takos et al. (2011) used the first gene for genomic colocalization of the second gene required for the synthesis of cyanogenic glucosides in *Lotus japonicus*. Thus, the order of genes in a genomic cluster enables the identification of related genes in the pathway.

The Regulation and Coordination of Gene Expression

The improvement in biotechnology and development of bioinformatics tools have provided easier prediction and development of plant-based gene expression systems. For example, artificial gene clusters or specific transcription factors can be used to coordinate the regulation of natural regulatory processes in plants (Jirschitzka et al., 2013). However, the engineering of metabolic processes for pathogen resistance and maximizing the production of the high-value compound differ in the fact that it requires metabolite production at the time of infection or pest attack. Constant production of defensive metabolites is undesirable as increased levels can have detrimental effects on the plant.

The extensive study on stilbene synthases demonstrates that pathogen-induced expression systems are beneficial over constitutive expression gene expression. When transgenic tobacco was introduced with constitutively expressing stilbene synthase gene, it led to undesirable effects on the plant such as fewer pollen production and dullness in floral colour were observed (Fischer et al., 1997). Transgenic strawberries and *Arabidopsis* have been reported with unintended modifications in phenylpropanoid metabolism arising as a result of stilbene synthase gene (Hanhineva et al., 2009; Park et al., 2021). However, the expression of metabolite production can be effectively controlled by the use of synthetic promoters, which can be selectively upregulated by using plant signalling molecules such as jasmonic acid, salicylic acid or ethylene (Liu et al., 2013b).

The simultaneous expression of numerous genes of pathway presents additional issues since the usage of the same promoter sequences might lead to silencing of the gene(s), thus affecting the metabolic pathway. The methods used to bypass this issue are by employing technologies such as synthetic promoters, transcription factors that regulate multiple genes for a pathway, synthetic gene clusters and polyprotein technology (Liu et al., 2013a). Polyproteins consist of distinct proteins covalently linked with each other, which have different functions. It contains a single promoter fused to multienzyme coding sequences separated by an oligopeptide sequence. During the translation of transcript, polyprotein is processed and contains individual enzyme subunits linked together covalently. This approach was used to successfully generated glucosinolate in the tobacco plant (Geu-Flores et al., 2009).

Transcription factors regulate the metabolic pathways by controlling gene expression. These are very much useful in the engineering of metabolic pathways and have been documented as a bioengineering tool. For example, phenylpropanoid metabolism was controlled using transcription factors such as AtMYB12 factor in *Arabidopsis* for the upregulation of multiple genes, and isoflavone synthase gene GmIFS1 soybean was also upregulated with a higher isoflavonoid production (Pandey et al., 2014). For certain biosynthetic gene clusters, highly coordinated gene expression systems have been discovered (Chu et al., 2011). A multigene cassette system has been employed to artificially facilitate the naturally occurring gene cluster sequences (Jirschitzka et al., 2013). The synthetic cluster of genes allows pathways to be introduced via single transformation process which is essential to create a genetically stable transgenic pathway regulated by non-homologous regulatory sequences.

Engineered Transportation and Metabolite Localization

Metabolism processes are compartmentalized within the plant cell; the localization of metabolites is highly specific and is controlled by the pathway system (Heinig et al., 2013; Jirschitzka et al., 2013). In plants, two different pathways for isoprenoid synthesis are present: cytosolic-mevalonic acid and plastid-localized methylerythritol phosphate pathways. The combination of farnesyl diphosphate synthase and patchoulol synthase in each route resulted in increased synthesis of terpenes in transgenic tobacco plants (Wu et al., 2006). Investigations demonstrated that the intermediates between the cytosolic and plastid routes had minimum exchanges between them. Specific enzymes in organelles and the presence of substrate or transporters are critical for the metabolic pathway. By engineering transporter proteins, crop improvement could be achieved by regulating the distribution of plan tissue-specific defensive phytometabolites. Studies on nitrate and peptide transporter gene family in *Arabidopsis* suggested that by impairing their function in source tissues, glucosinolates were not accumulated in seeds (Nour-Eldin et al., 2012). Accumulation of glucosinolates was also reduced in mustard (*Brassica napus*) seeds by disturbing specific transporter function. It is reported that such specific transporter, when engineered, might remove antinutrient compounds or metabolites from edible crops without affecting the defensive system present in plants.

New Technologies of Genome Editing

In recent years, various new technologies have arisen that permit accurate editing of the DNA of an organism. The targeted genome editing techniques take advantage of engineered nucleases. ZFNs and TALENs are engineered DNA-binding chimeric proteins which are designed accordingly to the sequence which needs to be edited. Upon specific binding dimerization, they have a nuclease domain attached to them which is responsible for cleavage (Gaj et al., 2013). CRISPR-Cas systems use guide RNA for specific binding with DNA and its removal by Cas nuclease. All these genetic modification tools have been used successfully for the desired editing in plant genome (Feng et al., 2014). Once the genocidal breaks are generated, they need to be repaired by the HDR pathway or homologous recombination, thus creating indels that allow to knock out the desired gene or can be modified according to the repair template provided.

These new approaches have now been used to provide resistance against diseases in plants. TALEN-based rice generation has been achieved. The promoter region of the sucrose-efflux transporter was deleted in the Os11N3 genes, thus *Xanthomonas oryzae* was unable to use its transporter for its entry into the rice plant (Li et al., 2012). ZFNs were used to target the TRANSPARENT TESTA 4 gene, which altered the plant metabolism to produce chalcone synthase to produce chalcones that act as a natural defence metabolite. Citrus resistance to *Xanthomonas citri* (citrus canker) was developed by CRISPR-Cas9 gene editing involving the editing of the promoter of the CsLOB1 gene leading to a more resistant citrus plant (Peng et al., 2017). Genome editing tools can be aimed to engineer metabolic pathway by altering biosynthetic enzymes involved in the cascade of reactions to redirect them according to their specificity and to change expression of metabolite synthesis. These genetic modifications imitate the natural development of plant chemicals in response to pathogenic stresses and herbivores. Change as tiny as single amino acid may result in new plant chemical defence responses, unproductive enzyme-substrate complex or decreased product specificity (Khersonsky & Tawfik, 2010; Lai et al., 2014).

Furthermore, enzymes may be explored by their mutational studies to alter enzymatic characteristics in order to develop new metabolites of interest (O'Maille et al., 2008). This indicates that extensive understanding of the biochemistry and regulation mechanism of desirable enzymes permit changes in the patterns of expression or the activity of enzymes and hence affect the chemical defence response of a plant. The technique for genome editing is considered a safe method for crop improvement. These systems require to produce marker-free and genetically improved plants. The current processes based on regulatory framework for the introduction of genetically engineered plants via genome-editing technologies into the EU have been discussed and suggest that risk analysis or regulations must be based on features introduced in transgenes and not on technologies used in their production (Hartung & Schiemann, 2014).

Trait Stacking

'Trait stacking' means the introduction of many desired genetic modifications or features into a single cultivar in a transgene crop (Halpin, 2005). The stacking characteristics of one single crop in the market are generally imparted by combining insect resistance, viral resistance and all the other important characteristics of individual genes. Trait stacking is useful as it defends against various infections. These restrictions in the protection are governed by resistance genes or by chemicals produced. It also reflects about the coevolution of pathogen host plants and also the disease-resistant features of the many classical breeds (Jones et al., 2014). The enhanced resistance against leaf pathogen black sigatoka or *Mycosphaerella fijiensis* by Cavendish bananas is an example of feature stacking involving plant chemical defence compounds (Vishnevetsky et al., 2011). In a field trial study conducted in Florida, black sigatoka was increased by a combined production of an endochitinase and stilbene synthase genes under the pathogenic-induced promoter and a superoxide dismutase gene (Vishnevetsky et al., 2011). Transgenic apple or peas were transformed with polygalacturonase-inhibiting protein and stilbene synthase which showed fungal resistance (Richter et al., 2006; Szankowski et al., 2003).

Position for Metabolic Engineering of Disease Resistance in Various Crops

Progress in plant science and technology improvements is permitted effective metabolic engineering of plant-based phytochemical disease resistance. Transgenic crops are placed on the market, with many technological hurdles, economic and political concerns, contentious public discussions concerning genetically engineered crops and worries about the company's business interests. Although the scientific agreement on biological safety for crop improvement was held (European

Academies Science Advisory Council, 2013), worldwide non-governmental organizations have substantially affected public opinion and government policy (Dunwell, 2014). The next generation is expected to support and provide public acceptability to genetically modified crops, as they bring benefits to consumers (Chen & Lin, 2013). The horticulture sector produces only commercially grown transgenic plants with changes to plant-specific metabolism that has a consumer-oriented characteristic.

Dianthus caryophyllus with different floral hues is vegetatively propagated and marketed in the European market. Many ornamental plants have blue-coloured anthocyanin derived from dihydroflavonol 4-reductase and flavonoid 3′5′hydroxylase coding genes of *Petunia* spp. (Fukui et al., 2003). Other features, in addition to flowering, floral aroma, are the morphology of the flora and plants, delayed senescence of cut flowers and resistance of diseases (Tanaka et al., 2005). It has been observed that features which help customers and not features that help producers in the horticulture business are profit-oriented and marketable. The use of crop protection chemicals is particularly expensive in the context of the prevention of plant diseases, which means that broad-spectrum compounds are preferred and employed in horticulture crops for many common pathogenic products (Tanaka et al., 2005).

Transgenic Crops

Their commercial production of crops with improved plant defence was a failure as it lacked offers of efficient degree of protection. Also, the easy availability and effectiveness of agrochemicals did not provide much attention for consumers. However, it is quite possible that because of the health benefits provided by such functional foods, it may rather bring customers' attention than disease prevention by these transgenic crops producing resveratrol (Baek et al., 2013; Giovinazzo et al., 2012). In addition, resveratrol has been related to the Sir2 activation of proteins and the increase of the lifespan of this antioxidant in the protection of cardiovascular disease (Wood et al., 2004; Sinclair & Guarente, 2014). The goal of metabolic engineering for polyphenols, like anthocyanins, to have positive effects on the prevention of cardiovascular and obesity-related chronic illnesses has also become a point of large interest. For example, increased anthocyanin content of transgenic tomato provided prolonged shelf-life and lowered the sensitivity to *Botrytis cinerea* infection after harvest along with the goodness of bioactive medicinal properties (Zhang et al., 2013). The metabolism of marine organisms has been entirely studied because of the benefits of omega-3 long-chain polyunsaturated fatty acids and has been successfully introduced to plant seed oil (Haslam et al., 2013). As studies on the metabolism of crops with beneficial properties for health are progressing recently, such transgenic plants might well lead the way for more acceptability in the public in the future years (Chen & Lin, 2013).

Conclusion

Plants have evolved their metabolism to assist them in survival such that they produce a wide diversity of phytochemicals having immense biological properties. These phytochemicals or metabolite-based stimuli have enhanced their biochemical properties, such as the production of many broad-spectrum biocidal chemicals to resist herbivores or pathogens. Therefore, the engineering of a specific metabolite and the approaches require to provide a long-term resistance pathogen. By gaining appropriate knowledge and increasing advances in modern technologies, one can engineer any desirable molecule. New technology options for biochemical analysis, DNA sequencing, bioinformatics, gene expression technique and, more recently, the possibility of genetic editing are progressively enabling the engineering of plant metabolism. The commercialization of bioengineered crops to create a plant defensive metabolite is primarily dependent on the economic benefits that these crops offer to farmers. In particular, the amount to which various cultures, such as those linked with agrochemicals, minimizes production and environmental costs, raises the total crop return or otherwise gives an added value to the product. Trait stacking provides scope and efficacy in every transgenic crop with multiple qualities such as resistance to many types of diseases and pests and also providing fitness to plants to resist abiotic stress. Additional advantages can be achieved by combining systems with both consumer- and farmer-oriented characteristics.

References

Ahuja, I., Kissen, R., & Bones, A. M. (2012). Phytoalexins in defense against pathogens. *Trends in Plant Science, 17*(2), 73–90.

Angelova, S., Buchheim, M., Frowitter, D., Schierhorn, A., & Roos, W. (2010). Overproduction of alkaloid phyto-alexins in California poppy cells is associated with the co-expression of biosynthetic and stress-protective enzymes. *Molecular Plant, 3*(5), 927–939.

Armah, C. N., Mackie, A. R., Roy, C., Price, K., Osbourn, A. E., Bowyer, P., & Ladha, S. (1999). The membrane-permeabilizing effect of avenacin A-1 involves the reorganization of bilayer cholesterol. *Biophysical Journal, 76*(1 Pt 1), 281–290.

Arneson, P. A., & Durbin, R. D. (1968). Sensitivity of fungi to a-tomatine. *Phytopathology, 58*, 536–537.

Ashihara, H., Sano, H., & Crozier, A. (2008). Caffeine and related purine alkaloids: Biosynthesis, catabolism, function and genetic engineering. *Phytochemistry, 69*(4), 841–856.

Baek, S. H., Shin, W.-C., Ryu, H.-S., Lee, D.-W., Moon, E., Seo, C.-S., Hwang, E., Lee, H. S., Ahn, M. H., Jeon, Y., & Kang, H. J. (2013). Creation of resveratrol-enriched rice for the treatment of metabolic syndrome and related diseases. *PLoS One, 8*(3), e579300.

Bednarek, P. (2012). Chemical warfare or modulators of defence responses – The function of secondary metabolites in plant immunity. *Current Opinion in Plant Biology, 15*, 407–414.

Beets, C. A., Huang, J. C., Madala, N. E., & Dubery, I. (2012). Activation of camalexin biosynthesis in *Arabidopsis thaliana* in response to perception of bacterial lipopolysaccharides: A gene-to-metabolite study. *Planta, 236*(1), 261–272.

Bouarab, K., Melton, R., Peart, J., Baulcombe, D., & Osbourn, A. (2002). A saponin detoxifying enzyme mediates suppression of plant defences. *Nature, 418*, 889–892.

Brown, K. K., & Hampton, M. B. (2011). Biological targets of isothiocyanates. *Biochimica et Biophysica Acta, 1810*(9), 888–894.

Chen, H., & Lin, Y. (2013). Promise and issues of genetically modified crops. *Current Opinion in Plant Biology, 16*(2), 255–260.

Chong, J., Poutaraud, A., & Hugueney, P. (2009). Metabolism and roles of stilbenes in plants. *Plant Science, 177*(3), 143–155.

Chu, H. Y., Wegel, E., & Osbourn, A. (2011). From hormones to secondary metabolism: The emergence of metabolic gene clusters in plants. *Plant Journal: For Cell and Molecular Biology, 66*(1), 66–79.

Dakora, F. D., & Phillips, D. A. (1996). Diverse functions of isoflavonoids in legumes transcend anti-microbial definitions of phytoalexins. *Physiological and Molecular Plant Pathology, 49*(1), 1–20.

De Luca, V., Salim, V., Atsumi, S. M., & Yu, F. (2012). Mining the biodiversity of plants: A revolution in the making. *Science, 336*(6089), 1658–1661.

Delaunois, B., Cordelier, S., Conreux, A., Clément, C., & Jeandet, P. (2009). Molecular engineering of resveratrol in plants. *Plant Biotechnology Journal, 7*(1), 2–12.

Desgagné-Penix, I. (2021). Biosynthesis of alkaloids in Amaryllidaceae plants: A review. *Phytochemistry Reviews, 20*, 409–431.

Dixon, D. P., Sellars, J. D., Kenwright, A. M., & Steel, P. G. (2012). The maize benzoxazinone DIMBOA reacts with glutathione and other thiols to form spirocyclic adducts. *Phytochemistry, 77*, 171–178.

Dunwell, J. M. (2014). Genetically modified (GM) crops: European and transatlantic divisions. *Molecular Plant Pathology, 15*(2), 119–121.

European Academies Science Advisory Council. (2013). Planting the future: Opportunities and challenges for using crop genetic improvement technologies for sustainable agriculture. *EASAC Policy Report, 21*. Retrieved from http://www.easac.eu

Feng, Z., Mao, Y., Xu, N., Zhang, B., Wei, P., Yang, D. L., Wang, Z., Zhang, Z., Zheng, R., Yang, L., Zeng, L., & Zhu, J. K. (2014). Multigeneration analysis reveals the inheritance, specificity, and patterns of CRISPR/Cas-induced gene modifications in Arabidopsis. *Proceedings of the National Academy of Sciences of the United States of America, 111*(12), 4632–4637.

Fischer, R., Budde, I., & Hain, R. (1997). Stilbene synthase gene expression causes changes in flower colour and male sterility in tobacco. *Plant Journal, 11*(3), 489–498.

Fukui, Y., Tanaka, Y., Kusumi, T., Iwashita, T., & Nomoto, K. (2003). A rationale for the shift in colour towards blue in transgenic carnation flowers expressing the flavonoid 3′,5′-hydroxylase gene. *Phytochemistry, 63*(1), 15–23.

Gaj, T., Gersbach, C. A., & Barbas, C. F. (2013). ZFN, TALEN, and CRISPR/Cas-based methods for genome engineering. *Trends in Biotechnology, 31*(7), 397–405.

Geu-Flores, F., Nielsen, M. T., Nafisi, M., Møldrup, M. E., Olsen, C. E., Motawia, M. S., & Halkier, B. A. (2009). Glucosinolate engineering identifies a γ-glutamyl peptidase. *Nature Chemical Biology, 5*(8), 575–577.

Giorcelli, A., Sparvoli, F., Mattivi, F., Tava, A., Balestrazzi, A., Vrhovsek, U., Calligari, P., Bollini, R., & Confalonieri, M. (2004). Expression of the stilbene synthase (StSy) gene from grapevine in transgenic white poplar results in high accumulation of the antioxidant resveratrol glucosides. *Transgenic Research, 13*(3), 203–214.

Giovinazzo, G., Ingrosso, I., Paradiso, A., De Gara, L., & Santino, A. (2012). Resveratrol biosynthesis: Plant metabolic engineering for nutritional improvement of food. *Plant Foods for Human Nutrition, 67*(3), 191–199.

Glawischnig, E., Hansen, B. G., Olsen, C. E., & Halkier, B. A. (2004). Camalexin is synthesized from indole-3-acetaldoxime, a key branching point between primary and secondary metabolism in Arabidopsis. *Proceedings of the National Academy of Sciences of the United States of America, 101*(21), 8245–8250.

Glenn, W. S., Runguphan, W., & O'Connor, S. E. (2013). Recent progress in the metabolic engineering of alkaloids in plant systems. *Current Opinion in Biotechnology, 24*(2), 354–365.

Großkinsky, D. K., van der Graaff, E., & Roitsch, T. (2012). Phytoalexin transgenics in crop protection—Fairy tale with a happy end? *Plant Science, 195*, 54–70.

Hain, R., Reif, H. J., Krause, E., Langebartels, R., Kindl, H., Vornam, B., & Stöcker, R. H. (1993). Disease resistance results from foreign phytoalexin expression in a novel plant. *Nature, 361*(6408), 153–156.

Halpin, C. (2005). Gene stacking in transgenic plants—The challenge for the 21st century plant biotechnology. *Plant Biotechnology Journal, 3*(2), 141–155.

Hanhineva, K., Kokko, H., Siljanen, H., Rogachev, I., Aharoni, A., & Kärenlampi, S. O. (2009). Stilbene synthase gene transfer caused alterations in the phenylpropanoid metabolism of transgenic strawberry (Fragaria × ananassa). *Journal of Experimental Botany, 60*(7), 2093–2106.

Hartung, F., & Schiemann, J. (2014). Precise plant breeding using new genome editing techniques: Opportunities, safety and regulation in the EU. *Plant Journal: For Cell and Molecular Biology, 78*(5), 742–752.

Haslam, R. P., Ruiz-Lopez, N., Eastmond, P., Moloney, M., Sayanova, O., & Napier, J. A. (2013). The modification of plant oil composition via metabolic engineering—Better nutrition by design. *Plant Biotechnology Journal, 11*(2), 157–168.

He, X. Z., & Dixon, R. A. (2000). Genetic manipulation of isoflavone 7-O-methyltransferase enhances biosynthesis of 4′-O-methylated isoflavonoid phytoalexins and decease resistance in alfalfa. *Plant Cell, 12*(9), 1689–1702.

Heinig, U., Gutensohn, M., Dudareva, N., & Aharoni, A. (2013). The challenges of cellular compartmentalization in plant metabolic engineering. *Current Opinion in Biotechnology, 24*(2), 239–246.

Inagaki, Y. S., Etherington, G., Geisler, K., Field, B., Dokarry, M., Ikeda, K., Mutsukado, Y., Dicks, J., & Osbourn, A. (2011). Investigation of the potential for triterpene synthesis in rice through genome mining and metabolic engineering. *New Phytologist, 191*(2), 432–448.

Iwuala, E., Odjegba, V., Unung, O., & Alam, A. (2020). Expression of stress responsive β-1,3-glucanase and chitinase genes in Arachis hypogaea seedlings against Macrophomina phaseolina. *Gene Reports*, 100693.

Jeandet, P., Hébrard, C., Deville, M. A., Cordelier, S., Dorey, S., Aziz, A., & Crouzet, J. (2014). Deciphering the role of phytoalexins in plant-microorganism interactions and human health. *Molecules (Basel, Switzerland), 19*(11), 18033–18056.

Jirschitzka, J., Mattern, D. J., Gershenzon, J., & D'Auria, J. C. (2013). Learning from nature: New approaches to the metabolic engineering of plant defense pathways. *Current Opinion in Biotechnology, 24*(2), 320–328.

Jones, J. D. G., Witek, K., Verweij, W., Jupe, F., Cooke, D., Dorling, S., Tomlinson, L., Smoker, M., Perkins, S., & Foster, S. (2014). Elevating crop disease resistance with cloned genes. *Philosophical Transactions of the Royal Society of London Series B, Biological Sciences, 369*(1639), 20130087.

Jørgensen, K., Rasmussen, A. V., Morant, M., Nielsen, A. H., Bjarnholt, N., Zagrobelny, M., Bak, S., & Møller, B. L. (2005). Metabolon formation and metabolic channeling in the biosynthesis of plant natural products. *Current Opinion in Plant Biology, 8*(3), 280–291.

Khersonsky, O., & Tawfik, D. S. (2010). Enzyme promiscuity: A mechanistic and evolutionary perspective. *Annual Review of Biochemistry, 79*, 471–505.

Kim, Y. S., & Sano, H. (2008). Pathogen resistance of trans-genic tobacco plants producing caffeine. *Phytochemistry, 69*(4), 882–888.

Kuśnierczyk, A., Winge, P., Jørstad, T. S., Troczyńska, J., Rossiter, J. T., & Bones, A. M. (2008). Towards global understanding of plant defence against aphids—Timing and dynamics of early Arabidopsis defence responses to cabbage aphid (Brevicoryne brassicae) attack. *Plant, Cell and Environment, 31*(8), 1097–1115.

Lai, D., Abou Hachem, M., Robson, F., Olsen, C. E., Wang, T. L., Møller, B. L., Takos, A. M., & Rook, F. (2014). The evolutionary appearance of non-cyanogenic hydroxynitrile glucosides in the Lotus genus is accompanied by the substrate specialisation of paralogous β- glucosidases resulting from a crucial amino acid substitution. *Plant Journal: For Cell and Molecular Biology, 79*(2), 299–311.

Li, T., Liu, B., Spalding, M. H., Weeks, D. P., & Yang, B. (2012). High-efficiency TALEN-based gene editing produces disease-resistant rice. *Nature Biotechnology, 30*(5), 390–392.

Liu, W., Mazarei, M., Rudis, M. R., Fethe, M. H., Peng, Y., Millwood, R. J., Schoene, G., Burris, J. N., & Stewart, C. N. (2013b). Bacterial pathogen phytosensing in trans- genic tobacco and Arabidopsis plants. *Plant Biotechnology Journal, 11*(1), 43–52.

Liu, W., Yuan, J. S., & Stewart, C. N. (2013a). Advanced genetic tools for plant biotechnology. *Nature Reviews. Genetics, 14*(11), 781–793.

Martin, C. (2013). The interface between plant metabolic engineering and human health. *Current Opinion in Biotechnology, 24*(2), 344–353.

Møldrup, M. E., Salomonsen, B., Geu-Flores, F., Olsen, C. E., & Halkier, B. A. (2013). De novo genetic engineering of the camalexin biosynthetic pathway. *Journal of Biotechnology, 167*(3), 296–301.

Muroi, A., Matsui, K., Shimoda, T., Kihara, H., Ozawa, R., Ishihara, A., Nishihara, M., & Arimura, G.-I. (2012). Acquired immunity of transgenic Torenia plants overexpressing agmatine coumaroyltransferase to pathogens and herbivore pests. *Scientific Reports, 2*, 689.

Nielsen, K. A., Gotfredsen, C. H., Buch-Pedersen, M. J., Ammitzbøll, H., Mattsson, O., Duus, J. Ø., & Nicholson, R. L. (2004). Inclusions of flavonoid 3-deoxyanthocyanidins in Sorghum bicolor self-organize into spherical structures. *Physiological and Molecular Plant Pathology, 65*(4), 187–196.

Nour-Eldin, H. H., Andersen, T. G., Burow, M., Madsen, S. R., Jørgensen, M. E., Olsen, C. E., Dreyer, I., Hedrich, R., Geiger, D., & Halkier, B. A. (2012). NRT/PTR transporters are essential for translocation of glucosinolate defence compounds to seeds. *Nature, 488*(7412), 531–534.

O'Maille, P. E., Malone, A., Dellas, N., Andes Hess, B. A., Smentek, L., Sheehan, I., Greenhagen, B. T., Chappell, J., Manning, G., & Noel, J. P. (2008). Quantitative exploration of the catalytic landscape separating divergent plant sesquiterpene synthases. *Nature Chemical Biology, 4*(10), 617–623.

Okazaki, Y., Isobe, T., Iwata, Y., Matsukawa, T., Matsuda, F., Miyagawa, H., Ishihara, A., Nishioka, T., & Iwamura, H. (2004). Metabolism of avenanthramide phytoalexins in oats. *The Plant Journal, 39*(4), 560–572.

Ökmen, B., Etalo, D. W., Joosten, M., Bouwmeester, H. J., de Vos, R., Collemare, J., & de Wit, P. (2013). Detoxification of α-tomatine by *Cladosporium fulvum* is required for full virulence on tomato. *The New Phytologist, 198*(4), 1203–1214.

Pan, L. P., Yu, S. L., Chen, C. J., Li, H., Wu, Y. L., & Li, H. H. (2012). Cloning a peanut resveratrol synthase gene and its expression in purple sweet potato. *Plant Cell Reports, 31*(1), 121–131.

Pandey, A., Misra, P., Khan, M. P., Swarnkar, G., Tewari, M. C., Bhambhani, S., Trivedi, R., Chattopadhyay, N., & Trivedi, P. K. (2014). Co-expression of Arabidopsis transcription factor, AtMYB12, and soybean isoflavone synthase, GmIFS1, genes in tobacco leads to enhanced biosynthesis of isoflavones and flavonols resulting in osteoprotective activity. *Plant Biotechnology Journal, 12*(1), 69–80.

Park, S. C., Pyun, J. W., Jeong, Y. J., Park, S. H., Kim, S., Kim, Y. H., Lee, J. R., Kim, C. Y., & Jeong, J. C. (2021). Overexpression of VlPRX21 and VlPRX35 genes in Arabidopsis plants leads to bioconversion of trans-resveratrol to δ-viniferin. *Plant Physiology and Biochemistry, 162*, 556–563.

Paxton, J. (1980). A new working definition of the term "phytoalexin". *Plant Disease, 64*, 734.

Peng, A., Chen, S., Lei, T., Xu, L., He, Y., Wu, L., Yao, L., & Zou, X. (2017). Engineering canker-resistant plants through CRISPR/Cas9-targeted editing of the susceptibility gene CsLOB1 promoter in citrus. *Plant Biotechnology Journal, 15*(12), 1509–1519. https://doi.org/10.1111/pbi.12733/full

Perez, F. J., & Niemeyer, H. M. (1989). Reaction of DIMBOA with amines. *Phytochemistry, 28*, 1831–1834.

Richter, A., Jacobsen, H. J., de Kathen, A., de Lorenzo, G., Briviba, K., Hain, R., Ramsay, G., & Kiesecker, H. (2006). Transgenic peas (Pisum sativum) expressing polygalacturonase inhibiting protein from raspberry (Rubus idaeus) and stilbene synthase from grape (*Vitis vinifera*). *Plant Cell Reports, 25*(11), 1166–1173.

Schmelz, E. A., Huffaker, A., Sims, J. W., Christensen, S. A., Lu, X., Okada, K., & Peters, R. J. (2014). Biosynthesis, elicitation and roles of monocot terpenoid phytoalexins. *Plant Journal: For Cell and Molecular Biology, 79*(4), 659–678.

Schwekendiek, A., Spring, O., Heyerick, A., Pickel, B., Pitsch, N. T., Peschke, F., De Keukeleire, D., & Weber, G. (2007). Constitutive expression of a grapevine stilbene synthase gene in transgenic hop (*Humulus lupulus* L.) yields resveratrol and its derivatives in substantial quantities. *Journal of Agricultural and Food Chemistry, 55*(17), 7002–7009.

Seppänen, S. K., Syrjälä, L., von Weissenberg, K., Teeri, T. H., Paajanen, L., & Pappinen, A. (2004). Antifungal activity of stilbenes in in vitro bioassays and in transgenic Populus expressing a gene encoding pinosylvin synthase. *Plant Cell Reports, 22*(8), 584–593.

Sinclair, D. A., & Guarente, L. (2014). Small-molecule allosteric activators of sirtuins. *Annual Review of Pharmacology and Toxicology, 54*, 363–380.

Singh, B., & Sharma, R. A. (2015). Plant terpenes: Defense responses, phylogenetic analysis, regulation and clinical applications. *3 Biotech, 5*(2), 129–151.

Smoliga, J. M., Baur, J. A., & Hausenblas, H. A. (2011). Resveratrol and health—A comprehensive review of human clinical trials. *Molecular Nutrition and Food Research, 55*(8), 1129–1141.

Szankowski, I., Briviba, K., Fleschhut, J., Schönherr, J., Jacobsen, H. J., & Kiesecker, H. (2003). Transformation of apple (Malus domestica Borkh.) with the stilbene synthase gene from grapevine (*Vitis vinifera* L.) and a PGIP gene from kiwi (*Actinidia deliciosa*). *Plant Cell Reports, 22*(2), 141–149.

Takos, A. M., Knudsen, C., Lai, D., Kannangara, R., Mikkelsen, L., Motawia, M. S., Olsen, C. E., Sato, S., Tabata, S., Jørgensen, K., Møller, B. L., & Rook, F. (2011). Genomic clustering of cyanogenic glucoside bio-synthetic genes aids their identification in Lotus japonicus and suggests the repeated evolution of this chemical defence pathway. *Plant Journal: For Cell and Molecular Biology, 68*(2), 273–286.

Takos, A. M., & Rook, F. (2012). Why biosynthetic genes for chemical defense compounds cluster. *Trends in Plant Science, 17*(7), 383–388.

Tanaka, Y., Katsumoto, Y., Brugliera, F., & Mason, J. (2005). Genetic engineering in floriculture. *Plant Cell, Tissue and Organ Culture, 80*(1), 1–24.

Thomzik, J. E., Stenzel, K., Stöcker, R., Schreier, P. H., Hain, R., & Stahl, D. J. (1997). Synthesis of a grapevine phytoalexin in transgenic tomatoes (Lycopersicon esculentum Mill.) conditions resistance against *Phytophthora infestans*. *Physiological and Molecular Plant Pathology, P51*, 265–278.

Tierens, K. F., Thomma, B. P., Brouwer, M., Schmidt, J., Kistner, K., Porzel, A., Mauch-Mani, B., Cammue, B. P., & Broekaert, W. F. (2001). Study of the role of antimicrobial glucosinolate-derived isothiocyanates in resistance of Arabidopsis to microbial pathogens. *Plant Physiology, 125*(4), 1688–1699.

VanEtten, H., Temporini, E., & Wasmann, C. (2001). Phytoalexin (and phytoanticipin) tolerance as a Virulence trait: Why is it not required by all pathogens? *Physiological and Molecular Plant Pathology, P59*, 83–93.

VanEtten, H. D., Mansfield, J. W., Bailey, J. A., & Farmer, E. E. (1994). Two classes of plant antibiotics: Phytoalexins versus "phytoanticipins". *Plant Cell, 6*(9), 1191–1192.

Vishnevetsky, J., White, T. L., Palmateer, A. J., Flaishman, M., Cohen, Y., Elad, Y., Velcheva, M., Hanania, U., Sahar, N., Dgani, O., & Perl, A. (2011). Improved tolerance toward fungal diseases in transgenic Cavendish banana (Musa spp. AAA group) cv. Grand Nain. *Transgenic Research, 20*(1), 61–72.

Voinnet, O., Rivas, S., Mestre, P., & Baulcombe, D. (2003). An enhanced transient expression system in plants based on suppression of gene silencing by the p19 protein of tomato bushy stunt virus. *Plant Journal: For Cell and Molecular Biology, 33*(5), 949–956.

Wittstock, U., & Burow, M. (2010). Glucosinolate breakdown in Arabidopsis: Mechanism, regulation and biological significance. *The Arabidopsis Book, 8*, e0134.

Wood, J. G., Rogina, B., Lavu, S., Howitz, K., Helfand, S. L., Tatar, M., & Sinclair, D. (2004). Sirtuin activators mimic caloric restriction and delay ageing in metazoans. *Nature, 430*(7000), 686–689.

Wu, S., Schalk, M., Clark, A., Miles, R. B., Coates, R., & Chappell, J. (2006). Redirection of cytosolic or plastic isoprenoid precursors elevates terpene production in plants. *Nature Biotechnology, 24*(11), 1441–1447.

Yang, T., Stoopen, G., Yalpani, N., Vervoort, J., de Vos, R., Voster, A., Verstappen, F. W., Bouwmeester, H. J., & Jongsma, M. A. (2011). Metabolic engineering of geranic acid in maize to achieve fungal resistance is compromised by novel glycosylation patterns. *Metabolic Engineering, 13*(4), 414–425.

Yu, O., Jung, W., Shi, J., Croes, R. A., Fader, G. M., McGonigle, B., & Odell, J. T. (2000). Production of the isoflavones genistein and daidzein in non-legume dicot and monocot tissues. *Plant Physiology, 124*(2), 781–794.

Zerbe, P., Hamberger, B., Yuen, M. M. S., Chiang, A., Sandhu, H. K., Madilao, L. L., Nguyen, A., Hamberger, B., Bach, S. S., & Bohlmann, J. (2013). Gene discovery of modular diterpene metabolism in nonmodel systems. *Plant Physiology, 162*(2), 1073–1091.

Zhang, Y., Butelli, E., De Stefano, R., Schoonbeek, H. J., Magusin, A., Pagliarani, C., Wellner, N., Hill, L., Orzaez, D., Granell, A., Jones, J. D., & Martin, C. (2013). Anthocyanins double the shelf life of tomatoes by delaying overripening and reducing susceptibility to gray mold. *Current Biology, 23*(12), 1094–1100.

Zhao, N., Wang, G., Norris, A., Chen, X., & Chen, F. (2013). Studying plant secondary metabolism in the age of genomics. *Critical Reviews in Plant Sciences, 32*(6), 369–382.

Ziegler, J., & Facchini, P. J. (2008). Alkaloid biosynthesis: Metabolism and trafficking. *Annual Review of Plant Biology, 59*, 735–769.

Chapter 6
Role of Metabolic Engineering in Enhancing Crop Nutritional Quality

Irshad Ahmad Bhat, Khushboo Guleria, Mudasir Fayaz, Roof-Ul-Qadir, Tareq Ahmad Wani, Jasfeeda Qadir, and Zahoor A. Kaloo

Abstract Human health faces one of the grave issues from the nutrition deficiency in crops. Therefore, it should be given prime importance to develop crops for subsiding this issue. In order to improve the nutritional quality of crops, biofortification aids in combatting nutrient deficiency by enhancing the nutrient-specific component levels. Plants have various types of complex metabolic pathways specific for the synthesis of particular nutrient compounds as well as other secondary metabolites. Plant metabolic engineering aids in the modulation of these specific biosynthetic pathways so as to introduce the new metabolic abilities by either downgrading or upgrading any particular enzyme according to the requirement for amplifying the nutritional quality. Progress and innovations made in the sequencing and bioinformatics will provide an ample opportunity about metabolic pathways and discover the best intervention point. Knowledge attained about vitamin metabolism, uptake of minerals, and relocation in plant makes it feasible today to increase micronutrient level in crop plants. While making comparison of different agronomic practices and conventional plant breeding with plant metabolic engineering and synthetic biological strategies, the latter ones have found to be more effective in synthesizing the specific nutrients, bioactive components in crops. Technological advancement and progresses made in the field of plant metabolic engineering particularly in view of research strategies of multigene stacking tool, complex metabolic engineering with prime focus on refining characters or traits associated with micronutrients, phytonutrients, and bioactive compounds will ultimately help the human race in developing the crops to meet the nutritional demand.

I. A. Bhat (✉) · M. Fayaz · T. A. Wani · J. Qadir · Z. A. Kaloo
Plant Tissue Culture Research Laboratory, Department of Botany, University of Kashmir, Hazratbal, Srinagar, Jammu and Kashmir, India
e-mail: bhatirshad.scholar@kashmiruniversity.net

K. Guleria
Department of Zoology, School of Bioscience and Bioengineering, Lovely Professional University, Jalandhar, Punjab, India

Roof-Ul-Qadir
Plant Reproductive Biology, Genetic Diversity and Phytochemistry Research Laboratory, University of Kashmir, Hazratbal, Srinagar, Jammu and Kashmir, India

T. Aftab, K. R. Hakeem (eds.), *Metabolic Engineering in Plants*,
https://doi.org/10.1007/978-981-16-7262-0_6

Keywords Human health · Nutritional quality · Biofortification · Micronutrients

Introduction

Inadequate consumption of minerals and vitamins comes under nutrient malnutrition or hidden hunger. This situation has posed a detrimental influence on human health across the globe and has influenced almost two billion people worldwide with the highest pervasiveness in developing countries (WHO, 2002). Women and children during their reproductive age have an utmost vulnerability to this hidden hunger which ultimately resulted in problems in health, stunted growth, and various birth-related issues. Crop plants have been subjected to metabolic engineering to increase the nutrient content to combat malnutrition worldwide. Employing biotechnology is one of the capable techniques in this regard, and biofortification through metabolic engineering in elevating levels of nutrients is advocated as a fundamental way to reduce hidden hunger (Garg et al., 2018). Remarkable achievements have been achieved in the biofortification of different single micronutrients across an array of food crops particularly staples (Bouis & Saltzman, 2017).

Conventional breeding and genetic engineering are utilized for biofortification; however, genetic engineering has not been given full endorsement yet for release to farmers. Biofortification done through genetic engineering has the potential for the increased accumulation of micronutrients and is not limited by available germplasm variations. Multiple nutrients can be simultaneously amplified by genetic engineering as well as refining the vitamins' stability postharvesting, while also including agronomically important traits, such as enhanced yield and stress resilience. Biofortification through metabolic engineering provides a solution, when the naturally occurring disparity in sexually companionable germplasm is inadequate to get desired micronutrient level in a particular crop via conventional breeding. Owing to the fact of growing to understand vitamin metabolism and uptake of minerals and relocation in plants, it is nowadays probable to increase nutrient concentration in crop plants thereby giving rise to a sustainable solution to humans with a suboptimal micronutrient intake, since the human body is in continuous need of nutrients for the sustenance of different physiological processes going on in the body to assure good mental and physical health. However, these nutrients cannot be produced by the human body itself and are therefore completely dependent on diet as a nutrient source. Although staple crop works as a fine source of daily calorie intake, the amount of micronutrient present in them does not suffice the requirement. Besides this storage of food, processing of food and preparation led to considerable losses of nutrients. For instance, the maximum amount of micronutrients present in rice are concentrated around the outer layers of the kernel (aleurone layer), which is unfortunately removed during milling to evade rancidification. Milling of rice is a simple process to enhance shelf life but it degrades nutritional value. Engineered crops fall under three categories and it relies on the fact of source of genes inserted such as transgenic, intragenic, and cisgenic. So far, transgenic crops are concerned they have transgene from an organism with sexual incompatibility, whereas intragenic crops have genes from sexually compatible individuals, thereby permitting recombination of genes with sexually compatible gene pool. Crops with cisgenic nature recombination are not endorsed and inserted gene needs to hold its original organization in normal orientation. By utilizing promoters with tissue-specific characters driving expression of focused (Trans) genes, metabolic engineering aids in micronutrient accumulation in particular tissues where they may be absent and if present but in very low concentration. During these particular cases, conventional breeding has no role or very little to play, and engineering of metabolites is the only way forward.

Iron paucity is one of the prime nutritional ailments in the world, and this is only micronutrient deficiency being one of the most prevalent in developed countries as well. It is approximately assessed that about one-third of the world's population suffers at the hands of anemia, affecting mostly children and women (http://www.who.int/vmnis/database/anemia/en/). In addition, developing countries are experiencing its intensified effects because of several infectious diseases like HIV, malaria, and tuberculosis. Unsatisfactory intake of iron during pregnancy causes obstructions in the normal physical and cognitive development (http://www.who.int/nutrition/topics/media/en/). Metabolic engineering can be put forward as the process to progess towards the production of target compounds in vivo via modulation of one or more genes or gene networks (Farré et al., 2014; Fu et al., 2018). Designing and establishment of innovative biological pathways for the biosynthesis of novel and new compounds in organisms come under synthetic biology (Liu & Stewart Jr., 2015; Hanson & Jez, 2018). There is correspondence among these two fields, and the current development and production of new bioactive compounds entail a combination of both synthetic biology and metabolic engineering. This combined method can be called "synthetic metabolic engineering" (Nielsen & Keasling, 2011; Ye et al., 2012; Pouvreau et al., 2018). Synthetic metabolic engineering encompasses majorly three steps to rely upon: obtaining knowledge about metabolic pathways or genetic pathways occurring naturally in organisms; reconstruction of the artificial pathway in terms of design, assemble, and transform in target host organism; and finally going for testing the practical aspect of the reconstructed pathway in the transformed target organisms (Pouvreau et al., 2018; Mortimer, 2019). After the first cycle, the "learn" step analyzes the results of the cycle to enhance system-level knowledge and drive subsequent learn/reconstruction/test (L-R-T) cycles. The iterative L-R-T cycles promote the development of better synthetic metabolic engineering approaches. The development of Astaxanthin rice is one good instance for the use of this cyclic approach. Due to molecular biotechnological advancements, various novel research tools and techniques have been utilized to analyze metabolic pathway reconstruction in target organisms and ultimately it has resulted in the development of synthetic metabolic engineering (García-Granados et al., 2019).

Approaches for Metabolic Engineering and Synthetic Biology in Plants

Metabolic engineering in plants encompasses manipulation of one or more prime genes or gene-related to rate-limiting enzyme in the metabolic pathway, and in some cases all those genes make up the entire pathways. Consequently, an assortment of different prime genes and the method involved to deliver and express multiple genes in host plants will affect the development of plant synthetic metabolic engineering, which is discussed in the following two sections.

Strategies for Synthetic Metabolic Engineering

A metabolic pathway is a set of a reaction involving a number of enzymes in series causing conversion of substrate into the target product and likely by-products in a certain order. To enhance the synthesis of a target product, different approaches are generally used:

(1) Enhancing the expression of those upstream genes involved in encoding rate-limiting enzymes or several key enzymes in the pathway of the target to make sure that there is an adequate supply of precursors and upsurge in the metabolic flux through target pathway.
(2) Stopping expression of those enzyme genes taking part in the competitive pathway of branch point so that one can sidestep intermediates being abstracted and prevent target metabolite decomposition.
(3) Transcription factors or certain crucial enzyme genes overexpression for simultaneously activating various pathways related to endogenous vital genes to increase metabolite synthesis.
(4) Besides this, CRISPR/dCas9-based activation/repression systems can also be used in metabolic engineering manipulation (Zalatan et al., 2015; Li et al., 2017).

Strategies for Multigene Transformation

Metabolic engineering of multifaceted metabolic pathways in plants frequently entails real-time expression of various target genes or sometimes whole genes in a particular pathway for safeguarding unobstructed metabolic flux. For this purpose, peculiar approaches have been created for multigene stacking such as sexual crossing and retransformation (time-consuming and labor-intensive iterative strategies) and cotransformation by using bombardment of particles and *Agrobacterium*-mediated transformation and the existing vectors for multigene vector transformation (multiple gene expression cassettes being linked in a single T-DNA region), polycistronic transgenes (using self-cleavage peptide 2A to link different protein sequences), and plastid transformation (Dafny-Yelin & Tzfira, 2007; Bock, 2013; Farré et al., 2014). From these different approaches, vector multigene transformation has a significant benefit over other strategies (Dafny-Yelin & Tzfira, 2007), like it makes multigene to be integrated as a single unit. This multigene vector strategy needs the gathering of multiple target genes in single T-DNA regions of binary vectors for *Agrobacterium*-mediated transformation or in single plasmid vectors for other transformation methods. Transformation by utilizing plastid is another operative and effective method but somehow it requires transferring multiple genes into the plastid genome. Since multigene vector transformation has apparent benefits, hence it is widely utilized in different projects for crop improvement (Boehm & Bock, 2019).

The following case studies demonstrate various aspects of metabolic engineering. In each case, we highlight the most significant candidate enzymes for engineering and the findings gained thus far, as well as views and ideas for future biofortification investigations to further improve the nutritional value of crops.

Identification of Genes with Potential to Improve the Nutritional Quality of Plants

The molecular biology approach has a lot of potential regarding the studies of genes in different organisms. They allow studying particular components of the genome and the whole interrelations between such components. It is now possible for researchers to identify any particular gene or multiple genes and their products involved in complex metabolic pathways. The basis for this analysis in plants is a genomic map, which can be either a genetic map relying upon information together from visible and molecular markers or a physical map, in which yeast artificial chromosomes (YACs) and bacterial artificial chromosomes (BACs) are aligned with the chromosomes to give a position to genes within the genome.

Nutritional genomics has been visualized as one of the prime approaches nowadays to gene discovery as it is most appropriate to compounds having nutritional importance, accumulated or biosynthesized by plants. It takes into account similarities or homologies between metabolic routes, giving rise to a final metabolic product in different organisms. From this very point of view, those genes which are already characterized by various metabolic pathways in a different organism can be easily found in databases, hence may be used as a source for the study of genetic information or modifications in target plant. The accomplishment of these modifications can be achieved by the use of genes having foreign origin or utilizing cloning genes from the recipient plant using as a molecular tool the information from these genes already isolated from other organisms. Application of nutritional genomics implication in the discovery of genes involved in the biosynthetic pathway of vitamin E as well as its modification is one of its prime example.

In *Arabidopsis thaliana*, the first gene involved in this pathway is taken from fungal and human orthologues as database queries. The purpose of the *Arabidopsis* gene sequence has aided to identify a 10-gene operon in the cyanobacterium *Synechocystis* PCC6803. Experiments done with gene disruption have posed that γ-tocopherol methyltransferase (γ-TMT) is also encoded by this operon, i.e., the ultimate step in the vitamin E synthesis. This γ-TMT gene of *Synechocystis* allowed the isolation of an orthologue from the *Arabidopsis* database, whose overexpression increased vitamin E levels ninefold in *Arabidopsis* seed oil.

However, when there is not sufficient knowledge about particular gene or genes of interest involved in a metabolic pathway, it becomes very much complicated for researchers to isolate the said gene, but there are some alternative ways like mutagenesis or genomic alterations producing variations in the phenotype due to the modification of the components of a given metabolic route. The technique of molecular marker mapping will help in the localization of modified genes through mutagenesis and may further be isolated and characterized. Seed development is affected by various mutant strains in maize, and the genes responsible for altered phenotype have been mapped with cloning technique. In the case of miniature-1, the altered phenotype is a reduction in the size of grain because of a lack of extracellular invertase activity. *Shrunken*, one of the mutant presents starchless grains and the affected gene was identified as the endosperm sucrose synthase.

Besides, other developments have been made for the identification and isolation of genes as well as sequencing of prokaryotic organism's entire genome. The development of new strategies has allowed mutagenesis for a large number of genes. These strategies fall into two categories: random insertional mutagenesis and targeted mutagenesis. As the name suggests in random mutagenesis, insertion mutations are randomly generated throughout the genome, followed by the identification of the gene(s) affected by comparing the sequence adjacent to the insertions with the genome sequence or expressed sequence tags (EST), while in targeted mutagenesis in which specific genes are deleted or analyzed. Transposons and retrotransposons use has abled one to carry out non-site directed mutagenesis. This strategy relies upon characteristics feature of natural transposable elements, found ubiquitously in eukaryotic organisms. Transposable elements' integration into the host genome disrupts genes, effectively producing a "tagged" mutation. A similar approach is to use *Agrobacterium tumefaciens* T-DNA insertions to produce mutations.

The above-designated methods are based on "one gene in one experiment" which indicates that the whole picture is difficult to get for the function of a gene, although different methods let researchers get a picture of a wider array of gene expression and their interaction. Proteomics has found its use in mapping translated genes and loci governing their expression, identifying deviations of multifaceted phenotypic traits. Among other techniques, the differential display is based on the analysis of differential mRNA populations produced in an organism as a result of its exposure to two or more different environmental conditions. Large-scale DNA sequencing and microarrays are new approaches for monitoring the genome on a single chip. This aids in the analysis of interaction among thousands of genes simultaneously.

Furthermore, combining these advanced methods with bioinformatics based and expression-based analysis to identify a limited set of candidate genes for a specific metabolic pathways (Table 6.1).

Vitamin A

As far as vitamin A is concerned, it is important for human growth and development, playing roles like differentiation of epithelial cells, functioning of the immune system, reproduction, and vision (Ross et al., 2000). According to the National Institute of Health, Office of Dietary Supplements, 900 μg and 700 μg for men and women, respectively, are recommended dietary allowances of retinol activity equivalents per day. The retinyl group is present as a general skeleton in retinoids (vitamin A), and it has components like the β-ionone ring and side chain of isoprenoids units. Carotenoids act as a provitamin A source in plants. These are tetraterpenoids having a C40 linear backbone and carotene is found to have the highest provitamin A activity. Biosynthesis of provitamin A proceeds with two molecules of geranyl-geranyl pyrophosphate condensation giving rise to 15-*cis*-phytoene via phytoene synthase (PSY, encoded by crtB in bacteria (Misawa et al., 1994). Geranyl-geranyl pyrophosphate molecule itself is the condensation product of three isopentenyl diphosphate and one dimethylallyl diphosphate molecule, proceeds via enzyme-GGPP synthase in plastids (Chappell, 1995). Four desaturation reactions are carried out to produce chromophores of carotenoid and in plants, and these reactions are because of phytoene desaturase and carotene desaturase. Subsequently, all-trans-lycopene is produced by carotenoid isomerase in nongreen tissue and by light and chlorophyll in green tissue (Isaacson et al., 2004; Li et al., 2010). 15-*Cis*-phytoene by the aid of enzyme CRT1 in bacteria is directly converted into all-trans-lycopene. All trans lycopene acts a substrate for two cyclases viz Lycopene ε-cyclase (LYCE) and lycopene-cyclase (LYCB; CRTY). To generate carotene cyclase, LYCB adds a ring to lycopene. The second addition of a ring by LYCB results in the formation of carotene.

Rice was the first a transgenic crop biofortified with provitamin A. In natural conditions, provitamin A is absent in rice endosperm. So, metabolic engineering was the only means to increase provitamin A amount. Owing to the fact of the transgenic rice kernel's yellow color, golden rice was the name given to it. For the first generation of golden rice, transgenes like psy/lycb and pacrt1 from daffodil and *Erwinia uredovora* (Bacteria), respectively, were utilized (Ye et al., 2000). The result obtained from the first generation was an accumulation of carotenoids by 1.6 μg/g dry weight. However, second-generation transgene from daffodil (psy) substituted with psy1 taken from corn enhanced carotene content to approximately 30 g/g dry weight (Paine et al., 2005a, 2005b). The progress made in the development of golden rice finally led to the development of other metabolic engineered crops like potato, sorghum, corn, wheat, and cassava. Transgenes employed in golden rice two, were also utilized in corn and wheat (Naqvi et al., 2009; Cong

Table 6.1 The list of plants and inserted transgene along with product for biofortification

S. No	Species	Transgene	Promoter or transgene assembly/ transformation	Product	Target tissue	Reference
1	Rice	*ZmR-S*, *ZmC1*	*npr33* Promoter endosperm specific	Flavonoids-methylquercetin	Pericarp	Shin et al. (2006)
2	Rice	*ZmLc*, *ZmPl*, *SsCHS*, *SsCHI*, *SsF3H*, *SsF3^{0}H*, *SsDFR*, *SsANS*	*GluC*, *GluB-1*, *GluB-4*, *Glb1*, *GluB-5*, *Npr33*, *10KDa*, *16KDa* specificity for endosperm	Anthocyanin (cyanidin 3-O-glucoside and peonidin 3-O-glucoside)	Endosperm	Zhu et al. (2017)
3	Rice	*OsPAL*, *OsCHS*	GluB-1 promoter endosperm specific	Flavonoids (Naringenin)	Endosperm	Ogo et al. (2013)
4	Rice	*OsPAL*, *OsCHS*	Promoter *Ole* seed specific	Flavonoids (Naringenin)	Embryo	Ogo et al. (2013)
5	Rice	*OsPAL*, *OsCHS*, *PcFNSI*, *GmFNSII*	*GluB-1* promoter endosperm specific	Flavone (Apigenin)	Embryo	
6	Rice	*OsPAL*, *OsCHS*, *PcFNSI*, *GmFNSII*	Seed-specific promoter *Ole*	Flavone (Apigenin)	Endosperm	
7	Rice	*OsPAL*, *OsCHS*, *PcFNSI*, *GmFNSII*, *OsOMT*, *ViolaF3^{0}5^{0}H*	*GluB-1* endosperm specificity	Flavone (Tricin)	Endosperm	
8	Maize	*ZmC1*, *ZmR2*, *ZmANS*, *ZmGST*	PZmBD1-embryo explicit promoter	Anthocyanin (cyanidin and pelargonidin)	Embryo	Liu et al. (2018a)
9	Maize	*ZmC1*, *ZmR2*, *ZmANS*, *ZmGST*	P2R5SGPA-bidirectional with seed explicit promoter	Anthocyanin (cyanidin, pelargonidin, and peonidin)	Endosperm	Liu et al. (2018b)
10	Tomato	*PhCHI*	*CaMV 35S*	Flavonols	Fruit peel	Muir et al. (2001)
11	Tomato	*ZmLc*, *ZmC1*	Fruit-specific promoter *E8*	Flavonol (Kaempferol)	Fruit	Bovy et al. (2002)
12	Tomato	*SlMYB75*	*CaMV 35S*	Anthocyanins	Fruit	Butelli et al. (2008)

13	Sweet Potato	*IbMYB1*	Storage root-specific promoter *SPO1*	Anthocyanins	Tuber	Jian et al. (2019a, 2019b)
14	Soybean	*CrtB*, *CrtW*	Cotransformation and particle bombardment	Canthaxanthin (carotene)	Seed	Pierce et al. (2015)
15	Soybean	*CrtB*, *BKT*	Cotransformation and particle bombardment	b-Carotene	Seed	Pierce et al. (2015)
16	Wheat	*CrtB*, RNAi-*BHY*	Particle bombardment	b-Carotene	Endosperm	Wang et al. (2014)
17	Wheat	RNAi-*BHY*	Particle bombardment	b-Carotene	Endosperm	Wang et al. (2014)
18	Sorghum	*ZmPSY1*, *PaCrtI*, *AtDXS*	Multiple gateway and transformation intermediated by *Agrobacterium*	b-Carotene	Endosperm	Che et al. (2016)
19	Sorghum	*ZmPSY1*, *PaCrtI*, *AtDXS*, *HGGT*	Multigene vectors (multiple gateways) and transformation via *Agrobacterium*.	b-Carotene	Endosperm	Che et al. (2016)
20	Rice	*ZmPSY1*, *PaCrtI*	Multigene vectors (restriction-ligation) and interceded transformation through *Agrobacterium*	b-Carotene	Endosperm	Paine et al. (2005a, 2005b)
21	Rice	*ZmPSY1*, *PaCrtI*, *tHMGR1*	Restriction-ligation multigene and *Agrobacterium*-arbitrated alteration	b-Carotene	Endosperm	Tian et al. (2019)
22	Rice	*sZmPSY1*, *sPaCrtI*, *sCrBKT*	Multigene vectors (Cre/*loxP*-based TGS II system) and *Agrobacterium*-mediated transformation	b-Carotene	Endosperm	Zhu et al. (2018)
23	Rice	*CaPSY*, *PaCrtI*, *stCaBch*, *stHpBKT*	Polycistronic transgene with 2A-multigene vector and transformation via *agrobacterium*	Zeaxanthin	Endosperm	Ha et al. (2019)
24	Maize	*ZmPSY1*, *PaCrtI*, *GlLycB*	Cotransformation and particle bombardment	b-Carotene	Endosperm	Zhu et al. (2008)
25	Maize	*ZmPSY1*, *PaCrtI*, *GlLycB*, *CrtW*	Cotransformation and particle bombardment	Astaxanthin	Endosperm	Zhu et al. (2008)

(continued)

Table 6.1 (continued)

S. No	Species	Transgene	Promoter or transgene assembly/ transformation	Product	Target tissue	Reference
26	Tomato	*CrBKT*	Transformation via agrobacterium and vectors of binary nature	Canthaxanthin, 338.4, Astaxanthin	Fruit	Huang et al. (2013)
27	Tomato	*CrBKT*, *HpBHY*	Binary vectors and transformation through *Agrobacterium*	Canthaxanthin, 338.4, Astaxanthin	Fruit	Huang et al. (2013)
28	Tomato	*AtOR*^{WT}	Transformation done via *Agrobacterium*	b-Carotene	Fruit	Yazdani et al. (2019)
29						

et al., 2009). Enhanced expression of AtDXS (*Arabidopsis* 1-deoxyxylulose-5-phosphates synthase) is an important protein toward GGPP flux, in association with PacrtB and corn psy1/PacrtI, which enriched carotene concentration in cassava and sorghum, respectively (Sayre et al., 2011; Lipkie et al., 2013).

Vitamin E

Vitamin E works as a strong antioxidant and is lipophilic. It is very much helpful in the prevention of various illnesses caused by oxidative degeneration (Ricciarelli et al., 2001). The National Institutes of Health, Office of Dietary Supplements has documented that the required daily dose of vitamin E for adults and children is 15 mg and 7 mg, respectively. Tocopherol biosynthesis involves five enzymes (VTE 1–5), and it starts with hydroxyphenylpyruvate getting converted into homogentisate (HGA) by the aid of enzyme hydroxyphenylpyruvate dioxygenase (HPPD). Upon HGA biosynthesis, tocochromanol production branches toward the production of tocotrienols and tocopherols. In the branch involving tocopherol, enzyme phytol kinase (VTE5) phosphorylates phytol to phytyl monophosphate, and another phosphorylation results in phytyl-diphosphate (PDP). 2-Methyl-6-phytyl-1, 4-benzoquinol (MPBQ) is produced by the condensation of PDP and HGA by enzyme homogentisate phytyltransferase (VTE2; HPT). MPBQ, the second methyl group, gets attached to it by enzyme MPBQ methyltransferase (VTE3; MPBQ-MT) to obtain 2, 3-dimethyl-6-phytyl-1 4-benzoquinol (DMPBQ). Tocopherol cyclase (VTE1) executes cyclization reaction to produce γ-tocopherol and δ-tocopherol from MPQB and DMPBQ, respectively. γ-Tocopherol methyltransferase (VTE4; γ-TMT) catalyzes second and third methylation reaction to get β-tocopherol from δ-tocopherol and α-tocopherol from γ-tocopherol. Tocotrienol branch involves the same enzymes involved in tocochromanol branch; however, geranyl-geranyl pyrophosphate (GGPP) is used in connotation with HGA by homogentisate geranyl-geranyl transferase (HGGT) to produce 2-methyl-6-geranyl-geranyl-plastoquinol, the tocotrienol equivalent of MPBQ. Numerous metabolic engineering efforts have been made to (1) enhance the concentration of tocochromanol amount in crops and (2) swing tocochromanols accumulation toward α-tocochromanols as it is highest in vitamin E activity. Overexpression of enzyme prephenate dehydrogenase (bacterial) in *Arabidopsis*, seeds of canola, and soybean has significant outcomes such as, conversion of prephenate to hydroxyphenylpyruvate in the association with AtHPPD (Hhydroxyphenylpyruvate dioxygenase), which results in enhanced content of tocochromanol level almost fourfold (Karunanandaa et al., 2005). Overexpression of AtHPPD in rice seedlings showed a slight increase in tocochromanol content and almost nothing increases in total tocopherol concentration; however, it caused a swing from γ-tocopherol to α-tocopherol (Farré et al., 2012). Overexpression of barley HGGT in corn enhanced tocochromanol concentration by six times (Butelli et al., 2003). Constitutive overexpression of AtHPPD and AtVTE3 tripled the amount of tocopherol in corn seeds (Naqvi et al., 2011). To enhance the vitamin E content effectively, a multitarget attitude is a prerequisite, boosting the production of tocopherol precursors HGA and PDP and diverting the flux towards alpha-tocopherol production.

Vitamin B6

Vitamin B6 is very much important for the reduction of risks involved in diabetes, kidney problems, hypertension, neurological disorders, cardiovascular diseases, and epilepsy (Hellmann & Mooney, 2010). The daily requirement for adults is 1.3 mg (National Institutes of Health, Office of Dietary Supplements). Vitamin B6 encompasses six vitamers, including pyridoxine, pyridoxal, and pyridoxamine and its phosphorylated derivatives. Among these pyridoxal-5-phosphate is regarded as the utmost important vitamer because it is involved in almost 140 chemical reactions in a cell as a cofactor (Hellmann & Mooney, 2010). So, for the biofortification approach amassing of pyridoxal-5-phosphate is of vital importance. In plants, vitamin B6 synthesis proceeds through only two proteins (Tambasco-Studart et al., 2005), PDX1 and PDX2. In rice (Ouyang et al., 2007), cassava (Prochnik et al., 2012), and *Arabidopsis* (Tambasco-Studart et al., 2005), three homologs of PDX1 have been found like PDX1.1, PDX1.2, and PDX1.3 and one PDX2 gene. From a vitamin B6 biosynthesis point of view, only PDX1.1 and PDX1.3 are important. Overexpression of PDX1 and PDX2 in tobacco from *Cercospora nicotianae* showed a minor 20% increase in the concentration of vitamin B6, but it also resulted in delayed germination and plant growth (Herrero & Daub, 2007). AtPDX1.3 and AtPDX2 constitutive overexpression in *Arabidopsis* enhanced vitamin B6 in seeds by 20%; however, seed-specific expression showed three times increase in seeds (Chen & Xiong, 2009) without upsetting the performance of plants. Constitutive expression of AtPDX1.1 in association with AtPDX2 promotes concentration of vitamin B6 as well as pyridoxal-5-phosphate specifically by five times in *Arabidopsis*, showing the importance of choosing the right homologues (Chen & Xiong, 2009). Excitingly, this very biofortification results in increased seed size owing to the fact of larger aerial organs and enlargement of an embryo (Raschke et al., 2011). AtPDX1.1 and AtPDX2 overexpression in cassava roots showed the enhanced level of vitamin B6 in leaves and roots (Li et al., 2015).

Vitamin B9

Vitamin B9 comprises a set of water-soluble B vitamins (folates), playing an important part as a cofactor in one-carbon (C1) metabolism in every organism (Blancquaert et al., 2010). During various metabolic processes like from methylation to replication of DNA, vitamin B9 provides or accepts the C1 in different forms of methylene, formyl groups, and methyl. Vitamin B9 possesses a backbone entailing three parts as a pterin group, apara-aminobenzoate (p-ABA) moiety, and a glutamate tail. Folates differ from each other in the glutamate tail length and C1 units attached. The daily requirement of folates is 400 μg and 600 μg in adults and pregnant women, respectively (National Institutes of Health, Office of Dietary Supplements). Deficiency of folates has serious results such as cardiovascular diseases (Scott & Weir, 1996), Alzheimer's diseases (Seshadri et al., 2002), coronary diseases (Stanger, 2004), cancers (Choi et al., 2015), and stroke (Endres et al., 2005). The active state of folic acids needs a reduction to dihydrofolate and THF (Blancquaert et al., 2014). Synthesis of folates takes place in three subcellular sections of cells, namely, plastids, cytosol, and mitochondria. In plastids p-ABA is synthesized, pterin in the cytosol, and both these precursors are abridged, reduced, and glutamylated in mitochondria. Initial metabolic engineering endeavors have shown interest toward two folate biosynthetic enzymes GTP cyclohydrolase I (GTPCHI, G) and aminodeoxychorismate synthase (ADCS, A), the first step in pterin branch and p-ABA branch, respectively. GTPCHI or ADCS overexpression in G lines of *Arabidopsis* (Hossain et al., 2004), rice (Storozhenko et al., 2007), tomato (Diaz et al., 2004), corn (Naqvi et al., 2009), and potato (Blancquaert et al., 2013) increases the content of pterin or p-ABA; however, it also leads to the depletion of nontargeted precursor. GTPCHI and ADCS synchronized overexpression in tomato by crossing G and A lines (Nunes et al., 2009). T-DNA insertion of GA lines in rice has shown a massive folate accumulation of 1723 g/g fresh weight in engineered crops. Although in potato tubers and *Arabidopsis*, this synchronized two-gene approach does not show desired results indicating that this is another bottleneck in the folate biosynthetic pathway (Blancquaert et al., 2013). Pterin and p-ABA concentration in these plants is high, so triple-gene strategy with bifunctional enzyme hydroxymethyl dihydropterin pyrophosphokinase/dihydropteroate synthase (HPPK/DHPS) was recommended for fruitful results (Blancquaert et al., 2014). Dong et al. (2014) documented that GTPCHI in arrangement with DHF synthetase overexpression on level of folate in rice seeds has shown only a little increment in folate amount because of p-ABA depletion in engineered lines. GTPCHI, ADCS, and FBP [folate-binding protein] fusion with carbonic anhydrase 2 from *Arabidopsis* combined overexpression in rice stemmed in an additional enhancement of folate levels by 50% as compared to original lines of GA (Blancquaert et al., 2015).

Iron

Various enzymes find iron as their very important part together with hemoglobin. Anemia is caused by an iron deficiency, which kills 0.8 million people each year (WHO). The RDA for iron is 8 mg for males and 18 mg for women (National Institutes of Health, Office of Dietary Supplements). Iron procurement is well-studied of all the metal uptake processes in plants. This is owing to the physiological features and effects of iron deficiency in plants being well defined. Iron is absorbed from the rhizosphere through the root epidermis, transferred to the xylem, and distributed throughout the plant to various tissues, where it can be stored until needed. Plants face a deficiency of iron (Fe (III), primarily because of solubility issues, particularly in alkaline and neutral soils, rather than the low amount of iron in the soil. Eudicots and non-graminaceous monocots have developed an approach for increment in Fe (III) solubility which is supported by the low iron presence in soil and is known to be a high-affinity uptake system. This system implicates soil acidification via H^+-ATPases proton pumping into rhizosphere and apoplast and the excretion of small chelating molecules, such as malate and citrate. Solubilization and chelation of Fe (III) are achieved in this way. Plasma membrane-bound Fe (III) chelate reductases then reduce bound Fe (III) at the root surface.

It has been found in *Arabidopsis* that FRO (ferric reductase oxidase) family has eight members. From these eight members, $AtFRO_2$ is known to have maximum activity for iron reduction. Additionally, it has a root-specific expression as well as is known to be a chief contributor to the reduction of Fe (III) in the rhizosphere (Wu et al., 2005). $AtFRO_2$ transformation in rice has been done but roots did not show any improvement in iron reductase activity. Nonetheless, during the transformation of the $AtFRO_2$ gene in rice, 0.6-kb upstream promoter of *Arabidopsis* was used in place of active rice promoter in the root epidermis, and detection of transcript for this transgene was absolutely nothing (Vasconcelos et al., 2004). In soybean $AtFRO_2$ genes constitutive overexpression has shown an almost five times higher amount of iron in leaves and pods; however, from the seeds point of view, 10% enhancement was spotted, indicating that some other factors are needed to transfer iron to seeds (Vasconcelos et al., 2014).

The study conducted by Beasley et al. (2019) on wheat endosperm having a low concentration of iron was subjected to constitutive expression of nicotianamine synthase 2 (OsNAS2) gene taken from rice to upregulate metal chelators such as nicotianamine (NA) and 20-deoxymugineic acid (DMA), playing an important role in the transportation of metals. Transgenic plants possessed grains having a higher content of iron in the endosperm.

Engineering Crops for Biofuel Production

Plants appear to be the ideal "invention" to address the combined issues of growing greenhouse gas emissions and the demand for green energy: they absorb CO_2 from the air and convert it to sugar, which is a great substrate for biofuel production. The majority of carbon is deposited in the form of dehydrated crystalline cellulose which is enfolded in a network of cross-linked phenylpropanoids, lignin (Sanderson, 2011). Biosynthesis of lignin has been subjected to genetic modification for altering its chemical configuration or to diminish its amount from tissues of plants for improvement in polysaccharides processing. For instance, a lignin biosynthetic gene caffeoyl shikimate esterase provided a fourfold enhancement in saccharification efficiency in an *Arabidopsis thaliana* knockout mutant relative to wild type (Vanholme et al., 2013). On the other hand, many lignin-altered plants have shown evident effects on growth and development; transgenic contains the amount of cellulose lessened by 25% as well as 40% lighter and smaller than wild type (Bonawitz & Chapple, 2013). Recently, the engineering of protein and expression of an enzyme 4-O-methyltransferase in *A. thaliana* significantly diminished the content of lignin by obstructing access to *p*-hydroxyls of lignin precursors needed for polymerization (Zhang et al., 2012). Excitingly, no noteworthy alterations in phenotype were detected and saccharification yields amended by 25%. To go further for lignin modifications, one relies mainly on the understanding of the lignin biosynthetic pathway as well as the physiological magnitudes of changing its structure. Now the question arises what if this job is done by plants on their own? Degradation of lignin from the plants on their own, liberation of cellulose, and subsequent degradation into glucose would be the ideal conditions for the crops grown for biofuel production. Metabolic engineering may be taken into consideration to engineer plants for cellulose breakdown on will, liberating biofuel-ready fermented sugars. However, the challenge posed to obtain this goal is degrading lignin enzymatically for the releasing of cellulose feasibility. No doubt it is a very difficult task, but white-rot fungi have the potential to degrade lignin, thereby proving the concept that there exists enzymes (e.g., lignin peroxidases) that can cleave the lignin meshwork into monomers and smaller polymers (Ten Have & Teunissen, 2001). There exists an alternative way besides lignin degradation: precipitation of lignin because of heavy cross-linkage and making it easy to separate cellulose. This process will be probably carried out by suites of rot fungi-degrading enzymes. However, the identification of enzymes that will be coexpressed remains an essential step for making required modifications to lignin. After harvesting it is very unlikely to continue transcription and translation for long, but the enzymes could be expressed in an inactive and caged form instead while the plant is still alive. Myrosinases in cruciferous plants provide an informative model by activating glucosinolates by hydrolysis of glycosides (Rask et al., 2000), although they are translated nonetheless physically sequestered in different cells from their substrate (a glucosinolate) and get activated only after an injury to tissues. Recently it was documented that there is a promising substitute approach grounded on integrating chemically labile bonds into the lignin backbone (Wilkerson et al., 2014).

Enhancing Photosynthetic Efficiency

Crop yields will need to rise to satisfy rising food demands as the world's population and urbanization expand quickly. Cereals will continue to play a significant role in the future food supply, both as a source of direct consumption and as a source of feed for animals. For green energy generation, crop yields for biofuel feedstocks must also be addressed. In calvin cycle Co_2 fixation being one of the prime step is catalyzed by enzyme Rubisco (Spreitzer & Salvucci, 2002). Rubisco uses oxygen [photorespiration] substrate as well besides Co_2 and owes to the fact of low turnover rate making it somewhat inefficient. The outcome of this issue makes Rubisco the most prevalent protein on earth. Alternative proficient systems have evolved to improve the photosynthetic efficiency via concentrating Co_2 actively and diminishing photorespiration (Hibberd et al., 2008; Von Caemmerer et al., 2012). C4 plants have been found to be efficient than C3 plants concerning nitrogen and water use, especially in a hot and dry environment. Additionally, they have a high ability for photon conversion into biomass almost by 50%. These properties have presented many suggestions for metabolic engineering in C3 plants to introduce C4 potential in them, and it could potentially lead to a 50% increase in crop yield. The International Rice Research Institute is currently working toward incorporating C4 photosynthesis into rice (http://c4rice.irri.org/). More than 60 distinct taxa have independently evolved from C3 to C4, signifying that this metabolic engineering endeavor may be more possible than it seems, though many new plant metabolic engineering methods will be prerequisites (Peterhansel, 2011). Enzymes involved in the C4 cycle are well documented and persist in C3 plants as well. However, simply expressing C4 cycle enzymes will not be adequate, as the plant's anatomy is vital to the pathway's success. The genes that govern C4 leaf architecture are mostly unknown, and mutant populations of model C4 plants like *Sorghum* are being used to find them (Zhu et al., 2010). To allow cell-type-specific expression in bundle sheath or mesophyll cells, cell-specific promoters must also be found. Finally, the installation of C4 photosynthesis in C3 plants will necessitate the development of more advanced transformation and genome editing technologies.

Metabolic Engineering of Vitamin Stability in Crop Plants

The stability of vitamins is very much important as it is subtle to variations in temperature, pH, humidity, and degradation by photooxidation (Fitzpatrick et al., 2012). Stability varies among different vitamers. Folic acid and 5-formyl THF from the vitamin B9 group are highly stable molecules (Rébeillé et al., 2006), but THF and DHF are most unstable due to degradation. It has been well documented that food processing affects the stability of vitamins but very little is known about the storage time effect as it is said commonly that with time degradation occurs. Vitamin E level in freeze-dried fortified apple plunged by 50% only kept for 6 months storage (Cortés et al., 2009), and in guava nectar vitamin C amount drops by 11% (Ordóñez-Santos & Vázquez-Riascos, 2010). Provitamin A concentration in biofortified corn gets minimized by 70% after half a year (Mugode et al., 2014). Speaking generally, those crops which are stapled in nature are kept stored until the next harvest. Storage time for corn is almost 9 months or longer after drying (Mejia, 2003). For crops like rice, corn, and cereal crops, vitamin stability must be ensured over time. Since then it will be fruitful to fight against deficiency of micronutrients through the implementation of biofortified crops. However, it is very unfortunate that this aspect of metabolic engineering is largely neglected. In recent times, it was revealed that the level of folate in metabolic engineered rice plunged by 50% after 4 months of storage at different storage temperatures (Blancquaert et al., 2015). Two approaches were taken into consideration to address the folate stability issue. In the first one, synthetic FBP (centered on FBP from bovine milk) expressed in association with GTPCHI and ADCS genes for folate enrichment. In-plant FBP has not been reported but is well studied in mammals. It occurs in milk, protects folates from degradation by forming complexes with them. The second strategy involves overexpressing FPGS (folylpolyglutamate synthetase), an enzyme in folate biosynthesis. The glutamate tail of this protein extends which aids folate-dependent enzyme binding and cellular retention through its anionic nature. Both approaches worked well in increasing the stability of folate in biofortified rice, and results have revealed that after a long storage period folate level was unchanged. Both approaches worked well in increasing the stability of folate in biofortified rice, and results have revealed that after a long storage period folate level was unbothered. One of the excellent vitamin stability engineering candidates is vitamin-binding protein. Carotenoids are susceptible to oxidation, which holds a nutritional value of foods having a rich amount of carotenoids such as golden rice, particularly through long-term storage. Hydroperoxy fatty acids have the potential to co-oxidize as well as decolorize carotenoids generated by lipoxygenase (LOX) (Casey, 1997; Wu et al., 1999). Silencing of r9- LOX1 in endosperm and aleurone activity by RNA interference got breakthrough by protecting carotenoids in golden rice from oxidation thereby stabilizing provitamin A amount upon storage (Gayen et al., 2015). For successful biofortification attempts, one should not only focus on the enhancement of nutrient levels but also on the stability of high nutrient levels particularly for those crops kept for storage. The above examples demonstrate that vitamin stability may be engineered. Vitamin stability can also be improved by directing vitamin biosynthesis to produce more stable vitamers or by investigating and regulating vitamin salvage mechanisms.

Bioavailability of Vitamins in Biofortified Crop Products

The amount present in a particular food product or diet that ultimately will be utilized by the human body during metabolic processing can be summarized as bioavailability. Studies related to bioavailability are very complex because it encompasses various factors one needs those to be considered. Some of those factors are vitamin release from the certain food matrix, uptake by the digestive system, as it is convoyed sometimes by alteration to a compound, loaded into the circulation system, and transported to different tissues and cells. A great bit of variation exists among vitamers from a bioavailability point of view. Synthetic folates have more bioavailability than natural folates (Food & Drug Administration, 1996; Bailey, 2004), and folate polyglutamates are less bioavailable than monoglutamates because it requires deconjugation reaction in gut mucosa before being released into the circulatory system (Melse-Boonstra et al., 2004). Glycosylation of vitamin B6 plunges its bioavailability almost by 50% (Gregory, 2012). Besides, different factors extant in diet have an impact on the bioavailability of vitamins through their interaction with micronutrients and have been shown for zinc, vitamin A, and iron (Dijkhuizen et al., 2004; Bloem, 1995). A study conducted on papaya revealed PACs bioavailability higher when supplied with zinc and iron after a 5-day PAC-free diet (Kana-sop et al., 2015). Vitamin bioavailability gets also influenced by the processing of foods by a change in the composition of the food matrix. Cryptoxanthin bioavailability in orange juice is enhanced by pasteurization as compared to fresh oranges (Aschoff et al., 2015), most likely because more citrus pectins are present in fresh oranges, which negatively affects carotenoid absorption. In humans, cell cultures, and murine models, bioavailability is assessed via rat feeding, which showed that folates in engineered rice have great bioavailability, making it a valuable source of dietary folate (Kiekens et al., 2015). The bioavailability aspect of vitamins related to studies of biofortification needs to be addressed. Metabolic engineering is one of the suitable ways to tackle the bioavailability problem of vitamins like antinutrients (phytate) to enhance compounds that promote bioavailability.

Flavonoid and Anthocyanin Biofortification in Crops

Anthocyanins can be described as a group of flavonoid compounds having strong antioxidant action and a wide distribution in vegetables and fruits promoting the health of humans (Deng et al., 2013). The presence of anthocyanins in cereals is confined to pericarp of particular variety like black corn, black rice, and purple wheat and in certain cereal endosperm completely absent. Besides this, in the East Asian region, people eat refined grains having no pericarp also diminishes its health-promoting properties. The biosynthesis of anthocyanins is a very well-understood pathway. It has numerous structural genes for encoding different metabolites for anthocyanin biosynthesis, decorating genes, transcription factors, as well as transporters. Transcription factors form a MYB-bHLH-WD40 (MBW) complex for controlling the expression of anthocyanin structural genes (Hichri et al., 2011; Dixon et al., 2013; Zhang et al., 2014; Yuan & Grotewold, 2015). Biofortification of flavonoids and anthocyanins in crops like rice, maize, and tomato is accomplished by making use of prime structural genes (Muir et al., 2001; Ogo et al., 2013), genes for transcription factors (Bovy et al., 2002; Butelli et al., 2008; Zhang et al., 2015; Jian et al., 2019a, 2019b), or utilizing both of them in combination (Zhu et al., 2017; Liu et al., 2018a, 2018b).

Fruits and vegetables have anthocyanins naturally present in them but not in the fruit of most tomato cultivars. One of the transcription factor genes, namely, petunia chalcone isomerase, one prime enzyme, regulator AtMYB12 (Muir et al., 2001; Zhang et al., 2015), or transcription factor of a regulatory complex like Lc and C1 of maize anthocyanin transcription factor overexpression may enhance flavonols concentration in tomato; however, it did not produce anthocyanins (Bovy et al., 2002). Nonetheless, coexpression of *AmDel* and *AmRos1* genes for anthocyanin regulator complex (Butelli et al., 2008) or expression of *SlMYB75* tomato regulatory complex gene (Jian et al., 2019a, 2019b) in fruits of tomato may attain good accumulation of anthocyanins. Results obtained showed that transcription having different sources has a different way of activation structural genes. However genetic manipulation is simple: the strategy of only using the transcription factor genes or their combination may not be successful. Metabolic engineering of anthocyanin in tomato is easy as compared to the cereal crops biofortification which is more complicated. For instance, in the endosperm of rice, certain transcriptional and structural genes are functionally defective, and those are involved in the biosynthetic pathway of anthocyanin. So, structural gene expression or coexpression of bHLH- and MYB-type regulatory genes (such as maize anthocyanin regulators *ZmR-S* and *ZmC1*) does not complete the biosynthesis of anthocyanins. However, the only intermediate product was produced (the anthocyanin upstream precursors) in the endosperm of rice (Shin et al., 2006; Ogo et al., 2013).

(*ZmLc* and *ZmPl*) regulatory genes of maize and all structural genes from the coleus biosynthetic anthocyanin pathway have been transformed recently driven by promoters having endosperm specificity (Zhu et al., 2017). Multigene stacking system TGS II has been utilized to assemble eight genes and subsequently transform them into rice (purple endosperm) for high anthocyanin biofortification amount and antioxidant action in the endosperm. Using the same approach, purple maize with is rich in anthocyanin endosperm has been developed via bidirectional seed-specific promoters to drive the target gene-coding sequences linked by the self-cleavage peptide 2A linker (Liu et al., 2018a, 2018b). Results have made any indication that using genes of transcription factors in combination with multiple structural genes possesses wider flexibility of anthocyanin metabolic engineering. With this strategy, it is also possible to obtain other purple endosperm cereals.

Conclusion

The engineering of metabolites has a potential role to play in the enhancement of the nutritional value of crop plants through genetic modifications. The approach utilized involves naturally occurring endogenous metabolic pathways amendment, providing the plant with one or more heterologous actors for increasing the amount of target product, molecules of undesirable nature reduction, or change the flux to introduce the accumulation of more bioavailable, stable, and active compound. For metabolic engineering, one should be equipped with very much knowledge of the metabolic pathways involved, so that efficacious engineering strategy will be employed in which major enzymes are targeted for overexpression or downregulation to enhance the biosynthesis or buildup of micronutrients without affecting crop development and yield.

References

Aschoff, J. K., Rolke, C. L., Breusing, N., Bosy-Westphal, A., Högel, J., Carle, R., & Schweiggert, R. M. (2015). Bioavailability of β-cryptoxanthin is greater from pasteurized orange juice than from fresh oranges: A randomized cross-over study. *Molecular Nutrition & Food Research, 59*(10), 1896–1904.

Bailey, L. B. (2004). Folate and vitamin B 12 recommended intakes and status in the United States. *Nutrition Reviews, 62*(Suppl. 1), S14–S20.

Beasley, J. T., Bonneau, J. P., Sánchez-Palacios, J. T., Moreno-Moyano, L. T., Callahan, D. L., Tako, E., Glahn, R. P., Lombi, E., & Johnson, A. (2019). Metabolic engineering of bread wheat improves grain iron concentration and bioavailability. *Plant Biotechnology Journal, 17*(8), 1514–1526.

Blancquaert, D., Storozhenko, S., Loizeau, K., De Steur, H., De Brouwer, V., Viaene, J., et al. (2010). Folates and folic acid: From fundamental research toward sustainable health. *Critical Reviews in Plant Science, 29*(1), 14–35.

Blancquaert, D., De Steur, H., Gellynck, X., & Van Der Straeten, D. (2014). Present and future of folate biofortification of crop plants. *Journal of Experimental Botany, 65*(4), 895–906.

Blancquaert, D., Storozhenko, S., Van Daele, J., Stove, C., Visser, R. G., Lambert, W., & Van Der Straeten, D. (2013). Enhancing pterin and para-aminobenzoate content is not sufficient to successfully biofortify potato tubers and Arabidopsis thaliana plants with folate. *Journal of Experimental Botany, 64*(12), 3899–3909.

Blancquaert, D., Van Daele, J., Strobbe, S., Kiekens, F., Storozhenko, S., De Steur, H., Gellynck, X., Lambert, W., Stove, C., & Van Der Straeten, D. (2015). Improving folate (vitamin B 9) stability in biofortified rice through metabolic engineering. *Nature Biotechnology, 33*(10), 1076–1078.

Bloem, M. W. (1995). Interdependence of vitamin A and iron: An important association for programmes of anaemia control. *Proceedings of the Nutrition Society, 54*(2), 501–508.

Bock, R. (2013). Strategies for metabolic pathway engineering with multiple transgenes. *Plant Molecular Biology, 83*, 21–31.

Boehm, C. R., & Bock, R. (2019). Recent advances and current challenges in synthetic biology of the plastid genetic system and metabolism. *Plant Physiology, 179*, 794–802.

Bonawitz, N. D., & Chapple, C. (2013). Can genetic engineering of lignin deposition be accomplished without an unacceptable yield penalty? *Current Opinion in Biotechnology, 24*(2), 336–343.

Bouis, H. E., & Saltzman, A. (2017). Improving nutrition through biofortification: A review of evidence from HarvestPlus, 2003 through 2016. *Global Food Security, 12*, 49–58.

Bovy, A., de Vos, R., Kemper, M., Schijlen, E., Almenar Pertejo, M., Muir, S., Collins, G., Robinson, S., Verhoeyen, M., Hughes, S., et al. (2002). High-flavonol tomatoes resulting from the heterologous expression of the maize transcription factor genes LC and C1. *Plant Cell, 14*, 2509–2526.

Butelli, E., Titta, L., Giorgio, M., Mock, H. P., Matros, A., Peterek, S., Cahoon, E. B., Hall, S. E., Ripp, K. G., Ganzke, T. S., Hitz, W. D., & Coughlan, S. J. (2003). Metabolic redesign of vitamin E biosynthesis in plants for tocotrienol production and increased antioxidant content. *Nature Biotechnology, 21*(9), 1082–1087.

Butelli, E., Titta, L., Giorgio, M., Mock, H. P., Matros, A., Peterek, S., Schijlen, E. G., Hall, R. D., Bovy, A. G., Luo, J., et al. (2008). Enrichment of tomato fruit with health-promoting anthocyanins by expression of select transcription factors. *Nature Biotechnology, 26*, 1301–1308.

Casey, R. (1997). Lipoxygenases and bread making. In S.A.G.F. Angelino, R.J. Hamer, W. van Hartingsfeld, F. Heidekamp & J. P. van der Lugt, (Eds.), *Proceedings of the First European Symposium on Enzymes and Grain Processing* (pp. 188–194).

Chappell, J. (1995). Biochemistry and molecular biology of the isoprenoid biosynthetic pathway in plants. *Annual Review of Plant Biology, 46*(1), 521–547.

Che, P., Zhao, Z. Y., Glassman, K., Dolde, D., Hu, T. X., Jones, T. J., Gruis, D. F., Obukosia, S., Wambugu, F., & Albertsen, M. C. (2016). Elevated vitamin E content improves all-*trans* b-carotene accumulation and stability in biofortified sorghum. *Proceedings of the National Academy of Sciences of the United States of America, 113*, 11040–11045.

Chen, H., & Xiong, L. (2009). Enhancement of vitamin B6 levels in seeds through metabolic engineering. *Plant Biotechnology Journal, 7*(7), 673–681.

Choi, S., Jun, H., Bang, J., Chung, S. H., Kim, Y., Kim, B. S., Kim, H., Beuchat, L. R., & Ryu, J. H. (2015). Behaviour of Aspergillus flavus and Fusarium graminearum on rice as affected by degree of milling, temperature, and relative humidity during storage. *Food Microbiology, 46*, 307–313.

Cong, L., Wang, C., Chen, L., Liu, H., Yang, G., & He, G. (2009). Expression of phytoene synthase1 and carotene desaturase crtI genes result in an increase in the total carotenoids content in transgenic elite wheat (*Triticum aestivum* L.). *Journal of Agricultural and Food Chemistry, 57*(18), 8652–8660.

Cortés, R. M., Chiralt, B. A., & Suarez, M. H. (2009). Influence of storage conditions on freeze-dried apple fortified with vitamin E. *Vitae, 16*(1), 31–41.

Dafny-Yelin, M., & Tzfira, T. (2007). Delivery of multiple transgenes to plant cells. *Plant Physiology, 145*, 1118–1128.

Deng, G. F., Xu, X. R., Zhang, Y., Li, D., Gan, R. Y., & Li, H. B. (2013). Phenolic compounds and bioactivities of pigmented rice. *Critical Reviews in Food Science and Nutrition, 53*, 296–306.

Diaz de la Garza, R., Quinlivan, E. P., Klaus, S. M., Basset, G. J., Gregory III, J. F., & Hanson, A. D. (2004). Folate biofortification in tomatoes by engineering the pteridine branch of folate synthesis.

Dijkhuizen, M. A., Wieringa, F. T., & West, C. E. (2004). Zinc plus β-carotene supplementation of pregnant women is superior to β-carotene supplementation alone in improving vitamin A status in both mothers and infants. *The American Journal of Clinical Nutrition, 80*(5), 1299–1307.

Dixon, R. A., Liu, C., & Jun, J. H. (2013). Metabolic engineering of anthocyanins and condensed tannins in plants. *Current Opinion in Biotechnology, 24*, 329–335.

Dong, W., Cheng, Z. J., Lei, C. L., Wang, J. L., Wang, J., Wu, F. Q., Zhang, X., Guo, X. P., Zhai, H. Q., & Wan, J. M. (2014). Overexpression of folate biosynthesis genes in rice (*Oryza sativa* L.) and evaluation of their impact on seed folate content. *Plant Foods for Human Nutrition, 69*(4), 379–385.

Endres, M., Ahmadi, M., Kruman, I., Biniszkiewicz, D., Meisel, A., & Gertz, K. (2005). Folate deficiency increases postischemic brain injury. *Stroke, 36*(2), 321–325.

Farré, G., Blancquaert, D., Capell, T., Van Der Straeten, D., Christou, P., & Zhu, C. (2014). Engineering complex metabolic pathways in plants. *Annual Review of Plant Biology, 65*, 187–223.

Farré, G., Sudhakar, D., Naqvi, S., Sandmann, G., Christou, P., Capell, T., & Zhu, C. (2012). Transgenic rice grains expressing a heterologous ρ-hydroxyphenylpyruvate dioxygenase shift tocopherol synthesis from the γ to the α isoform without increasing absolute tocopherol levels. *Transgenic Research, 21*(5), 1093–1097.

Fitzpatrick, T. B., Basset, G. J., Borel, P., Carrari, F., DellaPenna, D., Fraser, P. D., et al. (2012). Vitamin deficiencies in humans: Can plant science help? *The Plant Cell, 24*(2), 395–414.

Food and Drug Administration. (1996). Food standards: Amendment of standards of identity for enriched grain products to require addition of folic acid; final rule (21 CFR Parts 136, 137, and 139). *Federal Register, 61*, 8781–8797.

Fu, R., Martin, C., & Zhang, Y. (2018). Next-generation plant metabolic engineering, inspired by an Ancient Chinese irrigation system. *Molecular Plant, 11*, 47–57.

García-Granados, R., Lerma-Escalera, J. A., & Morones-Ramírez, J. R. (2019). Metabolic engineering and synthetic biology: Synergies, future, and challenges. *Frontiers in Bioengineering and Biotechnology, 7*, 36.

Garg, M., Sharma, N., Sharma, S., Kapoor, P., Kumar, A., Chunduri, V., & Arora, P. (2018). Biofortified crops generated by breeding, agronomy, and transgenic approaches are improving lives of millions of people around the world. *Frontiers in Nutrition, 5*, 12.

Gayen, D., Ali, N., Sarkar, S. N., Datta, S. K., & Datta, K. (2015). Down-regulation of lipoxygenase gene reduces degradation of carotenoids of golden rice during storage. *Planta, 242*(1), 353–363.

Gregory, J. F. (2012). Accounting for differences in the bioactivity and bioavailability of vitamers. *Food & Nutrition Research, 56*(1), 5809.

Ha, S. H., Kim, J. K., Jeong, Y. S., You, M. K., Lim, S. H., & Kim, J. K. (2019). Stepwise pathway engineering to the biosynthesis of zeaxanthin, astaxanthin and capsanthin in rice endosperm. *Metabolic Engineering, 52*, 178–189.

Hanson, A. D., & Jez, J. M. (2018). Synthetic biology meets plant metabolism. *Plant Science, 273*, 1–2.

Hellmann, H., & Mooney, S. (2010). Vitamin B6: A molecule for human health? *Molecules, 15*(1), 442–459.

Herrero, S., & Daub, M. E. (2007). Genetic manipulation of vitamin B-6 biosynthesis in tobacco and fungi uncovers limitations to up-regulation of the pathway. *Plant Science, 172*(3), 609–620.

Hibberd, J. M., Sheehy, J. E., & Langdale, J. A. (2008). Using C4 photosynthesis to increase the yield of rice—Rationale and feasibility. *Current Opinion in Plant Biology, 11*(2), 228–231.

Hichri, I., Barrieu, F., Bogs, J., Kappel, C., Delrot, S., & Lauvergeat, V. (2011). Recent advances in the transcriptional regulation of the flavonoid biosynthetic pathway. *Journal of Experimental Botany, 62*, 2465–2483.

Hossain, T., Rosenberg, I., Selhub, J., Kishore, G., Beachy, R., & Schubert, K. (2004). Enhancement of folates in plants through metabolic engineering. *Proceedings of the National Academy of Sciences, 101*(14), 5158–5163.

Huang, J. C., Zhong, Y. J., Liu, J., Sandmann, G., & Chen, F. (2013). Metabolic engineering of tomato for high-yield production of astaxanthin. *Metabolic Engineering, 17*, 59–67.

Isaacson, T., Ohad, I., Beyer, P., & Hirschberg, J. (2004). Analysis in vitro of the enzyme CRTISO establishes a poly-cis-carotenoid biosynthesis pathway in plants. *Plant Physiology, 136*(4), 4246–4255.

Jian, W., Cao, H., Yuan, S., Liu, Y., Lu, J., Lu, W., Li, N., Wang, J., Zou, J., Tang, N., et al. (2019a). SlMYB75, an MYB-type transcription factor, promotes anthocyanin accumulation and enhances volatile aroma production in tomato fruits. *Horticulture Research, 6*, 22.

Jian, W., Cao, H., Yuan, S., Liu, Y., Lu, J., Lu, W., Li, N., Wang, J., Zou, J., Tang, N., et al. (2019b). SlMYB75, an MYB-type transcription factor, promotes anthocyanin accumulation and enhances volatile aroma production in tomato fruits. *Horticulture Research, 6*, 22.

Kana-Sop, M. M., Gouado, I., Achu, M. B., Van Camp, J., Zollo, P. H. A., Schweigert, F. J., . Oberleas, D. & Ekoe, T. (2015). The influence of iron and zinc supplementation on the bioavailability of provitamin A carotenoids from papaya following consumption of a vitamin A-deficient diet. Journal of Nutritional Science and Vitaminology, 61(3), 205–214.

Karunanandaa, B., Qi, Q., Hao, M., Baszis, S. R., Jensen, P. K., Wong, Y. H. H., Jiang, J., Venkatramesh, M., Gruys, K. J., Moshiri, F., & Valentin, H. E. (2005). Metabolically engineered oilseed crops with enhanced seed tocopherol. *Metabolic Engineering, 7*(5–6), 384–400.

Kiekens, F., Blancquaert, D., Devisscher, L., Van Daele, J., Stove, V. V., Delanghe, J. R., Van Der Straeten, D., Lambert, W. E., & Stove, C. P. (2015). Folates from metabolically engineered rice: A long-term study in rats. *Molecular Nutrition & Food Research, 59*(3), 490–500.

Li, K. T., Moulin, M., Mangel, N., Albersen, M., Verhoeven-Duif, N. M., Ma, Q., Zhang, P., Fitzpatrick, T. B., Gruissem, W., & Vanderschuren, H. (2015). Increased bioavailable vitamin B 6 in field-grown transgenic cassava for dietary sufficiency. *Nature Biotechnology, 33*(10), 1029–1032.

Li, Q., Farre, G., Naqvi, S., Breitenbach, J., Sanahuja, G., Bai, C., Sandmann, G., Capell, T., Christou, P., & Zhu, C. (2010). Cloning and functional characterization of the maize carotenoid

isomerase and β-carotene hydroxylase genes and their regulation during endosperm maturation. *Transgenic Research, 19*(6), 1053–1068.
Li, Z., Zhang, D., Xiong, X., Yan, B., Xie, W., Sheen, J., & Li, J. F. (2017). A potent Cas9-derived gene activator for plant and mammalian cells. *Nature Plants, 3*, 930–936.
Lipkie, T. E., De Moura, F. F., Zhao, Z. Y., Albertsen, M. C., Che, P., Glassman, K., & Ferruzzi, M. G. (2013). Bioaccessibility of carotenoids from transgenic provitamin A biofortified sorghum. *Journal of Agricultural and Food Chemistry, 61*(24), 5764–5771.
Liu, W., & Stewart, C. N., Jr. (2015). Plant synthetic biology. *Trends in Plant Science, 20*, 309–317.
Liu, X., Li, S., Yang, W., Mu, B., Jiao, Y., Zhou, X., Zhang, C., Fan, Y., & Chen, R. (2018a). Synthesis of seed-specific bidirectional promoters for metabolic engineering of anthocyanin-rich maize. *Plant and Cell Physiology, 59*, 1942–1955.
Liu, X., Yang, W., Mu, B., Li, S., Li, Y., Zhou, X., Zhang, C., Fan, Y., & Chen, R. (2018b). Engineering of 'Purple Embryo Maize' with a multigene expression system derived from a bidirectional promoter and self-cleaving 2A peptides. *Plant Biotechnology Journal, 16*, 1107–1109.
Mejia, D. (2003). *Maize: Post-harvest operations*. FAO.
Melse-Boonstra, A., West, C. E., Katan, M. B., Kok, F. J., & Verhoef, P. (2004). Bioavailability of heptaglutamyl relative to monoglutamyl folic acid in healthy adults. *The American Journal of Clinical Nutrition, 79*(3), 424–429.
Misawa, N., Truesdale, M. R., Sandmann, G., Fraser, P. D., Bird, C., Schuch, W., & Bramley, P. M. (1994). Expression of a tomato cDNA coding for phytoene synthase in Escherichia coli, phytoene formation in vivo and in vitro, and functional analysis of the various truncated gene products. *The Journal of Biochemistry, 116*(5), 980–985.
Mortimer, J. C. (2019). Plant synthetic biology could drive a revolution in biofuels and medicine. *Experimental Biology and Medicine (Maywood), 244*, 323–331.
Mugode, L., Ha, B., Kaunda, A., Sikombe, T., Phiri, S., Mutale, R., Davis, C., Tanumihardjo, S., & De Moura, F. F. (2014). Carotenoid retention of biofortified provitamin A maize (Zea mays L.) after Zambian traditional methods of milling, cooking and storage. *Journal of Agricultural and Food Chemistry, 62*(27), 6317–6325.
Muir, S. R., Collins, G. J., Robinson, S., Hughes, S., Bovy, A., Ric De Vos, C. H., van Tunen, A. J., & Verhoeyen, M. E. (2001). Overexpression of petunia chalcone isomerase in tomato results in fruit containing increased levels of flavonols. *Nature Biotechnology, 19*, 470–474.
Naqvi, S., Farré, G., Zhu, C., Sandmann, G., Capell, T., & Christou, P. (2011). Simultaneous expression of Arabidopsis ρ-hydroxyphenylpyruvate dioxygenase and MPBQ methyltransferase in transgenic corn kernels triples the tocopherol content. *Transgenic Research, 20*(1), 177–181.
Naqvi, S., Zhu, C., Farre, G., Ramessar, K., Bassie, L., Breitenbach, J., Conesa, D. P., Ros, G., Sandmann, G., Capell, T., & Christou, P. (2009). Transgenic multivitamin corn through biofortification of endosperm with three vitamins representing three distinct metabolic pathways. *Proceedings of the National Academy of Sciences, 106*(19), 7762–7767.
Nielsen, J., & Keasling, J. D. (2011). Synergies between synthetic biology and metabolic engineering. *Nature Biotechnology, 29*, 693–695.
Nunes, A. C., Kalkmann, D. C., & Aragao, F. J. (2009). Folate biofortification of lettuce by expression of a codon optimized chicken GTP cyclohydrolase I gene. *Transgenic Research, 18*(5), 661.
Ogo, Y., Ozawa, K., Ishimaru, T., Murayama, T., & Takaiwa, F. (2013). Transgenic rice seed synthesizing diverse flavonoids at high levels: A new platform for flavonoid production with associated health benefits. *Plant Biotechnology Journal, 11*, 734–746.
Ordóñez-Santos, L. E., & Vázquez-Riascos, A. (2010). Effect of processing and storage time on the vitamin C and lycopene contents of nectar of pink guava (*Psidium guajava* L.). *Archivos latinoamericanos de nutricion, 60*(3), 280.

Ouyang, S., Zhu, W., Hamilton, J., Lin, H., Campbell, M., Childs, K., Thibaud-Nissen, F., Malek, R. L., Lee, Y., Zheng, L., Orvis, J., & Buell, C. R. (2007). The TIGR rice genome annotation resource: Improvements and new features. *Nucleic Acids Research, 35*(suppl_1), D883–D887.
Paine, J. A., Shipton, C. A., Chaggar, S., Howells, R. M., Kennedy, M. J., Vernon, G., Wright, S. Y., Hinchliffe, E., Adams, J. L., Silverstone, A. L., & Drake, R. (2005a). Improving the nutritional value of Golden Rice through increased pro-vitamin A content. *Nature Biotechnology, 23*(4), 482–487.
Paine, J. A., Shipton, C. A., Chaggar, S., Howells, R. M., Kennedy, M. J., Vernon, G., Wright, S. Y., Hinchliffe, E., Adams, J. L., Silverstone, A. L., et al. (2005b). Improving the nutritional value of Golden Rice through increased pro-vitamin A content. *Nature Biotechnology, 23*, 482–487.
Peterhansel, C. (2011). Best practice procedures for the establishment of a C4 cycle in transgenic C3 plants. *Journal of Experimental Botany, 62*(9), 3011–3019.
Pierce, E. C., LaFayette, P. R., Ortega, M. A., Joyce, B. L., Kopsell, D. A., & Parrott, W. A. (2015). Ketocarotenoid production in soybean seeds through metabolic engineering. *PLoS One, 10*, e0138196.
Pouvreau, B., Vanhercke, T., & Singh, S. (2018). From plant metabolic engineering to plant synthetic biology: The evolution of the design/build/test/learn cycle. *Plant Science, 273*, 3–12.
Prochnik, S., Marri, P. R., Desany, B., Rabinowicz, P. D., Kodira, C., Mohiuddin, M., Rodriguez, F., Fauquet, C., Tohme, J., Harkins, T., Rokhsar, D. S., & Rounsley, S. (2012). The cassava genome: Current progress, future directions. *Tropical Plant Biology, 5*(1), 88–94.
Raschke, M., Boycheva, S., Crèvecoeur, M., Nunes-Nesi, A., Witt, S., Fernie, A. R., Amrhein, N., & Fitzpatrick, T. B. (2011). Enhanced levels of vitamin B6 increase aerial organ size and positively affect stress tolerance in Arabidopsis. *The Plant Journal, 66*(3), 414–432.
Rask, L., Andréasson, E., Ekbom, B., Eriksson, S., Pontoppidan, B., & Meijer, J. (2000). Myrosinase: Gene family evolution and herbivore defense in Brassicaceae. *Plant Molecular Biology, 42*(1), 93–114.
Rébeillé, F., Ravanel, S., Jabrin, S., Douce, R., Storozhenko, S., & Van Der Straeten, D. (2006). Folates in plants: Biosynthesis, distribution, and enhancement. *Physiologia Plantarum, 126*(3), 330–342.
Ricciarelli, R., Zingg, J. M., & Azzi, A. (2001). Vitamin E: Protective role of a Janus molecule. *The FASEB Journal, 15*(13), 2314–2325.
Ross, S. A., McCaffery, P. J., Drager, U. C., & De Luca, L. M. (2000). Retinoids in embryonal development. *Physiological Reviews, 80*(3), 1021–1054.
Sanderson, K. (2011). Lignocellulose: A chewy problem. *Nature, 474*, S12–S14.
Sayre, R., Beeching, J. R., Cahoon, E. B., Egesi, C., Fauquet, C., Fellman, J., Fregene, M., Gruissem, W., Mallowa, S., Manary, M., Maziya-Dixon, B., & Zhang, P. (2011). The BioCassava plus program: Biofortification of cassava for sub-Saharan Africa. *Annual Review of Plant Biology, 62*, 251–272.
Scott, J. M., & Weir, D. G. (1996). Homocysteine and cardio-vascular disease. *The Quarterly Journal of Medicine, 89*, 561–563.
Seshadri, S., Beiser, A., Selhub, J., Jacques, P. F., Rosenberg, I. H., D'Agostino, R. B., et al. (2002). Plasma homocysteine as a risk factor for dementia and Alzheimer's disease. *New England Journal of Medicine, 346*(7), 476–483.
Shin, Y. M., Park, H. J., Yim, S. D., Baek, N. I., Lee, C. H., An, G., & Woo, Y. M. (2006). Transgenic rice lines expressing maize C1 and R-S regulatory genes produce various flavonoids in the endosperm. *Plant Biotechnology Journal, 4*, 303–315.
Spreitzer, R. J., & Salvucci, M. E. (2002). Rubisco: Structure, regulatory interactions, and possibilities for a better enzyme. *Annual Review of Plant Biology, 53*.
Stanger, O. (2004). The potential role of homocysteine in percutaneous coronary interventions (PCI): Review of current evidence and plausibility of action. *Cellular and Molecular Biology (Noisy-le-Grand, France), 50*(8), 953–988.

Storozhenko, S., De Brouwer, V., Volckaert, M., Navarrete, O., Blancquaert, D., Zhang, G. F., & Van Der Straeten, D. (2007). Folate fortification of rice by metabolic engineering. *Nature Biotechnology, 25*(11), 1277–1279.

Tambasco-Studart, M., Titiz, O., Raschle, T., Forster, G., Amrhein, N., & Fitzpatrick, T. B. (2005). Vitamin B6 biosynthesis in higher plants. *Proceedings of the National Academy of Sciences, 102*(38), 13687–13692.

Ten Have, R., & Teunissen, P. J. (2001). Oxidative mechanisms involved in lignin degradation by white-rot fungi. *Chemical Reviews, 101*(11), 3397–3414.

Tian, Y. S., Wang, B., Peng, R. H., Xu, J., Li, T., Fu, X. Y., Xiong, A. S., Gao, J. J., & Yao, Q. H. (2019). Enhancing carotenoid biosynthesis in rice endosperm by metabolic engineering. *Plant Biotechnology Journal, 17*, 849–851.

Vanholme, R., Cesarino, I., Rataj, K., Xiao, Y., Sundin, L., Goeminne, G., Kim, H., Cross, J., Morreel, K., Araujo, P., Welsh, L., & Boerjan, W. (2013). Caffeoyl shikimate esterase (CSE) is an enzyme in the lignin biosynthetic pathway in Arabidopsis. *Science, 341*(6150), 1103–1106.

Vasconcelos, M., Musetti, V., Li, C. M., Datta, S. K., & Grusak, M. A. (2004). Functional analysis of transgenic rice (Oryza sativa L.) transformed with an Arabidopsis thaliana ferric reductase (AtFR02). *Soil Science and Plant Nutrition, 50*(7), 1151–1157.

Vasconcelos, M. W., Clemente, T., & Grusak, M. A. (2014). Evaluation of constitutive iron reductase (AtFRO2) expression on mineral accumulation and distribution in soybean (Glycine max. L). *Frontiers in Plant Science, 5*, 112.

Von Caemmerer, S., Quick, W. P., & Furbank, R. T. (2012). The development of C4 rice: Current progress and future challenges. *Science, 336*(6089), 1671–1672.

Wang, C., Zeng, J., Li, Y., Hu, W., Chen, L., Miao, Y., Deng, P., Yuan, C., Ma, C., Chen, X., et al. (2014). Enrichment of provitamin A content in wheat (*Triticum aestivum* L.) by introduction of the bacterial carotenoid biosynthetic genes CrtB and CrtI. *Journal of Experimental Botany, 65*, 2545–2556.

WHO. (2002). *The world health report*. World Health Organization.

Wilkerson, C. G., Mansfield, S. D., Lu, F., Withers, S., Park, J. Y., Karlen, S. D., Gonzales-Vigil, E., Padmakshan, D., Unda, F., Rencoret, J., & Ralph, J. (2014). Monolignol ferulate transferase introduces chemically labile linkages into the lignin backbone. *Science, 344*(6179), 90–93.

Wu, H., Li, L., Du, J., Yuan, Y., Cheng, X., & Ling, H. Q. (2005). Molecular and biochemical characterization of the Fe (III) chelate reductase gene family in Arabidopsis thaliana. *Plant and Cell Physiology, 46*(9), 1505–1514.

Wu, Z., Robinson, D. S., Hughes, R. K., Casey, R., Hardy, D., & West, S. I. (1999). Co-oxidation of β-carotene catalyzed by soybean and recombinant pea lipoxygenases. *Journal of Agricultural and Food Chemistry, 47*(12), 4899–4906.

Yazdani, M., Sun, Z., Yuan, H., Zeng, S., Thannhauser, T. W., Vrebalov, J., Ma, Q., Xu, Y., Fei, Z., Van Eck, J., et al. (2019). Ectopic expression of ORANGE promotes carotenoid accumulation and fruit development in tomato. *Plant Biotechnology Journal, 17*, 33–49.

Ye, X., Al-Babili, S., Klöti, A., Zhang, J., Lucca, P., Beyer, P., & Potrykus, I. (2000). Engineering the provitamin A (β-carotene) biosynthetic pathway into (carotenoid-free) rice endosperm. *Science, 287*(5451), 303–305.

Ye, X., Honda, K., Sakai, T., Okano, K., Omasa, T., Hirota, R., Kuroda, A., & Ohtake, H. (2012). Synthetic metabolic engineering—A novel, simple technology for designing a chimeric metabolic pathway. *Microbial Cell Factories, 11*, 120.

Yuan, L., & Grotewold, E. (2015). Metabolic engineering to enhance the value of plants as green factories. *Metabolic Engineering, 27*, 83–91.

Zalatan, J. G., Lee, M. E., Almeida, R., Gilbert, L. A., Whitehead, E. H., La Russa, M., Tsai, J. C., Weissman, J. S., Dueber, J. E., Qi, L. S., et al. (2015). Engineering complex synthetic transcriptional programs with CRISPR RNA scaffolds. *Cell, 160*, 339–350.

Zhang, K., Bhuiya, M. W., Pazo, J. R., Miao, Y., Kim, H., Ralph, J., & Liu, C. J. (2012). An engineered monolignol 4-O-methyltransferase depresses lignin biosynthesis and confers novel metabolic capability in Arabidopsis. *The Plant Cell, 24*(7), 3135–3152.

Zhang, Y., Butelli, E., Alseekh, S., Tohge, T., Rallapalli, G., Luo, J., Kawar, P. G., Hill, L., Santino, A., Fernie, A. R., et al. (2015). Multi-level engineering facilitates the production of phenylpropanoid compounds in tomato. *Nature Communications, 6*, 8635.

Zhang, Y., Butelli, E., & Martin, C. (2014). Engineering anthocyanin biosynthesis in plants. *Current Opinion in Plant Biology, 19*, 81–90.

Zhu, C., Naqvi, S., Breitenbach, J., Sandmann, G., Christou, P., & Capell, T. (2008). Combinatorial genetic transformation generates a library of metabolic phenotypes for the carotenoid pathway in maize. *Proceedings of the National Academy of Sciences, 105*(47), 18232–18237.

Zhu, Q., Yu, S., Zeng, D., Liu, H., Wang, H., Yang, Z., Xie, X., Shen, R., Tan, J., Li, H., et al. (2017). Development of "purple endosperm rice" by engineering anthocyanin biosynthesis in the endosperm with a high-efficiency transgene stacking system. *Molecular Plant, 10*, 918–929.

Zhu, Q., Zeng, D., Yu, S., Cui, C., Li, J., Li, H., et al. (2018). From golden rice to aSTARice: Bioengineering astaxanthin biosynthesis in rice endosperm. *Molecular Plant, 11*(12), 1440–1448.

Zhu, X. G., Shan, L., Wang, Y., & Quick, W. P. (2010). C4 rice – An ideal arena for systems biology research. *Journal of Integrative Plant Biology, 52*(8), 762–770.

Chapter 7
miRNA- and RNAi-Mediated Metabolic Engineering in Plants

Deepu Pandita

Abstract MicroRNAs (miRNAs) are endogenous, highly conserved, small regulatory RNA molecules with a size of 19–24 nucleotides, micro in structure and macro in character and regulate gene expression in numerous developmental and signalling pathways. Transcribed from MIR genes, miRNAs act as key mediators of post-transcriptional gene regulation in eukaryotes. Biological functions of most miRNAs in plants include involvement in a huge range of biological processes such as development, signal transduction, biotic resistance, abiotic resistance, nutritional value and secondary metabolite biosynthesis. The miRNAs open a novel perspective to metabolic engineering, as miRNAs are involved in various biosynthetic pathways of secondary metabolites and their effects need to be explored. Current understanding about plant miRNAs, classical pathways of miRNA biogenesis and perspective analysis of RNAi and miRNA potent targets in metabolic engineering and potential in metabolic engineering in plants will be elucidated.

Keywords miRNA · RNAi · Post-translational gene regulation · Metabolic engineering · Secondary metabolites · Stress

Introduction

The plant microRNAs (miRNAs) for the first time reported in 2002 are the maximum investigated class of non-coding RNA (ncRNA) (Reinhart et al., 2002). The fundamental, endogenous, near-ubiquitous, micro in structure and macro in character and sequence-specific regulatory (posttranscriptional) miRNAs present throughout the plant kingdom have a size of 19–24 nucleotides (Yi et al., 2014; He et al., 2018a, 2018b; Pandita, 2018a; Xu et al., 2019). The miRNAs are transcribed from specific MIR genes to produce primary transcripts that fold into a step-loop structure. Step-loop structure is processed into a duplex with two mature miRNA sequences

D. Pandita (✉)
Government Department of School Education, Jammu, Jammu and Kashmir, India

T. Aftab, K. R. Hakeem (eds.), *Metabolic Engineering in Plants*,
https://doi.org/10.1007/978-981-16-7262-0_7

(Kamthan et al., 2015). Single-celled *Chlamydomonas reinhardtii* also produces miRNAs (Molnar et al., 2007; Zhao et al., 2007). A number of miRNA target transcription factors act as general growth regulators or regulate an immense array of biological functions such as control of plant hormones, fruit ripening, immune responses, development and stress adaptations (Wightman et al., 1993; Reinhart et al., 2000; Rhoades et al., 2002; Sunkar et al., 2007; Flynt & Lai, 2008; Pedersen & David, 2008; Garcia, 2008; Voinnet, 2008; Cedillo-Jimeneza et al., 2020). RNA interference (RNAi) is an umbrella term and promising gene regulatory approach involving sRNA-directed precise sequence-specific gene silencing/downregulation with potential for plant gene modification. RNA interference (RNAi), RNA silencing or RNA quelling is used to switch off undesirable genes (Hannon, 2002; Pandita, 2018b) and improve crops by improving the desired traits and decreasing the undesired traits (Gupta et al., 2017). Examples of RNAi-generated plants with improved traits include *Gossypium hirsutum* with increased content of cottonseed oil by downregulating phosphoenolpyruvate carboxylase 1 (GhPEPC1) (Xu et al., 2016), tolerance against leaf blight disease and rice blast in *Oryza sativa* by downregulation of rice fatty acid desaturase gene (OSSI2) (Jiang et al., 2009), production of seedless fruit by downregulation of gene encoding chalcone synthase (CHS) in *Zea mays*, petunia (Mahajan et al., 2011), and tomato (Schijlen et al., 2007), SQS inactivation enhanced artemisinin production, b- caryophyllene synthase (CPS) increased the content of artemisinin and dihydroartemisinic acid (DHAA) (Lu et al., 2016), trichome and artemisinin regulator 1 (TAR1) decreased accumulation of artemisinin (Tan et al., 2015), silencing of caffeine synthase (CS) reduced caffeine content (Mohanpuria et al., 2011) and silencing of ent-kaurenoic acid-13 hydroxylase (SrKA13H) and UDP glycosyltransferase (SrUGT85C2) restricted steviol glycoside synthesis (Guleria & Yadav, 2013). Plant miRNAs may act as important metabolic engineering targets in RNAi. The biosynthesis of some plant secondary metabolites is under regulation of miRNAs (Najafabadi & Naghavi, 2018). The miRNA-based metabolic engineering can be used in secondary metabolite production and in the designing of stress-tolerant crops (Chownk et al., 2018). MiRNA-based secondary metabolite regulation will help us in understanding avenues of miRNAs in plant metabolic engineering. Development of the miRNA-mediated transgenic plants through miRNA technology with enhanced secondary metabolites is vision and goal of plant researchers (Bulgakov & Avramenko, 2015; Zhang & Wang, 2015). This chapter deliberates the contemporary advances in the mechanisms of miRNA biogenesis and mode of action of plant miRNAs, roles of miRNAs in a variety of metabolic processes and miRNA-mediated metabolic engineering in plants.

Biogenesis of microRNAs

Primary miRNA transcript precursor known as pri-miRNAs (capped and polyadenylated imperfect fold-back structure) are mostly transcribed from miRNA-encoding (MIR) genes by RNA polymerase II (Pol II). MIR genes are located between non-protein-coding or protein-coding genes in plants. DAWDLE (DDL) encodes nuclear RNA-binding protein which stabilizes pri-miRNAs. Further direct interaction with DCL1 in nuclear processing centres called D-bodies converts pri-miRNA to stem-loop precursor (pre-miRNA) (Lee et al., 2004; Kim, 2005; Yu et al., 2008). The miRNA hairpins are present in genomic clusters and several hairpins are formed from a particular pri-miRNA (Rajagopalan et al., 2006; Axtell et al., 2007; Meng & Shao, 2012). Both the processes of conversion of pri-miRNA into pre-miRNA and processing of mature miRNAs are coordinated by the DCL1 in plants (Kurihara et al., 2006). The C_2H_2-zinc finger protein SERRATE (SE) and RNA-binding protein HYPONASTIC LEAVES 1 (HYL1) interact in a combined mode with Dicer-like 1 (DCL1) in D-bodies or SmD3/SmB-bodies (Kurihara et al., 2006; Fang & Spector, 2007) and nuclear cap-binding complex (CBC) (Laubinger et al., 2008; Gregory et al., 2008).

Pre- or mature miRNAs are exported to cytoplasm by HASTY which is a plant homologue of exportin-5 and other unidentified factors (Park et al., 2005). The precise miRNA form which is exported is ambiguous, as locations of miRNA/miRNA* strand separation and miRISC loading are indefinable. Mature miRNA duplexes (miRNA/miRNA*) excise from pre-miRNAs. In this duplex, miRNA is the guide strand and miRNA* is the degraded strand. The miRNA/miRNA* duplex is stabilized and methylated by S-adenosylmethionine-dependent methyltransferase Hua Enhancer 1 (HEN1) for their protection from degradation by the highly conserved small RNA-degrading nuclease (SDN) exonucleases. HEN1 methylates small RNAs at 3′ terminal nucleotide of each strand to avoid uridylation and subsequent degradation (Li et al., 2005; Yang et al., 2006). In *Arabidopsis thaliana*, four Dicer-like (DCL1, DCL2, DCL3, DCL4) proteins are present. DCL1 produces 18–21 nucleotides; DCL2 produces 22 nucleotides, DCL3 produces 24 nucleotides and DCL4 produces 21 nucleotide duplexes. Cleaved sRNA duplexes reserve inside the nucleus for chromatin level events or are exported to the cytoplasm for the post-transcriptional gene silencing (PTGS). The specific primary guide miRNA strand is then loaded into Argonaute (AGO) protein of RNA-induced silencing complexes (RISCs) also known as miRISC prior to binding their target gene(s) to carry out the silencing reactions. AGO has a sRNA-binding PAZ domain and a PIWI domain with endonucleolytic activity to cleave RNAs (Vaucheret, 2008; Winter & Diederichs, 2011; Sun et al., 2012; Iwakawa & Tomari, 2013). The detailed sketch of biogenesis and action of plant miRNAs can be found in various chapters (Pandita, 2019; Pandita & Wani, 2019). The other strand degrades by exosomes (Sun et al., 2012). The miRNA-based transcript repression takes place when mature miRNAs complementary to target mRNA boost site-specific exonucleolytic cleavage of target mRNA whereas miRNAs with impaired base pairing with target mRNA lead to translation

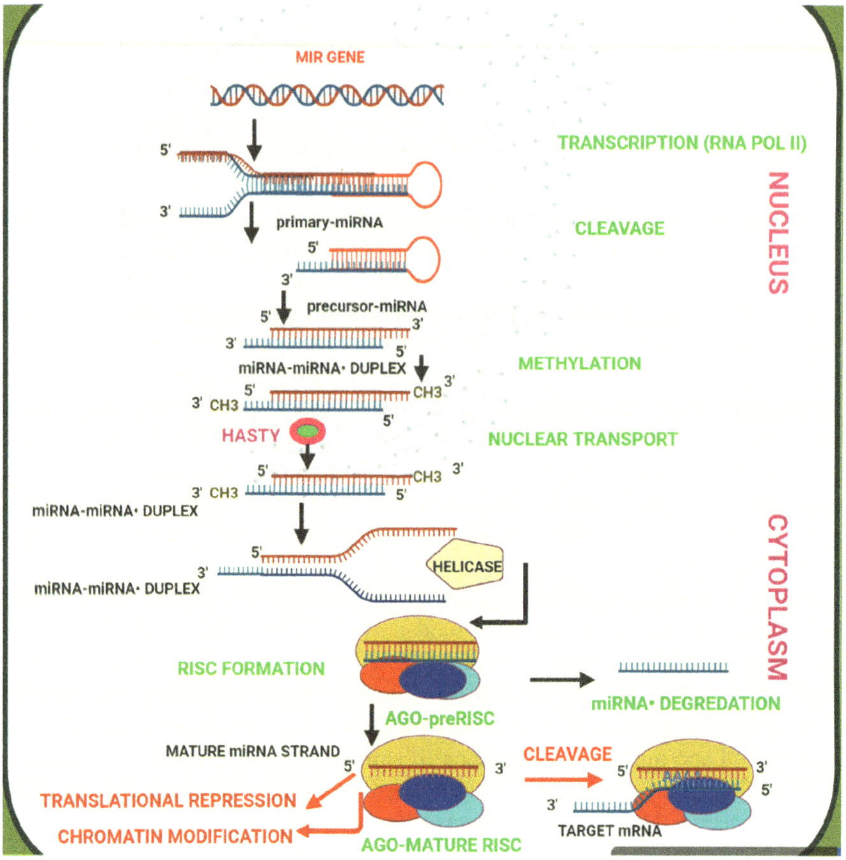

Fig. 7.1 The graphical representation of miRNA biogenesis and action in plants

inhibition or undergo chromatin modification (Eulalio et al., 2008; Sun et al., 2012; Sreekumar & Soniya, 2018). The graphic representation of miRNA biogenesis and action in plants is given in Fig. 7.1.

Mode of Action of Plant miRNA

RNA silencing includes induction and processing of double-stranded RNA (dsRNA), sRNA methylation and sRNA integration into miRISC complexes (Chapman and Carrington, 2007). Mature miRNA targets genes by cleavage or translation inhibition (Rogers & Chen, 2013). In RNA transcript cleavage or slicing mode of action, miRNA regulates target at mRNA levels. Extensively complementary conserved miRNAs pair with target sites of 9–11 nucleotides in open reading frames

(ORFs) of mRNA (Rhoades et al., 2002; Llave et al., 2002; Dunoyer et al., 2004; Souret et al., 2004).

RNA translational inhibition takes place on membrane-bound polysomes (MBPs). In this phenomenon, miRNA regulates target at protein level and translational repressors (TR) avoid slicing of miRNA-loaded AGO1 and AGO10. Translation inhibits by blocking read-through of protein making ribosomes (Wang et al., 2008) or their disturbed movement or inhibition of translation initiation (Iwakawa & Tomari, 2013).

AGO1 cleaves targets of the miRNA and also blocks their translation as in hypomorphic ago1–27 mutants (Baumberger & Baulcombe, 2005; Brodersen et al., 2008). MAD5-encoded microtubule-severing enzyme katanin (KTN) and Ge-1 orthologue varicose (VCS) also take part in translation inhibition, SUO and endoplasmic reticulum (ER)-localized altered meristem program 1 (AMP1) (Brodersen et al., 2008; Yang et al., 2012; Li et al., 2013).

miRNA-Mediated Metabolic Engineering in Plants

The small RNA inactivating tools including short tandem target mimic (STTM), miRNA target mimicry (TM) and miRNA sponge can be efficiently used for functional characterization of miRNAs and miRNA-based trait improvement (Franco-Zorrilla et al., 2007; Ebert et al., 2007; Tang et al., 2012; Yan et al., 2012). MiRNAs can be downregulated by amiRNAs and artificial/synthetic trans-acting siRNAs (atasiRNAs/syntasiRNAs) as well (Chang et al., 2006; Schwab et al., 2006; Eamens et al., 2011; Zhang, 2014). Metabolic engineering includes manipulation of complex biosynthetic metabolic pathways to boost novel beneficial metabolites, e.g. golden rice, artemisinin, flavonoids in plant and microbes and stress tolerance in crop plants, and decrease toxic metabolites in specific plants (Capell & Christou, 2004; Yuan & Grotewold, 2015; Chownk et al., 2018). Plant miRNA gene regulator tools will enable us to better comprehend and modify biosynthetic pathways for desired characters (Bulgakov & Avramenko, 2015).

Abiotic Stress-Resistant Crops

Seven miRNA families (miR2949, miR167, miR160, miR156, miR172, miR156, miR393 and miR3476) act as the main positive or negative regulators of heat stress (Chen et al., 2020). *Arabidopsis thaliana* miR156 and miR169 are important in plants for growth, development and responsiveness to stresses mainly high temperature stress memory (Zhang et al., 2011; Stief et al., 2014). Overexpressed rice miR156 transgenic showed sensitivity to high temperature stress (Wang et al., 2012). MiR169 transgenic plants showed better performance, reduced stomatal water loss and required less water (Zhang et al., 2011). The miR169 overexpression caused nitrogen starvation hypersensitivity in *A. thaliana* (Zhao et al., 2011). MiR169 target genes overcome deficiency of nitrogen and water. The miR319 expression increases during stress (Zhou et al., 2010). Transgenic creeping bentgrass showed higher miR319 expression and higher salt and drought tolerance (Zhou et al., 2013). The miR319 increases water retention and integrity of plasma membrane in transgenic (Pieczynski et al., 2013). Transgenic (miR319) rice boosts tolerance to cold stress (Yang et al., 2013). Rice and *Arabidopsis* miR393 show response to alkalinity and salt stress (Gao et al., 2011). Altered miR393 expression enhanced tolerance to salt, drought and heat stress in creeping bentgrass (Zhao et al., 2018). Rice knock-out mutants of miR166 showed increased tolerance to water deficiency (Zhang et al., 2018). Expression of premiR828 increased H_2O_2 production and biosynthesis of lignin in sweet potato (Lin et al., 2012). These studies are summarized in Table 7.1.

Secondary Metabolite (SM) Production

MicroRNAs are important players and regulatory factors in the field of secondary metabolite biosynthetic pathways (Gandhi et al., 2015). Various examples with insights of miRNA regulatory functions in production of SMs are discussed in coming paragraphs.

The miRNA-mediated regulation of alkaloid biosynthesis has been reported in *Papaver somniferum*. Pso-miR13, pso-miR2161 and psomiR408 which target 7-OMT, 4-OMT and reticuline oxidase-like protein, respectively, take part in the biosynthesis of benzylisoquinoline alkaloids (BIAs) (Boke et al., 2015). MiRNAs play roles in the regulation of nicotine biosynthetic pathway in *Nicotiana tabacum*. miRX17, miRX27, miRX20 and miRX19 target *quinolinate phosphoribosyltransferase* genes, namely, QPT1, QPT2, CYP82E4 and PMT2, respectively. Nicotine biosynthesis enhanced in treated tobacco plants by miRNA-eTM regulatory module involving nta-eTMX27-mediated inhibition of nta-miRX27 (Li et al., 2015). Similarly miR164 plays a role in the nicotine content in tobacco. In treated tobacco plants, miR164 shows downregulation and NtNAC-R1 shows upregulation resulting in increased nicotine content (Fu et al., 2013). Taxane 13α hydroxylase and taxane 2α-O-benzoyltransferase involved in the paclitaxel biosynthetic pathway are targeted and negatively regulated by miR164 and miR171, respectively, in *Taxus baccata* (Jennewein & Croteau, 2001; Hao et al., 2012; Ramirez-Estrada et al., 2016). Farnesyl diphosphate synthase (FPS), 3-hydroxy-3-methylglutaryl-coenzyme A reductase (HMGR), geranyl-diphosphate synthase, squalene synthase and squalene epoxidase (SE) genes of ginsenoside biosynthetic pathway are targeted by miR854b, miR854c, miR854e and other diverse miRNAs in *Panax ginseng* (Xie et al., 2011; Mathiyalagan et al., 2013). Mining of miRNAs in picroside biosynthetic pathway of *Picrorhiza kurroa* was done by Vashisht et al. (2015). The 3-deoxy-7-phosphoheptulonate synthase of cinnamic acid is targeted by miRNA4995, and thus it has a regulatory function in terpenoid biosynthesis and influences picroside-I production (Vashisht et al., 2015). Terpenoid indole alkaloid

Table 7.1 miRNA-mediated regulation of abiotic stress

Transgenic plants	microRNAs	Effects produced in transgenic plants	Reference/s
Arabidopsis thaliana	miR156	Boost in heat tolerance	Stief et al. (2014)
	miR169	Improved N-deprivation sensitivity	Zhao et al. (2011)
	miR173	Improved thermo tolerance	Li et al. (2014)
	miR393	Improved sensitivity to salinity and alkalinity	Gao et al. (2011)
	miR394	Boost in tolerance to water deficiency	Ni et al. (2012)
	miR395c/ miR395e	Decreased seed germination under salinity and drought	Kim et al. (2010a)
	miR396	Increased sensitivity to salinity and alkalinity	Gao et al. (2010)
	miR402	Improved salinity, drought and cold tolerance	Kim et al. (2010b)
	miR417	Increased sensitivity to salinity and ABA	Jung and Kang (2007)
	miR165/ 166	Drought tolerant mutants	Yang et al. (2019a, 2019b)
Agrostis spps.	miR319	Improved salinity and drought tolerance	Zhou et al. (2013)
	miR393a	Salinity, heat and drought tolerance	Zhao et al. (2018)
Brassica napus	miR395	Improved tolerance to oxidative stress and heavy metal stress	Zhang et al. (2013a, 2013b)
Cicer arietinum	MiR408	Improved tolerance to water deficiency	Hajyzadeh et al. (2015)
Ipomoea batatas	miR828	Oxidative stress	Lin et al. (2012)
Oryza sativa	miR156	Decreased cold tolerance	Cui et al. (2015)
	miR319	Improved tolerance to chilling stress	Yang et al. (2013)
	miR159	Improved tolerance to water deficiency	Zhao et al. (2017)
	miR166	Mutants were drought tolerant	Zhang et al. (2018)
	miR393	Knockdown mutants exhibited reduced tolerance to salt and water scarcity stresses	Xia et al. (2012)
Solanum lycopersicum	miR169	Improved water scarcity tolerance	Zhang et al. (2011)

(TIA) biosynthetic pathway of *Catharanthus roseus* produces antitumor phytochemicals of vinblastine and vincristine. GCPE protein and terpenoid cyclase of TIAs pathway are targeted by miR-5021 (El-Sayed & Verpoorte, 2007; Pani & Mahapatra, 2013). MiR-5021-mediated metabolic editing tool can enhance production of vinblastine and vincristine. MYB12 lipoxygenase of flavonoid biosynthetic pathway gets targeted by miR828a and miR948a in *Salvia sclarea* (Legrand et al., 2010). MiR2911 targets γ-tocopherol methyltransferase involved in γ-tocopherol production in tocopherol biosynthetic pathway in *Helianthus annuus* and thereby decreased α-tocopherol levels (Barozai et al., 2012). Production of artemisinin takes place in secretory cells of *Artemisia annua* and depends on the glandular trichome and trichome density miR390 targets gene of trichome development. MiR390 affected artemisinin synthesis indirectly (Pérez-Quintero et al., 2012; Zare-Mehrjerdi et al., 2013). MiR7539-, miR5021- and miR1134-based regulation of terpenoid biosynthesis in *Xanthium strumarium* takes place by targeting genes of terpenoid pathway. Gibberellin 3 oxidase, squalene epoxidase, beta-amyrin synthase and germacrene A oxidase, R-linalool synthase and ent-kaurene synthase were targeted by miR5183, miR5255, miR5491 and miR6435, miR7540 and miR6449, respectively. MiR6435 expresses in glandular trichome which is a primary site for xanthanolide biosynthesis in *X. strumarium* (Fan et al., 2015). The miR156a, miR166a, miR166b, miR168, miR11320 and miR11071 target the phosphoglycerate mutase (PGM), premnaspirodiene oxygenase (PSO), homeobox-leucine zipper protein (HD-ZIP), acetyl-CoA acetyltransferase (AACT), aspartate aminotransferase (PHAT) and ribulose-phosphate3epimerase (RPE), respectively, with roles in secondary metabolism, suggesting their roles in metabolic manipulation to enhance secondary metabolites of *Swertia chirayita* (Padhan et al., 2016). *Ferula gummosa* produces low levels of aromatic galbanum. MiR2919, miR838, miR5021, miR5251 and miR5658 are part of terpene biosynthetic pathway. The miR1533, miR5021 and miR5658 target SPL7, SPL11 and ATHB13 transcription factors and can be new contenders for metabolic engineering to enhance terpene quantity (Singh & Sharma, 2015; Najafabadi & Naghavi, 2018). MiR393 targets auxin receptors and halts auxin signalling. The miR393 overexpression enhances glucosinolate levels and declines camalexin levels and redirects metabolic flow towards antimicrobial compounds (Jones & Dangl, 2006; Robert-Seilaniantz et al., 2011). The other examples of miRNA-mediated regulation of biosynthesis of phytochemicals are summarized in Table 7.2.

Conclusion

Some challenges and advanced investigations are required to design fully miRNA-regulated plants. The miRNA- and RNAi based plant metabolic engineering adds to the improvement of crops not only at the quantity (grain yield) and quality (nutritional levels), and stress conditions but can also help in designing plants with significantly enhanced contents of secondary metabolites. These tailor made plants can change the face of future agriculture and drugs and human health sector.

Table 7.2 miRNA-mediated regulation of biosynthesis of phytochemicals

Plant species	miRNA/targets	Phytochemical biosynthesis	Reference/s
Arabidopsis thaliana	MiR156/SPL9	Anthocyanin	Gou et al. (2011)
	MiR858a/R2R3-MYB transcription factors	Flavonoid	Sharma et al. (2016)
Catharanthus roseus	miR5021/UDP-glucose iridoid glucosyl transferase	Indole alkaloids as well as quinolone alkaloids	Pani and Mahapatra (2013)
Papaver somniferum	miR13/7-O-metyltransferase (7-OMT)	BIA Biosynthesis	Boke et al. (2015)
Picrorhiza kurroa	miR4995/3-Deoxy-7 phosphoheptulonate synthase (DAHP synthase)	Picroside	Vashisht et al. (2015)
Podophyllum hexandrum	miR1438/Caffeoyl-CoAO-methyltransferase	Lignin	Biswas et al. (2016)
	miR5532/2-hydroxy isoflavanone dehydratase	Isoflavonoid	Biswas et al. (2016)
	miR5538/Protein-S-isoprenylcysteine O-methyl transferase	Terpenoid backbone biosynthesis	Biswas et al. (2016)
Pogostemon cablin	miR156/SPL9	Sesquiterpenoid and triterpenoid	Yu et al. (2015)
Salvia miltiorrhiza	miR5072/Acetyl-CoAC-acetyl transferase	Tanshinones	Xu et al. (2014)
Xanthium strumarium	miR1134/3-hydroxy-3-methylglutaryl coenzyme A reductase (HMGR)	Terpenoid backbone biosynthesis	Fan et al. (2015)
Zingiber officinale	miR838/CYP71	Terpenoid metabolism	Singh et al. (2016)

Acknowledgements DP acknowledges Dr. Saurabh Raghuvanshi, Associate Professor, Department of Plant Molecular Biology, South Campus University of Delhi, New Delhi, India, for introducing her to the world of plant microRNAs.

References

Axtell, M. J., Snyder, J. A., & Bartel, D. P. (2007). Common functions for diverse small RNAs of land plants. *Plant Cell, 19*, 1750–1769. https://doi.org/10.1105/tpc.107.051706

Barozai, M. Y. K., Baloch, I. A., & Din, M. (2012). Identification of microRNAs and their targets in *Helianthus*. *Molecular Biology Reports, 39*, 2523–2532.

Baumberger, N., & Baulcombe, D. C. (2005). Arabidopsis ARGONAUTE1 is an RNA slicer that selectively recruits microRNAs and short interfering RNAs. *Proceedings of the National Academy of Sciences of the United States of America, 102*, 11,928–11,933.

Biswas, S., Hazra, S., & Chattopadhyay, S. (2016). Identification of conserved miRNAs and their putative target genes in *Podophyllum hexandrum* (Himalayan Mayapple). *Plant Gene, 6*, 82–89. https://doi.org/10.1016/j.plgene.2016.04.002

Boke, H., Ozhuner, E., Turktas, M., Parmaksiz, I., Ozcan, S., & Unver, T. (2015). Regulation of the alkaloid biosynthesis by miRNA in opium poppy. *Plant Biotechnology Journal, 13*, 409–420. https://doi.org/10.1111/pbi.12346

Brodersen, P., Sakvarelidze-Achard, L., Bruun-Rasmussen, M., Dunoyer, P., Yamamoto, Y. Y., Sieburth, L., & Voinnet, O. (2008). Widespread translational inhibition by plant miRNAs and siRNAs. *Science, 320*, 1185–1190.

Bulgakov, V. P., & Avramenko, T. V. (2015). New opportunities for the regulation of secondary metabolism in plants: Focus on microRNAs. *Biotechnology Letters, 37*, 1719–1727.

Capell, T., & Christou, P. (2004). Progress in plant metabolic engineering. *Current Opinion in Biotechnology, 15*, 148–154.

Cedillo-Jimeneza, C. A., Feregrino-Pereza, A. A., Guevara-Gonzáleza, R. G., & Cruz-Hernándezb, A. (2020). MicroRNA regulation during the tomato fruit development and ripening: A review. *Scientia Horticulturae, 270*, 109435. https://doi.org/10.1016/j.scienta.2020.109435

Chang, K., Elledge, S. J., & Hannon, G. J. (2006). Lessons from nature: MicroRNA-based shRNA libraries. *Nature Methods, 3*, 707–714.

Chapman, E. J., & Carrington, J. C. (2007). Specialization and evolution of endogenous small RNA pathways. *Nature Reviews Genetics, 8*(11), 884–896. https://doi.org/10.1038/nrg2179

Chen, J., Pan, A., He, S., Su, P., Yuan, X., Zhu, S., & Liu, Z. (2020). Different microRNA families involved in regulating high temperature stress response during cotton (*Gossypium hirsutum* L.) anther development. *International Journal of Molecular Sciences, 21*, 1280. https://doi.org/10.3390/ijms21041280

Chownk, M., Thakur, K., & Sudesh Kumar Yadav, S. K. (2018). Retrospect and prospects of plant metabolic engineering. *Journal of Plant Biochemistry and Biotechnology*. https://doi.org/10.1007/s13562-018-0473-7

Cui, N., Sun, X., Sun, M., Jia, B., Duanmu, H., Lv, D., et al. (2015). Overexpression of OsmiR156k leads to reduced tolerance to cold stress in rice (*Oryza sativa*). *Molecular Breeding, 35*, 214. https://doi.org/10.1007/s11032-015-0402-6

Dunoyer, P., Lecellier, C. H., Parizotto, E. A., Himber, C., & Voinnet, O. (2004). Probing the microRNA and small interfering RNA pathways with virus-encoded suppressors of RNA silencing. *Plant Cell, 16*, 1235–1250.

Eamens, A. L., Agius, C., Smith, N. A., Waterhouse, P. M., & Wang, M. B. (2011). Efficient silencing of endogenous microRNAs using artificial microRNAs in *Arabidopsis thaliana*. *Molecular Plant, 4*, 157–170. https://doi.org/10.1093/mp/ssq061

Ebert, M. S., Neilson, J. R., & Sharp, P. A. (2007). MicroRNA sponges: competitive inhibitors of small RNAs in mammalian cells. *Nature Methods, 4*, 721. https://doi.org/10.1038/nmeth1079

El-Sayed, M., & Verpoorte, R. (2007). Catharanthus terpenoid indole alkaloids: biosynthesis and regulation. *Phytochemistry Reviews, 6*, 277–305.

Eulalio, A., Huntzinger, E., & Izaurralde, E. (2008). Getting to the root of miRNA-mediated gene silencing. *Cell, 132*, 9–14.

Fan, R., Li, Y., Li, C., & Zhang, Y. (2015). Differential microRNA analysis of glandular trichomes and young leaves in *Xanthium strumarium* L. reveals their putative roles in regulating terpenoid biosynthesis. *PLoS One, 10*, e0139002. https://doi.org/10.1371/journal.pone.0139002

Fang, Y., & Spector, D. L. (2007). Identification of nuclear dicing bodies containing proteins for microRNA biogenesis in living Arabidopsis plants. *Current Biology, 17*, 818–823.

Flynt, A. S., & Lai, E. C. (2008). Biological principles of microRNA-mediated regulation: shared themes amid diversity. *Nature Reviews. Genetics, 9*, 831–842.

Franco-Zorrilla, J. M., Valli, A., Todesco, M., Mateos, I., Puga, M. I., Rubio-Somoza, I., Leyva, A., Weigel, D., García, J. A., & Paz-Ares, J. (2007). Target mimicry provides a new mechanism for regulation of microRNA activity. *Nature Genetics, 39*(8), 1033–1037. https://doi.org/10.1038/ng2079

Fu, Y., Guo, H., Cheng, Z., Wang, R., Li, G., Huo, G., & Liu, W. (2013). NtNAC-R1, a novel NAC transcription factor gene in tobacco roots, responds to mechanical damage of shoot meristem. *Plant Physiology and Biochemistry, 69*, 74–81.

Gandhi, S. G., Mahajan, V., & Bedi, Y. S. (2015). Changing trends in biotechnology of secondary metabolism in medicinal and aromatic plants. *Planta, 241*, 303–317.

Gao, P., Bai, X., Yang, L., Lv, D., Li, Y., Cai, H., Ji, W., Guo, D., & Zhu, Y. (2010). Over-expression of Osa-MIR396c decreases salt and alkali stress tolerance. *Planta, 231*, 991–1001. https://doi.org/10.1007/s00425010-1104-2

Gao, P., Bai, X., Yang, L., Lv, D., Pan, X., Li, Y., Cai, H., Ji, W., Chen, Q., & Zhu, Y. (2011). Osa-MIR393: a salinity- and alkaline stress-related microRNA gene. *Molecular Biology Reports, 38*, 237–242. https://doi.org/10.1007/s11033-010-0100-8

Garcia, D. (2008). A miRacle in plant development: Role of microRNAs in cell differentiation and patterning. *Seminars in Cell & Developmental Biology, 19*, 586–595.

Gou, J. Y., Felippes, F. F., Liu, C. J., Weigel, D., & Wang, J. W. (2011). Negative regulation of anthocyanin biosynthesis in Arabidopsis by a miR156-targeted SPL transcription factor. *Plant Cell, 23*, 1512–1522. https://doi.org/10.1105/tpc.111.084525

Gregory, B. D., O'Malley, R. C., Lister, R., Urich, M. A., Tonti-Filippini, J., Chen, H., Millar, A. H., & Ecker, J. R. (2008). A link between RNA metabolism and silencing affecting Arabidopsis development. *Developmental Cell, 14*, 854–866.

Guleria, P., & Yadav, S. K. (2013). Agrobacterium mediated transient gene silencing (AMTS) in *Stevia rebaudiana*: Insights into steviol glycoside biosynthesis pathway. *PLoS One, 8*, e74731.

Gupta, O. P., Karkute, S. G., Banerjee, S., Meena, N. L., & Dahuja, A. (2017). Contemporary understanding of miRNA-based regulation of secondary metabolites biosynthesis in plants. *Frontiers in Plant Science, 8*, 374.

Hajyzadeh, M., Turktas, M., Khawar, K. M., & Unver, T. (2015). miR408 overexpression causes increased drought tolerance in chickpea. *Gene, 555*, 186–193. https://doi.org/10.1016/j.gene.2014.11.002

Hannon, G. J. (2002). RNA interference. *Nature, 418*, 244–251.

Hao, D. C., Yang, L., Xiao, P. G., & Liu, M. (2012). Identification of Taxus microRNAs and their targets with high-throughput sequencing and degradome analysis. *Physiologia Plantarum, 146*, 388–403.

He, J., Xu, M., Willmann, M. R., McCormick, K., Hu, T., Yang, L., Starker, C. G., Voytas, D. F., Meyers, B. C., & Poethig, R. S. (2018a). Threshold-dependent repression of SPL gene expression by miR156/miR157 controls vegetative phase change in *Arabidopsis thaliana*. *PLoS Genetics, 14*, e1007337. https://doi.org/10.1371/journal.pgen.1007337

He, Q., Liu, Y., & Sun, W. (2018b). Statistical analysis of non-coding RNA data. *Cancer Letters, 417*(161). https://doi.org/10.1016/j.canlet.2017.12.029

Iwakawa, H. O., & Tomari, Y. (2013). Molecular insights into microRNA mediated translational repression in plants. *Molecular Cell, 52*, 591–601. https://doi.org/10.1016/j.molcel.2013.10.033

Jennewein, S., & Croteau, R. (2001). Taxol: biosynthesis, molecular genetics and biotechnological applications. *Applied Microbiology and Biotechnology, 57*, 13–19.

Jiang, C. J., Shimono, M., Maeda, S., Inoue, H., Mori, M., Hasegawa, M., Sugano, S., & Takatsuji, H. (2009). Suppression of the rice fatty-acid desaturase gene OsSSI2 enhances resistance to blast and leaf blight diseases in rice. *Molecular Plant-Microbe Interactions, 22*(7), 820–829.

Jones, J. D., & Dangl, J. L. (2006). The plant immune system. *Nature, 444*, 323–329.

Jung, H. J., & Kang, H. (2007). Expression and functional analyses of microRNA417 in *Arabidopsis thaliana* under stress conditions. *Plant Physiology and Biochemistry, 45*, 805–811. https://doi.org/10.1016/j.plaphy.2007.07.015

Kamthan, A., Chaudhuri, A., Kamthan, M., & Datta, A. (2015). Small RNAs in plants: recent development and application for crop improvement. *Frontiers in Plant Science, 06*, 208. https://doi.org/10.3389/fpls.2015.00208

Kim, J., Lee, H., Jung, H., Maruyama, K., Suzuki, N., & Kang, H. (2010a). Overexpression of microRNA395c or 395e affects differently the seed germination of *Arabidopsis thaliana* under stress conditions. *Planta, 232*, 1447–1454. https://doi.org/10.1007/s00425-010-1267-x

Kim, J. Y., Kwak, K. J., Jung, H. J., Lee, H. J., & Kang, H. (2010b). MicroRNA402 affects seed germination of *Arabidopsis thaliana* under stress conditions via targeting DEMETER-LIKE Protein3 mRNA. *Plant & Cell Physiology, 51*, 1079–1083. https://doi.org/10.1093/pcp/pcq072

Kim, V. N. (2005). MicroRNA biogenesis: coordinated cropping and dicing. *Nature Reviews. Molecular Cell Biology, 6*, 376–385.

Kurihara, Y., Takashi, Y., & Watanabe, Y. (2006). The interaction between DCL1 and HYL1 is important for efficient and precise processing of pri-miRNA in plant microRNA biogenesis. *RNA, 12*, 206–212.

Laubinger, S., Sachsenberg, T., Zeller, G., Busch, W., Lohmann, J. U., Ratsch, G., & Weigel, D. (2008). Dual roles of the nuclear cap-binding complex and SERRATE in pre-mRNA splicing and microRNA processing in *Arabidopsis thaliana*. *Proceedings of the National Academy of Sciences USA, 105*, 8795–8800.

Lee, Y., Kim, M., Han, J., Yeom, K. H., Lee, S., Baek, S. H., & Kim, V. N. (2004). MicroRNA genes are transcribed by RNA polymerase II. *The EMBO Journal, 23*, 4051–4060.

Legrand, S., Valot, N., Nicolé, F., Moja, S., Baudino, S., Jullien, F., Magnard, J. L., Caissard, J. C., & Legendre, L. (2010). Onestep identification of conserved miRNAs, their targets, potential transcription factors and effector genes of complete secondary metabolism pathways after 454 pyrosequencing of calyx cDNAs from the Labiate *Salvia sclarea* L. *Gene, 450*, 55–62.

Li, F., Wang, W., Zhao, N., Xiao, B., Cao, P., Wu, X., Ye, C., Shen, E., Qiu, J., Zhu, Q. H., Xie, J., Zhou, X., & Fan, L. (2015). Regulation of nicotine biosynthesis by an endogenous target mimicry of microRNA in tobacco. *Plant Physiology, 169*, 1062–1071.

Li, J., Yang, Z., Yu, B., Liu, J., & Chen, X. (2005). Methylation protects miRNAs and siRNAs from a 30-enduridylation activity in Arabidopsis. *Current Biology, 15*, 1501–1507.

Li, S., Liu, J., Liu, Z., Li, X., Wu, F., & He, Y. (2014). HEAT-INDUCED TAS1 TARGET1 mediates thermostolerance via HEAT STRESS TRANSCRIPTION FACTOR A1a-directed pathways in Arabidopsis. *Plant Cell, 26*, 1764–1780. https://doi.org/10.1105/tpc.114.124883

Li, S., Liu, L., Zhuang, X., Yu, Y., Liu, X., Cui, X., Ji, L., Pan, Z., Cao, X., Mo, B., et al. (2013). MicroRNAs inhibit the translation of target mRNAs on the endoplasmic reticulum in Arabidopsis. *Cell, 153*, 562–574.

Lin, J. S., Lin, C. C., Lin, H. H., Chen, Y. C., & Jeng, S. T. (2012). MicroR828 regulates lignin and H_2O_2 accumulation in sweet potato on wounding. *The New Phytologist, 196*, 427–440. https://doi.org/10.1111/j.1469-8137.2012.04277.x

Llave, C., Xie, Z., Kasschau, K. D., & Carrington, J. C. (2002). Cleavage of Scarecrow-like mRNA targets directed by a class of Arabidopsis miRNA. *Science, 297*, 2053–2056.

Lu, X., Tang, K., & Li, P. (2016). Plant metabolic engineering strategies for the production of pharmaceutical terpenoids. *Frontiers in Plant Science, 7*, 1647.

Mahajan, M., Ahuja, P. S., & Yadav, S. K. (2011). Post-transcriptional silencing of flavonol synthase mRNA in Tobacco leads to fruits with arrested seed set. *PLoS One, 6*, e28315.

Mathiyalagan, R., Subramaniyam, S., Natarajan, S., Kim, Y. J., Sun, M. S., Kim, Y., Kim, Y. J., & Yang, D. C. (2013). In silico profiling of microRNAs in Korean ginseng (*Panax ginseng* Meyer). *Journal of Ginseng Research, 37*, 227–247.

Meng, Y., & Shao, C. (2012). Large-scale identification of mirtrons in Arabidopsis and rice. *PLoS One, 7*, e31163. https://doi.org/10.1371/journal.pone.0031163

Mohanpuria, P., Kumar, V., & Ahuja, P. S. (2011). Yadav SK (2011) Producing low-caffeine tea through post-transcriptional silencing of caffeine synthase mRNA. *Plant Molecular Biology, 76*, 523–534.

Molnar, A., Schwach, F., Studholme, D. J., Thuenemann, E. C., & Baulcombe, D. C. (2007). miRNAs control gene expression in the single-cell alga *Chlamydomonas reinhardtii*. *Nature, 447*, 1126–1129.

Najafabadi, A. S., & Naghavi, M. R. (2018). Mining *Ferula gummosa* transcriptome to identify miRNAs involved in the regulation and biosynthesis of terpenes. *Gene, 645*, 41–47.

Ni, Z., Hu, Z., Jiang, Q., & Zhang, H. (2012). Overexpression of gma-MIR394a confers tolerance to drought in transgenic *Arabidopsis thaliana*. *Biochemical and Biophysical Research Communications, 427*, 330–335. https://doi.org/10.1016/j.bbrc.2012.09.055

Padhan, J. K., Kumar, P., Sood, H., & Chauhan, R. (2016). Prospecting NGS-transcriptomes to assess regulation of miRNA-mediated secondary metabolites biosynthesis in *Swertia chirayita*, a medicinal herb of the North-Western Himalayas. *International Journal of Phytomedicines and Related Industries, 8*, 219–228.

Pandita, D. (2018a). Plant miRNAs: micro structure and macro character. *Research & Reviews: Journal of Agriculture and Allied Sciences, 7*(1), 83–84.

Pandita, D. (2018b). RNA interference: What and why? *Journal of Genetics and Molecular Biology, 2*(1), 1–3.

Pandita, D. (2019). Plant MIRnome: miRNA biogenesis and abiotic stress response. In M. Hasanuzzaman, K. Hakeem, K. Nahar, & H. Alharby (Eds.), *Plant abiotic stress tolerance*. Springer. https://doi.org/10.1007/978-3-030-06118-0_18.

Pandita, D., & Wani, S. H. (2019). MicroRNA as a tool for mitigating abiotic stress in rice (*Oryza sativa* L.). In S. Wani (Ed.), *Recent approaches in omics for plant resilience to climate change*. Springer. https://doi.org/10.1007/978-3-030-21687-0_6

Pani, A., & Mahapatra, R. N. (2013). Computational identification of microRNAs and their targets in *Catharanthus roseus* expressed sequence tags. *Genomics Data, 1*, 2–6. https://doi.org/10.1016/j.gdata.2013.06.001

Park, M. Y., Wu, G., Gonzalez-Sulser, A., Vaucheret, H., & Poethig, R. S. (2005). Nuclear processing and export of microRNAs in Arabidopsis. *Proc. Natl. Acad. Sci. USA, 102*, 3691–3696.

Pedersen, I., & David, M. (2008). MicroRNAs in the immune response. *Cytokine, 43*, 391–394.

Pérez-Quintero, A. L., Sablok, G., Tatarinova, T. V., Conesa, A., Kuo, J., & López, C. (2012). Mining of miRNAs and potential targets from gene oriented clusters of transcripts sequences of the anti-malarial plant *Artemisia annua*. *Biotechnology Letters, 34*, 737–745.

Pieczynski, M., Marczewski, W., Hennig, J., et al. (2013). Down-regulation of CBP80 gene expression as a strategy to engineer a drought-tolerant potato. *Plant Biotechnology Journal, 11*, 459–469. https://doi.org/10.1111/pbi.12032

Rajagopalan, R., Vaucheret, H., Trejo, J., & Bartel, D. P. (2006). A diverse and evolutionarily fluid set of microRNAs in *Arabidopsis thaliana*. *Genes & Development, 20*, 3407–3425. https://doi.org/10.1101/gad.1476406

Ramirez-Estrada, K., Vidal-Limon, H., Hidalgo, D., Moyano, E., Golenioswki, M., Cusidó, R. M., & Palazon, J. (2016). Elicitation, an effective strategy for the biotechnological production of bioactive high-added value compounds in plant cell factories. *Molecules, 21*(2), 182. https://doi.org/10.3390/molecules21020182

Reinhart, B. J., Slack, F. J., Basson, M., Pasquinelli, A. E., Bettinger, J. C., Rougvie, A. E., Horvitz, H. R., & Ruvkun, G. (2000). The 21-nucleotide let-7 RNA regulates developmental timing in *Caenorhabditis elegans*. *Nature, 403*, 901–906.

Reinhart, B. J., Weinstein, E. G., Rhoades, M., Bartel, B., & Bartel, D. P. (2002). MicroRNAs in plants. *Genes & Development, 16*, 1616–1626.
Rhoades, M. W., Reinhart, B. J., Lim, L. P., Burge, C. B., Bartel, B., & Bartel, D. P. (2002). Prediction of plant microRNA targets. *Cell, 110*, 513–520.
Robert-Seilaniantz, A., MacLean, D., Jikumaru, Y., Hill, L., Yamaguchi, S., Kamiya, Y., & Jones, J. D. G. (2011). The microRNA miR393 re-directs secondary metabolite biosynthesis away from camalexin and towards glucosinolates. *The Plant Journal., 67*, 218–231.
Rogers, K., & Chen, X. (2013). Biogenesis, turnover, mode of action of plant microRNAs. *Plant Cell, 25*, 2383–2399.
Schijlen, E. G., de Vos, C. H., Martens, S., Jonker, H. H., Rosin, F. M., Molthoff, J. W., Tikunov, Y. M., Angenent, G. C., van Tunen, A. J., & Bovy, A. G. (2007). RNA interference silencing of chalcone synthase, the first step in the flavonoid biosynthesis pathway, leads to parthenocarpic tomato fruits. *Plant Physiology, 144*(3), 1520–1530. https://doi.org/10.1104/pp.107.100305
Schwab, R., Ossowski, S., Riester, M., Warthmann, N., & Weigel, D. (2006). Highly specific gene silencing by artificial microRNAs in Arabidopsis. *Plant Cell, 18*, 1121–1133. https://doi.org/10.1105/tpc.105.039834
Sharma, D., Tiwari, M., Pandey, A., Bhatia, C., Sharma, A., & Trivedi, P. K. (2016). MicroRNA858is a potential regulatorof phenylpropanoid pathway and plant development. *Plant Physiology, 171*, 944–959. https://doi.org/10.1104/pp.15.01831
Singh, B., & Sharma, R. A. (2015). Plant terpenes: defense responses, phylogenetic analysis, regulation and clinical applications. *3 Biotech, 5*, 129–151.
Singh, N., Srivastava, S., & Sharma, A. (2016). Identification and analysis of miRNAs and their targets in ginger using bioinformatics approach. *Gene, 575*, 570–576. https://doi.org/10.1016/j.gene.2015.09.036
Souret, F. F., Kastenmayer, J. P., & Green, P. J. (2004). AtXRN4 degrades mRNA in Arabidopsis and its substrates include selected miRNA targets. *Molecular Cell, 15*, 173–183.
Sreekumar, S., & Soniya, E. V. (2018). Artificial microRNAs promote high level production of biomolecules through metabolic engineering of phenylpropanoid pathway. *Critical Reviews in Plant Sciences*. https://doi.org/10.1080/07352689.2018.1444361
Stief, A., Altmann, S., Hoffmann, K., Pant, B. D., Scheible, W. R., & Bäurle, I. (2014). Arabidopsis miR156 regulates tolerance to recurring environmental stress through SPL transcription factors. *Plant Cell, 26*, 1792–1807. https://doi.org/10.1105/tpc.114.123851
Sun, G., Stewart, C. N., Jr., Xiao, P., & Zhang, B. (2012). MicroRNA expression analysis in the cellulosic biofuel crop switchgrass (*Panicum virgatum*) under abiotic stress. *PLoS One, 7*(3), e32017. https://doi.org/10.1371/journal.pone.0032017
Sunkar, R., Chinnusamy, V., Zhu, J., & Zhu, J. K. (2007). Small RNAs as big players in plant abiotic stress responses and nutrient deprivation. *Trends in Plant Science, 12*, 301–309.
Tan, H., Xiao, L., Gao, S., Li, Q., Chen, J., Xiao, Y., Ji, Q., Chen, R., Chen, W., & Zhang, L. (2015). Trichome and artemisinin regulator 1 is required for trichome development and artemisinin biosynthesis in *Artemisia annua*. *Molecular Plant, 8*(9), 1396–1411. https://doi.org/10.1016/j.molp.2015.04.002
Tang, G., Yan, J., Gu, Y., Qiao, M., Fan, R., Mao, Y., & Tang, X. (2012). Construction of short tandem target mimic (STTM) to block the functions of plant and animal microRNAs. *Methods, 58*, 118–125. https://doi.org/10.1016/j.ymeth.2012.10.006
Vashisht, I., Mishra, P., Pal, T., Chanumolu, S., Singh, T. R., & Chauhan, R. S. (2015). Mining NGS transcriptomes for miRNAs and dissecting their role in regulating growth, development, and secondary metabolites production in different organs of a medicinal herb, *Picrorhiza kurroa*. *Planta, 241*, 1255–1268. https://doi.org/10.1007/s00425-015-2255-y
Vaucheret, H. (2008). Plant ARGONAUTES. *Trends in Plant Science, 13*, 350–358.
Voinnet, O. (2008). Post-transcriptional RNA silencing in plant-microbe interactions: A touch of robustness and versatility. *Current Opinion in Plant Biology, 11*, 464–470.
Wang, B., Malik, R., Nigg, E. A., & Korner, R. (2008). Evaluation of the low-specificity protease elastase for large-scale phosphoproteome analysis. *Analytical Chemistry, 80*, 9526–9533.

Wang, Y., Sun, F., Cao, H., Peng, H., Ni, Z., Sun, Q., & Yao, Y. (2012). TamiR159 directed wheat TaGAMYB cleavage and its involvement in another development and heat response. *PLoS One, 7*, e48445. https://doi.org/10.1371/journal.pone.0048445

Wightman, B., Ha, I., & Ruvkun, G. (1993). Posttranscriptional regulation of the heterochronic gene lin-14 by lin-4 mediates temporal pattern formation in *C. elegans*. *Cell, 75*, 855–862.

Winter, J., & Diederichs, S. (2011). Argonaute proteins regulate microRNA stability. *RNA Biology, 8*, 1149–1157. https://doi.org/10.4161/rna.8.6.17665

Xia, K., Wang, R., Ou, X., Fang, Z., Tian, C., & Duan, J. (2012). OsTIR1 and OsAFB2 down regulation via OsmiR393 over expression leads to more tillers, early flowering and less tolerance to salt and drought in rice. *PLoS One, 7*, e30039.

Xie, F., Frazier, T. P., & Zhang, B. (2011). Identification, characterization and expression analysis of microRNAs and their targets in the potato (*Solanum tuberosum*). *Gene, 473*, 8–22.

Xu, J., Hou, Q.-M., Khare, T., Verma, S. K., & Kumar, V. (2019). Exploring miRNAs for developing climate-resilient crops: a perspective review. *Science of the Total Environment, 653*, 91–104. https://doi.org/10.1016/j.scitotenv.2018.10.340

Xu, X., Jiang, Q., Ma, X., Ying, Q., Shen, B., Qian, Y., et al. (2014). Deep sequencing identifies tissue-specific microRNAs and their target genes involving in the biosynthesis of tanshinones in *Salvia miltiorrhiza*. *PLoS One, 9*, e111679. https://doi.org/10.1371/journal.pone.0111679

Xu, Z., Li, J., Guo, X., Jin, S., & Zhang, X. (2016). Metabolic engineering of cottonseed oil biosynthesis pathway via RNA interference. *Scientific Reports, 6*, 33,342.

Yan, J., Gu, Y., Jia, X., Kang, W., Pan, S., Tang, X., Chen, X., & Tang, G. (2012). Effective small RNA destruction by the expression of a short tandem target mimic in Arabidopsis. *Plant Cell, 24*, 415–427. https://doi.org/. https://doi.org/10.1105/tpc.111.094144

Yang, B., Tang, J., Yu, Z., Khare, T., Srivastav, A., Datir, S., & Kumar, V. (2019a). Light stress responses and prospects for engineering light stress tolerance in crop plants. *Journal of Plant Growth Regulation*. https://doi.org/10.1007/s00344-019-09951-8

Yang, C., Li, D., Mao, D., Liu, X. U. E., Ji, C., Li, X., Zhao, X., Cheng, Z., Chen, C., & Zhu, L. (2013). Overexpression of microRNA319 impacts leaf morphogenesis and leads to enhanced cold tolerance in rice (*Oryza sativa* L.). *Plant, Cell & Environment, 36*, 2207–2218. https://doi.org/10.1111/pce.12130

Yang, L., Wu, G., & Poethig, R. S. (2012). Mutations in the GW-repeat protein SUO reveal a developmental function for microRNA-mediated translational repression in Arabidopsis. *Proceedings of the National Academy of Sciences of the United States of America, 109*, 315–320.

Yang, T., Wang, Y., Teotia, S., Wang, Z., Shi, C., Sun, H., Gu, Y., Zhang, Z., & Tang, G. (2019b). The interaction between miR160 and miR165/166 in the control of leaf development and drought tolerance in Arabidopsis. *Scientific Reports, 9*, 2832. https://doi.org/10.1038/s41598-019-39397-7

Yang, Z., Ebright, Y. W., Yu, B., & Chen, X. (2006). HEN1 recognizes 21–24 nt small RNA duplexes and deposits a methyl group onto the 20 OH of the 30 terminal nucleotide. *Nucleic Acids Research, 34*, 667–675.

Yi, X., Zhang, Z., Ling, Y., Xu, W., & Su, Z. (2014). PNRD: A plant non-coding RNA database. *Nucleic Acids Research, 43*, D982–D989. https://doi.org/10.1093/nar/gku1162

Yu, B., Bi, L., Zheng, B., Ji, L., Chevalier, D., Agarwal, M., Ramachandran, V., Li, W., Lagrange, T., Walker, J. C., & Chen, X. (2008). The FHA domain proteins DAWDLE in Arabidopsis and SNIP1 in humans act in small RNA biogenesis. *Proc. Natl. Acad. Sci. USA, 105*, 10,073–10,078.

Yu, Z. X., Wang, L. J., Zhao, B., Shan, C. M., Zhang, Y. H., Chen, D. F., et al. (2015). Progressive regulation of sesquiterpene biosynthesis in Arabidopsis and Patchouli (*Pogostemon cablin*) by the miR156-targeted SPL transcription factors. *Molecular Plant, 8*, 98–110. https://doi.org/10.1016/j.molp.2014.11.002

Yuan, L., & Grotewold, E. (2015). Metabolic engineering to enhance the value of plants as green factories. *Metabolic Engineering, 27*, 83–91.

Zare-Mehrjerdi, M., Bihamta, M. R., Omidi, M., Naghavi, M. R., Soltanloo, H., & Ranjbar, M. (2013). Effects of exogenous methyl jasmonate and 2-isopentenyladenine on artemisinin production and gene expression in Artemisia annua. *Turkish Journal of Botany, 37*, 499–505.
Zhang, B., & Wang, Q. (2015). MicroRNA-based biotechnology for plant improvement. *Journal of Cellular Physiology, 230*, 1–15.
Zhang, J., Zhang, H., Srivastava, A. K., Pan, Y., Bai, J., Fang, J., Shi, H., & Zhu, J.-K. (2018). Knockdown of rice microRNA166 confers drought resistance by causing leaf rolling and altering stem xylem development. *Plant Physiology, 176*, 2082–2094. https://doi.org/10.1104/pp.17.01432
Zhang, L. W., Song, J. B., Shu, X. X., Zhang, Y., & Yang, Z. M. (2013a). miR395 is involved in detoxification of cadmium in *Brassica napus*. *Journal of Hazardous Materials, 250–251*, 204–211. https://doi.org/10.1016/j.jhazmat.2013.01.053
Zhang, J., Zhang, S., Li, S., Han, S., Wu, T., Li, X., & Qi, L. (2013b). A genome-wide survey of microRNA truncation and 3' nucleotide addition events in larch (*Larix leptolepis*). *Planta, 237*, 1047–1056. https://doi.org/10.1007/s00425-012-1823-7
Zhang, X., Zou, Z., Gong, P., Zhang, J., Ziaf, K., Li, H., Xiao, F., & Ye, Z. (2011). Over-expression of microRNA169 confers enhanced drought tolerance to tomato. *Biotechnology Letters, 33*, 403–409. https://doi.org/10.1007/s10529-010-0436-0
Zhang, Z. J. (2014). Artificial trans-acting small interfering RNA: A tool for plant biology study and crop improvements. *Planta, 239*, 1139–1146. https://doi.org/10.1007/s00425-014-2054-x
Zhao, J., Yuan, S., Zhou, M., Yuan, N., Li, Z., Hu, Q., Bethea, F. G., Liu, H., Li, S., & Luo, H. (2018). Transgenic creeping bentgrass overexpressing OsamiR393aexhibits altered plant development and improved multiple stress tolerance. *Plant Biotechnology Journal, 17*, 233–251. https://doi.org/10.1111/pbi.12960
Zhao, M., Ding, H., Zhu, J. K., Zhang, F., & Li, W. X. (2011). Involvement of miR169 in the nitrogen-starvation responses in Arabidopsis. *The New Phytologist, 190*, 906–915. https://doi.org/10.1111/j.1469-8137.2011.03647.x
Zhao, T., Li, G., Mi, S., Li, S., Hannon, G. J., Wang, X. J., & Qi, Y. (2007). A complex system of small RNAs in the unicellular green alga *Chlamydomonas reinhardtii*. *Genes & Development, 21*, 1190–1203.
Zhao, Y., Wen, H., Teotia, S., Du, Y., Zhang, J., Li, J., Sun, H., Tang, G., Peng, T., & Zhao, Q. (2017). Suppression of microRNA159 impacts multiple agronomic traits in rice (*Oryza sativa* L.). *BMC Plant Biology, 17*(1), 215. https://doi.org/10.1186/s12870-017-1171-7
Zhou, L., Liu, Y., Liu, Z., Kong, D., Duan, M., & Luo, L. (2010). Genome-wide identification and analysis of drought-responsive microRNAs in *Oryza sativa*. *Journal of Experimental Botany, 61*, 4157–4168. https://doi.org/10.1093/jxb/erq237
Zhou, M., Li, D., Li, Z., Hu, Q., Yang, C., Zhu, L., & Luo, H. (2013). Constitutive expressionofamiR319genealtersplantdevelopmentandenhances salt and drought tolerance in transgenic creeping bentgrass. *Plant Physiology, 161*, 1375–1391. https://doi.org/10.1104/pp.112.208702

Chapter 8
CRISPR/Cas Genome Editing in Engineering Plant Secondary Metabolites of Therapeutic Benefits

Tuyelee Das, Mimosa Ghorai, Devendra Kumar Pandey, Radha, Mamta Thakur, Sonia Rathour, Abdel Rahman Al-Tawaha, Ercan Bursal, Vinay Kumar, Potshangbam Nongdam, Mahipal S. Shekhawat, Gaber El-Saber Batiha, Arabinda Ghosh, Padmanabh Dwivedi, Vinay Kumar, Manoj Kumar, and Abhijit Dey

Abstract Plants hold the ability to produce wide types of bioactive secondary metabolites. Having emerged in the pregenomic era, increasingly more biosynthetic genes are being discovered in plants, leading to the discovery of new types of

---

T. Das · M. Ghorai · A. Dey (✉)
Department of Life Sciences, Presidency University, Kolkata, West Bengal, India
e-mail: abhijit.dbs@presiuniv.ac.in

D. K. Pandey
Department of Biotechnology, Lovely Professional University, Phagwara, Punjab, India

Radha · M. Thakur · S. Rathour
School of Biological and Environmental Sciences, Shoolini University of Biotechnology and Management Sciences, Solan, Himachal Pradesh, India

A. R. Al-Tawaha
Department of Biological Sciences, Al-Hussein Bin Talal University, Maan, Jordon

E. Bursal
Department of Biochemistry, Mus Alparslan University, Mus, Turkey

V. Kumar
Department of Biotechnology, Modern College, Ganeshkhind, Pune, India

P. Nongdam
Department of Biotechnology, Manipur University, Canchipur, Imphal, Manipur, India

M. S. Shekhawat
Plant Biotechnology Unit, KM Government Institute for Postgraduate Studies and Research, Puducherry, India

G. E.-S. Batiha
Department of Pharmacology and Therapeutics, Faculty of Veterinary Medicine, Damanhour University, Damanhour, AlBeheira, Egypt

A. Ghosh
Department of Botany, Gauhati University, Guwahati, Assam, India

T. Aftab, K. R. Hakeem (eds.), *Metabolic Engineering in Plants*,
https://doi.org/10.1007/978-981-16-7262-0_8

bioactive secondary metabolites. Utilisation of classical techniques is limited that hampers the discovery of pharmacologically important secondary metabolites. However, the development of CRISPR (clustered regularly interspaced short palindromic repeats)/Cas (CRISPR associated protein)-based tools may alleviate this impasse. This chapter briefly presents existing information about the CRISPR/Cas9 system, and by what implies it was engineered to enhance important secondary metabolite production in plants. CRISPR/Cas systems have been among the most versatile genome editing tools available, revolutionising molecular biology. This chapter intends to highlight and discuss the lasting challenges of CRISPR/Cas-based genome editing and the improvement of secondary metabolite amount in plant natural product engineering. The plants canvassed in this chapter include *Atropa belladonna*, *Brassica napus*, *Camelina sativa*, *Dendrobium officinale*, *Dioscorea zingiberensis*, *Glycine max*, *Humulus lupulus*, *Papaver somniferum* and *Salvia miltiorrhiza*. Additionally, we highlight the prospects of using CRISPR/Cas in plant secondary metabolite engineering.

Keywords CRISPR/Cas9 · Metabolites · Plant natural products · Medicinal plants genome editing

Introduction

For centuries, medicinal plants have been used by us for the production of a plethora of unique metabolites (Calabrò, 2015). Metabolites are biomolecules which formed during the metabolism of an organism. Metabolites have crucial roles in intracellular and intercellular signalling pathways, growth, development and defence against pathogens. Moreover, some metabolites are economically important for the production of drugs, flavourings, pesticides, aromatic compounds etc. (Hussain et al., 2012). There has been evidence that secondary metabolites of plants could serve as potential pharmaceutical leads for reducing human deaths (Dey et al., 2017b). Various natural compounds have been implicated in the reversibility of drug resistance (Das et al., 2021; Guo et al., 2017; Wang et al., 2015). The efficacy of secondary metabolites in medicinal herbalism has also been demonstrated by extensive clinical and preclinical studies (Anand et al., 2020; Kaur et al., 2020a, 2020b).

P. Dwivedi
Department of Plant Physiology, Institute of Agricultural Sciences, Banaras Hindu University, Varanasi, India

V. Kumar
Department of Biotechnology, Modern College (Savitribai Phule Pune University), Ganeshkhind, Pune, India

M. Kumar
Chemical and Biochemical Processing Division, ICAR—Central Institute for Research on Cotton Technology, Mumbai, India

For enhancing the production of valuable plant secondary metabolites, several biotechnological interventions have been applied (Dey et al., 2017a).

Plants produce different secondary metabolites that are restricted to specific groups of plants within the plant kingdom. In recent times, the world has shifted towards natural products due to their significant role in medicine without severe side effects. For that, evolving demand for secondary metabolites resulted in boundless interest mainly in the possibility of changing the production of bioactive plant metabolites via targeted metabolic engineering, elicitation, endophytes, bioreactors, precursor feeding, nanotechnology and other specialised biotechnological approaches (Calabrò, 2015; Hussain et al., 2012; Narayani & Srivastava, 2017; Verpoorte & Memelink, 2002). Metabolic engineering can be applied in several ways to increase metabolite yields, including by blocking catabolism, reducing competitive pathways and disabling rate-limiting steps (Verpoorte et al., 1999). Modulations in the production of plant metabolites can be accomplished by targeting regulatory genes for their biosynthetic pathways. Enhancement of secondary metabolite production can sometimes be achieved by targeting a single pathway gene that can act as a transcription factor involved in regulating multiple pathway genes or through knocking out the gene which is responsible for the suppression of natural product production (Staniek et al., 2013; Verpoorte & Memelink, 2002).

There is an increasing demand for genetic engineering strategies; however, conventional genetic engineering approaches have numerous shortcomings, such as being unable to manipulate the genomes of large plants. A variety of tools for genome editing at the transcriptional level such as zinc-finger nucleases (ZFNs), meganucleases (MNs), transcription activator-like endonucleases (TALENs) and CRISPR are helping to resolve the difficulties of specific genome editing of plants (Rehman et al., 2021; Chen & Gao, 2015). ZFNs and TALENs initiate genome editing by employing Fok-I, a protein-based endonuclease for gene knockout through producing double-stranded breaks (DSBs) in the DNA target which further arouse error-prone non-homologous end joining (NHEJ) or homology-directed repair (HDR) at precise genomic positions (Gaj et al., 2013; Khan, 2019). However, in order to utilise the TALENs and ZFNs techniques, two distinct protein hybrid designs are required which can identify sparsely existing DNA flanks (Li et al., 2011). Furthermore, these techniques are lengthy and less precise, which prompted scientists to develop an efficient genome editing method. Successful gene manipulation of plants has been achieved with CRISPR and associated proteins, for example, CRISPR-associated protein Cas9 (Upadhyay & Sharma, 2014). Figure 8.1 presents the key steps of genome editing through the CRISPR tool in plants.

Background Study of CRISPR-Cas9

The CRISPR/Cas is part of the adaptive immune system of bacteria and a few archaea. Foreign particles like viruses and plasmids are repelled by the CRISPR/Cas mechanism. The foreign genetic material (DNA) will be sliced and reorganised

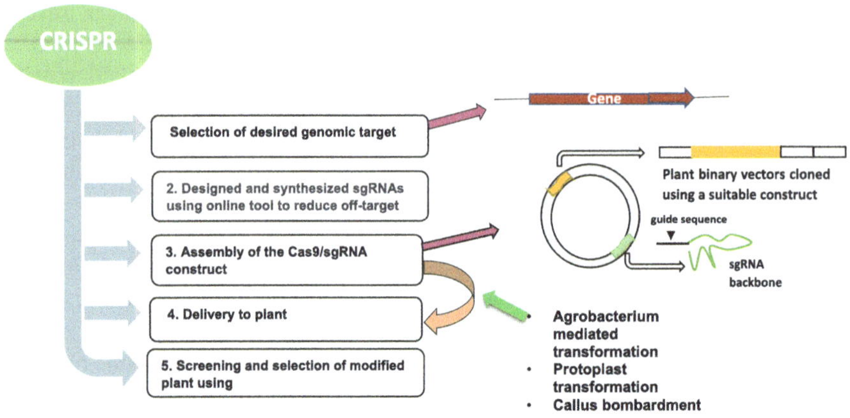

Fig. 8.1 The key steps of genome editing through the CRISPR technique in plants

and its genetic sequence preserved for when the same virus or plasmid invades again, the genetic material will be recognised (Koonin & Makarova, 2013; Koonin & Makarova, 2009). CRISPR/Cas system contains sgRNA (single guide RNA) and Cas9 endonuclease which form a complex that can cleave target DNA sites. CRISPR-Cas9 genome editing system is functioning by introducing DSBs (Jinek et al., 2012). Previous gene manipulating nuclease platforms such as ZFNs and TALENS, could initiate DNA DSBs; however CRISPR-Cas9 differs from ZFNs and TALENS as Cas9 nucleases form a complex with the sgRNAs that guide specific DNA sequences within host nuclei and initiate DSB induction (Jinek et al., 2012). The most likely outcome of a DSB is the initiation of NHEJ mechanisms producing various mutations such as indel and substitutions. By introducing frameshift mutations, NHEJ mechanisms can be used in breeding, mutant library construction and high-throughput mutational screening (Barakate & Stephens, 2016). DSB formation can also initiate HDR, although that is less likely. The CRISPR/Cas technology is a new, popular and powerful gene editing tool and has wide application in gene manipulation in the field of genomic research due to its cost-effectiveness and efficiency (Cui et al., 2018). There are numerous uses of CRISPR/Cas9 in genetically modifying both medicinal and food plants and model organisms (Alagoz et al., 2016; Malnoy et al., 2016; Ricroch et al., 2017; Shan et al., 2014; Singh et al., 2020). CRISPR/Cas9 has successfully edited approximately 32 different plants, but the application of this technology to medicinal plants is still limited. Therefore, this chapter intends to summarise data available in the past pertaining to ten medicinal plant genomes edited with CRISPR. CRISPR-Cas9 toolkit limitations and prospective are also discussed in this chapter.

Secondary Metabolite Engineering of Plants with CRISPR/Cas9

It is believed that there are over lakhs of secondary metabolites, which are classified by structure, function and biosynthesis. Secondary metabolites are divided into alkaloids, terpenoids and steroids, fatty acid-derived compounds, polyketides and nonribosomal polypeptides and enzyme cofactors (McMurry, 2015). CRISPR/Cas offers an alternative production strategy of commercially important plant-derived natural products by modulating the phytochemical profile of medicinal plants with CRISPR/Cas9 technology. Table 8.1 presents application of gene editing with CRISPR/Cas in medicinal plants for the production of valuable secondary metabolites. In Fig. 8.2 we have highlighted different plant metabolites and their genome editing by CRISPR/Cas9 action.

Atropa belladonna L.

Atropa belladonna L., alternatively referred to as deadly nightshade, is used extensively against anti-inflammatory and neuropharmacological disorders. Hyoscyamine is the major tropane alkaloid compound produced by *A. belladonna* which has activity against arrhythmias and organophosphate poisoning. This plant also produces the derivatives of hyoscyamine such as anisodamine and scopolamine (Hashimoto, 1992). Therefore, it is necessary to separate one compound from the others. However, due to their similar structures, this can be difficult to isolate compounds autonomously. So, it is crucial to develop *A. belladonna* without anisodamine and scopolamine, in order to achieve low costs for hyoscyamine separation. Additionally, hyoscyamine content in *A. belladonna* is very low and commercial demand for this compound is high; therefore, Zeng et al. (2021) develop high yields of hyoscyamine from *A. belladonna* plants for the first time, using CRISPR/Cas9 tool. The editing method, CRISPR/Cas9, disrupted the hyoscyamine 6β-hydroxylase (*AbH6H*) gene, initiated homozygous mutations in *AbH6H* and led to elevated hyoscyamine production, as well as decreased hyoscyamine 6β-hydroxylase function without anisodamine or scopolamine (Zeng et al., 2021). The editing method, CRISPR/Cas9, disrupted the hyoscyamine gene, initiated homozygous mutations in *AbH6H* and led to elevated hyoscyamine production, as well as reduced gene function. Hyoscyamine 6β-hydroxylase catalyses the 6β-hydroxylation of hyoscyamine which further produces anisodamine and subsequently converts it to scopolamine (Hashimoto & Yamada, 1986).

Table 8.1 Gene editing with CRISPR/Cas in medicinal plants for the production of valuable secondary metabolites

Plant species	Family	Targeted gene	Promoter for sgRNA	Secondary metabolite(s)	Editing type	Method of Cas9 system delivery	Reference
Atropa bella-donna L.	Solanaceae	*AbH6H*	U6-26	Hyoscyamine	CRISPR/Cas9, targeted mutagenesis	*A. tumefaciens* EHA105-mediated freeze-thaw method	Zeng et al. (2021)
Brassica napus L.	Brassicaceae	*BnaFAD2*	U3	Oleic acid linoleic acid, linolenic acid	CRISPR/Cas9, knockout	*A. tumefaciens*-mediated hypocotyl method	Huang et al. (2020)
		BnaA. FAD2	AtU6	Oleic acid	CRISPR/Cas9, knockout	*A. tumefaciens* GV3101 mediated	Okuzaki et al. (2018)
Camelina sativa (L.) Crntz.	Brassicaceae	*FAD2*	U3, U6	Oleic acid, PUFA	CRISPR/Cas9, targeted mutagenesis	*A. tumefaciens*-mediated floral-dip method	Morineau et al. (2017)
		FAD2	U6	Oleic acid, PUFA	CRISPR/Cas9, knockout	*A. tumefaciens*-mediated floral-dip method	Jiang et al. (2017)
		FAE1	U6- 26	Oleic acid or α-linolenic acid	CRISPR/Cas9, knockout	Floral vacuum infiltration method	Ozseyhan et al. (2018)
Dendrobium officinale Kimura & Migo	Orchidaceae	*C3H, C4H, 4CL, CCR, IRX*	OsU3	Alkaloids, phenanthrenes, polysaccharides, bibenzyls, essential oils, glycosides	CRISPR/Cas9, knockout	*Agrobacterium* mediated	Kui et al. (2017)
Dioscorea zingiberensis C. H. Wright	Dioscoreaceae	*Dzfps*	OsU3	Diosgenin	CRISPR/Cas9, targeted mutagenesis	*A. tumefaciens* GV3101 mediated	Feng et al. (2018)
Dioscorea alata L.	Dioscoreaceae	*DrPDS*	DaU6.3	–	CRISPR/Cas9, knockout	*Agrobacterium* mediated	Syombua et al. (2021)

Glycine max (L.) Merr.	Fabaceae	*IFS*	GmU3 or GmU6 promoter	Isoflavonoids	CRISPR/Cas9, knock-out	*A. rhizogenes*-mediated hairy root culture	Zhang et al. (2020)
Humulus lupulus L.	Cannabaceae	*PDS*	U6-626p and U6-29p	Carotenoid	CRISPR/Cas9, targeted mutagenesis	*A. tumefaciens* mediated	Awasthi et al. (2021)
Nicotiana tabacum L.	Solanaceae	*FucT, XylT*	U6	Alkaloids, flavonoids, terpenoids, phenylpropanoids	CRISPR/Cas9, knockout	*A. tumefaciens* (EHA105, LBA4404) mediated	Mercx et al. (2017)
Papaver somniferum L.	Papaveraceae	*4′OMT2*	AtU6p	Alkaloid	CRISPR/Cas9, knockout	*A. tumefaciens*-mediated leaf infiltration	Alagoz et al. (2016)
Salvia miltiorrhiza Bunge	Lamiaceae	*SmRAS*	U3	Rosmarinic acid phenolic acids, diterpenoids	CRISPR/Cas9, knockout	*A. rhizogenes*-mediated hairy root culture	Zhou et al. (2018)

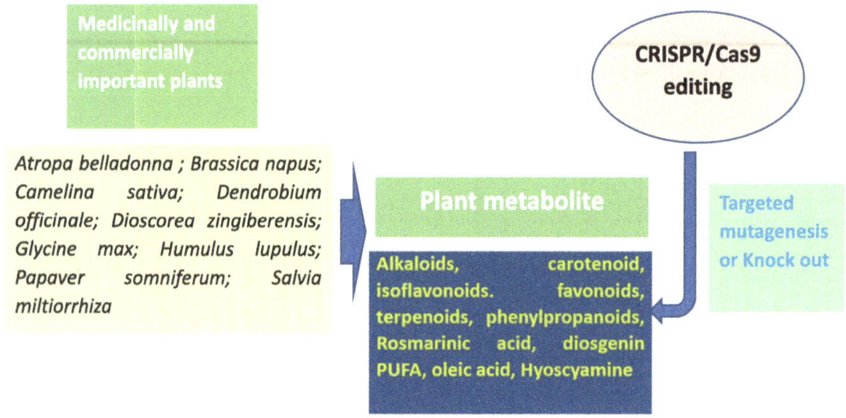

Fig. 8.2 Plant metabolites and their genome editing by CRISPR/Cas9

Brassica napus L.

Rapeseed oil ((*B. napus)* contains three major unsaturated fatty acids, including linoleic acid (18:2), linolenic acid (18:3) and oleic acid (18:1), as well as palmitic acid (16:0) and stearic acid (18:0), which are of high nutritional value (Nesi et al., 2008; Peng et al., 2010). Higher oleic acid content enhances thermal stability of the oil. Oleic acid content of napus is regulated by *FAD2* (fatty acid desaturase 2 gene) loci. Standardised CRISPR/Cas9 technology has generated mutated *BnaA.FAD2* through *Agrobacterium*-mediated transformation in *B. napus*. In an attempt to boost the content of oleic acid in *B. napus* genes using the CRISPR/Cas9 genome editing tool, mutated *BnaFAD2* copies were successfully introduced. Their result suggested that modification by CRISPR/Cas9 system showed significant increase in the content of oleic acid by 80% and a notable decrease in linoleic acid and linolenic acid (Huang et al., 2020). The report also suggested that genome editing of polyploidy species is quite feasible and does not lead to chimeric modifications. One study using CRISPR/Cas9 technology to drive knockout mutations in *FAD2* genes in *B. napus* resulted in an increase in the oleic acid (73%–80%) and reduction in linoleic acid (16%–9%). Though, 4-bp deletion of mutant fad2_Aa allele exhibited normal plant growth. Thus, knocking out via CRISPR/Cas9 technology can also be used to produce mutant plants that are agronomically more advantageous, such as mutant plants that contain higher oleic acid content important (Okuzaki et al., 2018).

Camelina sativa (L.) Crantz

CRISPR/Cas9-based gene editing was effective in enhancing the levels of oleic acid in the hexaploid *C. sativa* plant. Knockout of the *FAD2* genes successfully increased seed oil. A significant increase in oleic acid content (16 % to over 50%) was observed, accompanied by a drop in linoleic acid (~16% to <4%) and linolenic acid (~35% to <10%). gRNAs targeted all three homologous *FAD 2* genes simultaneously in *Camelina* seeds of T3 and T4 generations (Jiang et al., 2017). An alternative study used CRISPR/Cas9 genome editing technology to improve oleic acid content (from 10 to 62%) and reduce PUFA in the *C. sativa* seed oil using three closely related *FAD 2* genes through targeted mutagenesis. The results of evaluating the mutations in three isologous CsFAD 2 genes over four generations have confirmed a large heritability of mutations (Morineau et al., 2017). In another experiment in allohexaploid *C. sativa*, CRISPR/Cas9 employed knockout mutagenesis by ethyl methanesulfonate which formed fatty acid elongase 1 (*FAE1*) gene mutants that resulted in the reduction in the C20–C24 very long-chain fatty acids (over 60%). Therefore, the inactivation of *FAE1* genes enabled the improvement of oleic acid content or α-linolenic acid in seed oil. Interestingly, knocking out mutant fae1 showed normal growth and seed development (Ozseyhan et al., 2018). Others altered three conserved homologous genes (*CsDGAT1* or *CsPDAT1*) via sgRNA to manipulate TAG (triacylglycerol) composition in the seeds of *C. sativa*. In addition, the seed of both *CsDGAT1* and *CsPDAT1* targeted lines shows wrinkled and shrinking surface (Aznar-Moreno & Durrett, 2017).

Dendrobium officinale Kimura and Migo

One of the largest groups of Angiosperm monocotyledon family, Orchidaceae, finds its uses not only in cosmetics, perfumes and for decoration purposes but also as natural remedies for the treatment of numerous diseases (Sut et al., 2017). The Chinese medicinal herb *Dendrobium officinale*, a species belonging to this family, produces many useful primary and secondary metabolites. There is also evidence to suggest that it heals yin deficiency disorders (Guo et al., 2020; Tang et al., 2017). Major bioactive compounds of *D. officinale* are polysaccharides, besides other compounds such as alkaloids, bibenzyls, essential oils, phenanthrenes and glycosides which have anticancer, antibacterial, antioxidant, anti-inflammatory, antidiabetic, antiviral, anti-ageing and hair growth-promoting properties (Tang et al., 2017; Teixeira et al., 2015). Metabolic profiling of *D. officinale* has led to the identification of several pharmaceutically active ingredients (Jin et al., 2016). The protocol for genome editing through CRISPR/Cas9 is well established for this orchid (Kui et al., 2017). This has not only eased the process of precise genetic manipulation of this plant but also helped in eliminating the controversies associated with transgenic acceptability. This knockout system reached 10–100% editing efficiency by using highly effective promoter (*Cauliflower mosaic virus* 35S), reporter genes β-glucuronidase (GUS), superfolder green fluorescence protein (SG) and vector (pCambia-1301-35SN) by using *Agrobacterium*-mediated transformation system. Five lignocellulose biosynthesis pathway genes, *C3H* (coumarate 3-hydroxylase), *C4H* (cinnamate 4-hydroxylase), *4CL* (4-coumarate), *CCR* (cinnamoyl coenzyme A reductase) and *IRX* (irregular xylem5), were used to determine the efficiency of genome editing with CRISPR/Cas9 in *D. officinale*. The high efficiency of genome editing marks the entry of *D. officinale* in the new age of reverse genetics to decipher orchid gene functions which will further lead to the sustainable exploitation of this plant for our benefit (Kui et al., 2017).

Dioscorea zingiberensis C. H. Wright

Dioscorea spp. are medicinally important species of plants that are used for the extraction of diosgenin. The rhizomes of this species are used for the isolation of steroidal saponin, i.e. diosgenin, which has therapeutic significance (Pandey et al., 2017). Syombua et al. (2021), for the first time, evaluated the efficiency of CRISPR/Cas9 genome editing tool by targeting the phytoene desaturase gene (DrPDS) of *Dioscorea alata*. PDS (phytoene desaturase) catalyses the conversion of phytoene into both phytofluene and f-carotene (carotenoid precursors) (Mann et al., 1994). Results of this study proved that CRISPR/Cas9 efficiently induced site-specific disruption of the PDS gene with 83.3% editing efficiency and also induced phenotypical changes in yam. It is expected that the established CRISPR/Cas9 system in this species will serve as basal information for establishing editing protocols in other *Dioscorea* species. This will assist in functional genomics studies' trait improvement in other related species like *D. alata*. *D. zingiberensis* (a perennial vine, commonly known as 'yellow ginger') is extensively used in Chinese medicines (Zhang et al., 2018). A range of biological effects like anti-inflammatory, anthelmintic, anti-thrombosis, cardiovascular, hyperlipidaemia and neuroprotection activity have been confirmed. In order to target the farnesyl pyrophosphate synthase gene (Dzfps) (the gene controlling diosgenin biosynthesis), the CRISPR/Cas9 system was used via an Agrobacterium tumefaciens-facilitated transfection method in D. zingiberensis. Farnesyl pyrophosphate (FPP) synthase is important for the production of isoprenoids such as carotenoids, ubiquinones, sterols etc. (Dhar et al., 2013). Cas9 and sgRNA expression cassettes are controlled by the OsU3 and 35S promoters, respectively, which are designed to target the Dzfps gene. Dzfps gene is involved in sequentially catalysing dimethylallyl diphosphate (DMAPP), geranyl diphosphate (GPP) to produce FPP and finally squalene which is the precursor of diosgenin. Both Dzfps transcript level and squalene content were relatively decreased in mutants with genome editing compared to wild type (Feng et al., 2018).

Glycine max (L.) Merr.

Glycine max (soybean) contains important vegetable oils, proteins and bioactive secondary metabolites. Remarkably, isoflavonoids content in *G. max* is almost 100 times higher as compared to other leguminous plants, which have significant functions in plant disease resistance and human health (Budryn et al., 2018; Kant et al., 2019; Yu et al., 2003). As a signal molecule, isoflavonoids are involved in activating nod genes (Subramanian et al., 2006). Soybean phenylpropanoid pathway produces isoflavonoids by a sequential but complex process (Zhang et al., 2016). Briefly, the metabolic pathway for isoflavonoid biosynthesis starts by hydroxylation of flavanone which is catalysed by isoflavone synthase (*IFS*) which shares common substrate with flavanone-3-hydroxylase (*F3H*) and flavone synthase II (*FNS II*). *GmIFS1* and *GmIFS2* have been isolated from *G. max* (Jung et al., 2000). Zhang et al. (2020) studied multiplex CRISPR/Cas9 genome editing that was used to target *GmF3H1*, *GmF3H2* and *GmFNSII-1* in the hairy roots of *G. max*. Metabolomic analysis of *GmF3H1*, *GmF3H2* and *GmFNSII-1* triple mutants showed significant enhancement in isoflavone content. Additionally, the viral titre of *Soybean mosaic virus* (SMV) was significantly decreased. According to these findings, improved isoflavone content significantly enhanced soybean's resistance to SMV (Zhang et al., 2020).

Humulus lupulus L.

Humulus lupulus, also called hops, contains phenols, bitter acids, prenylated flavonoids etc., which are considered to be medically significant for those suffering from diseases such as prostate and breast cancer, osteoporosis, menopause and anxiety (Mishra et al., 2020; Srečec et al., 2012). CRISPR/Cas9 was used for the first time to manipulate the gene expression of *H. lupulus*. *Agrobacterium*-mediated transformation [using the binary pKSE401] successfully edited 33.3% of the transformed plants. Targeted editing of a *PDS* gene in carotenoid biosynthesis pathway resulted in endogenous edited genes in *H. lupulus* and decreased concentrations of chlorophyll a/b and carotenoid pigments. The CRISPR/Cas9 system was cited as a precise way to target the genome sequence of hops (Awasthi et al., 2021).

Nicotiana tabacum L.

Nicotiana tabacum, a perennial herbaceous plant that produces tobacco, is reported to be a storehouse of important secondary metabolites such as alkaloids, terpenoids, flavonoids, phenylpropanoids etc. A number of recombinant products have also been produced using *N. tabacum*, including glycoenzymes α-galactosidase, glycohormone erythropoietin, proteases and xylanase (Jutras et al., 2020; Pantaleoni et al., 2014). Plants have been used for the production of pharmacological glycoproteins. These glycoproteins carry N-glycans with β(1,2)-xylose and a core α(1,3)-fucose, which have a significant influence on immunogenicity and allergenicity. CRISPR/Cas9 has been successfully used to knockout six glycosyltransferase genes in *Nicotiana* sp. It was used to produce recombinant proteins without core sugar α-1,3-fucose and β-1,2-xylose (Jansing et al., 2019). Gao et al. (2015) have projected that CRISPR/Cas9 system is a useful tool for targeted mutagenesis of *N. tabacum* genome mainly due to the high efficiency of this editing system in this species (Gao et al., 2015). Mercx et al. (2017) conducted the study on the knockout lines of *XylT* (β(1,2)-xylosyltransferase) and *FucT* (α(1,3)-fucosyltransferase) in suspended *N. tabacum* BY-2 cells. Expression cassettes for Cas9 and gRNA were driven by the 35S-PPDK and U6 transcriptional promoters, respectively, which knockout four *FucT* and two *XylT* genes. The IgG glycosylation profile was screened by mass spectrometry that displayed the presence of GnGn (69%), Man7 (9.3%) and GnM (4.8%) structure and lack of β(1,2)-xylose or α(1,3)-fucose on the glycosylation moiety (Mercx et al., 2017).

Papaver somniferum L

The benzylisoquinoline alkaloids produced by *Papaver somniferum* have clinical significance in biomedicine (Labanca et al., 2018). A genome editing strategy based on CRISPR/Cas9 genome editing has been applied to this plant successfully. CRISPR-SpCas9 type II was used to knockout *4′OMT2* (gene controlling benzylisoquinoline alkaloids biosynthesis). A resulting DSB was repaired by NHEJ, causing short indels to be introduced, resulting in gene dysfunction. *Agrobacterium*-mediated transformation was carried out utilising TRV (*Tobacco rattle virus*)-based synthetic binary plasmids expressing sgRNA and hCas9 (human-codon optimised Cas9)-encoding synthetic vector to inactivate *4′OMT2*, thus regulating the biosynthesis of the BIAs. An HPLC-ToF/MS (high-performance liquid chromatography-time-of-flight mass spectrometry) study showed that *P. somniferum* plants with gene knockouts had significantly lower production of benzylisoquinoline alkaloids (Alagoz et al., 2016).

Salvia miltiorrhiza Bunge

Salvia miltiorrhiza, a traditional medicinal herb from China, contains diterpene compounds such as tanshinones and rosmarinic acid (RA) (Li et al., 2017a, 2017b; Shi et al., 2019). *S. miltiorrhiza* is extensively used against cardiovascular and cerebrovascular diseases and diabetes. Moreover, this plant has antioxidant and anti-inflammatory properties as well as cardioprotective and anticancer properties. According to earlier reports, an inactivation with the CRISPR/Cas9 system was shown to effectively target the *SmCPS1* (diterpene synthase) gene from the tanshinone biosynthesis pathway. Rosmarinic acid synthase (RAS), an enzyme which catalyses RA biosynthesis, also accumulates lithospermic acid B (LAB) (Sander & Petersen, 2011). *Agrobacterium rhizogenes*-mediated transformation of hairy roots using CRISPR-Cas9 inactivated the *SmCPS1* gene implicated in the production of tanshinone. An analysis of metabolite profiles in homozygous mutants revealed that tanshinone synthesis is completely absent, but these mutants produce the other phenolics. However, the tanshinone content of the chimeric mutants was reduced (Li et al., 2017a, 2017b). *S. miltiorrhiza* suspension cells were successfully transfected with *Agrobacterium rhizogenes* in order to knockout the *SmRAS* gene using CRISPR/Cas9 genome editing. One homozygous mutant, two heterozygous mutants and five biallelic mutants were generated using the *Arabidopsis* U6 promoter. Mutation in the *SmRAS* genes resulted in a decline in RA and LAB and an increase in 3,4-dihydroxyphenyllactic acid, a precursor of RA (Zhou et al., 2018). *SmPAL1* was targeted and edited in the phenylpropane metabolic pathway using a software designed for CRISPR/Cas9 in the *S. miltiorrhiza*. SmPAL1-g1, SmPAL1-g2 and SmPAL1-g3 are three sequences which are likely CRISPR targets. These sequences achieved 53.3%, 76.6% and 10.0% enzyme digestion efficiencies, respectively (Qiu et al., 2018).

Challenges Associated with the Use of CRISPR/Cas Tools for Plant Natural Product Research

Off-Target Effects

Despite CRISPR/Cas9 being a popular tool among other genome editing tools due to its affordability, precision and simplicity, some negative consequences are also associated with it. The off-target effect is one such type. There are unintentional translocations, insertions, deletions, inversions and point mutations associated with off-target effects. Utilising the CRISPR/Cas9 tool, off-target mutations limit therapeutic implications. In lower group of organisms such as bacteria, a smaller number of off-target effects are reported because of a lower frequency of a spacer-PAM combination in their genome. There are a number of effective solutions to fighting off-target effects, such as mutations in the nuclease domain of Cas9 (Cho et al., 2014), dimerisation of nucleases (Fok1-dCas9) (Tsai et al., 2014) and CRISPR interference (Gilbert et al., 2014). There is considerable effort underway in higher eukaryotes to minimise off-target effects. Improved sgRNAs need to reduce mismatches (Doench et al., 2016) and restrict Cas9 levels in the cells to lessen off-target effects of these tools (Shen et al., 2019).

Editing Accuracy

The dominant error-prone NHEJ pathway for DSB repair is randomly generated small indels at the DSB locus. Unpredictable precision size of the indel leads to frameshift mutations. HDR creates more precise gene editing products. The efficiency of HDR is also reduced multiple folds due to the dominance of NHEJ pathway, regardless of whether an editing template is accommodated for HDR (Yang et al., 2020). In mammalian cells the efficiency of HDR is much lower (25%) than NHEJ (75%). To conquer this limitation, a few strategies have been effectively utilised to suppress the NHEJ pathway, for example, utilising the small molecule Scr7 on promoting HDR efficiency, as Scr7 binds to the DNA binding domain of ligase IV, KU70 or KU80 and thus inhibits with the of NHEJ pathway events (Li et al., 2017a, 2017b).

Efficient Delivery

The large size and positive charge of Cas9 protein and negative charge of sgRNAs make difficult the delivery of this protein into cell. The efficacious CRISPR/Cas9 application needs strong Cas9 and sgRNA delivery either by plasmid or by mRNA or by ribonucleoprotein (RNP) complex (Rahimi et al., 2020). Inappropriately, few medicinal plants discussed in this chapter lack efficient transformation methods. Nanoparticle-based delivery (polymeric nanocarriers, gold-based nanomaterials, metal-organic frameworks) is fetching a more striking approach due to its specific targeting and minimal exposure to nucleases (Carboni et al., 2019; Chen et al., 2020).

Genome Instability

DSBs may cause chromosomal translocations, instability of the genome which is a dangerous cellular event. CRISPR/Cas9 tool initiates DSB that may put the cells under severe stress. Cas9 expression in *Streptomycetes* leads to deletions in linear chromosomes (Hoff et al., 2018). CRISPR-based editing system (CRISPR-BEST), *CRISPR* interference (CRISPRi) and deaminase-based DNA base editors, that are not based on DSBs, have been conveyed as an effective genome editing tool (Eid et al., 2018; Tong et al., 2020).

Toxicity of Cas9

CRISPR-Cas9 tool is for high-throughput approach, but can be **astonished** by nuclease-induced toxicity because of DNA damage. Toxicity due to Cas9 expression has been reported in bacteria and green algae but has not yet been reported in plants. The presence of high intercellular Cas9 leads to toxicity. Reducing Cas9 levels using weak promoters or by controlling the size and frequency of target gene edits lead to a decreased toxicity (Dow et al., 2015; Morgens et al., 2017).

Indigenous Vs Introduced CRISPR's Role

Bacteria and *Streptomyces* have evolved CRISPR/Cas systems which provide protection against foreign nucleic acids. A future research area of interest could be the interaction between native and introduced CRISPR/Cas.

Perspectives of CRISPR/Cas Genetic Engineering on Plant Natural Products

CRISPR-based genome editing has become a major breakthrough in the twenty-first century due to its effectiveness in knocking out. The success of this application prompted biochemical research companies to completely shift their focus to using CRISPR tools in plants for the production of sustainable agricultural crops and valuable plant natural products (Brinegar et al., 2017). It is crucial to efficaciously edit the genomes of medicinal plants. It enables accurate control of secondary metabolite pathways through regulating metabolic routes and removing rate-limiting constraints and feedback inhibitions. In order to meet the ever-expanding demand of biomedical industries, CRISPR/Cas genome editing may prove to be an exciting tool in metabolite engineering. For plants with a long life cycle, direct CRISPR/Cas-mediated editing can be ineffective. It may be extremely useful to employ microbial cell factories to substitute heterologous expression of secondary metabolic pathways. Earlier, CRISPR/Cas tool was mostly used to modify just one trait in somatic cells of higher plants through NHEJ. However, in recent years, CRISPR/Cas research mainly focused on the utilisation of homologous recombination for chromosomal rearrangement driven genetic traits in plants (Schindele et al., 2020). Incorporating artificial DNA sequences into plant genomes altering plant behaviour due to novel functions by CRISPR/Cas system is a promising approach to improve plant design and synthetic biology (Chen et al., 2019). Genome editing using CRISPR/Cas has immense potential to create biochemical factories, clone long DNA segments and manipulate biosynthetic pathways (Bennett-Baker & Mueller, 2017; Liu et al., 2015). Targeting long genomes with Cas9-aided targeting of CHROMOSOME (CATCH) and isolating those megabase-sized genomic fragments with CRISPR-mediated isolation (CISMR) are powerful tools for synthetic and chemical biology (Bennett-Baker & Mueller, 2017; Liu et al., 2015). Recent studies have shown that inducing artificial polyploidy via the CRISPR/Cas9 mechanism on *A. rhizogenes* hairy root cultures is a novel and promising way to increase secondary metabolite production in a wide range of medicinal plants (Niazian, 2019). As *Agrobacterium*-mediated transformation is expensive and time-consuming, tissue culture-free genome editing offers an alternative, which will likely be more efficient. The specificity of the Cas9-linked base editors has also been improvised through the expansion of the guide sequences sgRNAs and RNP-mediated delivery of the base editors. A genomic editing process has also been improved by enhancing the frequency of HDR to improve the efficiency and precision of the process. Despite receiving a lot of attention during the past several years, CRISPR/Cas has only just begun to make its way into natural product discovery. Few applications have been reported, and most of the work has focused on proving the feasibility of CRISPR/Cas systems in plants. The current CRISPR/Cas technology should be optimised, developed and innovated further, as no technology is perfect.

Conclusion

A number of plant secondary metabolites could potentially be used to treat several life-threatening human diseases. They provide protection against a wide range of diseases with a wide array of defence properties without causing harmful side effects. They are also relatively affordable and less costly than other methods of treatment. CRISPR/Cas9 has appeared as a breakthrough tool for metabolic engineering in plants. CRISPR/Cas9 modifies desired genomic parts by knocking out, knocking in, mutations etc. It is still in the early stages of CRISPR/Cas9-based genome editing in most medicinal plants as information about whole genomes and mRNAs sequences is lacking. Due to insufficient sequence information, this tool cannot edit key genes involved in plant secondary metabolite pathways. The limitations of this tool and strategic approaches to overcome them have already been discussed. However, a large number of secondary metabolites producing genes have not been modified yet with CRISPR/Cas9 for enhancing metabolite production. The use of higher plants as renewable sources of bioactive metabolites will be extended and enhanced through CRISPR/Cas9. In the future, we expect to achieve controllable and successful biotechnological production of valuable and as yet unknown plant phytochemicals through the continuation of efforts in this area.

References

Alagoz, Y., Gurkok, T., Zhang, B., & Unver, T. (2016). Manipulating the biosynthesis of bioactive compound alkaloids for next-generation metabolic engineering in opium poppy Using CRISPR-Cas 9 genome editing technology. *Scientific Reports, 6*(1), 1–9. https://doi.org/10.1038/srep30910

Anand, U., Nandy, S., Mundhra, A., Das, N., Pandey, D. K., & Dey, A. (2020). A review on antimicrobial botanicals, phytochemicals and natural resistance modifying agents from Apocynaceae family: Possible therapeutic approaches against multidrug resistance in pathogenic microorganisms. *Drug Resistance Updates, 51*, 100695. https://doi.org/10.1016/j.drup.2020.100695

Awasthi, P., Kocábek, T., Mishra, A. K., Nath, V. S., Shrestha, A., & Matoušek, J. (2021). Establishment of CRISPR/Cas9 mediated targeted mutagenesis in hop (*Humulus lupulus*). *Plant Physiology and Biochemistry, 160*, 1–7. https://doi.org/10.1016/j.plaphy.2021.01.006

Aznar-Moreno, J. A., & Durrett, T. P. (2017). Simultaneous targeting of multiple gene homeologs to alter seed oil production in *Camelina sativa*. *Plant and Cell Physiology, 58*(7), 1260–1267.

Barakate, A., & Stephens, J. (2016). An overview of crispr-based tools and their improvements: New opportunities in understanding plant-pathogen interactions for better crop protection. *Frontiers in Plant Science, 7*, 1–7. https://doi.org/10.3389/fpls.2016.00765

Bennett-Baker, P. E., & Mueller, J. L. (2017). CRISPR-mediated isolation of specific megabase segments of genomic DNA. *Nucleic Acids Research, 45*(19), 1–8. https://doi.org/10.1093/nar/gkx749

Brinegar, K., Yetisen, K., Choi, S., Vallillo, E., Ruiz-Esparza, G. U., Prabhakar, A. M., Khademhosseini, A., & Yun, S. H. (2017). The commercialization of genome-editing technologies. *Critical reviews in biotechnology, 37*(7), 924–932.

Budryn, G., Grzelczyk, J., & Pérez-Sánchez, H. (2018). Binding of red clover isoflavones to actin as a potential mechanism of anti-metastatic activity restricting the migration of cancer cells. *Molecules, 23*(10), 1–11. https://doi.org/10.3390/molecules23102471

Calabrò, S. (2015). Plant secondary metabolites. *Rumen microbiology: from evolution to revolution, October* (pp. 153–159). doi: https://doi.org/10.1007/978-81-322-2401-3_11

Carboni, V., Maaliki, C., Alyami, M., Alsaiari, S., & Khashab, N. (2019). Synthetic vehicles for encapsulation and delivery of CRISPR/Cas9 gene editing machinery. *Advanced Therapeutics, 2*(4), 1800085. https://doi.org/10.1002/adtp.201800085

Chen, F., Alphonse, M., & Liu, Q. (2020). Strategies for nonviral nanoparticle-based delivery of CRISPR/Cas9 therapeutics. *Wiley Interdisciplinary Reviews: Nanomedicine and Nanobiotechnology, 12*(3), 1–14. https://doi.org/10.1002/wnan.1609

Chen, K., & Gao, C. (2015). Targeted gene mutation in plants. In *Somatic genome manipulation* (pp. 253–272). Springer.

Chen, K., Wang, Y., Zhang, R., Zhang, H., & Gao, C. (2019). CRISPR/Cas genome editing and precision plant breeding in agriculture. *Annual Review of Plant Biology, 70*, 667–697. https://doi.org/10.1146/annurev-arplant-050718-100049

Cho, S. W., Kim, S., Kim, Y., Kweon, J., Kim, H. S., Bae, S., & Kim, J. S. (2014). Analysis of off-target effects of CRISPR/Cas-derived RNA-guided endonucleases and nickases. *Genome Research, 24*(1), 132–141. https://doi.org/10.1101/gr.162339.113

Cui, Y., Xu, J., Cheng, M., Liao, X., & Peng, S. (2018). Review of CRISPR/Cas9 sgRNA design tools. *Interdisciplinary Sciences: Computational Life Sciences, 10*(2), 455–465. https://doi.org/10.1007/s12539-018-0298-z

Das, T., Anand, U., Pandey, S. K., Ashby, C. R., Assaraf, Y. G., Chen, Z. S., & Dey, A. (2021). Therapeutic strategies to overcome taxane resistance in cancer. *Drug Resistance Updates, 55*, 100754. https://doi.org/10.1016/j.drup.2021.100754

Dey, A., Bhattacharya, R., Mukherjee, A., & Pandey, D. K. (2017a). Natural products against Alzheimer's disease: Pharmaco-therapeutics and biotechnological interventions. *Biotechnology Advances, 35*(2), 178–216. https://doi.org/10.1016/j.biotechadv.2016.12.005

Dey, A., Gorai, P., Mukherjee, A., Dhan, R., & Modak, B. K. (2017b). Ethnobiological treatments of neurological conditions in the Chota Nagpur Plateau, India. *Journal of Ethnopharmacology, 198*, 33–44. https://doi.org/10.1016/j.jep.2016.12.040

Dhar, M. K., Koul, A., & Kaul, S. (2013). Farnesyl pyrophosphate synthase: A key enzyme in isoprenoid biosynthetic pathway and potential molecular target for drug development. *New biotechnology, 30*(2), 114–123. https://doi.org/10.1016/j.nbt.2012.07.001

Doench, J. G., Fusi, N., Sullender, M., Hegde, M., Vaimberg, E. W., Donovan, K. F., Smith, I., Tothova, Z., Wilen, C., Orchard, R., Virgin, H. W., Listgarten, J., & Root, D. E. (2016). Optimized sgRNA design to maximize activity and minimize off-target effects of CRISPR-Cas9. *Nature Biotechnology, 34*(2), 184–191. https://doi.org/10.1038/nbt.3437

Dow, L. E., Fisher, J., Rourke, K. P. O., Muley, A., Edward, R., Livshits, G., Tschaharganeh, D. F., Socci, N. D., & Lowe, W. (2015). Inducible in vivo genome editing with CRISPR/Cas9. *Nature Biotechnology, 33*(4), 390–394. https://doi.org/10.1038/nbt.3155.Inducible

Eid, A., Alshareef, S., & Mahfouz, M. M. (2018). CRISPR base editors: Genome editing without double-stranded breaks. *Biochemical Journal, 475*(11), 1955–1964. https://doi.org/10.1042/BCJ20170793

Feng, S., Song, W., Fu, R., Zhang, H., Xu, A., & Li, J. (2018). Application of the CRISPR/Cas9 system in *Dioscorea zingiberensis*. *Plant Cell, Tissue and Organ Culture, 135*(1), 133–141. https://doi.org/10.1007/s11240-018-1450-5

Gaj, T., Gersbach, C. A., & Barbas, C. F. (2013). ZFN, TALEN, and CRISPR/Cas-based methods for genome engineering. *Trends in Biotechnology, 31*(7), 397–405. https://doi.org/10.1016/j.tibtech.2013.04.004

Gao, J., Wang, G., Ma, S., Xie, X., Wu, X., Zhang, X., Wu, Y., Zhao, P., & Xia, Q. (2015). CRISPR/Cas9-mediated targeted mutagenesis in *Nicotiana tabacum* CRISPR/Cas9-mediated targeted mutagenesis in Nicotiana tabacum. *Plant Molecular Biology, 87*(1–2), 99–110. https://doi.org/10.1007/s11103-014-0263-0

Gilbert, L. A., Horlbeck, M. A., Adamson, B., Villalta, J. E., Chen, Y., Whitehead, E. H., Guimaraes, C., Panning, B., Ploegh, H. L., Bassik, M. C., Qi, L. S., Kampmann, M., & Weissman, J. S. (2014). Genome-scale CRISPR-mediated control of gene repression and activation. *Cell, 159*(3), 647–661. https://doi.org/10.1016/j.cell.2014.09.029

Guo, L., Qi, J., Du, D., Liu, Y., & Jiang, X. (2020). Current advances of *Dendrobium officinale* polysaccharides in dermatology: A literature review. *Pharmaceutical Biology, 58*(1), 664–673. https://doi.org/10.1080/13880209.2020.1787470

Guo, Q., Cao, H., Qi, X., Li, H., Ye, P., Wang, Z., Wang, D., & Sun, M. (2017). Research progress in reversal of tumor multi-drug resistance via natural products. *Anti-Cancer Agents in Medicinal Chemistry, 17*(11), 1466–1476. https://doi.org/10.2174/1871520617666171016105704

Hashimoto, T. (1992). Metabolic engineering of medicinal plants: Transgenic Atropa belladonna with an improved alkaloid composition. *Proceedings of the National Academy of Sciences of the United States of America, 89*, 11,799–11,803.

Hashimoto, T., & Yamada, Y. (1986). Hyoscyamine 6β-hydroxylase, a 2-oxoglutarate-dependent dioxygenase, in alkaloid-producing root cultures. *Plant Physiology, 81*(2), 619–625. https://doi.org/10.1104/pp.81.2.619

Hoff, G., Bertrand, C., Piotrowski, E., Thibessard, A., & Leblond, P. (2018). Genome plasticity is governed by double strand break DNA repair in *Streptomyces*. *Scientific Reports, 8*(1), 1–11. https://doi.org/10.1038/s41598-018-23622-w

Huang, H., Cui, T., Zhang, L., Yang, Q., Yang, Y., Xie, K., Fan, C., & Zhou, Y. (2020). Modifications of fatty acid profile through targeted mutation at BnaFAD2 gene with CRISPR/Cas9-mediated gene editing in *Brassica napus*. *Theoretical and Applied Genetics, 133*(8), 2401–2411. https://doi.org/10.1007/s00122-020-03607-y

Hussain, M. S., Fareed, S., Ansari, S., Rahman, M. A., Ahmad, I. Z., & Saeed, M. (2012). Current approaches toward production of secondary plant metabolites. *Journal of Pharmacy and Bioallied Sciences, 4*(1), 10–20. https://doi.org/10.4103/0975-7406.92725

Jansing, J., Sack, M., Augustine, S. M., Fischer, R., & Bortesi, L. (2019). CRISPR/Cas9-mediated knockout of six glycosyltransferase genes in *Nicotiana benthamiana* for the production of recombinant proteins lacking β-1,2-xylose and core α-1,3-fucose. *Plant Biotechnology Journal, 17*(2), 350–361. https://doi.org/10.1111/pbi.12981

Jiang, W. Z., Henry, I. M., Lynagh, P. G., Comai, L., Cahoon, E. B., & Weeks, D. P. (2017). Significant enhancement of fatty acid composition in seeds of the allohexaploid, *Camelina sativa*, using CRISPR/Cas9 gene editing. *Plant Biotechnology Journal, 15*(5), 648–657. https://doi.org/10.1111/pbi.12663

Jin, Q., Jiao, C., Sun, S., Song, C., Cai, Y., Lin, Y., & Fan, H. (2016). Metabolic analysis of medicinal *Dendrobium officinale* and *Dendrobium huoshanense* during different growth years. *PLoS ONE, 11*(1), e0146607. https://doi.org/10.1371/journal.pone.0146607

Jinek, M. K., Chylinski, I., Fonfara, M., Hauer, J., Doudna, & Charpentier, E. (2012). A programmable dual-RNA-guided DNA endonuclease in adaptive bacterial immunity. *Science*, (337), 816–821.

Jung, W., Yu, O., Lau, S. M. C., O'Keefe, D. P., Odell, J., Fader, G., & McGonigle, B. (2000). Identification and expression of isoflavone synthase, the key enzyme for biosynthesis of isoflavones in legumes. *Nature Biotechnology, 18*(5), 559. https://doi.org/10.1038/75448

Jutras, P. V., Dodds, I., & van der Hoorn, R. A. L. (2020). Proteases of *Nicotiana benthamiana*: an emerging battle for molecular farming. *Current Opinion in Biotechnology, 61*, 60–65. https://doi.org/10.1016/j.copbio.2019.10.006

Kant, R., Tyagi, K., Ghosh, S., & Jha, G. (2019). Host alternative NADH:Ubiquinone oxidoreductase serves as a susceptibility factor to promote pathogenesis of *Rhizoctonia solani* in plants. *Phytopathology, 109*(10), 1741–1750. https://doi.org/10.1094/PHYTO-02-19-0055-R

Kaur, N., Alok, A., Shivani, K. P., Kaur, N., Awasthi, P., Chaturvedi, S., Pandey, P., Pandey, A., Pandey, A. K., & Tiwari, S. (2020a). CRISPR/Cas9 directed editing of lycopene epsilon-cyclase modulates metabolic flux for β-carotene biosynthesis in banana fruit. *Metabolic Engineering, 59*, 76–86. https://doi.org/10.1016/j.ymben.2020.01.008

Kaur, P., Gupta, R. C., Dey, A., Malik, T., & Pandey, D. K. (2020b). Validation and quantification of major biomarkers in 'Mahasudarshan Churna'—an ayurvedic polyherbal formulation through high-performance thin-layer chromatography. *BMC Complementary Medicine and Therapies, 20*(1), 184. https://doi.org/10.1186/s12906-020-02970-z

Khan, S. H. (2019). Genome-editing technologies: Concept, pros, and cons of various genome-editing techniques and bioethical concerns for clinical application. *Molecular Therapy - Nucleic Acids, 16*, 326–334. https://doi.org/10.1016/j.omtn.2019.02.027

Koonin, E. V., & Makarova, K. S. (2009). CRISPR-Cas: an adaptive immunity system in prokaryotes. *F1000 Biology Reports, 1*, 1–6. https://doi.org/10.3410/b1-95

Koonin, E. V., & Makarova, K. S. (2013). CRISPR-Cas: Evolution of an RNA-based adaptive immunity system in prokaryotes. *RNA Biology, 10*(5), 679–686. https://doi.org/10.4161/rna.24022

Kui, L., Chen, H., Zhang, W., He, S., Xiong, Z., Zhang, Y., Yan, L., Zhong, C., He, F., Chen, J., Zeng, P., Zhang, G., Yang, S., Dong, Y., Wang, W., & Cai, J. (2017). Building a genetic manipulation tool box for orchid biology: Identification of constitutive promoters and application of CRISPR/Cas9 in the orchid, dendrobium officinale. *Frontiers in Plant Science, 7*, 1–13. https://doi.org/10.3389/fpls.2016.02036

Labanca, F., Ovesnà, J., & Milella, L. (2018). Papaver somniferum L. taxonomy, uses and new insight in poppy alkaloid pathways. *Phytochemistry Reviews, 17*(4), 853–871. https://doi.org/10.1007/s11101-018-9563-3

Li, B., Cui, G., Shen, G., Zhan, Z., Huang, L., Chen, J., & Qi, X. (2017a). Targeted mutagenesis in the medicinal plant *Salvia miltiorrhiza*. *Scientific Reports, 7*(January), 1–9. https://doi.org/10.1038/srep43320

Li, G., Zhang, X., Zhong, C., Mo, J., Quan, R., Yang, J., Liu, D., Li, Z., Yang, H., & Wu, Z. (2017b). Small molecules enhance CRISPR/Cas9-mediated homology-directed genome editing in primary cells. *Scientific Reports, 7*(1), 1–11. https://doi.org/10.1038/s41598-017-09306-x

Li, T., Huang, S., Jiang, W. Z., Wright, D., Spalding, M. H., Weeks, D. P., & Yang, B. (2011). TAL nucleases (TALNs): Hybrid proteins composed of TAL effectors and FokI DNA-cleavage domain. *Nucleic Acids Research, 39*(1), 359–372. https://doi.org/10.1093/nar/gkq704

Liu, Y., Tao, W., Wen, S., Li, Z., Yang, A., Deng, Z., & Sun, Y. (2015). In vitro CRISPR/cas9 system for efficient targeted DNA editing. *MBio, 6*(6). https://doi.org/10.1128/mBio.01714-15

Malnoy, M., Viola, R., Jung, M. H., Koo, O. J., Kim, S., Kim, J. S., Velasco, R., & Kanchiswamy, C. N. (2016). DNA-free genetically edited grapevine and apple protoplast using CRISPR/Cas9 ribonucleoproteins. *Frontiers in Plant Science, 7*, 1–9. https://doi.org/10.3389/fpls.2016.01904

Mann, V., Pecker, I., & Hirschberg, J. (1994). Cloning and characterization of the gene for phytoene desaturase (Pds) from tomato (*Lycopersicon esculentum*). *Plant Molecular Biology, 24*(3), 429–434.

McMurry, J. E. (2015). Organic chemistry with biological applications. In *Secondary metabolites: An introduction to natural products chemistry* (pp. 1016–1046). Cengage Learning Ltd.

Mercx, S., Smargiasso, N., Chaumont, F., De Pauw, E., Boutry, M., & Navarre, C. (2017). Inactivation of the β(1,2)-xylosyltransferase and the α(1,3)-fucosyltransferase genes in Nicotiana tabacum BY-2 cells by a multiplex CRISPR/Cas9 strategy results in glycoproteins without plant-specific glycans. *Frontiers in Plant Science, 8*, 1–11. https://doi.org/10.3389/fpls.2017.00403

Mishra, A. K., Kocábek, T., Nath, V. S., Awasthi, P., Shrestha, A., Killi, U. K., Jakse, J., Patzak, J., Krofta, K., & Matoušek, J. (2020). Dissection of dynamic transcriptome landscape of leaf, bract, and lupulin gland in hop (*Humulus lupulus* L.). *International Journal of Molecular Sciences, 21*(1). https://doi.org/10.3390/ijms21010233

Morgens, D. W., Wainberg, M., Boyle, E. A., Ursu, O., Araya, C. L., Kimberly Tsui, C., Haney, M. S., Hess, G. T., Han, K., Jeng, E. E., Li, A., Snyder, M. P., Greenleaf, W. J., Kundaje, A., & Bassik, M. C. (2017). Genome-scale measurement of off-target activity using Cas9 toxicity in high-throughput screens. *Nature Communications, 8*, 1–8. https://doi.org/10.1038/ncomms15178

Morineau, C., Bellec, Y., Tellier, F., Gissot, L., Kelemen, Z., Nogué, F., & Faure, J. D. (2017). Selective gene dosage by CRISPR-Cas9 genome editing in hexaploid *Camelina sativa*. *Plant Biotechnology Journal, 15*(6), 729–739. https://doi.org/10.1111/pbi.12671

Narayani, M., & Srivastava, S. (2017). Elicitation: A stimulation of stress in in vitro plant cell/tissue cultures for enhancement of secondary metabolite production. *Phytochemistry Reviews, 16*(6), 1227–1252. https://doi.org/10.1007/s11101-017-9534-0

Nesi, N., Delourme, R., Brégeon, M., Falentin, C., & Renard, M. (2008). Genetic and molecular approaches to improve nutritional value of *Brassica napus* L. seed. *Comptes Rendus—Biologies, 331*(10), 763–771. https://doi.org/10.1016/j.crvi.2008.07.018

Niazian, M. (2019). Application of genetics and biotechnology for improving medicinal plants. *Planta, 249*(4), 953–973. https://doi.org/10.1007/s00425-019-03099-1

Okuzaki, A., Ogawa, T., Koizuka, C., Kaneko, K., Inaba, M., Imamura, J., & Koizuka, N. (2018). CRISPR/Cas9-mediated genome editing of the fatty acid desaturase 2 gene in *Brassica napus*. *Plant Physiology and Biochemistry, 131*, 63–69. https://doi.org/10.1016/j.plaphy.2018.04.025

Ozseyhan, M. E., Kang, J., Mu, X., & Lu, C. (2018). Mutagenesis of the FAE1 genes significantly changes fatty acid composition in seeds of *Camelina sativa*. *Plant Physiology and Biochemistry, 123*, 1–7. https://doi.org/10.1016/j.plaphy.2017.11.021

Pandey, D. K., Nazir, A., & Dey, A. (2017). Isolation and characterization of phosphate solubilizing bacteria from rhizosphere of *Dioscorea alata* stimulating growth and diosgenin production. *Proceedings of the National Academy of Sciences India Section B—Biological Sciences, 87*(4), 1143–1152. https://doi.org/10.1007/s40011-015-0670-2

Pantaleoni, L., Longoni, P., Ferroni, L., Baldisserotto, C., Leelavathi, S., Siva Reddy, V., Pancaldi, S., & Cella, R. (2014). Chloroplast molecular farming: Efficient production of a thermostable xylanase by *Nicotiana tabacum* plants and long-term conservation of the recombinant enzyme. *Protoplasma, 251*(3), 639–648. https://doi.org/10.1007/s00709-013-0564-1

Peng, Q., Hu, Y., Wei, R., Zhang, Y., Guan, C., Ruan, Y., & Liu, C. (2010). Simultaneous silencing of FAD2 and FAE1 genes affects both oleic acid and erucic acid contents in *Brassica napus* seeds. *Plant Cell Reports, 29*(4), 317–325. https://doi.org/10.1007/s00299-010-0823-y

Qiu, J. R., Su, Y. K., Song, Z. Q., Fang, X. S., Li, J. Y., Zhang, J., & Wang, J. H. (2018). Directing construction of CRISPR/Cas9 vector of SmPAL1 in *Salvia miltiorrhiza* by target efficiency detection in vitro. *Zhongguo Zhong yao za zhi=Zhongguo zhongyao zazhi=China journal of Chinese materia medica, 43*(21), 4226–4230.

Rahimi, H., Salehiabar, M., Charmi, J., Barsbay, M., Ghaffarlou, M., Roohi Razlighi, M., Davaran, S., Khalilov, R., Sugiyama, M., Nosrati, H., Kaboli, S., Danafar, H., & Webster, T. J. (2020). Harnessing nanoparticles for the efficient delivery of the CRISPR/Cas9 system. *Nano Today, 34*, 100895. https://doi.org/10.1016/j.nantod.2020.100895

Rehman, S., Rehman, I. U., Jan, B., Rashid, I., Reshi, Z. A., & Ganie, A. H. (2021). Genome editing: applications for medicinal and aromatic plants. In T. Aftab & K. R. Hakeem (Eds.), *Medicinal and aromatic plants* (pp. 119–144). Academic Press.

Ricroch, A., Clairand, P., & Harwood, W. (2017). Use of CRISPR systems in plant genome editing: Toward new opportunities in agriculture. *Emerging Topics in Life Sciences, 1*(2), 169–182. https://doi.org/10.1042/ETLS20170085

Sander, M., & Petersen, M. (2011). Distinct substrate specificities and unusual substrate flexibilities of two hydroxycinnamoyltransferases, rosmarinic acid synthase and hydroxycinnamoyl-CoA: Shikimate hydroxycinnamoyl-transferase, from Coleus blumei Benth. *Planta, 233*, 1157–1171. https://doi.org/10.1007/s00425-011-1367-2

Schindele, A., Dorn, A., & Puchta, H. (2020). CRISPR/Cas brings plant biology and breeding into the fast lane. *Current Opinion in Biotechnology, 61*, 7–14. https://doi.org/10.1016/j.copbio.2019.08.006

Shan, Q., Wang, Y., Li, J., & Gao, C. (2014). Genome editing in rice and wheat using the CRISPR/Cas system. *Nature Protocols, 9*(10), 2395–2410. https://doi.org/10.1038/nprot.2014.157

Shen, C. C., Hsu, M. N., Chang, C. W., Lin, M. W., Hwu, J. R., Tu, Y., & Hu, Y. C. (2019). Synthetic switch to minimize CRISPR off-target effects by self-restricting Cas9 transcription and translation. *Nucleic Acids Research, 47*(3), 1–11. https://doi.org/10.1093/nar/gky1165

Shi, M., Huang, F., Deng, C., Wang, Y., & Kai, G. (2019). Bioactivities, biosynthesis and biotechnological production of phenolic acids in *Salvia miltiorrhiza*. *Critical Reviews in Food Science and Nutrition, 59*(6), 953–964. https://doi.org/10.1080/10408398.2018.1474170

Singh, R. K., Prasad, A., Muthamilarasan, M., Parida, S. K., & Prasad, M. (2020). Breeding and biotechnological interventions for trait improvement: Status and prospects. *Planta, 252*(4), 1–18. https://doi.org/10.1007/s00425-020-03465-4

Srečec, S., Zechner-Krpan, V., Petravić-Tominac, V., Košir, I. J., & Čerenak, A. (2012). Importance of medical eff ects of Xanthohumol, Hop (*Humulus lupulus* L.) biofl avonoid in restructuring of world hop industry. *Agriculturae Conspectus Scientificus, 77*(2), 61–67.

Staniek, A., Bouwmeester, H., Fraser, P. D., Kayser, O., Martens, S., Tissier, A., van der Krol, S., Wessjohann, L., & Warzecha, H. (2013). Natural products—Modifying metabolite pathways in plants. *Biotechnology Journal, 8*(10), 1159–1171. https://doi.org/10.1002/biot.201300224

Subramanian, S., Stacey, G., & Yu, O. (2006). Endogenous isoflavones are essential for the establishment of symbiosis between soybean and *Bradyrhizobium japonicum*. *Plant Journal, 48*(2), 261–273. https://doi.org/10.1111/j.1365-313X.2006.02874.x

Sut, S., Maggi, F., & Dall, S. (2017). Bioactive secondary metabolites from orchids (Orchidaceae). *Chemistry & Biodiversity, 14*(11), e1700172. https://doi.org/10.1002/cbdv.201700172

Tang, H., Zhao, T., Sheng, Y., Zheng, T., Fu, L., & Zhang, Y. (2017). *Dendrobium officinale* Kimura et Migo: A review on its ethnopharmacology, phytochemistry, pharmacology, and industrialization. *Evidence-Based Complementary and Alternative Medicine, 2017*, 7436259.

Teixeira, J. A., Tsavkelova, E. A., Zeng, S., Cardoso, C., & Cardoso, J. C. (2015). Symbiotic in vitro seed propagation of *Dendrobium*: Fungal and bacterial partners and their influence on plant growth and development. *Planta, 242*(1), 1–22. https://doi.org/10.1007/s00425-015-2301-9

Tong, Y., Whitford, C. M., Blin, K., Jørgensen, T. S., Weber, T., & Lee, S. Y. (2020). CRISPR–Cas9, CRISPRi and CRISPR-BEST-mediated genetic manipulation in *Streptomycetes*. *Nature Protocols, 15*(8), 2470–2502. https://doi.org/10.1038/s41596-020-0339-z

Tsai, S. Q., Wyvekens, N., Khayter, C., Foden, J. A., Thapar, V., Reyon, D., Goodwin, M. J., Aryee, M. J., & Joung, J. K. (2014). Dimeric CRISPR RNA-guided FokI nucleases for highly specific genome editing. *Nature Biotechnology, 32*(6), 569–576. https://doi.org/10.1038/nbt.2908

Upadhyay, S. K., & Sharma, S. (2014). SSFinder: High throughput CRISPR-Cas target sites prediction tool. *BioMed Research International, 2014*. https://doi.org/10.1155/2014/742482

Verpoorte, R., & Memelink, J. (2002). Engineering secondary metabolite production in plants. *Current Opinion in Biotechnology, 13*(2), 181–187. https://doi.org/10.1016/S0958-1669(02)00308-7

Verpoorte, R., Van Der Hejden, R., Ten Hoopen, H. J. G., & Memelink, J. (1999). Metabolic engineering of plant secondary metabolite pathways for the production of fine chemicals. *Biotechnology Letters, 21*(6), 467–479. https://doi.org/10.1023/A:1005502632053

Wang, P., Yang, H. L., Yang, Y. J., Wang, L., & Lee, S. C. (2015). Overcome cancer cell drug resistance using natural products. *Evidence-Based Complementary and Alternative Medicine, 2015*. https://doi.org/10.1155/2015/767136

Yang, H., Ren, S., Yu, S., Pan, H., Li, T., Ge, S., Zhang, J., & Xia, N. (2020). Methods favoring homology-directed repair choice in response to crispr/cas9 induced-double strand breaks. *International Journal of Molecular Sciences, 21*(18), 1–20. https://doi.org/10.3390/ijms21186461

Yu, O., Shi, J., Hession, A. O., Maxwell, C. A., McGonigle, B., & Odell, J. T. (2003). Metabolic engineering to increase isoflavone biosynthesis in soybean seed. *Phytochemistry, 63*(7), 753–763. https://doi.org/10.1016/S0031-9422(03)00345-5

Zeng, L., Zhang, Q., Jiang, C., Zheng, Y., Zuo, Y., Qin, J., Liao, Z., & Deng, H. (2021). Development of atropa belladonna L. Plants with high-yield hyoscyamine and without its derivatives using the CRISPR/Cas9 system. *International Journal of Molecular Sciences, 22*(4), 1–10. https://doi.org/10.3390/ijms22041731

Zhang, P., Du, H., Wang, J., Pu, Y., Yang, C., Yan, R., Yang, H., Cheng, H., & Yu, D. (2020). Multiplex CRISPR/Cas9-mediated metabolic engineering increases soya bean isoflavone content and resistance to soya bean mosaic virus. *Plant Biotechnology Journal, 18*(6), 1384–1395. https://doi.org/10.1111/pbi.13302

Zhang, X., Jin, M., Tadesse, N., Dang, J., Zhou, T., Zhang, H., Wang, S., Guo, Z., & Ito, Y. (2018). *Dioscorea zingiberensis* C. H. Wright: An overview on its traditional use, phytochemistry, pharmacology, clinical applications, quality control, and toxicity. *Journal of Ethnopharmacology, 220*, 283–293. https://doi.org/10.1016/j.jep.2018.03.017

Zhang, Z., Mao, Y., Ha, S., Liu, W., Botella, J. R., & Zhu, J. K. (2016). A multiplex CRISPR/Cas9 platform for fast and efficient editing of multiple genes in *Arabidopsis*. *Plant Cell Reports, 35*(7), 1519–1533. https://doi.org/10.1007/s00299-015-1900-z

Zhou, Z., Tan, H., Li, Q., Chen, J., Gao, S., Wang, Y., Chen, W., & Zhang, L. (2018). CRISPR/Cas9-mediated efficient targeted mutagenesis of RAS in *Salvia miltiorrhiza*. *Phytochemistry, 148*, 63–70. https://doi.org/10.1016/j.phytochem.2018.01.015

Chapter 9
Enhanced Production of Plant Aromatic Compounds Through Metabolic Engineering

Ravikant Singh, Saket Jha, Ashutosh Pathak, Gunja Jha, Parul Singh, and Anupam Dikshit

Abstract Aromatic compounds are cyclic six carbon structures possessing sweet or pleasant aroma. Arene or aromatic compounds are considered to be important to chemical industry as well as bio-chemistry fields. These aromatic compounds (mostly derived from petroleum industry) were used in various biological field and need to be explored more. The production via chemically causes more global warming and becomes alarming for alternative methods to produce these compounds. The metabolic engineering technique provides the easy method to produce these chemicals as per their increasing demands which is not fulfilled by oceanic fuel. Tremendous efforts using microbial inoculants have led to the high yield of these aroma compounds through interaction of the plant secondary metabolites. Metabolic engineering to design and construct microorganisms suitable for the production of aromatic amino acids and derivatives thereof requires control of a complicated network of metabolic reactions that partly entertain in equivalent and recurrently are in hasty steadiness.

Keywords Aromatic compounds · Metabolic engineering · Shikimate pathway · Microbes · Oceanic fuels

---

R. Singh
Department of Biotechnology, Swami Vivekanand University, Sagar, Madhya Pradesh, India

S. Jha (✉) · A. Dikshit
Biological Product Lab, Department of Botany, University of Allahabad, Prayagraj, Uttar Pradesh, India

A. Pathak
Forest Botany Division, Forest Research Institute, Dehradun, Uttarakhand, India

G. Jha · P. Singh
Department of Biological Sciences, GD Goenka University, Gurugram, Haryana, India

T. Aftab, K. R. Hakeem (eds.), *Metabolic Engineering in Plants*,
https://doi.org/10.1007/978-981-16-7262-0_9

Introduction

The synthesis of protein is very essential for a living cell. There are some aromatic amino acids (AAA), viz., L-tryptophan (Trp), L-phenylalanine (Phe), and L-tyrosine (Tyr), that play a very important role in the protein synthesis of living cells. In plat, these AAA also serve as a predecessor of a broad diversity of plant natural products that play important role in plant's development, growth, reproduction, defense, and response (Fig. 9.2). Trp is an amino acid essential to biosynthesis of plant hormone auxin, as well as a precursor of alkaloids, phyto-alexins, and indole glucosinolates, while Tyr is a precursor of iso-quinoline alkaloids, pigment betalains, and quinones (tocochromanols and plastoquinone) (Fig. 9.1) (Radwanski & Last, 1995; Kutchan, 1995). Phe is a usual precursor of many phenolic compounds, which contain flavonoids, liquidize tannins, lignans, lignin, and phenyl-propanoid/benzenoid volatiles (Vogt, 2010). Among all three AAA, the biggest carbon flux is frequently administered to Phe, as Phe-derived compounds can compose up to 30% of organic matter in some plant species (Pribat et al., 2010). All three AAA are produced from the final product (chorismate) of the shikimate pathway. Chorismate is also a precursor for vitamins K1 and B9 and the plant protection hormone salicylic acid (Maeda & Dudareva, 2012).

The AAA channel comprises of the Shikimate pathway (the pre-chorismate pathway) and single post-chorismate pathways leading to the production of Trp, Phe, and Tyr (the Trp, Phe, and Tyr pathways, appropriately; Fig. 9.1). These pathways are present in fungi, bacteria, plants, and some protists but are missing in the human diet, while in animal Try can be made from Phe by Phe hydroxylase (Fitzpatrick, 1999). In animal, Try and Trp are also precursors of the serotonin and catecholamine neurotransmitters, appropriately (Fitzpatrick, 1999; Fernstrom & Fernstrom, 2007). The lack of the AAA channel in animals also makes these pathways striking targets for antimicrobial agents and herbicides (Baylis, 2000). Despite the importance of the AAA, the channel has frequently been reason from microbial studies, and as a result, the management of AAA biosynthesis is badly understood in a plant (Maeda & Dudareva, 2012). This basic understanding gap also creates a constriction in functional plant breeding and metabolic engineering for the better construction of AAA-derived target compounds. In this book chapter, our focal point is on the current identification and genetic investigation of enzymes that participate in plant AAA biosynthesis.

By combining current genetic and previous biochemical data, we present an overview of the transcriptional and posttranscriptional regulations of the AAA channel. We also highlight the understanding of the gap in the transfer of AAA and pathways in between across plastid membranes. Finally, we talk over metabolic engineering efforts and outlook for building on the production of AAA and their downstream metabolites in plants (Maeda & Dudareva, 2012).

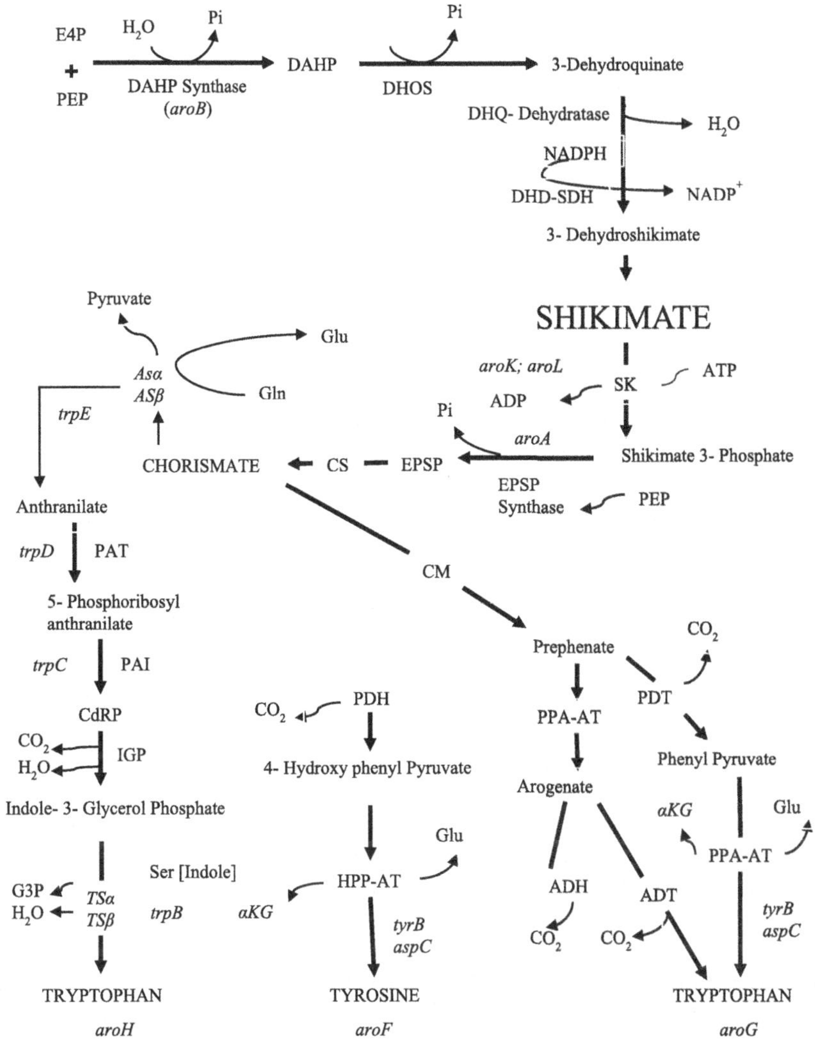

Fig. 9.1 Pathway of aromatic amino acid biosynthesis and its regulation

The Shikimate Pathway

The seven enzymatic reactions of the shikimate channel join central carbon metabolism and the AAA matrix by changing phosphoenolpyruvate (PEP) and D-erythrose 4-phosphate (E4P)—intermediary in the pentose phosphate and glycolysis pathways, appropriately—to chorismate, the global precursor for all AAA (Figs. 9.1 and 9.2) (Bentley & Haslam, 1990; Schmid & Amrhein, 1995). The channel was named after the first recognized intermediate, shikimate, which was

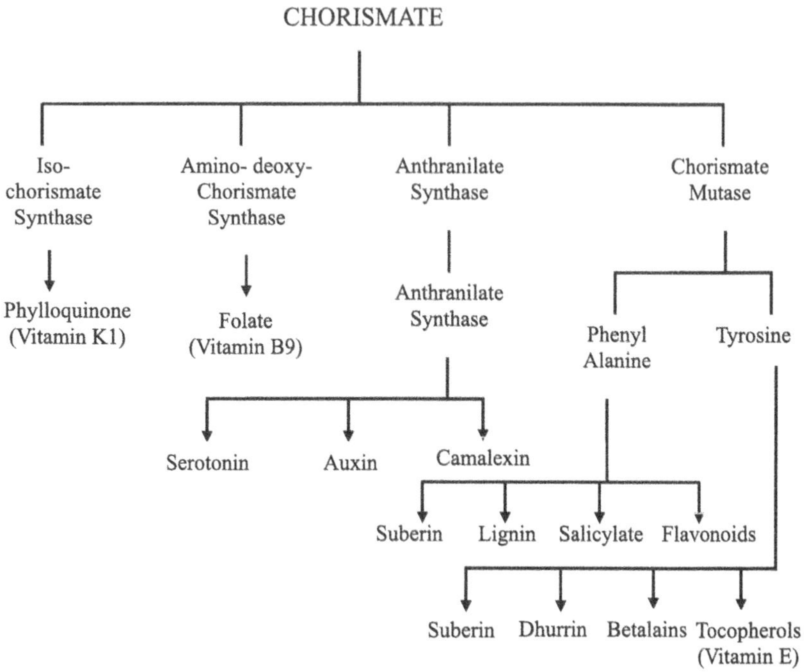

Fig. 9.2 The aromatic amino acid pathways support the formation of numerous natural products in plants

isolated from the fruit of *Illicium religiosum* (commonly known as the Japanese star anise, shikimi) (Bohm, 1965). The corresponding genes have been identified from both plants and microbes (Table 9.1), and all the enzymes involved in the shikimate pathway have been biochemically characterized. In plants, an entire set of the shikimate channel enzymes survive in the plastids, based on experimental confirmation and assumption of their subcellular localization (Table 9.1). While microbial enzymes have been subjected to considerable genetic research that has given an extensive understanding of shikimate pathway directives, only a few genetic studies have been carrying out with plant enzymes, and in plants, this regulation is always badly understood.

3-Deoxy-D-Arabino-Heptulosonate 7-Phosphate (DAHP) Synthase

The first performed step of the shikimate pathway is 3-Deoxy-D-arabinose-heptulosonate7-phosphate (DAHP) synthase catalyzes and an aldol condensation

Table 9.1 Shikimate pathway (enzymes and genes involved in AAA biosynthetic pathways)

S. No.	Enzyme name	Abbreviation	*E. coli* genes	Number of isolated genes (plant species)
1.	3-deoxy-D-arabinoheptulosonate 7-phosphate synthase	DAHP synthase	*AroF* *AroG* *AroH*	2 (*Solanum tuberosum*) 2 (*Arabidopsis thaliana*) 2 (*Lycopersicon esculentum*)
2.	3-Dehydroquinate synthase	DHQS	*AroB*	1 (*Lycopersicon esculentum*)
3.	3-Dehydroquinate dehydratase	DHD^{c}	*AroD*	1 (*Lycopersicon esculentum*) 1 (*Arabidopsis thaliana*)
4.	Shikimate dehydrogenase	SDH^{c}	*AroE*	2 (*Nicotiana tabacum*)
5.	Shikimate kinase	SK	*AroL* *AroK*	1(*Lycopersicon esculentum*) 2 (*Arabidopsis thaliana*) 3 (*Oryza sativa*)
6.	5-Enolpyruvylshikimate-3-phosphate synthase	EPSP synthase	*AroA*	1 (*Arabidopsis thaliana*) 1 (*Solanum lycopersicum*) 2 (*Petunia hybrida*)
7.	Chorismate synthase	CS	*AroC*	1 (*Corydalis sempervirens*) 2 (*Solanum lycopersicum*)

of PEP and E4P to produce DAHP (Fig. 9.1). DAHP synthases are metallo-enzymes accommodating a divalent metal cation (e.g., Mn^{2+} or Co^{2+}). All DAHP synthases have a basic (β/α) 8-barrel monomer structure and show a large diversity in their peripheral small kingdom responsible for AAA moderate allosteric regulation depending on sequence affinity in their DAHP synthases and are divided into two types, which contribute <10% amino acid sequence identity. An example of type one enzyme comprises DAHP synthases from Saccharomyces cerevisiae (*Aro3* and *Aro4*) containing an N-terminal Phe/Tyr-binding domain (Bentley & Haslam, 1990), bi-functional chorismate mutase (CM)–, and *Escherichia coli* (*AroF*, *AroG*, and *AroH*) (Maeda & Dudareva, 2012).

DAHP synthase enzyme (e.g., from *Bacillus subtilis*) (Wu & Woodard, 2006) and deregulated DAHP synthase have only the basic barrel structure, e.g., from *Pyrococcus furiosus* (Schofield et al., 2005). In-plant type II DAHP synthases are present and some microbes (e.g., *Mycobacterium tuberculosis*) and have special AAA-binding elements as adding to the basic barrel structure (Webby et al., 2005). It is fact that plants have DAHP synthases that are nearby related to *E. coli* and yeast enzymes; two copies of DAHP synthase genes have been extracted from many plant species through the accompaniment of *E. coli* or yeast mutant defective in the correlated activity (Dyer et al., 1990; Keith et al., 1991). The two DAHP synthase genes (*dahps1* and *dahps2*) show different expressions in *Arabidopsis thaliana* (Arabidopsis), *Solanum tuberosum* (potato), and *Solanum lycopersicum* (tomato). In potato, *dahps2* (*shkB*) gene is continually expressed, while the *dahps1* (*shkA* in potato) is strongly persuaded in response to wounding and pathogen infection (Görlach et al., 1995; Keith et al., 1991). In *Arabidopsis*, DAHP synthase I related to redox moderate by a ferredoxin-thioredoxin system that connects carbon flow into the shikimate pathway with photosynthetic electron flow (Entus et al., 2002). Despite plant DAHP synthases 1 and 2, both utilized Mn^{2+} (Pinto et al., 1986), and further Co^{2+}-dependent DAHP synthase activity has been identified in the cytosol of many plant tissues (Rubin & Jensen, 1985). Although the corresponding genes have not been identified, their physiological task is currently mysterious.

3-Dehydroquinate Synthase (DHQS)

3-Dehydroquinate synthase (DHQS) transforms DAHP to 3-dehydroquinate utilizing a divalent cation (e.g., Co^{2+}) and NAD+ cofactors by five following chemical reactions: β-elimination of inorganic phosphate, alcohol oxidation, carbonyl reduction, ring-opening, and intramolecular aldol condensation (Srinivasan et al., 1963). Elicited from crystal structure studies, DHQS executes these in one active site without establishing by-products (Carpenter et al., 1998). A plant DHQS gene has been extracted from tomato through the accompaniment of an *E. coli* mutant (Bischoff et al., 1996). This single-duplicate gene is highly expressed in tomato roots, and its expression is instigated by elicitor treatment in suspension cell culture (Bischoff et al., 1996). In *E. coli* (*AroB*) and plants, DHQS genes are monofunctional enzymes, while those in fungi are part of a penta-functional enzyme (AROM complex) that catalyzes five consecutive reactions transferring DAHP to 5-enolpyruvylshikimate 3-phosphate (EPSP) in the shikimate channel (Duncan et al., 1987).

3-Dehydroquinate Dehydratase (DHD)–Shikimate Dehydrogenase (SDH)

The third and fourth enzymatic reactions in the shikimate pathway comprise (a) the dehydration of 3-dehydroquinate into 3-dehydroshikimate to establish the first double bond in the ring and (b) the reversible reduction of 3-dehydroshikimate into shikimate utilizing NADPH (Fig. 9.2). 3-dehydroquinate dehydratase (DHD) and shikimate dehydrogenase (SDH also known as shikimate: $NADP^+$ oxidoreductase) catalyze the appropriate reactions and have pointedly different enzyme organizations in three domains. These enzymes are mono-functional in *E. coli* (*AroD* and *AroE*, respectively) (Duncan et al., 1986) and part of the AROM complex in fungi, while in plants they are combined to form a bi-functional DHD-SDH enzyme (Bischoff et al., 2001). The crystal structure of the *Arabidopsis* enzyme represents that the active parts of DHD and SDH are restricted in close presence and face each other, which make possible an optimal, local 3-dehydroshikimate concentration for productive SDH catalysis (Singh & Christendat, 2006). In plant tissues, approximately 10 times higher activity of SDH than that of DHD additionally certifies that the 3-dehydroshikimate intermediate will be strongly converted to shikimate, therefore increasing metabolic flux through the shikimate pathway (Fiedler & Schultz, 1985; Mousdale & Coggins, 1985). A single gene encoding DHD-SDH, which contains an assumed plastid transit peptide, has been recognized in many plant species, including *Arabidopsis* (Table 9.1) (Singh & Christendat, 2006). An assumption is the genome of *Nicotiana tabacum* (tobacco), which carries two genes encoding both plastidic and cytosolic DHD-SDHs (Ding et al., 2007). RNA interference (RNAi) defeating of the plastidic DHD-SDH in tobacco leaves develops in an aggregation of both 3-dehydroquinate and shikimate—the substrate and product of the enzyme, respectively—together with decreased levels of Phe, Tyr, lignin, and chlorogenic acid (Ding et al., 2007). Therefore, the underlying mechanisms for the increased shikimate level in the RNAi lines as well as the role(s) of the cytosolic DHD-SDH in tobacco plants persist to be investigated.

Shikimate Kinase (SK)

Shikimate kinase (SK), the fifth enzyme of the shikimate channel, catalyzes the phosphorylation of the C3 hydroxyl class of shikimate utilizing ATP as a co-substrate to give shikimate 3-phosphate (Fig. 9.2). SK needs a divalent cation (e.g., Mg^{2+} orMn^{2+}) for its activity (Koshiba, 1979) and comprises a base, lid, shikimate-binding, and nucleotide-binding domains, which undergo substantial conformational changes upon binding of shikimate and ATP (Gan et al., 2006). While plants have different numbers of isozymes depending on species: one in tomato, two in *Arabidopsis*, and three in *Oryza sativa* (rice), *E. coli* has two isozymes, *AroL* and *AroK* (DeFeyter & Pittard, 1986) (Table 9.1) (Fucile et al., 2011). The purified *Spinacia oleracea* (spinach) SK is regulated by the rank of the energy charge (the respective concentrations of ATP, ADP, and AMP) (Schmidt et al., 1990), close to *B. subtilis* SK and other ATP-using enzymes (Pacold & Anderson, 1973). Therefore, SK may give a modulate link between the energy-need shikimate pathway and cellular energy stable in plants.

5-Enolpyruvylshikimate 3-Phosphate (EPSP) Synthase

EPSP synthase (also known as 3-phosphoshikimate 1-carboxyvinyltranferase) produces the penultimate step of the shikimate pathway, the development of EPSP, by moving the enolpyruvyl moiety of phosphoenolpyruvate of its 5-hydroxyl position of shikimate 3 phosphate PEP to the 5-hydroxyl position of shikimate 3-phosphate (Fig. 9.2). This C3 enolpyruvyl unit, in the end, becomes the side chain of Phe and Tyr and is separate during the biosynthesis of Trp. Crystalline structure studies have proven that the binding of the first substrate, shikimate 3-phosphate, activates a global conformational change to form the active site in the inter-domain cleft of EPSP synthase (Schönbrunn et al., 2001). EPSP synthase is the first target of the non-selective, broad-spectrum herbicide glyphosate [N-(phosphonomethyl) glycine. Glyphosate insistently inhibits EPSP synthase with esteem to the second substrate, PEP, by covering the PEP binding site of the enzyme-shikimate 3-phosphate complex (Schönbrunn et al., 2001).

EPSP synthases from different organisms have been classifying into two classes depending on glyphosate sensitivity: some bacteria, such as *Agrobacterium* sp. strain CP4, have class II EPSP synthases that are relatively resistant to glyphosate and therefore have been used to generate glyphosate-resistant crops, whereas all plants and most bacteria, including *E. coli*, have glyphosate-sensitive (class I EPSP synthases) (Funke et al., 2006). The genes encoding for EPSP synthases have been extracted from many plant species (Table 9.1) (Shah et al., 1986). Compatible with the purification of two isozymes with very similar kinetic function in *Zea mays* (maize) (Forlani et al., 1994), the *Arabidopsis* genome consists of two genes encoding one functional (Klee et al., 1987) and one putative EPSP synthase (Kaul et al., 2000). The EPSP synthase gene is the main constituent expressed at low levels but displays tissue-specific and developmentally moderate expression in *Petunia hybrida* (petunia) flowers, likely for the production of Phe-derived volatiles (Gasser et al., 1988).

Chorismate Synthase (CS)

Chorismate synthase (CS) produces the final reaction of the shikimate channel, the 1,4-antielimination of the 3-phosphate, and C6-pro-R hydrogen from EPSP, introducing the second double bond in the ring to make chorismate (Macheroux et al., 1999). Therefore, the reaction is redox neutral, CS needs lower flavin mononucleotide (FMN) as a cofactor that transiently donates one electron to the EPSP substrate to make possible the phosphate break, and it is engaging in the C6 hydrogen abstraction (Maclean & Ali, 2003). While CSs from a different domain are highly homologous and have the same structural fold kinetic properties and cofactor specificity (Maclean & Ali, 2003), two classes have been differentiating based on their ability to lower the oxidized FMN. CSs from fungi are connected with

NADPH-dependent flavin reductase as a part of a bi-functional enzyme (Schaller et al., 1991a), while most bacteria and plants have mono-functional enzymes based on an external source of lower FMN. In plants, the lower FMN may be given by blue-light-mediated FMN photoreduction or by flavin reductase activity not physically connected with the enzyme (Schaller et al., 1991b). Further examination into the sources of lower FMN in the plastids will give us to understand of the function of redox potential in moderating CS activity in plants. To date, plant genes encoding CSs have been extracted from *Corydalis sempervirens* and tomato, the latter having two different expressed genes (Görlach et al., 1993).

The Tryptophan Pathway

In plants, chorismate is a usual precursor for small four branches of metabolic pathways leading to the production of Trp, Phe/Tyr, salicylate/phylloquinone, and folate (Fig. 9.1). Four enzymes—CM, isochorismate synthase (ICS), anthranilate synthase (AS), and amino-deoxy-chorismate synthase (ADCS)—catalyze the carry-out step of the specific pathways and compete for chorismate (Fig. 9.1). The Trp pathway changes from chorismate to Trp by six enzymatic reactions (Fig. 9.2) (Siehl, 1999). All enzymes involved in the Trp biosynthetic pathway are exhibit or divine to confine in the plastids (Table 9.2) (Kriechbaumer et al., 2008). In comparison to diverse enzymes often found in the microbial Trp biosynthetic pathway (Crawford, 1989), all plant enzymes are mono-functional (Radwanski & Last,

Table 9.2 Tryptophan pathway (enzymes and genes involved in AAA biosynthetic pathways)

S. No.	Enzyme name	Abbreviation	*E. coli* genes	Number of isolated genes (plant species)
1.	Anthranilate synthetase α subunit	ASα	*TrpE*	2 (*Arabidopsis thaliana*) 2 (*Ruta graveolens*)
2.	Anthranilate synthetase β subunit	ASβ	*TrpG*	1 (*Arabidopsis thaliana*)
3.	Phosphoribosyl-anthranilate transferase	PAT	*TrpD*	1 (*Arabidopsis thaliana*)
4.	Phosphoribosyl-anthranilate isomerase	PAI	*TrpF*	3 (*Arabidopsis thaliana*)
5.	Indole-3-glycerol phosphate synthase	IGPS	*TrpC*	1 (*Arabidopsis thaliana*)
6.	Tryptophan synthase α subunit	TSα	*TrpA*	1 (*Arabidopsis thaliana*) 1 (*Zea mays*)
7.	Tryptophan synthase β subunit	TSβ	*TrpB*	2 (*Arabidopsis thaliana*) 2 (*Zea mays*)

1995). Therefore, the first and last reactions are catalyzed by noncovalent enzyme complexes. Isolation and investigation of Trp biosynthetic mutants have provided genetic proof for the participation of these enzymes in Trp biosynthesis and its moderation in plants and have advanced our understanding for the development of other indole compounds.

Anthranilate Synthase (AS)

The first step in Trp biosynthesis is the formation of anthranilate which is catalyzed or lyase by an amino-accepting chorismate-pyruvate. It is comprised of large α and small β subunits (*ASα* and *ASβ*, respectively), which form an α/β heterodimer or an α2/β2 tetramer (Romero et al., 1995). *Asα* binds to chorismate and catalyzes the amination and pyruvate exclusion reactions, while *ASβ* hydrolysis glutamine gives ammonia to *ASα* (Knöchel et al., 1999). The binding of chorismate to *ASα* activates a conformational change to an active state and produces an intermolecular pathway for ammonia transfer from *ASβ* to *Asα* (Morollo & Eck, 2001). The AS enzyme is allosterically inhibited by Trp, which binds to *ASα* and stops its conformational change (Morollo & Eck, 2001), indicating that *ASα* (but not *ASβ*) is responsible for reaction inhibition by Trp. Most *ASα* subunits in plants are feedback-sensitive to Trp, with the known prediction of feedback-insensitive tobacco and *Ruta graveolens ASα* (Bohlmann et al., 1996). Higher plants consider to date contain two small genes encoding *ASα* and one gene encoding *Asβ* (Table 9.1) (Song et al., 1998). One of the *Asα* genes is rapidly expressed, while the other is moderate developmentally and induced in response to wounding and pathogens (Tozawa et al., 2001), suggesting its participation in the formation of Trp pathway made from natural products as a part of plant defense.

Phosphoribosylanthranilate Transferase (PAT)

Phosphoribosylanthranilate transferase (PAT) transfers the phosphoribosyl element from phosphoribosyl pyrophosphate to anthranilate and forms 5-phosphoribosylanthranilate. The first separated *Arabidopsis* mutant reduced in Trp biosynthesis, *trp1* (Last & Fink, 1988), contains a mutation in a single-copy gene encoding PAT (Rose et al., 1992). The PAT gene is rapidly expressed in *Arabidopsis*, and the first two introns were shown to improve the abundance of PAT messenger RNA (mRNA) posttranscriptionally (Rose & Last, 1997). The *trp1* mutant shows a series of auxin-deficient phenotypes (e.g., reduced size and apical dominance) as well as a blue fluorescence phenotype owing to the aggregation of anthranilate glucosides (Quiel & Bender, 2003). Feeding of Trp did not reinstate the auxin-deficient phenotype of *trp1* (Rose et al., 1997), highlighting the importance of the Trp-independent channel of auxin biosynthesis.

Phosphoribosylanthranilate Isomerase (PAI)

Phosphoribosylanthranilate isomerase (PAI) catalyzes the irreversible disarray of 5-phosphoribosylanthranilate to 1-(o-carboxy phenyl amino)-1-deoxy-ribulose 5-phosphate (CdRP), a reaction that can also take place non-enzymatically. *Arabidopsis* has three or four highly same *PAI* genes depending on ecotype (Bender & Fink, 1995). *PAI1* and *PAI2* encode useful PAI enzymes (Melquist et al., 1999), and their mRNAs show main transcripts differentially exhibit in response to environmental pressure and in a cell type-specific mode (He & Li, 2001). Similar to the *PAT* mutant (*trp1*), lower in PAI activity in PAI antisense plants or deficiency in the tandem *PAI1* and *PAI4* genes in the *trp6* mutant outcome in the blue fluorescence phenotype (He & Li, 2001; Bender & Fink, 1995). The respective contribution of each isogeny to Trp biosynthesis and other indole complex has not been fully solved owing to their high chain similarities (Li et al., 1995b) and concert epigenetic control of the PAI gene family in some *Arabidopsis* ecotypes (Melquist et al., 1999).

Indole-3-Glycerol Phosphate Synthase (IGPS)

Indole-3-glycerol phosphate synthase (IGPS) catalyzes the irreversible transformation of CdRP to indole-3-glycerol phosphate. One of two *Arabidopsis* genes encoding *IGPSs* (*igps1*) has been extracted through complementation in an *E. coli trpC* mutant defective in *IGPS* (Li et al., 1995c). Antisense suppression of *igps1* lowers the levels of both *Trp* and auxin, while the *trp2* and *trp3* mutants defective in Trp synthase have less Trp but aggregate more auxin, indicating that indole-3-glycerol phosphate acts as a key branch-point intermediate in Trp-independent auxin biosynthesis (Ouyang et al., 2000).

Tryptophan Synthase (TS)

The final two reactions of the Trp channel are catalyzed by the Trp synthase α subunit (*TSα*) and β subunit (*TSβ*). *TSα* catalyzes the reversible retro-aldol break of indole-3-glycerol phosphate to indole and glyceraldehyde 3-phosphate (G3P), and *TSβ* eventually condenses indole and serine to build Trp utilizing pyridoxal 5-phosphate (PLP) as a cofactor (Barends et al., 2008). *TSα* and *TSβ* form an $\alpha 2\beta 2$ hetero-complex, and indole is converted from the active site of *TSα* to that of *TSβ* through a 25-A°-long intermolecular hole (Barends et al., 2008). While fungi carry a single gene encoding a bi-functional *TSα-TSβ* enzyme, bacteria such as *E. coli* have two different genes encoding *TSα* and *TSβ* (Crawford, 1989). The *Arabidopsis* genome comprises at least two and three putative genes encoding *TSα* and *TSβ*, as result (Kaul et al., 2000; Radwanski et al., 1995).

The essential roles of *Arabidopsis TSα1* and *TSβ1* in Trp biosynthesis have been genetically proven to depend on the Trp auxotrophic phenotypes of the *trp3* and *trp2* mutants, respectively, while the functions of the other genes remain to be analyzed (Radwanski et al., 1996). In comparison to *Arabidopsis*, maize comprises two highly same and redundant *TSβ* genes, and only simultaneous loss of both utility leads to Trp auxotroph (Wright et al., 1992). Plants carry other enzymes in addition to *TSα* that transfer indole-3-glycerol phosphate to indole, a usual precursor of not only Trp but also other plant natural products (Fig. 9.1, Table 9.2). In maize, indole-3-glycerol phosphate lyase (IGL) constructs explosive indole under herbivore attack (Frey et al., 2000), while *BX1* (*benzoxazineless1*) catalyzes the first process in the biosynthesis of the natural pesticides benzoxazinoids (Frey et al., 2009). In comparison to maize *IGL* and *BX1*, which have a predicted plastid transit peptide (Frey et al., 2000), the *Arabidopsis* genome also carries an indole-producing *TSα*-like enzyme that localizes in the cytosol (Zhang et al., 2008). Therefore, all three genes are *TSα* paralogs; the correlate with enzymes catalyzes indole production independent of a *TSβ*-like subunit (Frey et al., 2000). Because the formation of indole or benzoxazinoids sometimes exceeds that of Trp (Frey et al., 2009), the hetero-complex production of *TSα* and *TSβ* stops the release of the indole intermediate and makes sure a basal level of Trp production for protein and auxin biosynthesis.

The Phenylalanine and Tyrosine Pathways

This pathway has been discovered recently. As genes encoding the pathway enzymes have only recently been identified, therefore knowledge about this pathway is very new for us. Phe and Tyr are derived from chorismate, the final product of the shikimate pathway. Conversion of chorismate into Phe and Tyr may take place via two alternative pathways (Fig. 9.2) (Siehl, 1999).

In the arogenate pathway, transamination of prephenate to L-arogenate takes place which is followed by dehydration/decarboxylation to Phe or dehydrogenation/decarboxylation to Tyr catalyzed by arogenate dehydratase (ADT) or arogenate dehydrogenase (ADH), respectively. In the phenylpyruvate or 4-hydroxyphenylpyruvate pathway, the prephenate in the presence of prephenate dehydratase (PDT) is first subjected to dehydration/decarboxylation or to dehydrogenation/decarboxylation by prephenate dehydrogenase (PDH). This process is followed by transamination of phenylpyruvate to Phe and 4-hydroxyphenylpyruvate to Tyr (Fig. 9.2). For Phe biosynthesis in plants, arogenate pathway is the predominant route, which is now clear from recent genetic studies (Maeda et al., 2010), whereas the major pathway for Tyr biosynthesis remains to be determined. Both pathways are important for the better understanding and determination of actual biosynthetic routes of Tyr and Phe (Table 9.3).

Table 9.3 Phe/Tyr pathway (enzymes and genes involved in the biosynthetic pathways)

S. No.	Enzyme name	Abbreviation	*E. coli* genes	Number of isolated genes (plant species)
1.	Chorismate mutase	CM	*AroQ* *PheA*[e] *TyrA*[e]	2 (*Petunia hybrida*) 3 (*Arabidopsis thaliana*)
2.	Prephenate aminotransferase	PPA-AT	Absent	1 (*Arabidopsis thaliana*) 1 (*Petunia hybrida*) 1 (*Lycopersicon esculentum*)
3.	Arogenate/prephenate dehydratase	ADT/PDT	*PheA*[e]	6 (*Arabidopsis thaliana*) 3 (*Petunia hybrida*) 1 (*Oryza sativa*)
4.	Arogenate/prephenate dehydrogenase	ADH/PDH	*TyrA*[e]	2 (*Arabidopsis thaliana*) 4 (*Zea mays*)
5.	Phenylpyruvate/4-hydroxyphenylpyruvate aminotransferase	PPY-AT/ HPP-AT	*TyrB* *AspC* *IlvE*	1 (*Cucumis melo*) 1 (*Papaver somniferum*) 1 (*Arabidopsis thaliana*)
6.	Phenylalanine hydroxylase	Phe hydroxylase	Absent	1 (*Pinus taeda*) 1 (*Physcomitrella patens*) 1 (*Chlamydomonas reinhardtii*)

Chorismate Mutase (CM)

Conversion of chorismate to prephenate is the first step in Phe and Tyr biosynthesis and is catalyzed by CM. CMs exist in various forms. Two classes of CMs with α-helical or α/β-barrel structures (*AroQ* or *AroH* types, respectively) have been identified on the basis of protein folds (Lee et al., 1995). Like CM1, CM3 contains a putative plastid transit peptide and is subject to allosteric regulation, but its affinity toward chorismate is closer to that of CM2 (Mobley et al., 1999). All three Arabidopsis genes display differential expression, with CM1 and, to a lesser extent, CM3 being inducible in response to elicitors and pathogen treatments (Eberhard et al., 1996), suggesting their distinct physiological functions under different developmental and environmental conditions.

Prephenate Aminotransferase (PPA-AT)

Prephenate aminotransferase (*PPA-AT*) catalyzes the initial step of the arogenate pathway in Phe and Tyr biosynthesis, a reversible transamination between prephenate and arogenate using PLP as a cofactor (Fig. 9.2, Table 9.3). *PPA-AT* activities have been detected in some bacteria (Fazel & Jensen, 1979) as well as in many plant species (Bonner & Jensen, 1985; Siehl et al., 1986). Despite extensive biochemical characterization of PPA-ATs, the corresponding genes had not been identified in any organisms until recently. Identification of *PPA-AT* genes from Arabidopsis, petunia, and tomato (Maeda et al., 2011) revealed that *PPA-ATs* belong to the class Ib aspartate aminotransferases, which contain a lysine residue (De la Torre et al., 2009). Similarly, to most aminotransferases, which often exhibit broad substrate specificity (Wightman & Forest, 1978), plant *PPA-ATs* that are purified from plant tissues or recombinantly expressed display aspartate aminotransferase activity as well (Maeda et al., 2011).

Prephenate and Arogenate Dehydratases (PDT and ADT)

PDT catalyzes the decarboxylation and dehydration of prephenate to phenylpyruvate, the initial step in the phenylpyruvate route of Phe biosynthesis; whereas in the final step, ADT changed the arogenate to phenyl alanine (Figs. 9.1 and 9.2). Prephenate and arogenate can be spontaneously converted to phenylpyruvate and Phe, respectively, under acidic conditions in vitro (Zamir et al., 1983). However, crystal-structure and site-directed mutagenesis studies of PDTs showed that the acidic residues of the enzyme are not directly involved in the catalysis (Zhang et al., 2000). Genes homologous to bacterial PDTs have recently been identified in *Arabidopsis*, rice, and petunia with six, four, and three genes, respectively, though additional genes might exist in the rice and petunia genomes (Yamada et al., 2008).

Moreover, comparative analysis of PPA-AT and ADT activities in petunia petals revealed that PPA-AT activity is at least three orders of magnitude higher than ADT activity (Maeda et al., 2011), signifying that ADT catalyzes a rate-limiting step within the arogenate biosynthesis pathway of phenyl alanine. In general, PDT activity and phenylpyruvate have rarely been detectable in plant tissues (Maeda et al., 2010). The role of the phenylpyruvate route in plant Phe biosynthesis under physiological conditions remains to be determined. Given that both the arogenate and phenylpyruvate pathways can potentially be active in plants (Maeda et al., 2010), an open question is why the major flux goes via the arogenate route in plant Phe biosynthesis. ADTs that can use prephenate in addition to arogenate compete with PPA-ATs for the prephenate substrate illustrated in Fig. 9.1 (Maeda et al., 2011).

Arogenate and Prephenate Dehydrogenases (ADH and PDH)

ADH and PDH catalyze the oxidative decarboxylation of arogenate and prephenate to Tyr and 4-hydroxyphenylpyruvate, respectively, using an NAD^+ or $NADP^+$ cofactor. Similar to PDT enzymes, PDHs exist as either mono-functional or bi-functional enzymes in which the PDH domain is fused with other enzymes such as CM (*AroQ*), EPSP synthase (PDH-*AroF*), or AAA aminotransferase (*HisH*) (Song et al., 2005). PDHs/ADHs are active as tetramers, and each monomer consists of an N-terminal nucleotide-binding domain and a C-terminal substrate-binding domain (Legrand et al., 2006). Plant ADHs specifically use $NADP^+$ as a cofactor (Rippert & Matringe, 2002), with the exception of the NAD^+-dependent ADH in maize (Byng et al., 1981).

Phenylpyruvate and 4-Hydroxyphenylpyruvate Aminotransferases (PPY-AT and HPP-AT)

Phenylpyruvate aminotransferase (PPY-AT) catalyzes a reversible transamination between phenylpyruvate and Phe using a PLP cofactor (Figs. 9.1 and 9.2). Similarly, 4-hydroxyphenylpyruvate aminotransferase (HPP-AT) converts 4-hydroxyphenylpyruvate and Tyr. In *E. coli*, two distinct aminotransferases [an AAA aminotransferase (*AroAT/TyrB*) and one aspartate aminotransferase (*AspC*)] catalyze the phenylpyruvate/4-hydroxyphenylpyruvate transamination to Phe/Tyr. However, an additional enzyme [a branched-chain amino acid aminotransferase (*IlvE*)] can also convert phenylpyruvate to Phe (Gelfand & Steinberg, 1977). Similarly, to those of microbes, AAA aminotransferase activities detected in a number of plant species exhibit broad substrate specificity with respect to amino donors and keto acid acceptors (Gonda et al., 2010). Seven genes encoding putative Tyr aminotransferases exist in the *Arabidopsis* genome, and two of these enzymes have been biochemically characterized (Prabhu & Hudson, 2010).

Phenylalanine Hydroxylase

In animals, Phe is an essential amino acid and can be converted to Tyr by Phe hydroxylase (Fitzpatrick, 1999). Protists and some bacteria also contain iron-dependent monooxygenases, which hydroxylate the aromatic ring of Phe to Tyr. Recently, Phe-specific AAA hydroxylases have been identified in nonflowering plants, including gymnosperms, mosses, and algae, but not in angiosperms (Pribat et al., 2010).

Regulation of Aromatic Amino Acid Biosynthesis

Both plant and microorganisms control carbon flux toward AAA biosynthesis at the transcriptional and posttranscriptional levels (Bentley & Haslam, 1990). Apart from basal levels of AAA production for protein biosynthesis, plants have to balance their production for the biosynthesis of downstream natural products, including the main cell wall component lignin and defense complex, the levels of which usually awfully change under specific developmental and environmental conditions. Therefore, the moderation of AAA biosynthesis in plants should be coordinated with the activities of the downstream metabolic channel and different from those of microorganisms.

Transcriptional Regulation

In microbes, the expression of the first gene in the shikimate pathway (DAHP synthase) is regulated in response to the cellular levels of AAA, playing a key role in controlling the carbon flux into the pathway. The expression of DAHP synthase genes (*Aro3p* and *Aro4p*) is regulated in fungi through a master transcriptional activator (*Gcn4p*), which is induced under general amino acid starvation (Natarajan et al., 2001). In plants, there is finite information about the effect of AAA levels on the expression of the shikimate pathway genes. Reduction of AAA biosynthesis through the glyphosate-mediated prohibition of EPSP synthase induces DAHP synthase protein level and activity in potato cells either transcriptionally or translationally (Pinto et al., 1988). In addition, reduced Phe levels in petunia flowers of ADT1-RNAi lines increase the expression of the shikimate pathway genes (Maeda et al., 2010). Although the underlying molecular mechanisms are currently unknown, these results suggest that reduced levels of AAA or their downstream products may act as a signal to induce the expression of the shikimate pathway genes and reinstitute the carbon flux through the pathway in plants.

In *Arabidopsis*, methyl jasmonate treatment induces the expression of genes encoding DAHP synthase and Trp pathway enzymes via a COI1 (coronatine insensitive 1)-dependent signaling pathway (Devoto et al., 2005). In addition, salicylate is involved in the transcriptional activation of the Trp pathway, as demonstrated by the fact that salicylate-deficient *Arabidopsis* NahG plants show a reduction in pathogen-induced expression of the Trp pathway genes (Zhao & Last, 1996). The *Arabidopsis ups1* (under inducer after pathogen and stress 1) mutant exhibits reduced PAT expression and a compromised salicylate-, jasmonate, and reactive oxygen species-mediated induction of defense gene expressions (Denby et al., 2005), implying that *ups1* is an upstream component involved in transcriptional activation of the Trp pathway and plant defense (Figs. 9.1 and 9.2).

Many transcription factors that regulate the biosynthesis of AAA-derived natural products have been isolated, and some have been shown to co-regulate the expression of genes in the AAA pathway as well (Figs. 9.1 and 9.2). In *Arabidopsis*, MYB transcription factors regulating indole glucosinolate biosynthesis (*hig1/myb51* and *atr1/myb34*) also activate genes encoding DAHP synthase 1 and *ASα1*, respectively (Gigolashvili et al., 2007). RNAi suppression of *myb8* in *Nicotiana* attenuate significantly reduces the expression of all seven shikimate pathway genes, resulting in complete elimination of phenylpropanoid-polyamine conjugates (Kaur et al., 2010). Expression of genes encoding 6 out of 10 enzymes essential for Phe biosynthesis was reduced in an *Arabidopsis* double mutant defective in *nst1* and *nst3*, NAC transcription factors regulating secondary wall formation (Mitsuda et al., 2007).

These results indicate that in plants the expression of genes encoding enzymes in the AAA and their downstream pathways are coordinately regulated, often by the same transcription factor, to achieve the required production of AAA-derived natural products. However, it remains to be determined whether these transcription factors activate the target promoters directly or indirectly and whether additional factors are involved. Three transcription factors controlling shikimate pathway genes—a C_2H_2-type zinc finger DNA-binding protein, *epf1*, and two R_2R_3-type MYB transcription factors, *odorant1* and *eobII*—have been isolated in petunia flowers and provided new intuition into this transcriptional regulatory network (Verdonk et al., 2005). *epf1* directly binds to the EPSP synthase promoter and controls its spatial and developmental (Takatsuji et al., 1992). Current studies have shown that *eobII* binds to the *odorant1* promoter via a putative MYB binding site (Van Moerkercke et al., 2011) and that *eobII* suppression results in partial *odorant1* downregulation (Spitzer-Rimon et al., 2010). Because the overexpression of *eobII* did not significantly alter the *odorant1* and CM expressions (Spitzer-Rimon et al., 2010), one or more additional factors are likely involved in the *eobII*-mediated activation of these promoters.

Posttranscriptional Regulation

In addition to the transcriptional regulation, AAA biosynthesis is subject to complex posttranscriptional regulations, which control carbon flux into the shikimate pathway as well as the carbon allocation toward individual AAA. Within the pathway, the partitioning of carbon flux between the Trp and Phe/Tyr pathways is controlled at the level of two enzymes, AS and CM, both of which compete for chorismate as a substrate. AS and CM are feedback inhibited by the final product(s) of the corresponding pathways (i.e., Trp and Phe/Tyr, respectively) in both microbes and plants (Knöchel et al., 1999). In addition, Trp activates CM to redirect flux from Trp to Phe/Tyr biosynthesis (Kuroki & Conn, 1988). Likewise, the enzymes localized at the branch points of Phe and Tyr biosynthesis, ADT and ADH (or PDT and PDH, depending on the organism), are feedback inhibited by Phe and Tyr, respectively (Yamada et al., 2008).

In contrast to microbial enzymes (Wu & Woodard, 2006), plant DAHP synthases are not inhibited by AAA (Herrmann, 1995). Although arogenate has been shown to inhibit spinach and mung bean DAHP synthases in vitro (Rubin & Jensen, 1985), such arogenate feedback regulation has not been demonstrated in vivo. Considering the expression of genes encoding the shikimate and Phe pathway enzymes (i.e., DAHP synthase, EPSP synthase, and CM) were upregulated and feeding of exogenous shikimate restored the reduced Trp and Phe (Maeda et al., 2010).

E4P and PEP Precursor Supply to the Shikimate Pathway

The supply of the DAHP synthase substrates, E4P and PEP, can also play an important role in the regulation of the carbon flux into the shikimate pathway. In photosynthetic tissues, transketolase (TK) in the Calvin cycle converts G3P and fructose 6-phosphate (F6P) to xylose5-phosphate (X5P) and E4P (Figs. 9.1 and 9.2). A light decrease in TK activity in transgenic tobacco leaves leads to a substantial reduction in the levels of AAA and their downstream metabolites (Henkes et al., 2001), suggesting that the E4P supply via TK can be a limiting factor for AAA biosynthesis. In non-photosynthetic tissues, transaldolase (TA) and TK in the oxidative pentose phosphate pathway (OPPP) likely play key roles in E4P supply to the shikimate pathway (Fig. 9.2 and Table 9.3) (Kruger & von Schaewen, 2003). In bacteria, overexpression of TK rather than TA was found to be more effective in directing the carbon flux into the AAA pathways (Bongaerts et al., 2001); however, the relative contributions of these enzymes to E4P supply for the shikimate pathway in plants have not been investigated. The broad substrate specificity of TK and TA and the presence of additional intermediates (e.g., octulose 8-phosphate) potentially involved in the OPPP (Van Winden et al., 2001) suggest that the OPPP and its regulation may be much more dynamic and complex than currently thought to meet the high demand of E4P for the biosynthesis of AAA, especially in plants.

Plastidic PEP can be derived from (a) plastidic glycolysis via phosphor glycerolmutase (PGyM) and enolase (*eno1*) (Prabhakar et al., 2009), (b) import from the cytosol along the PEP phosphate translocator (PPT) (Streatfield et al., 1999), and/or (c) phosphorylation of pyruvate which is catalyzed by plastidic pyruvate orthophosphate di-kinase (Flügge et al., 2011). Although ENO1, PGyMs, PPTs, and PPDK expression analysis revealed that the relative contributions of different pathways are tissue specific (Parsley & Hibberd, 2006), the mutant analysis showed that multiple pathways can simultaneously contribute to the plastidic PEP internal pool in plants. The *Arabidopsis ppt1* (*Cue1*) knockout mutant displays a mesophyll-specific defect in chloroplast development (Li et al., 1995a) that can be rescued by the constitutive overexpression of PPDK in the plastids (Voll et al., 2003), indicating that sufficient levels of pyruvate exist in the chloroplasts to compensate for the shortage of PEP transport from the cytosol (Voll et al., 2003). The gametophyte-lethal phenotype of the *ppt1-eno1* double-knockout mutant indicates that PEP generation through both the plastidic glycolysis and its import from the cytosol is necessary during early *Arabidopsis* development (Prabhakar et al., 2010). However, considering that plastidic PEP is also used for the biosynthesis of fatty acids, branch-chained amino acids, and isoprenoids (via the MEP pathway), the mutant phenotypes described above cannot be attributed to a single metabolic pathway (e.g., the shikimate pathway), and the major route for PEP supply for the shikimate pathway remains to be determined.

Subcellular Localization of Aromatic Amino Acid Biosynthesis in Plants

Several lines of the evidence conclusively indicate that the plastids contain a full set of biosynthetic enzymes for the production of all AAA from PEP and E4P:

(i) Feeding of radiolabeled precursors (e.g., PEP) to isolated chloroplasts led to the production of labeled Trp, Phe, and Tyr (Homeyer & Schultz, 1988).
(ii) Almost all enzymatic activities involved in AAA biosynthesis have been detected in the plastid fractions of plant tissue extracts (Zhao & Last, 1995).
(iii) The amino acid sequence of at least one isozyme responsible for each biochemical step in the plant AAA pathway involves a plastid transit peptide.
(iv) Plastidic localization for most channel enzymes has been confirmed by utilizing GFP (green fluorescent protein) fusion proteins or protein import assays (Maeda & Dudareva, 2012), except CS, *ASβ*, IGPS, and PPY-AT/HPP-AT (Tables 9.1, 9.2, and 9.3).

An open question is whether some of the biochemical steps are also present outside of the plastids. Genes encoding cytosolic tobacco DHD/SDH2 using GFP fusion protein (Ding et al., 2007) and *Arabidopsis* and petunia CM2 (Eberhard et al., 1996) have been isolated, and the extra plastidic localization of the corresponding proteins was shown using a protein import assay (Colquhoun et al., 2010). A gene encoding cytosolic DAHP synthase has not been identified; however, a PPY-AT recently isolated from melon lacks a plastid transit peptide and is likely localized in the cytosol (Gonda et al., 2010). Further investigation of molecular identity, subcellular localization, and physiological functions of potential cytosolic enzymes will clarify whether the AAA pathways exist in more than one cellular compartment.

Transport of Aromatic Amino Acids and Pathway Intermediates across Plastid Membranes

Shikimate is conjugated with p-coumaroyl-coenzyme A (CoA) by hydroxy cinnamoyl-CoA: shikimate hydroxy cinnamoyl transferase in the cytosol, forming p-coumaroyl shikimate, an intermediate in the monolignol biosynthesis (Hoffmann et al., 2003). Thus, shikimate has to be exported from the plastids. Most plant enzymes utilizing chorismate are localized in the plastids (Fig. 9.1) (Colquhoun et al., 2010); however, CM2 resides in the cytosol (Colquhoun et al., 2010) and is inhibited in tobacco by caffeic acid (Goers & Jensen, 1984). Future identifications of plastid membrane transporters will allow us to understand the roles of shikimate and chorismate in the crosstalk between the plastidic AAA pathway and cytosolic metabolic pathways.

Plant-Based Biosynthesis of Aromatic Amino Acid

To increase Phe production and elucidate the effect of high Phe levels on the AAA metabolic network, a bacterial bi-functional CM/PDT that converts chorismate into phenyl pyruvate via prephenate was recently overexpressed in *Arabidopsis* plastids (Tzin et al., 2009). The C-terminal allosteric domain was removed from the introduced CM/PDT to prevent its feedback inhibition by Phe. Interestingly, these transgenic lines contain increased levels of Phe- and Tyr-derived metabolites (e.g., phenyl propanoids, glucosinolates, and vitamin E), whereas the levels of Trp-derived secondary metabolites were reduced (Tzin et al., 2009). These examples show that the overexpression of feedback-insensitive *ASα* is generally effective in enhancing Trp accumulation, but the degree of enhancement depends on the biological system. However, *Arabidopsis* Trp-overproducing lines show high levels of Phe and Tyr (Ishihara et al., 2006), which may be attributed to an activation of CM by elevated levels of Trp in transgenics (Benesova & Bode, 1992). Thus, in contrast to the upregulation of Phe biosynthesis, an increase in carbon flux toward Trp does not compromise the biosynthesis of other AAA and their derived compounds.

Trp is a precursor of auxin as well as a diverse range of secondary metabolites, including indole alkaloids, glucosinolates, and phytoalexins (Fig. 9.1). In contrast, a 200-fold increase in Trp in the *Arabidopsis* transgenic lines led to only a twofold increase in the levels of indole-3-ylmethyl glucosinolate and unaltered levels of the indole phytoalexin camalexin after pathogenic fungus inoculation (Ishihara et al., 2006). All these results exhibited that the level of Trp is not a major limiting factor for indole glucosinolates and phytoalexin biosynthesis, as had been previously suggested (Zhao & Last, 1996). Much less carbon is generally incorporated into Tyr, which is further converted to 4-hydroxyphenylpyruvate and homogentisate, the aromatic precursors of tocochromanols collectively known as vitamin E. To increase flux toward 4-hydroxyphenyl pyruvate and production of vitamin E, prephenate was directly converted into 4-hydroxyphenylpyruvate by the expression of Tyr-insensitive *S. cerevisiae* PDH in the plastids of tobacco plants (Rippert et al., 2004). Although tobacco plants expressing only the yeast PDH transgene accumulate trace amounts of tocotrienols, the co-expression of the yeast PDH with an Arabidopsis 4-hydroxyphenylpyruvate dioxygenase that converts 4-hydroxyphenylpyruvate to homogentisate resulted in a massive accumulation of tocotrienols (Rippert et al., 2004).

Because an overexpression of *Arabidopsis* 4-hydroxyphenylpyruvate dioxygenase alone had only a limited effect on tocopherol production (Tsegaye et al., 2002), both supplies of 4-hydroxyphenylpyruvate and its subsequent conversion to homogentizate represent bottlenecks in vitamin E accumulation in plants. Taken together, these results show that the elimination of posttranscriptional feedback regulatory mechanisms by introducing feedback-insensitive enzymes can lead to an increase in AAA production; however, upregulation of additional downstream biochemical steps is likely required for effective production of target AAA-derived metabolites.

References

Barends, T. R., Dunn, M. F., & Schlichting, I. (2008). Tryptophan synthase, an allosteric molecular factory. *Current Opinion in Chemical Biology, 12*(5), 593–600. https://doi.org/10.1016/j.cbpa.2008.07.011

Baylis, A. D. (2000). Why glyphosate is a global herbicide: Strengths, weaknesses and prospects. *Pest Management Science: Formerly Pesticide Science, 56*(4), 299–308. https://doi.org/10.1002/(SICI)1526-4998(200004)56:4<299::AID-PS144>3.0.CO;2-K

Bender, J., & Fink, G. R. (1995). Epigenetic control of an endogenous gene family is revealed by a novel blue fluorescent mutant of *Arabidopsis*. *Cell, 83*(5), 725–734. https://doi.org/10.1016/0092-8674(95)90185-X

Benesova, M., & Bode, R. (1992). Chorismate mutase isoforms from seeds and seedlings of *Papaver somniferum*. *Phytochemistry, 31*(9), 2983–2987. https://doi.org/10.1016/0031-9422(92)83431-W

Bentley, R., & Haslam, E. (1990). The shikimate pathway—A metabolic tree with many branche. *Critical Reviews in Biochemistry and Molecular Biology, 25*(5), 307–384. https://doi.org/10.3109/10409239009090615

Bischoff, M., Rösler, J., Raesecke, H. R., Görlach, J., Amrhein, N., & Schmid, J. (1996). Cloning of a cDNA encoding a 3-dehydroquinate synthase from a higher plant, and analysis of the organ-specific and elicitor-induced expression of the corresponding gene. *Plant Molecular Biology, 31*(1), 69–76. https://doi.org/10.1007/BF00020607

Bischoff, M., Schaller, A., Bieri, F., Kessler, F., Amrhein, N., & Schmid, J. (2001). Molecular characterization of tomato 3-dehydroquinate dehydratase-shikimate: NADP oxidoreductase. *Plant Physiology, 125*(4), 1891–1900. https://doi.org/10.1104/pp.125.4.1891

Bohlmann, J., Lins, T., Martin, W., & Eilert, U. (1996). Anthranilate synthase from *Ruta graveolens* (duplicated ASα-genes encode tryptophan-sensitive and tryptophan-insensitive isoenzymes specific to amino acid and alkaloid biosynthesis). *Plant Physiology, 111*(2), 507–514. https://doi.org/10.1104/pp.111.2.507

Bohm, B. A. (1965). Shikimic acid (3, 4, 5-trihydroxy-1-cyclohexene-1-carboxylic acid). *Chemical Reviews, 65*(4), 435–466.

Bongaerts, J., Krämer, M., Müller, U., Raeven, L., & Wubbolts, M. (2001). Metabolic engineering for microbial production of aromatic amino acids and derived compounds. *Metabolic Engineering, 3*(4), 289–300. https://doi.org/10.1006/mben.2001.0196

Bonner, C. A., & Jensen, R. A. (1985). Novel features of prephenate aminotransferase from cell cultures of *Nicotiana silvestris*. *Archives of Biochemistry and Biophysics, 238*(1), 237–246. https://doi.org/10.1016/0003-9861(85)90161-4

Byng, G., Whitaker, R., Flick, C., & Jensen, R. A. (1981). Enzymology of l-tyrosine biosynthesis in corn (*Zea mays*). *Phytochemistry, 20*(6), 1289–1292. https://doi.org/10.1016/0031-9422(81)80023-4

Carpenter, E. P., Hawkins, A. R., Frost, J. W., & Brown, K. A. (1998). Structure of dehydroquinate synthase reveals an active site capable of multistep catalysis. *Nature, 394*(6690), 299–302. https://doi.org/10.1038/28431

Colquhoun, T. A., Schimmel, B. C., Kim, J. Y., Reinhardt, D., Cline, K., & Clark, D. G. (2010). A petunia chorismate mutase specialized for the production of floral volatiles. *The Plant Journal, 61*(1), 145–155. https://doi.org/10.1111/j.1365-313X.2009.04042.x

Crawford, I. P. (1989). Evolution of a biosynthetic pathway: The tryptophan paradigm. *Annual Review of Microbiology, 43*(1), 567–600.

De la Torre, F., Moya-García, A. A., Suárez, M. F., Rodríguez-Caso, C., Cañas, R. A., Sánchez-Jiménez, F., & Cánovas, F. M. (2009). Molecular modeling and site-directed mutagenesis reveal essential residues for catalysis in a prokaryote-type aspartate aminotransferase. *Plant Physiology, 149*(4), 1648–1660. https://doi.org/10.1104/pp.108.134510

DeFeyter, R. C., & Pittard, J. A. M. E. S. (1986). Purification and properties of shikimate kinase II from *Escherichia coli* K-12. *Journal of Bacteriology, 165*(1), 331–333. https://doi.org/10.1128/jb.165.1.331-333.1986

Denby, K. J., Jason, L. J., Murray, S. L., & Last, R. L. (2005). ups1, an *Arabidopsis thaliana* camalexin accumulation mutant defective in multiple defence signaling pathways. *The Plant Journal, 41*(5), 673–684. https://doi.org/10.1111/j.1365-313X.2005.02327.x

Devoto, A., Ellis, C., Magusin, A., Chang, H. S., Chilcott, C., Zhu, T., & Turner, J. G. (2005). Expression profiling reveals COI1 to be a key regulator of genes involved in wound-and methyl jasmonate-induced secondary metabolism, defence, and hormone interactions. *Plant Molecular Biology, 58*(4), 497–513. https://doi.org/10.1007/s11103-005-7306-5

Ding, L., Hofius, D., Hajirezaei, M. R., Fernie, A. R., Börnke, F., & Sonnewald, U. (2007). Functional analysis of the essential bifunctional tobacco enzyme 3-dehydroquinate dehydratase/shikimate dehydrogenase in transgenic tobacco plants. *Journal of Experimental Botany, 58*(8), 2053–2067. https://doi.org/10.1093/jxb/erm059

Duncan, K., Chaudhuri, S., Campbell, M. S., & Coggins, J. R. (1986). The overexpression and complete amino acid sequence of *Escherichia coli* 3-dehydroquinase. *Biochemical Journal, 238* (2), 475–483. https://doi.org/10.1042/bj2380475

Duncan, K., Edwards, R. M., & Coggins, J. R. (1987). The pentafunctional arom enzyme of *Saccharomyces cerevisiae* is a mosaic of monofunctional domains. *Biochemical Journal, 246* (2), 375–386. https://doi.org/10.1042/bj2460375

Dyer, W. E., Weaver, L. M., Zhao, J. M., Kuhn, D. N., Weller, S. C., & Herrmann, K. M. (1990). A cDNA encoding 3-deoxy-D-arabino-heptulosonate 7-phosphate synthase from *Solanum tuberosum* L. *Journal of Biological Chemistry, 265*(3), 1608–1614. https://doi.org/10.1016/S0021-9258(19)40060-4

Eberhard, J., Ehrler, T. T., Epple, P., Felix, G., Raesecke, H. R., Amrhein, N., & Schmid, J. (1996). Cytosolic and plastidic chorismate mutase isozymes from *Arabidopsis thaliana*: Molecular characterization and enzymatic properties. *The Plant Journal, 10*(5), 815–821. https://doi.org/10.1046/j.1365-313X.1996.10050815.x

Entus, R., Poling, M., & Herrmann, K. M. (2002). Redox regulation of *Arabidopsis* 3-deoxy-D-arabino-heptulosonate 7-phosphate synthase. *Plant Physiology, 129*(4), 1866–1871. https://doi.org/10.1104/pp.002626

Fazel, A. M., & Jensen, R. A. (1979). Obligatory biosynthesis of L-tyrosine via the pretyrosine branchlet in coryneform bacteria. *Journal of Bacteriology, 138*(3), 805–815. https://doi.org/10.1128/jb.138.3.805-815.1979

Fernstrom, J. D., & Fernstrom, M. H. (2007). Tyrosine, phenylalanine, and catecholamine synthesis and function in the brain. *The Journal of Nutrition, 137*(6), 1539S–1547S. https://doi.org/10.1093/jn/137.6.1539S

Fiedler, E., & Schultz, G. (1985). Localization, purification, and characterization of shikimate oxidoreductase-dehydroquinate hydrolyase from stroma of spinach chloroplasts. *Plant Physiology, 79*(1), 212–218. https://doi.org/10.1104/pp.79.1.212

Fitzpatrick, P. F. (1999). Tetrahydropterin-dependent amino acid hydroxylases. *Annual Review of Biochemistry, 68*(1), 355–381. https://doi.org/10.1146/annurev.biochem.68.1.355

Flügge, U. I., Häusler, R. E., Ludewig, F., & Gierth, M. (2011). The role of transporters in supplying energy to plant plastids. *Journal of Experimental Botany, 62*(7), 2381–2392. https://doi.org/10.1093/jxb/erq361

Forlani, G., Parisi, B., & Nielsen, E. (1994). 5-enol-pyruvyl-shikimate-3-phosphate synthase from *Zea mays* cultured cells (purification and properties). *Plant Physiology, 105*(4), 1107–1114. https://doi.org/10.1104/pp.105.4.1107

Frey, M., Schullehner, K., Dick, R., Fiesselmann, A., & Gierl, A. (2009). Benzoxazinoid biosynthesis, a model for evolution of secondary metabolic pathways in plants. *Phytochemistry, 70* (15–16), 1645–1651. https://doi.org/10.1016/j.phytochem.2009.05.012

Frey, M., Stettner, C., Paré, P. W., Schmelz, E. A., Tumlinson, J. H., & Gierl, A. (2000). An herbivore elicitor activates the gene for indole emission in maize. *Proceedings of the National Academy of Sciences, 97*(26), 14801–14806. https://doi.org/10.1073/pnas.260499897

Fucile, G., Garcia, C., Carlsson, J., Sunnerhagen, M., & Christendat, D. (2011). Structural and biochemical investigation of two Arabidopsis shikimate kinases: The heat-inducible isoform is thermostable. *Protein Science, 20*(7), 1125–1136. https://doi.org/10.1002/pro.640

Funke, T., Han, H., Healy-Fried, M. L., Fischer, M., & Schönbrunn, E. (2006). Molecular basis for the herbicide resistance of roundup ready crops. *Proceedings of the National Academy of Sciences, 103*(35), 13,010–13,015. https://doi.org/10.1073/pnas.0603638103

Gan, J., Gu, Y., Li, Y., Yan, H., & Ji, X. (2006). Crystal structure of Mycobacterium tuberculosis shikimate kinase in complex with shikimic acid and an ATP analogue. *Biochemistry, 45*(28), 8539–8545. https://doi.org/10.1021/bi0606290

Gasser, C. S., Winter, J. A., Hironaka, C. M., & Shah, D. M. (1988). Structure, expression, and evolution of the 5-enolpyruvylshikimate-3-phosphate synthase genes of petunia and tomato. *Journal of Biological Chemistry, 263*(9), 4280–4287. https://doi.org/10.1016/S0021-9258(18)68922-7

Gelfand, D. H., & Steinberg, R. A. (1977). Escherichia coli mutants deficient in the aspartate and aromatic amino acid aminotransferases. *Journal of Bacteriology, 130*(1), 429–440. https://doi.org/10.1128/jb.130.1.429-440.1977

Gigolashvili, T., Berger, B., Mock, H. P., Müller, C., Weisshaar, B., & Flügge, U. I. (2007). The transcription factor HIG1/MYB51 regulates indolic glucosinolate biosynthesis in *Arabidopsis thaliana*. *The Plant Journal, 50*(5), 886–901. https://doi.org/10.1111/j.1365-313X.2007.03099.x

Goers, S. K., & Jensen, R. A. (1984). The differential allosteric regulation of two chorismate-mutase isoenzymes of *Nicotiana silvestris*. *Planta, 162*(2), 117–124. https://doi.org/10.1007/BF00410207

Gonda, I., Bar, E., Portnoy, V., Lev, S., Burger, J., Schaffer, A. A., Tadmor, Y. A., Gepstein, S., Giovannoni, J. J., Katzir, N., & Lewinsohn, E. (2010). Branched-chain and aromatic amino acid catabolism into aroma volatiles in *Cucumis melo* L. fruit. *Journal of Experimental Botany, 61* (4), 1111–1123. https://doi.org/10.1093/jxb/erp390

Görlach, J., Raesecke, H. R., Rentsch, D., Regenass, M., Roy, P. H. I. L. I. P. P. E., Zala, M., Keel, C., Boller, T., Amrhein, N., & Schmid, J. (1995). Temporally distinct accumulation of transcripts encoding enzymes of the prechorismate pathway in elicitor-treated, cultured tomato cells. *Proceedings of the National Academy of Sciences, 92*(8), 3166–3170. https://doi.org/10.1073/pnas.92.8.3166

Görlach, J., Schmid, J., & Amrhein, N. (1993). Differential expression of tomato (*Lycopersicon esculentum* L.) genes encoding shikimate pathway isoenzymes. II. Chorismate synthase. *Plant Molecular Biology, 23*(4), 707–716. https://doi.org/10.1007/BF00021526

He, Y., & Li, J. (2001). Differential expression of triplicate phosphoribosylanthranilate isomerase isogenes in the tryptophan biosynthetic pathway of *Arabidopsis thaliana* (L.) Heynh. *Planta, 212*(5), 641–647. https://doi.org/10.1007/s004250000452

Henkes, S., Sonnewald, U., Badur, R., Flachmann, R., & Stitt, M. (2001). A small decrease of plastid transketolase activity in antisense tobacco transformants has dramatic effects on photosynthesis and phenylpropanoid metabolism. *The Plant Cell, 13*(3), 535–551. https://doi.org/10.1105/tpc.13.3.535

Herrmann, K. M. (1995). The shikimate pathway: Early steps in the biosynthesis of aromatic compounds. *The Plant Cell, 7*(7), 907. https://doi.org/10.1105/tpc.7.7.907

Hoffmann, L., Maury, S., Martz, F., Geoffroy, P., & Legrand, M. (2003). Purification, cloning, and properties of an acyltransferase controlling shikimate and quinate ester intermediates in phenylpropanoid metabolism. *Journal of Biological Chemistry, 278*(1), 95–103. https://doi.org/10.1074/jbc.M209362200

Homeyer, U., & Schultz, G. (1988). Activation by light of plastidic shikimate pathway in spinach. *Plant Physiology and Biochemistry (Paris), 26*(3), 365–370.

Ishihara, A., Asada, Y., Takahashi, Y., Yabe, N., Komeda, Y., Nishioka, T., Miyagawa, H., & Wakasa, K. (2006). Metabolic changes in *Arabidopsis thaliana* expressing the feedback-resistant anthranilate synthase α subunit gene OASA1D. *Phytochemistry, 67*(21), 2349–2362. https://doi.org/10.1016/j.phytochem.2006.08.008

Kaul, S., Koo, H. L., Jenkins, J., Rizzo, M., Rooney, T., Tallon, L. J., Feldblyum, T., Nierman, W., Benito, M. I., Lin, X., & Town, C. D. (2000). Analysis of the genome sequence of the flowering plant *Arabidopsis thaliana. Nature, 408*(6814), 796–815. https://doi.org/10.1038/35048692

Kaur, H., Heinzel, N., Schöttner, M., Baldwin, I. T., & Gális, I. (2010). R2R3-NaMYB8 regulates the accumulation of phenylpropanoid-polyamine conjugates, which are essential for local and systemic defense against insect herbivores in *Nicotiana attenuata. Plant Physiology, 152*(3), 1731–1747. https://doi.org/10.1104/pp.109.151738

Keith, B., Dong, X. N., Ausubel, F. M., & Fink, G. R. (1991). Differential induction of 3-deoxy-D-arabino-heptulosonate 7-phosphate synthase genes in *Arabidopsis thaliana* by wounding and pathogenic attack. *Proceedings of the National Academy of Sciences, 88*(19), 8821–8825. https://doi.org/10.1073/pnas.88.19.8821

Klee, H. J., Muskopf, Y. M., & Gasser, C. S. (1987). Cloning of an *Arabidopsis thaliana* gene encoding 5-enolpyruvylshikimate-3-phosphate synthase: Sequence analysis and manipulation to obtain glyphosate-tolerant plants. *Molecular and General Genetics MGG, 210*(3), 437–442. https://doi.org/10.1007/BF00327194

Knöchel, T., Ivens, A., Hester, G., Gonzalez, A., Bauerle, R., Wilmanns, M., Kirschner, K., & Jansonius, J. N. (1999). The crystal structure of anthranilate synthase from *Sulfolobus solfataricus*: Functional implications. *Proceedings of the National Academy of Sciences, 96* (17), 9479–9484. https://doi.org/10.1073/pnas.96.17.9479

Koshiba, T. (1979). Alicyclic acid metabolism in plants 12. Partial purification and some properties of shikimate kinase from *Phaseolus mungo* seedlings. *Plant and Cell Physiology, 20*(4), 803–809. https://doi.org/10.1093/oxfordjournals.pcp.a075872

Kriechbaumer, V., Weigang, L., Fießelmann, A., Letzel, T., Frey, M., Gierl, A., & Glawischnig, E. (2008). Characterization of the tryptophan synthase alpha subunit in maize. *BMC Plant Biology, 8*(1), 1–11. https://doi.org/10.1186/1471-2229-8-44

Kruger, N. J., & von Schaewen, A. (2003). The oxidative pentose phosphate pathway: Structure and organization. *Current Opinion in Plant Biology, 6*(3), 236–246. https://doi.org/10.1016/S1369-5266(03)00039-6

Kuroki, G. W., & Conn, E. E. (1988). Purification and characterization of an inducible aromatic amino acid-sensitive form of chorismate mutase from *Solanum tuberosum* L. tubers. *Archives of Biochemistry and Biophysics, 260*(2), 616–621. https://doi.org/10.1016/0003-9861(88)90489-4

Kutchan, T. M. (1995). Alkaloid biosynthesis [mdash] the basis for metabolic engineering of medicinal plants. *The Plant Cell, 7*(7), 1059. https://doi.org/10.1105/tpc.7.7.1059

Last, R. L., & Fink, G. R. (1988). Tryptophan-requiring mutants of the plant *Arabidopsis thaliana. Science, 240*(4850), 305–310. https://doi.org/10.1126/science.240.4850.305

Lee, A. Y., Karplus, P. A., Ganem, B., & Clardy, J. (1995). Atomic structure of the buried catalytic pocket of *Escherichia coli* chorismate mutase. *Journal of the American Chemical Society, 117* (12), 3627–3628. https://doi.org/10.1021/ja00117a038

Legrand, P., Dumas, R., Seux, M., Rippert, P., Ravelli, R., Ferrer, J. L., & Matringe, M. (2006). Biochemical characterization and crystal structure of *Synechocystis* arogenate dehydrogenase provide insights into catalytic reaction. *Structure, 14*(4), 767–776. https://doi.org/10.1016/j.str.2006.01.006

Li, H. M., Culligan, K., Dixon, R. A., & Chory, J. (1995a). CUE1: A mesophyll cell-specific positive regulator of light-controlled gene expression in *Arabidopsis*. *The Plant Cell, 7*(10), 1599–1610. https://doi.org/10.1105/tpc.7.10.1599

Li, J., Chen, S., Zhu, L., & Last, R. L. (1995b). Isolation of cDNAs encoding the tryptophan pathway enzyme indole-3-glycerol phosphate synthase from *Arabidopsis thaliana*. *Plant Physiology, 108*(2), 877. https://doi.org/10.1104/pp.108.2.877

Li, J., Zhao, J., Rose, A. B., Schmidt, R., & Last, R. L. (1995c). Arabidopsis phosphoribosylanthranilate isomerase: Molecular genetic analysis of triplicate tryptophan pathway genes. *The Plant Cell, 7*(4), 447–461. https://doi.org/10.1105/tpc.7.4.447

Macheroux, P., Schmid, J., Amrhein, N., & Schaller, A. (1999). A unique reaction in a common pathway: Mechanism and function of chorismate synthase in the shikimate pathway. *Planta, 207*(3), 325–334. https://doi.org/10.1007/s004250050489

Maclean, J., & Ali, S. (2003). The structure of chorismate synthase reveals a novel flavin binding site fundamental to a unique chemical reaction. *Structure, 11*(12), 1499–1511. https://doi.org/10.1016/j.str.2003.11.005

Maeda, H., & Dudareva, N. (2012). The shikimate pathway and aromatic amino acid biosynthesis in plants. *Annual Review of Plant Biology, 63*, 73–105. https://doi.org/10.1146/annurev-arplant-042811-105439

Maeda, H., Shasany, A. K., Schnepp, J., Orlova, I., Taguchi, G., Cooper, B. R., Rhodes, D., Pichersky, E., & Dudareva, N. (2010). RNAi suppression of Arogenate Dehydratase1 reveals that phenylalanine is synthesized predominantly via the arogenate pathway in petunia petals. *The Plant Cell, 22*(3), 832–849. https://doi.org/10.1105/tpc.109.073247

Maeda, H., Yoo, H., & Dudareva, N. (2011). Prephenate aminotransferase directs plant phenylalanine biosynthesis via arogenate. *Nature Chemical Biology, 7*(1), 19–21. https://doi.org/10.1038/nchembio.485

Melquist, S., Luff, B., & Bender, J. (1999). Arabidopsis PAI gene arrangements, cytosine methylation and expression. *Genetics, 153*(1), 401–413. https://doi.org/10.1093/genetics/153.1.401

Mitsuda, N., Iwase, A., Yamamoto, H., Yoshida, M., Seki, M., Shinozaki, K., & Ohme-Takagi, M. (2007). NAC transcription factors, NST1 and NST3, are key regulators of the formation of secondary walls in woody tissues of *Arabidopsis*. *The Plant Cell, 19*(1), 270–280. https://doi.org/10.1105/tpc.106.047043

Mobley, E. M., Kunkel, B. N., & Keith, B. (1999). Identification, characterization and comparative analysis of a novel chorismate mutase gene in *Arabidopsis thaliana*. *Gene, 240*(1), 115–123. https://doi.org/10.1016/S0378-1119(99)00423-0

Morollo, A. A., & Eck, M. J. (2001). Structure of the cooperative allosteric anthranilate synthase from *Salmonella typhimurium*. *Nature Structural Biology, 8*(3), 243–247. https://doi.org/10.1038/84988

Mousdale, D. M., & Coggins, J. R. (1985). Subcellular localization of the common shikimate-pathway enzymes in *Pisum sativum* L. *Planta, 163*(2), 241–249. https://doi.org/10.1007/BF00393514

Natarajan, K., Meyer, M. R., Jackson, B. M., Slade, D., Roberts, C., Hinnebusch, A. G., & Marton, M. J. (2001). Transcriptional profiling shows that Gcn4p is a master regulator of gene expression during amino acid starvation in yeast. *Molecular and Cellular Biology, 21*(13), 4347–4368. https://doi.org/10.1128/MCB.21.13.4347-4368.2001

Ouyang, J., Shao, X., & Li, J. (2000). Indole-3-glycerol phosphate, a branchpoint of indole-3-acetic acid biosynthesis from the tryptophan biosynthetic pathway in *Arabidopsis thaliana*. *The Plant Journal, 24*(3), 327–334. https://doi.org/10.1046/j.1365-313x.2000.00883.x

Pacold, I., & Anderson, L. E. (1973). Energy charge control of the Calvin cycle enzyme 3-phosphoglyceric acid kinase. *Biochemical and Biophysical Research Communications, 51* (1), 139–143. https://doi.org/10.1016/0006-291X(73)90519-6

Parsley, K., & Hibberd, J. M. (2006). The *Arabidopsis* PPDK gene is transcribed from two promoters to produce differentially expressed transcripts responsible for cytosolic and plastidic proteins. *Plant Molecular Biology, 62*(3), 339–349. https://doi.org/10.1007/s11103-006-9023-0

Pinto, J. E., Dyer, W. E., Weller, S. C., & Herrmann, K. M. (1988). Glyphosate induces 3-deoxy-D-arabino-heptulosonate 7-phosphate synthase in potato (*Solanum tuberosum* L.) cells grown in suspension culture. *Plant Physiology, 87*(4), 891–893. https://doi.org/10.1104/pp.87.4.891

Pinto, J. E., Suzich, J. A., & Herrmann, K. M. (1986). 3-Deoxy-D-arabino-heptulosonate 7-phosphate synthase from potato tuber (*Solanum tuberosum* L.). *Plant Physiology, 82*(4), 1040–1044. https://doi.org/10.1104/pp.82.4.1040

Prabhakar, V., Löttgert, T., Geimer, S., Dörmann, P., Krüger, S., Vijayakumar, V., Schreiber, L., Göbel, C., Feussner, K., Feussner, I., & Marin, K. (2010). Phosphoenolpyruvate provision to plastids is essential for gametophyte and sporophyte development in *Arabidopsis thaliana*. *The Plant Cell, 22*(8), 2594–2617. https://doi.org/10.1105/tpc.109.073171

Prabhakar, V., Löttgert, T., Gigolashvili, T., Bell, K., Flügge, U. I., & Häusler, R. E. (2009). Molecular and functional characterization of the plastid-localized phosphoenolpyruvate enolase (ENO1) from *Arabidopsis thaliana*. *FEBS Letters, 583*(6), 983–991. https://doi.org/10.1016/j.febslet.2009.02.017

Prabhu, P. R., & Hudson, A. O. (2010). Identification and partial characterization of an L-tyrosine aminotransferase (TAT) from *Arabidopsis thaliana*. *Biochemistry Research International, 2010*. https://doi.org/10.1155/2010/549572

Pribat, A., Noiriel, A., Morse, A. M., Davis, J. M., Fouquet, R., Loizeau, K., Ravanel, S., Frank, W., Haas, R., Reski, R., & Bedair, M. (2010). Nonflowering plants possess a unique folate-dependent phenylalanine hydroxylase that is localized in chloroplasts. *The Plant Cell, 22*(10), 3410–3422. https://doi.org/10.1105/tpc.110.078824

Quiel, J. A., & Bender, J. (2003). Glucose conjugation of anthranilate by the *Arabidopsis* UGT74F2 glucosyltransferase is required for tryptophan mutant blue fluorescence. *Journal of Biological Chemistry, 278*(8), 6275–6281. https://doi.org/10.1074/jbc.M211822200

Radwanski, E. R., Barczak, A. J., & Last, R. L. (1996). Characterization of tryptophan synthase alpha subunit mutants of *Arabidopsis thaliana*. *Molecular and General Genetics MGG, 253*(3), 353–361. https://doi.org/10.1007/PL00008602

Radwanski, E. R., & Last, R. L. (1995). Tryptophan biosynthesis and metabolism: Biochemical and molecular genetics. *The Plant Cell, 7*(7), 921. https://doi.org/10.1105/tpc.7.7.921

Radwanski, E. R., Zhao, J., & Last, R. L. (1995). Arabidopsis thaliana tryptophan synthase alpha: Gene cloning, expression, and subunit interaction. *Molecular and General Genetics MGG, 248* (6), 657–667. https://doi.org/10.1007/BF02191705

Rippert, P., & Matringe, M. (2002). Purification and kinetic analysis of the two recombinant arogenate dehydrogenase isoforms of *Arabidopsis thaliana*. *European Journal of Biochemistry, 269*(19), 4753–4761. https://doi.org/10.1046/j.1432-1033.2002.03172.x

Rippert, P., Scimemi, C., Dubald, M., & Matringe, M. (2004). Engineering plant shikimate pathway for production of tocotrienol and improving herbicide resistance. *Plant Physiology, 134*(1), 92–100. https://doi.org/10.1104/pp.103.032441

Romero, R. M., Roberts, M. F., & Phillipson, J. D. (1995). Anthranilate synthase in microorganisms and plants. *Phytochemistry, 39*(2), 263–276. https://doi.org/10.1016/0031-9422(95)00010-5

Rose, A. B., Casselman, A. L., & Last, R. L. (1992). A phosphoribosylanthranilate transferase gene is defective in blue fluorescent *Arabidopsis thaliana* tryptophan mutants. *Plant Physiology, 100* (2), 582–592. https://doi.org/10.1104/pp.100.2.582

Rose, A. B., & Last, R. L. (1997). Introns act post-transcriptionally to increase expression of the *Arabidopsis thaliana* tryptophan pathway gene PAT1. *The Plant Journal, 11*(3), 455–464. https://doi.org/10.1046/j.1365-313X.1997.11030455.x

Rose, A. B., Li, J., & Last, R. L. (1997). An allelic series of blue fluorescent trp1 mutants of *Arabidopsis thaliana*. *Genetics, 145*(1), 197–205.
Rubin, J. L., & Jensen, R. A. (1985). Differentially regulated isozymes of 3-deoxy-D-arabino-heptulosonate-7-phosphate synthase from seedlings of *Vigna radiata* [L.]. *Wilczek. Plant Physiology, 79*(3), 711–718. https://doi.org/10.1104/pp.79.3.711
Schaller, A., Schmid, J., Leibinger, U., & Amrhein, N. (1991a). Molecular cloning and analysis of a cDNA coding for chorismate synthase from the higher plant *Corydalis sempervirens* Pers. *Journal of Biological Chemistry, 266*(32), 21,434–21,438. https://doi.org/10.1016/S0021-9258(18)54657-3
Schaller, A., van Afferden, M., Windhofer, V., Bülow, S., Abel, G., Schmid, J., & Amrhein, N. (1991b). Purification and characterization of chorismate synthase from *Euglena gracilis*: Comparison with chorismate synthases of plant and microbial origin. *Plant Physiology, 97*(4), 1271–1279. https://doi.org/10.1104/pp.97.4.1271
Schmid, J., & Amrhein, N. (1995). Molecular organization of the shikimate pathway in higher plants. *Phytochemistry, 39*(4), 737–749. https://doi.org/10.1016/0031-9422(94)00962-S
Schmidt, C. L., Danneel, H. J., Schultz, G., & Buchanan, B. B. (1990). Shikimate kinase from spinach chloroplasts: Purification, characterization, and regulatory function in aromatic amino acid biosynthesis. *Plant Physiology, 93*(2), 758–766. https://doi.org/10.1104/pp.93.2.758
Schofield, L. R., Anderson, B. F., Patchett, M. L., Norris, G. E., Jameson, G. B., & Parker, E. J. (2005). Substrate ambiguity and crystal structure of *Pyrococcus furiosus* 3-deoxy-D-arabino-heptulosonate-7-phosphate synthase: An ancestral 3-deoxyald-2-ulosonate-phosphate synthase? *Biochemistry, 44*(36), 11,950–11,962. https://doi.org/10.1021/bi050577z
Schönbrunn, E., Eschenburg, S., Shuttleworth, W. A., Schloss, J. V., Amrhein, N., Evans, J. N., & Kabsch, W. (2001). Interaction of the herbicide glyphosate with its target enzyme 5-enolpyruvylshikimate 3-phosphate synthase in atomic detail. *Proceedings of the National Academy of Sciences, 98*(4), 1376–1380. https://doi.org/10.1073/pnas.98.4.1376
Shah, D. M., Horsch, R. B., Klee, H. J., Kishore, G. M., Winter, J. A., Tumer, N. E., Hironaka, C. M., Sanders, P. R., Gasser, C. S., Aykent, S., & Siegel, N. R. (1986). Engineering herbicide tolerance in transgenic plants. *Science, 233*(4762), 478–481. https://doi.org/10.1126/science.233.4762.478
Siehl, D. (1999). *The biosynthesis of tryptophan, tyrosine and phenylalanine from chorismate. Plant amino acids*. BK Singh.
Siehl, D. L., Singh, B. K., & Conn, E. E. (1986). Tissue distribution and subcellular localization of prephenate aminotransferase in leaves of *Sorghum bicolor*. *Plant Physiology, 81*(2), 711–713. https://doi.org/10.1104/pp.81.2.711
Singh, S. A., & Christendat, D. (2006). Structure of *Arabidopsis* dehydroquinate dehydratase-shikimate dehydrogenase and implications for metabolic channeling in the shikimate pathway. *Biochemistry, 45*(25), 7787–7796. https://doi.org/10.1021/bi060366
Song, H. S., Brotherton, J. E., Gonzales, R. A., & Widholm, J. M. (1998). Tissue culture-specific expression of a naturally occurring tobacco feedback-insensitive anthranilate synthase. *Plant Physiology, 117*(2), 533–543. https://doi.org/10.1104/pp.117.2.533
Song, J., Bonner, C. A., Wolinsky, M., & Jensen, R. A. (2005). The TyrA family of aromatic-pathway dehydrogenases in phylogenetic context. *BMC Biology, 3*(1), 1–30. https://doi.org/10.1186/1741-7007-3-13
Spitzer-Rimon, B., Marhevka, E., Barkai, O., Marton, I., Edelbaum, O., Masci, T., Prathapani, N. K., Shklarman, E., Ovadis, M., & Vainstein, A. (2010). EOBII, a gene encoding a flower-specific regulator of phenylpropanoid volatiles' biosynthesis in petunia. *The Plant Cell, 22*(6), 1961–1976. https://doi.org/10.1105/tpc.109.067280
Srinivasan, P. R., Rothschild, J., & Sprinson, D. B. (1963). The enzymic conversion of 3-deoxy-D-arabino-heptulosonic acid 7-phosphate to 5-dehydroquinate. *Journal of Biological Chemistry, 238*(10), 3176–3182.
Streatfield, S. J., Weber, A., Kinsman, E. A., Häusler, R. E., Li, J., Post-Beittenmiller, D., Kaiser, W. M., Pyke, K. A., Flügge, U. I., & Chory, J. (1999). The phosphoenolpyruvate/phosphate

translocator is required for phenolic metabolism, palisade cell development, and plastid-dependent nuclear gene expression. *The Plant Cell, 11*(9), 1609–1621. https://doi.org/10.1105/tpc.11.9.1609

Takatsuji, H., Mori, M., Benfey, P. N., Ren, L., & Chua, N. H. (1992). Characterization of a zinc finger DNA-binding protein expressed specifically in *Petunia* petals and seedlings. *The EMBO Journal, 11*(1), 241–249. https://doi.org/10.1002/j.1460-2075.1992.tb05047.x

Tozawa, Y., Hasegawa, H., Terakawa, T., & Wakasa, K. (2001). Characterization of rice anthranilate synthase α-subunit genes OASA1 and OASA2. Tryptophan accumulation in transgenic rice expressing a feedback-insensitive mutant of OASA1. *Plant Physiology, 126*(4), 1493–1506. https://doi.org/10.1104/pp.126.4.1493

Tsegaye, Y., Shintani, D. K., & DellaPenna, D. (2002). Overexpression of the enzyme p-hydroxyphenolpyruvate dioxygenase in *Arabidopsis* and its relation to tocopherol biosynthesis. *Plant Physiology and Biochemistry, 40*(11), 913–920. https://doi.org/10.1016/S0981-9428(02)01461-4

Tzin, V., Malitsky, S., Aharoni, A., & Galili, G. (2009). Expression of a bacterial bi-functional chorismate mutase/prephenate dehydratase modulates primary and secondary metabolism associated with aromatic amino acids in *Arabidopsis*. *The Plant Journal, 60*(1), 156–167. https://doi.org/10.1111/j.1365-313X.2009.03945.x

Van Moerkercke, A., Haring, M. A., & Schuurink, R. C. (2011). The transcription factor EMISSION OF BENZENOIDS II activates the MYB ODORANT1 promoter at a MYB binding site specific for fragrant petunias. *The Plant Journal, 67*(5), 917–928. https://doi.org/10.1111/j.1365-313X.2011.04644.x

Van Winden, W., Verheijen, P., & Heijnen, S. (2001). Possible pitfalls of flux calculations based on 13C- labeling. *Metabolic Engineering, 3*(2), 151–162. https://doi.org/10.1006/mben.2000.0174

Verdonk, J. C., Haring, M. A., Van Tunen, A. J., & Schuurink, R. C. (2005). ODORANT1 regulates fragrance biosynthesis in petunia flowers. *The Plant Cell, 17*(5), 1612–1624. https://doi.org/10.1105/tpc.104.028837

Vogt, T. (2010). Phenylpropanoid biosynthesis. *Molecular plant, 3*(1), 2–20. https://doi.org/10.1093/mp/ssp106

Voll, L., Häusler, R. E., Hecker, R., Weber, A., Weissenböck, G., Fiene, G., Waffenschmidt, S., & Flügge, U. I. (2003). The phenotype of the *Arabidopsis* cue1 mutant is not simply caused by a general restriction of the shikimate pathway. *The Plant Journal, 36*(3), 301–317. https://doi.org/10.1046/j.1365-313X.2003.01889.x

Webby, C. J., Baker, H. M., Lott, J. S., Baker, E. N., & Parker, E. J. (2005). The structure of 3-deoxy-D-arabino-heptulosonate 7-phosphate synthase from *Mycobacterium* tuberculosis reveals a common catalytic scaffold and ancestry for type I and type II enzymes. *Journal of Molecular Biology, 354*(4), 927–939. https://doi.org/10.1016/j.jmb.2005.09.093

Wightman, F., & Forest, J. C. (1978). Properties of plant aminotransferases. *Phytochemistry, 17*(9), 1455–1471. https://doi.org/10.1016/S0031-9422(00)94622-3

Wright, A. D., Moehlenkamp, C. A., Perrot, G. H., Neuffer, M. G., & Cone, K. C. (1992). The maize auxotrophic mutant orange pericarp is defective in duplicate genes for tryptophan synthase beta. *The Plant Cell, 4*(6), 711–719. https://doi.org/10.1105/tpc.4.6.711

Wu, J., & Woodard, R. W. (2006). New insights into the evolutionary links relating to the 3-deoxy-D-arabino-heptulosonate 7-phosphate synthase subfamilies. *Journal of Biological Chemistry, 281*(7), 4042–4048. https://doi.org/10.1074/jbc.M512223200

Yamada, T., Matsuda, F., Kasai, K., Fukuoka, S., Kitamura, K., Tozawa, Y., Miyagawa, H., & Wakasa, K. (2008). Mutation of a rice gene encoding a phenylalanine biosynthetic enzyme results in accumulation of phenylalanine and tryptophan. *The Plant Cell, 20*(5), 1316–1329. https://doi.org/10.1105/tpc.107.057455

Zamir, L. O., Tiberio, R., & Jensen, R. A. (1983). Differential acid-catalyzed aromatization of prephenate, arogenate, and spiro-arogenate. *Tetrahedron Letters, 24*(28), 2815–2818. https://doi.org/10.1016/S0040-4039(00)88031-4

Zhang, R., Wang, B., Ouyang, J., Li, J., & Wang, Y. (2008). *Arabidopsis* indole synthase, a homolog of tryptophan synthase alpha, is an enzyme involved in the trp-independent indole-containing metabolite biosynthesis. *Journal of Integrative Plant Biology, 50*(9), 1070–1077. https://doi.org/10.1111/j.1744-7909.2008.00729.x

Zhang, S., Wilson, D. B., & Ganem, B. (2000). Probing the catalytic mechanism of prephenate dehydratase by site-directed mutagenesis of the *Escherichia coli* P-protein dehydratase domain. *Biochemistry, 39*(16), 4722–4728. https://doi.org/10.1021/bi9926680

Zhao, J., & Last, R. L. (1995). Immunological characterization and chloroplast localization of the tryptophan biosynthetic enzymes of the flowering plant *Arabidopsis thaliana (∗). Journal of Biological Chemistry, 270*(11), 6081–6087. https://doi.org/10.1074/jbc.270.11.6081

Zhao, J., & Last, R. L. (1996). Coordinate regulation of the tryptophan biosynthetic pathway and indolic phytoalexin accumulation in *Arabidopsis. The Plant Cell, 8*(12), 2235–2244. https://doi.org/10.1105/tpc.8.12.2235

Chapter 10
Molecular Farming: Sustainable Manufacturing of Vaccines, Antibodies, and Other Therapeutic Substances

Kaisar Ahmad Bhat, Lubna Tariq, Aadil Ayaz, Madhiya Manzoor, S. M. Zargar, and Ali Asghar Shah

Abstract Many important and potential benefits have been claimed from plants since the moment these were considered as production systems for various recombinant pharmaceutical proteins. With the advent of technological developments, vital benefits have been reaped from the plant system in the form of production of many important therapeutic proteins, including antibodies, blood products, cytokines, growth factors, hormones, recombinant enzymes, and different vaccines. The genetic alteration or gene editing in various plant species (both monocot as well as dicot) has led to the production of different important plant products. The past decade saw the resurgence in the production of antibodies, vaccines, and other therapeutic substances which have direct impact on the theragnosis of human and other animal diseases treatment. Plant systems have some of the important and reliable advantages over other systems for the production of these different molecules. These advantages include the safety; like that of absence of human replicating pathogens; unparalleled potential for scalability, diversity, and range in the synthesis of the recombinant molecules; and potential to produce small peptides, polypeptides, and complex polymeric proteins; many of which cannot be made at cell or microbial

K. A. Bhat (✉)
School of Biosciences and Biotechnology, Baba Ghulam Shah Badshah University, Rajouri, Jammu and Kashmir, India

Division of Plant Biotechnology, Sher-e-Kashmir University of Agricultural Sciences & Technology Kashmir, Srinagar, Jammu and Kashmir, India
e-mail: bhatk2@bgsbu.ac.in

L. Tariq · A. A. Shah
School of Biosciences and Biotechnology, Baba Ghulam Shah Badshah University, Rajouri, Jammu and Kashmir, India

A. Ayaz
Department of Microbiology, SKIMS Medical College, Bemina, Srinagar, Jammu and Kashmir, India

M. Manzoor · S. M. Zargar
Division of Plant Biotechnology, Sher-e-Kashmir University of Agricultural Sciences & Technology Kashmir, Srinagar, Jammu and Kashmir, India

T. Aftab, K. R. Hakeem (eds.), *Metabolic Engineering in Plants*,
https://doi.org/10.1007/978-981-16-7262-0_10

level. This book chapter provides an overview on different aspects of plant-derived products for the development of vaccines, antibodies, and other therapeutic substances using the plant system.

Keywords Molecular farming · Antibodies · Vaccines · Proteins · Therapeutic substances

Introduction

While plants have been used medicinally for thousands of years, still microbial and chemical medications are more widely used. In higher plants, recombinant DNA technology has created a novel area of fundamental research and role in plant science. Foreign genes can be inserted into any plant, allowing therapeutic proteins like growth factors, blood products, antibodies, hormones, cytokines, enzymes, and vaccines to be produced. Plant molecular farming (PMF), elucidated as the technique of employing plants to create human remedial proteins, has garnered international attention. Over the last two decades, PMF has grown and evolved significantly. Numerous remedial proteins have been generated in plants, with various undergoing preclinical or clinical studies and being considered for commercialization. PMF is the process of mass producing and purifying pharmaceutically significant proteins in plant systems (Fig. 10.1). The age of molecular farming started in 1989 with the unprecedented expression of complete antibodies in transgenic plants. In 1992, the

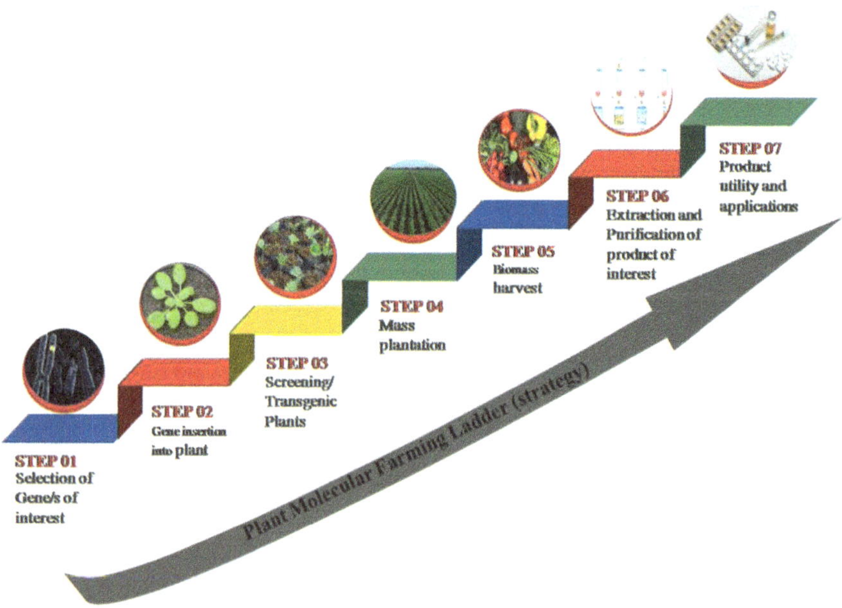

Fig. 10.1 General strategy for plant molecular farming

league was succeeded by the development of plant-based vaccines. Nevertheless, numerous different products have been generated from the plant system occasionally, including antibodies and fragments of antibodies with medicinal and veterinary applications. In the 1990s, the first ever transgenic plant-derived pharmaceutical protein (PDP) was synthesized from transgenic potato and tobacco plants (Sijmons et al., 1990). These products are currently in the process of commercialization for the therapy of human ailments. Fifteen years down the lane, the first plant transgenic proteins are commercially available, and evidence of concept for the development of numerous therapeutic proteins has been established, including growth factors, blood products, antibodies, transgenic enzymes, hormones, human and veterinary vaccines, and cytokines (Twyman et al., 2005). Additionally, numerous PDP products for the diagnosis of human diseases are nearing commercialization, like antibodies for tooth decay deterrence and non-Hodgkin's lymphoma treatment recombinant and gastric lipase for cystic fibrosis (Ma et al., 2003). Several veterinary vaccines are also in development as Dow AgroSciences (Indianapolis, IN, USA) has declared its desire to develop plant-based vaccinations for the animal health business. The focus has shifted away from fundamental research and toward commercialization, and molecular farming has advanced to the point where it could confront conventional making technologies based on yeast, bacteria, and grown mammalian cells. Numerous potential implications have been asserted since plants were initially studied as a mechanism for producing transgenic medicinal proteins. However, as technology has advanced, so has public awareness garnished regarding true benefits that plants bring. There is also a debate that recombinant plants offer unmatched scaling potential. Plants grown in the field have a practically limitless growth potential, and even if production sites are adequately segregated to avoid cross-pollination or mixing with several other crops, there are still many locations on the world where huge growth is possible. Even underneath confinement, a significant number of pharmaceutical plants might be grown; enormous greenhouse services are frequently utilized by the food industries and for horticultural activities. Human growth hormone was synthesized in 1986 in tobacco and sunflower, just a few years after the first plant-derived transgenic remedial protein (Barta et al., 1986). Mason et al. (1992) later on demonstrated that transgenic tobacco can express the hepatitis B surface antigen (HBsAg). This plant-based antigen was physically and immunologically identical to recombinant yeast HBsAg, and human serum HBsAg produced from yeast is used clinically to immunize against HBV. Since 1994, plants have been used to generate and describe over 100 pharmacological proteins. By 2011, there were over 20 PMF medicines in preclinical or clinical testing (Paul & Ma, 2011). Numerous PMF products have successfully concluded phase II trial. Table 10.1 provides the list of plant-derived pharmaceuticals in clinical trials. Numerous plant-derived medicines have been commercialized as diagnostic and research reagents, e.g. rice-derived lysozyme and tobacco-derived aprotinin from Sigma-Aldrich Company (St. Louis, MO, USA), or have recognized USDA authorization for use as a poultry vaccine additive (Dow Agro Sciences, Indianapolis, IN, USA) (Arntzen, 2015). Plants constitute an exciting approach for the large-scale manufacturing of human medicinal proteins at reasonable costs. Numerous manufacturing challenges, including plant glycosylation (Ford et al., 2015), low yield (Paul & Ma, 2011;

Table 10.1 Plant-derived pharmaceuticals in clinical trials

Drug/Product	Application	Company	Host plant
Vaccine	Norwalk virus infection	Arntzen Group	Potato
	Rabies	Yusibov (2002)	Spinach
	Hepatitis B	Arntzen Group	Potato
	Diarrhoea	ProdiGene Inc	Maize
	Malaria	Center for Molecular Biotechnology, Plymouth, MI, USA	Tobacco
	Anthrax	Center for Molecular Biotechnology, Plymouth, MI, USA	Tobacco
	H5N1 Vaccine	Center for Molecular Biotechnology, Plymouth, MI, USA	Tobacco
	Vaccine growth	PhycoBiologics Inc. (n.d.)	Algae
Antibody	Non-Hodgkin lymphoma	Large Scale Biology Corporation	Tobacco
	Dental caries	Plant Biotechnology Inc	Tobacco
	HIV	University of Surrey, Guildford, UK	Tobacco
	Ebola Virus	National Institute of Allergy and Infectious Diseases (NIAID), Bethesda, MD, USA	Tobacco
Dietary	Vitamin B12 deficiency	Cobanto Biotech AS	Arabidopsis
	Gastrointestinal infections	Meristem Therapeutics	Maize
Therapeutic enzyme	Cystic fibrosis pancreatitis	Meristem Therapeutics	Maize

Boothe et al., 2010; Sack et al., 2015; Abiri et al., 2015), downstream processing, and purification obstacles (Fischer et al., 2015), have hampered the clinical expansion of PMF-based human drugs. Vaccines and monoclonal antibodies (mAbs) derived from plants have been developed for a range of chronic diseases prevalent in developing countries, such as hepatitis B virus (HBV), tetanus, human papillomavirus (HPV), human immunodeficiency virus (HIV), malaria, enterotoxigenic *Escherichia coli*, rabies, and *Vibrio cholerae*. Although an effective HPV vaccine exists, it is prohibitively expensive for widespread use in many underdeveloped countries. There is no effective vaccination against HIV infection, and the expensive cost of antiretrovirals prohibits them from being accessible to millions of infected people living in developing nations. Vaccines derived from plants may potentially be beneficial for treating tropical diseases that are hard to manage, such as tuberculosis, malaria, and newly developing infectious pathogens such as the Ebola virus.

Plant-Derived Vaccines

Vaccine Antigens from Plant Suspension Cell Cultures

To date, research has established that cultures of plant cell (primarily tobacco NT-1 and BY-2) are able to produce a variety of pharmacologically active proteins, like immunoglobulins (e.g. IgG) (Girard et al., 2006), immunotoxins (Francisco et al., 1997), hGM-CSF, IL-12 (Kwon et al., 2003), and human IFN-a (Xu et al., 2007), with few of these (e.g. Protalix BioTherapeutics, human GCD) (Huang et al., 2010a, 2010b) approaching the stage of commercial development. While the transgenic rotavirus VP6 antigen was among the first antigens expressed in recombinant cells of tomato (with an intracellular yield of 0.15–0.33 mg/L) (Chung et al., 2000), the majority of research on vaccine antigens manufactured in vitro by plant cell cultures has focused on the HBsAgI (Hepatitis B surface antigen), a complicated, widely disulphide cross-linked lipoprotein unit. It was initially expressed in suspension cultures of tobacco and soybean NT-1 cells (Smith et al., 2002). Constantly, expression levels were accomplished in the cell line of soybean, with antigen titres approaching 65 mg/g fresh weight, compared to about tenfold lower titres in the most active tobacco cell line. This difference was most likely owing to the greater transgenic copy number in soybean, 4–6, compared to 1 in the tobacco cell line, and/or to a substantial downregulation of antigen expression throughout consecutive subculture in tobacco cultures. Sojikul and co-workers (2003) synthesized HBsAg protein in tobacco NT-1 cell line culture with a eukaryotic ER retention signal (Ser-Glu-Lys-Asp-Glu-Leu) or an ER signal peptide from VSPaS (soybean vegetative storage protein) at the C- and N-terminal ends, respectively. The signal peptide directed the protein directly to the endoplasmic reticulum (ER) and remained intact, enhancing the aggregation and stability of the VSPaS–HBsAg fusion protein in tobacco plant cells. Kumar and co-workers confirmed the existence of HBsAg in suspension cultures of the tobacco NT-1 cell line (Kumar et al., 2005). Preclinical investigations established the vaccine's ability to induce an antigen-specific immune reaction in mice.

Vaccine Antigens from Hairy Root Cultures

The majority of research involving hairy roots has concentrated on bioactive compounds. Hairy roots have been used to create functional and medicinal proteins, with the goal of integrating the benefit of transgenic protein-based vaccinations with possible benefits of confined plant synthesis. Hairy root cultures can be generated from recombinant plants by inoculation or co-cultivation with *A. rhizogenes* for the goal of producing useful heterologous proteins. Hairy roots from recombinant tobacco plants have been effectively employed to manufacture the immune modulator ricin B (the nontoxic galactose/N-acetylgalactosamine-binding subunit of ricin) fused to a model antigen (green fluorescent protein) and tested for mucosal vaccine adjuvancy/delivery (Medina-Bolivar et al., 2003). The results demonstrated for the first time that a recombinant ricin B genetically merged to an antigen and expressed in roots is capable of delivering the antigen to the mucosa surface and augmenting the immunogenicity of the fused antigen following intranasal immunization in mice with greater adjuvancy to cholera toxin in evoking a Th2-type systemic and mucosal response. As a result, the amount of ricin B–GFP required to generate comparable antibody responses was tenfold less than the amount of cholera toxin. In a mist bioreactor, tobacco lines with high levels of murine IL-12 were further transformed with *A. rhizogenes* to obtain hairy roots with high levels of IL-12, demonstrating the successful yield of therapeutic protein in a mist bioreactor with the capacity for large-scale applications in root cultures (Liu et al., 2009). Kumar et al. (2005) demonstrated for the first time the presence of HBsAg in roots produced from transgenic potatoes. When a C-terminal ER retention signal was used, higher amounts of HBsAg were obtained (Kumar et al., 2006). Genetically modified hairy roots expressing a surface protective antigen (SpaA) from *Erysipelothrix rhusiopathiae* (swine erysipelas) fused to the cholera toxin B subunit (CTB) and containing an ER retention signal were collected from untransformed tobacco leaf discs employing *A. rhizogenes* expressing the CTB–SpaA fusion (Ko et al., 2006).

Vaccine Antigens from Microalgae

PhycoBiologics Inc. (n.d.) has leveraged the capability of *Chlamydomonas* chloroplast transformation to develop oral aquaculture vaccines (live algae or freeze-dried) capable of inducing antigen-specific immune responses in mucus and serum (Dohi et al., 2006). In preclinical experiments, lyophilized microalgae encoding the *Staphylococcus aureus* fibronectin-binding domain D2 linked to the CTB mucosal adjuvant were employed for oral vaccination (Dreesen et al., 2010). Immunized mice generated mucosal and systemic immune reactions; the pathogen burden in the spleen and gut was decreased, and 80% of them were shielded against deadly dosages of *S. aureus*. At room temperature, this vaccination remained active for more than 1.5 years. Additionally, scientists from the Scripps Research Institute (La Jolla, CA, USA) have harnessed the potency of this system to generate human therapeutic proteins. CTB was utilized to express the foot-and-mouth disease virus VP1 protein. The fusion protein was produced well in *Chlamydomonas* chloroplasts, yielding 3–4% of the TSP (Sun et al., 2003). Another work demonstrated the validity of employing *Chlamydomonas* chloroplast as a vaccine making base when glutamic acid decarboxylase-65, an insulin-dependent auto-antigen, was produced at 0.3% of the TSP, isolated, and identified by diabetic serum samples taken from NOD mice (Wang et al., 2008). By replacing the chloroplast psbA sequence with the bovine mammary-associated serum amyloid sequence (encoding a protein found in mammalian colostrum that protects newborns against intestinal bacteria and viruses), the maximum concentration of recombinant antigen accumulation has been achieved (Manuell et al., 2007). Presently, novel platforms for pharmaceutical synthesis in confined systems like higher aquatic plants, Lemnaceae, and the moss *Physcomitrella patens*. Lemnaceae are palatable freshwater plants that float freely in water. They are currently employed for animal and human nutrition. They have higher protein content (up to 45% dry weight) and can be produced in aseptic modules without agitation with minimal expenditure in full containment (Vunsh et al., 2007; Stomp, 2005). Due to the swift clonal development (doubling times of as little as 1.5–2 days) and a typical plant architecture (which requires vascular tissue or some mechanical support), this system scales up quickly and accumulates a substantial quantity of biomass (Stomp, 2005). The adjuvant characteristics of *Lemna*, a pectic polysaccharide isolated from *Lemna minor*, were explained by inducing Th1- and Th2-type responses (Popov et al., 2006), suggesting an advantage for oral vaccine formulation. Agrobacterium-mediated gene transfer or the biolistic technique has been employed to accomplish stable nuclear transformation of various Lemnaceae species (Yamamoto et al., 2001), with expression levels changing depending to the protein and the species utilized. At the moment, *Spirodela oligorrhiza* is the suitable example of foreign protein production in a higher plant via nuclear transformation, accounting for 25% of the TSP accumulation for the GFP (Vunsh et al., 2007). Numerous beneficial proteins have been successfully displayed in this system, and Biolex Therapeutics Inc. (Xu et al., 2007), which acquired the French company Lemnagene in 2005, now uses the two genera *Spirodela* and *Lemna* as a portal for the manufacturing of drugs for human and animal use, including IFN-a2b (which is currently in phase III clinical trials), plasmin, and the monoclonal antibody antiCD20 for the treatment of rheumatoid arthritis. Additionally, *Wolffia* species are being developed for molecular farming objectives (Boehm, 2007). The glycosylation of an anti-human CD30 monoclonal antibody was improved by co-expressing the heavy and light chains of the antibody with an RNA interference construct targeting the endogenous a1,3-fucosyltransferase and b1,2-xylosyltransferase genes. The resulting mAb included no detectable plant-specific N-glycans and performed even better than the mAb produced in cultivated CHO cells (cell-mediated cytotoxicity and effector cell receptor binding) (Cox et al., 2006). The moss *P. patens* can be grown in confined circumstances throughout its life, providing unique benefits for biopharmaceutical development (Decker & Reski, 2007, 2008). While sexual reproduction is favoured when moss is grown on agar plates, vegetative growth is favoured when moss is grown on liquid media (flasks and bioreactors). *Physcomitrella* can be cultured at this stage ('protonema') in basic settings using swirling glass tanks or modular glass tube photobioreactors (e.g. inorganic salts and airborne CO_2 as a carbon source). Several photobioreactors for moss cultivation have been created, and a 100-L modular photobioreactor is actively being employed for GMP-compliant biopharmaceutical making. Due to *Physcomitrella's* extensive pH tolerance, this factor can be changed to correspond to any protein that must be released into the growth medium (Decker & Reski, 2004). The whole *Physcomitrella* genome has been sequenced and a second, 'more user-friendly' version is scheduled for release by the end of this year (Rensing et al., 2008). Contrary to higher plants, there is no considerable bias for codon usage in expressed genes, implying that transgenes can be expressed without codon optimization (Rensing et al., 2005). With *Physcomitrella*'s nuclear DNA, precise genetic alterations like base-specific targeted gene knockouts may be easily performed at up to 100% frequency (Kamisugi et al., 2005), whereas HR occurs at considerably lower frequency (10-4–10-5) in flowering plants (Britt & May, 2003).

Plant-Based Production Systems and Plant-Derived Protein Products

Numerous scientists have desired to create transgenic proteins in plants since the 1990s. Typically, they chose plants that had already been utilized for various scientific goals due to the ease with which gene transfer techniques could be used. This resulted in the development of a plethora of production systems, including various tissue, whole plants, and cell systems (hairy roots and cell suspension cultures), and a variety of expression strategies (stably transformed transgenic and transplastomic plants, various protein targeting strategies, inducible expression, and transient expression systems) Twyman et al., 2003, 2005; Schillberg et al., 2013; Spiegel et al., 2018). While a viable platform is likely to exist for any potential product, the lack of a potential platform splits and hinders efforts to gear up efficiency and makes defining industrial production standards more challenging. When it comes to product options, study has primarily focused on biopharmaceuticals with a higher efficacy than diagnostic and technical proteins. Three distinct classes of protein products have started to emerge in this context: vaccine candidates, antibodies, and replacement human proteins like blood products (human serum albumin), replacement proteins for rare and common diseases (glucocerebrosidase for Gaucher's disease, growth factors and cytokine gastric lipase for cystic fibrosis, and insulin for diabetes) (Spiegel et al., 2018). Transgenic antibody fragments, antibodies, and antibody fusion proteins have emerged the most frequently expressed plant products (Nölke et al., 2003; Vasilev et al., 2016), owing to their economic value as pharmaceuticals (Walsh, 2018) and their relative stability and ease of characterization. This implies they accrue high concentrations (>100 mg/kg fresh plant weight or >100 mg/L culture media) and are simple to purify even from complex plant matrices, and their performance can be determined using simple binding tests. Furthermore, antibody titres in plants remain significantly lower than those obtained in CHO cells, making it doubtful that plants will ever become a commercially viable system for antibody products on a routine basis. Numerous research on the synthesis of transgenic proteins in plants have skirted the harsh constraints of commercial development, concentrating instead on early-stage objectives like confirming expression, improving production and filtration, and performing preliminary functional tests. Few research findings have included translational research indicating economic viability, in part due to the economic and organizational hurdles associated with testing plant-derived biopharmaceuticals in clinical testing. It is nearly impractical to obtain financial and business backing when the potential of market and intellectual property rights is unknown, as is the case with the majority of plant-based protein products. However, without a compelling financial justification, the industry will continue to use current microbial and mammalian production techniques, as the risk is not warranted. Nonetheless, a few plant-derived biopharmaceutical potential candidates have entered clinical evaluations aiding in the development of GMP-compliant production processes approved by regulatory bodies (Fischer et al., 2012; Sack et al., 2015). Only a handful has been commercialized, the first being transgenic glucocerebrosidase (prGCD), also known as taliglucerase alfa and marketed by Protalix BioTherapeutics as Elelyso (Rup et al., 2017; Zimran et al., 2018). To establish commercial viability of plant-derived biopharmaceuticals, extensive periods and large expenditures may be necessary. Goods with less stringent regulatory criteria, such as diagnostic, technical, and cosmetic products, can be brought to market faster. The diagnostic reagent avidin was first marketed in maize 20 years ago (Hood et al., 1997) and is still sold by Sigma-Aldrich, as well as human epidermal growth factor, which is produced in barley and distributed by Sif Cosmetics (Iceland). However, poor yields and high downstream processing costs must be solved before plants can compete more extensively with alternative expression methods.

Plant-Derived Veterinary Vaccines

The phenomenal advancements in plant-based veterinary vaccinations are largely attributable to the capacity to execute challenge tests on selected animal species (Santi 2009). Black-eyed bean generated a brief, linear, and neutralizing epitope from the viral VP2 capsid protein to show the vaccine's protection against MEV (Dalsgaard et al., 1997). On the surface of plant chimeric virus particles (CVPs) after infection, the short epitope was inserted into the cowpea mosaic virus. Subcutaneous injection of 1 mg MEV peptide-encoding chimeric viral particles shielded mink from clinical disease and MEV virulence challenge. Another research demonstrated that the VP60 protein of the rabbit haemorrhagic disease virus generated from recombinant potatoes protected rabbits against infection following parenteral vaccination (Castanon et al., 1999). Additionally, it was immunogenic and gave temporary relief in rabbits following oral administration of the vaccine (Martín-Alonso et al., 2003). Similarly, the most convincing proof of concept for a palatable plant-based animal vaccination was against swine TGEV (Howard, 2004). The S protein of TGEV was produced in maize and given orally to 10-day-old pigs (Streatfield et al., 2001). Several economically significant diseases have been the focus of research on plant-based vaccinations in the chicken vaccine arena. The NDV is one of them, with expression directed toward the virus's surface glycoprotein and/or hemagglutinin-neuraminidase. Along with the first authorized plant-produced NDV vaccine (Vermij & Waltz, 2006), different plant systems were used to express NDV viral proteins. The IBV S1 glycoprotein also contains vaccine-targeted virus-neutralizing and hemagglutination-inhibiting epitopes. Transgenic potato tuber extracts were utilized for immunization and protective effectiveness investigations in chicken (Zhou et al., 2004a, 2004b).

Therapeutical Applications in Terms of Health Supplements and Topical Utility

It has been demonstrated that subcutaneous infusions of plant-derived proteins induce an immunogenic response to plant-specific glycans (Costa et al., 2014). External applications of plant-derived glycoproteins in humans, on the other hand, have not been associated with adverse consequences, suggesting a viable path for PMF-based goods. External application of monoclonal antibodies (mAbs) generated from soybeans diffused rapidly into human cervical mucus and inhibited herpes simplex virus 2 (HSV-2) infection. Tan et al. (2014) demonstrated that the medicinal herb *Salvia miltiorrhiza* expressed human acidic fibroblast growth factor 1 (FGF-1). The product incorporated both FGF-1 and bioactive chemicals found in the medicinal plant. In mice with burn wounds, topical application of extracts from the transgenic medicinal plant stimulated fibroblast cells, enhanced blood vessel formation, and accelerated the healing process. This is an illustration of how PMF may be utilized to integrate a recombinant protein's therapeutic activity with the natural characteristics of a medicinal plant. The expense of purification and downstream processing would be considerably reduced if a plant extract were applied topically. By and large, topical treatment is safer than oral ingestion or injection, which alleviates public safety concerns. PMF's major goal is to minimize the expense of developing new remedial proteins. Utilizing PMF to generate a fruit, seed, or vegetable, health supplement may be a more viable option than utilizing PMF to generate a processed pharmaceutical medication. Guan et al. (2014) expressed lumbrokinase, an anti-thrombotic enzyme derived from earthworms, in sunflower kernels. Rats and mice given transgenic kernels demonstrated a significant reduction of blood clots (Guan et al., 2014). Neither a vaccine nor a therapeutic protein, lumbrokinase has been extensively marketed and utilized as a health supplement to break blood clots and preserve normal circulatory function in humans. This makes lumbrokinase an excellent option for PMF, as health supplements often do not require a medical prescription and are subject to less rules for commercialization (Wang et al., 2013).

Plant Molecular Farming as a Strategy Against COVID -19

As of July 28, 2021, the severe acute respiratory syndrome coronavirus 2 (SARS-CoV-2) has killed 41,80,161 persons worldwide (WHO) and caused massive socio-economic damage. This led to the race of development of vaccine and test kits for identification, monitoring, and medications to reduce mortality. Excellent achievements have been so for recorded in production of COVID vaccine. More efforts are being carried out to use plants (pertinent to their higher efficacy in generating antibodies) to provide an alternative method for producing recombinant anti-COVID-19 antibodies.

Neutralizing Antibodies

Numerous studies show that passive immunotherapy with COVID-19 neutralizing antibodies may be effective against SARS-CoV-2 (Tortorici et al., 2020). Many drug manufacturers are previously disclosed phase I studies of monoclonal neutralizing antibodies that prevented against SARS-CoV-2 infection in animal models (Ren et al., 2020). One of these trials is testing REGN-COV2, a mixture of two antibodies that recognize areas of the RBD that do not overlap (Baum et al., 2020; Hansen et al., 2020). Eli/Lilly recently announced the initial results of the double-blind, randomized, placebo-controlled phase II trial BLAZE-1, which evaluated the combination of LY-CoV016 and LY-CoV555, two SARS-CoV-2 neutralizing antibodies, in the outpatient environment for the cure of symptomatic COVID-19. Given the high efficiency with which antibodies are produced in plants, particularly via transitory expression utilizing viral vectors, plants may provide an alternative method for producing recombinant anti-COVID-19 antibodies (Donini & Marusic, 2019).

Other Proteins

Along with active and passive immunisation, antiviral proteins may provide an additional line of defence towards the virus throughout the infection phase, while other biologics may address COVID-19 symptoms including the cytokine storm caused by the preliminary wave of contagion. Numerous murine, avian, and swine coronaviruses' S proteins, as well as the N-terminal fragment of the SARS-CoV S protein, have been successfully synthesized in recombinant tobacco, maize, tomato, or potato plants via conventional Agrobacterium-mediated transformation or viral display on the exterior of plant viruses, and in all cases, the products elicit an immune response. Due to the extensive presence of glycans on the surface of viruses, lectins have been studied as antivirals and have been shown to disrupt the infectious cycle of influenza A, HIV, CMV, respiratory syncytial virus, and a variety of coronaviruses. Certainly, over 20 distinct plant lectins have been found to inhibit SARS-CoV infection, most likely by their preferential binding to glycans on the S protein (Keyaerts et al., 2007).

Antigen-Based Vaccines

Antibodies that neutralize coronaviruses frequently disrupt connections between the S protein and its receptor. This is ACE2 (angiotensin converting enzyme 2), which is predominantly present on the surface of epithelial cells of lung but is also detected in other tissues in the case of SARS-CoV-2 (Liu & Li, 2020; Long et al., 2020). As a result, the protein S or fragments thereof are preferred as vaccine contender, but the

N and M proteins have been investigated as well. In animal models, earlier coronavirus vaccines in opposition to MERS-CoV and SARS-CoV caused Th2-mediated immunopathology, and scientists are concentrating on protein engineering techniques to mitigate this impact (Koirala et al., 2020). One potential strategy is to express the RBD or S1 regions of the S protein, which directs the immune reaction to the regions of the S protein that link up with ACE2 and therefore increases the probability of generating neutralising antibodies. Numerous murine, avian, and swine coronaviruses' S proteins, as well as the N-terminal fragment of the SARS-CoV S protein, have been successfully synthesized in recombinant tobacco, maize, tomato, or potato plants via conventional Agrobacterium-mediated transformation or viral display on the exterior of plant viruses, and in all cases, the products elicit an immune response (Zhou et al., 2004a, 2004b) or nasal delivery (Koo et al., 1999).

PMF Production Platforms

Transient Expression Platform

Transgenic plants are frequently produced in 6 months to a year or more. Multiple generations are necessary to create homozygous plants for the transgene. Additionally, the majority of transformation procedures culminate in the random introduction of the gene into the genome of plant. Because of the random insertion factor, as well as the requirement to find regulatory areas (promoters) capable of driving high levels of foreign gene expression, recombinant protein output is generally modest, even in stable transformed plants. The time required to implement a typical PMF strategy is insufficient for dealing with unexpected viral epidemics like an Ebola, MERS-CoV, or severe acute respiratory syndrome (SARS) outbreak. Alternatively, transient expression methods may be used to rapidly generate recombinant protein within 3–5 days (Sainsbury & Lomonossoff, 2014; Peyret & Lomonossoff, 2015; Catrice & Sainsbury, 2015). Numerous viral vectors are being designed for the generation of PMF on a small or medium scale. Huang et al. (2009) created a very effective single-vector DNA replicon system relying on the bean yellow dwarf virus (BeYDV), which contained several DNA replicon cassettes. Using a similar approach, Mapp Biopharmaceutical Inc. transiently produced humanized antibodies, ZMab, and MB-003 on leaves of tobacco. Later, MB-003 and ZMab were merged to become ZMapp. The pharmaceutical usage of these antibodies was able to treat 100 percent of Ebola-infected rhesus macaques monkeys (Qiu et al., 2014). The failure to acquire enough Ebola vaccine is due to Agrobacterium penetration of leaves for transient expression. The US Department of Health and Human Services (n.d.) (HHS) awarded Mapp Biopharmaceutical Inc. a $42.2 million grant in September 2014 to design a technique for large-scale manufacturing of ZMapp. Future advances in the temporary expression technique may enable large-scale manufacturing in a short time span (www.hhs.gov.).

Bioreactor-Based Platforms

Plant-Cell-Culture System

At the moment, plant-cell culture-based bioreactors appear highly promising than typical PMFs that utilize entire plants to manufacture medicines (Schillberg et al., 2013; Magy et al., 2014; Raven et al., 2015). As is the case with microbial or mammalian cell bioreactors, plant cells are cultivated in a closed, sterile cargo environment devoid of human pathogens and soil contamination. Additionally, biosafety problems related with accidental pollen dispersion and cross-fertilization are avoided. Because cultured plant cells involve just basic ingredients to thrive, they have a significantly lower operating cost than microbial or mammalian bioreactors. Processing and purification of the transgenic protein downstream are simplified due to the lack of complex plant fibres and a variety of secondary metabolites, resulting in considerable cost savings. As stated previously, the first FDA-approved PMF-based medicine, taliglucerase alfa, was generated in carrot cell suspension cultures to treat Gaucher's illness (Fox, 2012). Due to the rarity of Gaucher's disease, therapy with soul medication is prohibitively expensive ($200,000 US yearly per patient for life). However, by utilising a carrot cell manufacturing system, the expense per patient per year is reduced to $150,000. Over 20 transgenic proteins have been generated using culture methods of plant cells (Spök et al., 2008). Tobacco strains NT-1 and BY-2 are the most often utilized plant cell culture-based bioreactors in PMF. To facilitate downstream purification, the transgenic protein can be produced to be released into the culture medium. However, depending on the size and folded architecture of particular foreign proteins, the pore size in plant cell walls may prohibit them from being released (Schillberg et al., 2013). Additionally, proteolytic activity in cultured cells may end in a deficient generation of antibodies. Magy et al. (2014) established that the proteolytic profile varies according to host species. They evaluated several combinations of culture conditions, isotype, and host species and discovered that the optimal combination that resulted in a ten-fold increase in expression level. In optimal circumstances, more than 30 mg/L intact antibody was generated. ProCellExTM (Protalix BioTherapeutics, Karmiel, Israel) is a commercially accessible, cost-effective plant cell culture system that significantly reduces the expense of recombinant protein synthesis on an industrial scale.

Moss Culture

Presently, the utilization of moss, a non-seed plant, is actively being studied as a viable option for protein synthesis in bioreactors (Huang & McDonald, 2012; Kircheis et al., 2012; Reski et al., 2015). Moss cell's capacity to photosynthesize greatly decreases the culture nutrients cost. As with plant and yeast cell suspension cultures, transgenic proteins may be engineered to secrete into the moss culture media, enabling downstream processing and filtration. Moss cells may also produce a humanized form of a glycosylated protein, the Y-specific monoclonal antibody MB314 (Kircheis et al., 2012), which circumvents the problems associated with plant-specific glycosylation, as mentioned below. Several recombinant therapeutic proteins have been generated in moss cultures, including epidermal growth factor (Niederkrüger et al., 2014), amylase (Anterola, et al., 2009), and galactosidase (Niederkrüger et al., 2014). The most often used plant material in bioreactors for protein synthesis is *Physcomitrella patens* cultures. Greenovation Biotech GmbH, a German biopharmaceutical firm, has created a platform (Bryo-Technology) based on *P. patens* for significant, first-class transgenic protein synthesis. Moss-GAA is recommended for the treatment of Pompe disease, Moss-GBA is indicated for the treatment of Gaucher's disease, and moss-aGal is indicated for the treatment of Fabry disease. Moss-AGAL has completed preliminary testing satisfactorily and is progressing to phase I testing.

Algal Bioreactors

For many years, microalgae cultures have been utilized to produce foreign proteins and biofuel (Anterola et al., 2009). Microalgae are unicellular, colonial, or filamentous in structure. Due to their brief life cycle, algae may create a high quantity of biomass in a short period of time. Purification of transgenic proteins downstream in algae is comparable to that in yeast and bacteria and is hence typically less costly than whole plant synthesis. Nevertheless, because recombinant proteins derived from algae may not experience certain post-translational modifications, they may be incompatible with the synthesis of some glycoproteins. For example, algae may be unable to generate glycosylated proteins in the human form due to an absence of the necessary enzyme machinery (Mathieu-Rivet et al., 2014). Moreover, a range of diagnostic and therapeutic recombinant proteins have been generated in algal systems, comprising vaccines, enzymes, and antibodies (Rasala & Mayfield, 2011). Furthermore, in certain situations, foreign genes are expressed only transiently in algae (Teng et al., 2002). The employment of strong promoters, appropriate codon use, intron integration, and specialized transformation techniques significantly improves the efficiency and stability of foreign gene expression in algae (Eichler-Stahlberg et al., 2009). PhycoBiologics, a firm located in the United States of America, has created an efficient microalgae manufacturing system. Up to 20% total soluble protein transgenic yields have been obtained, making the algal platform viable for industrial pharmaceutical production (https://www.phycotransgenics.com.)

Seed-Based Platforms

Stability of protein is another critical factor to consider when storing harvested PMF-based recombinant products. At the moment, the majority of medicinal proteins are produced in leafy crops for maximum biomass production. Leaf proteins, on the other hand, undergo fast proteolytic breakdown upon harvest (Spok et al., 2008). Additionally, long-time preservation of leaf material is quite difficult. Foreign protein overexpression in leaf cells can also cause necrosis and eventual cell death (Phoolcharoen et al., 2011). Some preliminary findings revealed that temporary production of a variety of blood clot-dissolving serine proteinases in leaves, including vampire bat plasminogen activator (DSPA1), lumbrokinase, and nattokinase, results in leaf necrosis 4 days after penetration. Additionally, no necrosis was observed when these proteins were directed to seeds, and the separated proteins exhibited the ability to dissolve fibrin and blood clots. As a result, manufacturing PMF-based solutions specifically for seeds is becoming an interesting option (Ou et al., 2014). PMF systems based on seeds have been established in a variety of plant species, such as tobacco (Hsu et al., 2013), Arabidopsis (De Jaeger et al., 2002), corn (Zhong et al., 2006), and rice (Ou et al., 2014). It has been found that recombinant proteins directed targeting seeds accumulate to extremely high quantities. Using a seed-specific regulatory sequence from *Phaseolus vulgaris* to induce transcription of a murine scFV (single-chain variable fragment) in Arabidopsis, a considerable amount of transgenic protein was produced. Human lysozyme was found in recombinant rice grains at a concentration of 1% of grain weight. When single-chain variable fragment (scFV) was fused with elastin-like polypeptides, the amount of scFV accumulation in tobacco seeds exceeded 25% of total protein (Scheller et al., 2006). The application of PMF-based solutions to seeds ensures long-lasting protein stability while also facilitating storage, harvesting, and transportation. Compared to leaves, seeds contain less secondary metabolites, phenolic compounds, and natural proteins. Seeds are quite easy to clean and sanitize on their surface, which facilitates commercial manufacture. Taken together, the advantages of seed-based recombinant protein synthesis make it an economically viable method of producing PMF-based products (Khan et al., 2012).

Advantages

A significant benefit of PMF studies is that they do not require a substantial financial commitment to get started. Plants can be grown in greenhouses or, if necessary, in biosafety laboratories. Plant maintenance is inexpensive in contrast to mammalian cell, *E. coli*, or yeast expression techniques, and the supply of recombinant protein (plant leaves or seeds) is theoretically infinite (Aboul-Ata et al., 2014). Plant expression systems surpass prokaryotic and other eukaryotic cell systems in terms of production speed, expense, and safety. Plants are capable of correctly folding and assembling complex proteins appropriately, including full-length immunoglobulins, secretory antibodies, and the homodimeric vascular endothelial growth factor (VEGF). Commercialization of human VEGF generated in barley grain for scientific purposes has occurred (Antibodies-online Inc., 2015; PromoKine Home Page, 2015). Human VEGF expressed in plants is used to cure hair loss (UNICO Enterprises, Pasadena, CA, USA). Moreover, plants are capable of introducing post-translational changes. Additionally, the use of plants reduces the possibility of the medicinal medication being contaminated with animal diseases (prions, viruses, and mycoplasmas), therefore enhancing safety. Generally, PMF-derived products cost less than 0.1% of cell culture systems of mammals and between 2% and 10% of microbes.

Limitations

Plant biotechnology techniques so far available are unable to accurately regulate the expression level of recombinants in plants in a reliable way, and not all plant species are easily converted. This implies that the quantity of medicine generated by a plant species, or even by individual plant sections, may vary (i.e. seeds, fruit, and leaves). Subsequent generation levels of expression may likewise fluctuate. Owing to this fact, properly quantifying the optimum amount of edible vaccines for children and adults is extremely difficult. Edible vaccines, when administered orally, have the ability to establish immunological tolerance. Finally, the majority of consumed protein will be destroyed throughout the digestive process. Taken together, these limitations severely limit the practical utilization of consumable vaccines (Barzegari et al., 2014). While the research and studies in the field of PMF is quite young, the industry has established conventional and high-throughput purification procedures for over 30 years using cell expression systems of microbes and animals. In comparison, purifying procedures for therapeutic proteins produced from plants vary considerably. While rice, tobacco, and maize have been used to generate PMF products in greenhouses or open fields, each plant species includes its own unique collection of metabolites and proteins. As a result, each PMF platform requires a unique purification technique that is customized to the product and plant production system (Fischer et al., 2015). Factors like plant phenolic compounds, infections, secondary metabolites, fertilizers, and pesticides all contribute to the difficulties of industrially purifying a PMF product. Pollen contamination problems with field crop-based PMF platforms, like rice or maize, present biosafety concerns, since pollen may infect non-transgenic crops grown in regular production of agriculture (MacDonald et al., 2015). Presently, the FDA has a limited regulation regarding the use of food crops to produce recombinant medicinal molecules (FDA, 2015). ProdiGene Inc. (College Station, TX, USA) has begun the commercialization of recombinant maize that generates trypsin on a huge scale. The USDA, on the other hand, determined that remains of a previous ProdiGene study polluted a neighbouring conventional farm. ProdiGene was fined $250,000 and assessed a

$3 million penalty for mishandling the field test. ProdiGene was driven into bankruptcy as a result of the punitive action. This incident was a severe setback for the commercialization of PMF (Strauss, 2012). PMF must create both standard biosafety protocols and purification techniques. Tobacco is an ideal choice for PMF production since it is not a food crop and hence cannot harm other crops (Spök et al., 2008). Genome editing and expression is possibly the best mechanism in tobacco. In 6 months, a transgenic tobacco plant capable of producing the required protein in both leaves and seeds could be produced.

Conclusion

Plants have the capabilities to efficiently generate huge quantities of transgenic proteins at a price that is affordable in comparison to other manufacturing techniques. PMF-based pharmaceutical, topical, nutritional supplement, antibody, and vaccine generation for human and veterinary applications is practicable due to the significant benefits related with plant cell cultures for the manufacturing of commercially beneficial foreign proteins over mammalian cell cultures. Nevertheless, issues concerning biosafety, human health, including the allergic reaction to plant-specific glycans, and other aspects must be addressed appropriately. At the moment, downstream processing and filtration of PMF products is time consuming and expensive. Yet, two key problems need be addressed before completely commercializing plant systems for vaccine development: poor expression of recombinant vaccine proteins and undesired plant-specific glycosylation. Hence, plants may be regarded a promising alternative to the present mammalian-based vaccine manufacturing method. Significant methods must be created to improve the purification process and enable industrial scale manufacturing of PMF-based products viable and cost effective. The proper candidate genes, a compelling industrial need, and an efficient manufacturing method will facilitate the transition from fundamental research on PMF to commercial use.

References

Abiri, R., Valdiani, A., Maziah, M., Shaharuddin, N. A., Sahebi, M., Yusof, Z. N. B., & Talei, D. (2015). A critical review of the concept of transgenic plants: Insights into pharmaceutical biotechnology and molecular farming. *Current Issues in Molecular Biology, 18*(1), 21–42.

Aboul-Ata, A. A. E., Vitti, A., Nuzzaci, M., El-Attar, A. K., Piazzolla, G., Tortorella, C., & Piazzolla, P. (2014). Plant-based vaccines: Novel and low-cost possible route for Mediterranean innovative vaccination strategies. *Advances in Virus Research, 89*, 1–37.

Anterola, A., Shanle, E., Perroud, P. F., & Quatrano, R. (2009). Production of taxa-4 (5), 11 (12)-diene by transgenic *Physcomitrella patens*. *Transgenic Research, 18*(4), 655–660.

Antibodies-online Inc. (2015). Home page. Retrieved from http://www.antibodies-online.com

Arntzen, C. (2015). Plant-made pharmaceuticals: From 'edible vaccines' to Ebola therapeutics. *Plant Biotechnology Journal, 13*(8), 1013–1016.

Barta, A., Sommergruber, K., Thompson, D., Hartmuth, K., Matzke, M. A., & Matzke, A. J. (1986). The expression of a nopaline synthase—Human growth hormone chimaeric gene in transformed tobacco and sunflower callus tissue. *Plant Molecular Biology, 6*(5), 347–357.

Barzegari, A., Saeedi, N., Zarredar, H., Barar, J., & Omidi, Y. (2014). The search for a promising cell factory system for production of edible vaccine: Spirulina as a robust alternate to plants. *Human Vaccines & Immunotherapeutics, 10*(8), 2497–2502.

Baum, A., Fulton, B. O., Wloga, E., Copin, R., Pascal, K. E., Russo, V., et al. (2020). Antibody cocktail to SARS-CoV-2 spike protein prevents rapid mutational escape seen with individual antibodies. *Science, 369*(6506), 1014–1018.

Boehm, R. (2007). Bioproduction of therapeutic proteins in the 21st century and the role of plants and plant cells as production platforms. *Annals of the New York Academy of Sciences, 1102*(1), 121.

Boothe, J., Nykiforuk, C., Shen, Y., Zaplachinski, S., Szarka, S., Kuhlman, P., & Moloney, M. M. (2010). Seed-based expression systems for plant molecular farming. *Plant Biotechnology Journal, 8*(5), 588–606.

Britt, A. B., & May, G. D. (2003). Re-engineering plant gene targeting. *Trends in Plant Science, 8*(2), 90–95.

Castanon, S., Marin, M. S., Martin-Alonso, J. M., Boga, J. A., Casais, R., Humara, J. M., et al. (1999). Immunization with potato plants expressing VP60 protein protects against rabbit hemorrhagic disease virus. *Journal of Virology, 73*(5), 4452–4455.

Catrice, E. V., & Sainsbury, F. (2015). Assembly and purification of polyomavirus-like particles from plants. *Molecular Biotechnology, 57*(10), 904–913.

Chung, I. S., Kim, C. H., Kim, K. I., Hong, S. H., Park, J. H., Kim, J. K., & Kim, W. Y. (2000). Production of recombinant rotavirus VP6 from a suspension culture of transgenic tomato (*Lycopersicon esculentum* Mill.) cells. *Biotechnology Letters, 22*(4), 251–255.

Costa, A. R., Rodrigues, M. E., Henriques, M., Oliveira, R., & Azeredo, J. (2014). Glycosylation: Impact, control and improvement during therapeutic protein production. *Critical Reviews in Biotechnology, 34*(4), 281–299.

Cox, K. M., Sterling, J. D., Regan, J. T., Gasdaska, J. R., Frantz, K. K., Peele, C. G., et al. (2006). Glycan optimization of a human monoclonal antibody in the aquatic plant *Lemna minor*. *Nature Biotechnology, 24*(12), 1591–1597.

Dalsgaard, K., Uttenthal, Å., Jones, T. D., Xu, F., Merryweather, A., Hamilton, W. D., et al. (1997). Plant-derived vaccine protects target animals against a viral disease. *Nature Biotechnology, 15*(3), 248–252.

De Jaeger, G., Scheffer, S., Jacobs, A., Zambre, M., Zobell, O., Goossens, A., et al. (2002). Boosting heterologous protein production in transgenic dicotyledonous seeds using *Phaseolus vulgaris* regulatory sequences. *Nature Biotechnology, 20*(12), 1265–1268.

Decker, E. L., & Reski, R. (2004). The moss bioreactor. *Current Opinion in Plant Biology, 7*(2), 166–170.

Decker, E. L., & Reski, R. (2007). Moss bioreactors producing improved biopharmaceuticals. *Current Opinion in Biotechnology, 18*(5), 393–398.

Decker, E. L., & Reski, R. (2008). Current achievements in the production of complex biopharmaceuticals with moss bioreactors. *Bioprocess and Biosystems Engineering, 31*(1), 3–9.

Dohi, K., Nishikiori, M., Tamai, A., Ishikawa, M., Meshi, T., & Mori, M. (2006). Inducible virus-mediated expression of a foreign protein in suspension-cultured plant cells. *Archives of Virology, 151*(6), 1075–1084.

Donini, M., & Marusic, C. (2019). Current state-of-the-art in plant-based antibody production systems. *Biotechnology Letters, 41*(3), 335–346.

Dreesen, I. A., Charpin-El Hamri, G., & Fussenegger, M. (2010). Heat-stable oral alga-based vaccine protects mice from *Staphylococcus aureus* infection. *Journal of Biotechnology, 145*(3), 273–280.

Eichler-Stahlberg, A., Weisheit, W., Ruecker, O., & Heitzer, M. (2009). Strategies to facilitate transgene expression in *Chlamydomonas reinhardtii*. *Planta, 229*(4), 873–883.
FDA. (2015). FDA guidance for industry: Drugs, biologics, and medical devices derived from bioengineered plants for use in humans and animals (draft guidance). *FDA website*. Retrieved November 27, 2015, from http://www.fda.gov/downloads/AnimalVeterinary/GuidanceComplianceEnforcement/GuidanceforIndustry/ucm055424.pdf.
Fischer, R., Schillberg, S., Hellwig, S., Twyman, R. M., & Drossard, J. (2012). GMP issues for recombinant plant-derived pharmaceutical proteins. *Biotechnology Advances, 30*(2), 434–439.
Fischer, R., Vasilev, N., Twyman, R. M., & Schillberg, S. (2015). High-value products from plants: The challenges of process optimization. *Current Opinion in Biotechnology, 32*, 156–162.
Ford, K., Zeng, W., Heazlewood, J. L., & Bacic, A. (2015). Characterization of protein N-glycosylation by tandem mass spectrometry using complementary fragmentation techniques. *Frontiers in Plant Science, 6*, 674.
Fox, J. L. (2012). First plant-made biologic approved. *Nature Biotechnology, 30*(6), 472–473.
Francisco, J. A., Gawlak, S. L., Miller, M., Bathe, J., Russell, D., Chace, D., et al. (1997). Expression and characterization of bryodin 1 and a bryodin 1-based single-chain immunotoxin from tobacco cell culture. *Bioconjugate Chemistry, 8*(5), 708–713.
Girard, L. S., Fabis, M. J., Bastin, M., Courtois, D., Pétiard, V., & Koprowski, H. (2006). Expression of a human anti-rabies virus monoclonal antibody in tobacco cell culture. *Biochemical and Biophysical Research Communications, 345*(2), 602–607.
Guan, C., Du, X., Wang, G., Ji, J., Jin, C., & Li, X. (2014). Expression of biologically active anti-thrombosis protein lumbrokinase in edible sunflower seed kernel. *Journal of Plant Biochemistry and Biotechnology, 23*(3), 257–265.
Hansen, J., Baum, A., Pascal, K. E., Russo, V., Giordano, S., Wloga, E., et al. (2020). Studies in humanized mice and convalescent humans yield a SARS-CoV-2 antibody cocktail. *Science, 369*(6506), 1010–1014.
Hood, E. E., Witcher, D. R., Maddock, S., Meyer, T., Baszczynski, C., Bailey, M., et al. (1997). Commercial production of avidin from transgenic maize: Characterization of transformant, production, processing, extraction and purification. *Molecular Breeding, 3*(4), 291–306.
Howard, J. A. (2004). Commercialization of plant-based vaccines from research and development to manufacturing. *Animal Health Research Reviews, 5*(2), 243–245.
Huang, T. K., & McDonald, K. A. (2012). Bioreactor systems for in vitro production of foreign proteins using plant cell cultures. *Biotechnology Advances, 30*(2), 398–409.
Huang, Z., Chen, Q., Hjelm, B., Arntzen, C., & Mason, H. (2009). A DNA replicon system for rapid high-level production of virus-like particles in plants. *Biotechnology and Bioengineering, 103*(4), 706–714. https://doi.org/10.1002/bit.22299
Huang, T. K., Plesha, M. A., & McDonald, K. A. (2010a). Semicontinuous bioreactor production of a recombinant human therapeutic protein using a chemically inducible viral amplicon expression system in transgenic plant cell suspension cultures. *Biotechnology and Bioengineering, 106*(3), 408–421.
Huang, Z., Phoolcharoen, W., Lai, H., Piensook, K., Cardineau, G., Zeitlin, L., et al. (2010b). High-level rapid production of full-size monoclonal antibodies in plants by a single-vector DNA replicon system. *Biotechnology and Bioengineering, 106*(1), 9–17.
Kamisugi, Y., Cuming, A. C., & Cove, D. J. (2005). Parameters determining the efficiency of gene targeting in the moss *Physcomitrella patens*. *Nucleic Acids Research, 33*(19), e173–e173.
Keyaerts, E., Vijgen, L., Pannecouque, C., Van Damme, E., Peumans, W., Egberink, H., et al. (2007). Plant lectins are potent inhibitors of coronaviruses by interfering with two targets in the viral replication cycle. *Antiviral Research, 75*(3), 179–187.
Khan, I., Twyman, R. M., Arcalis, E., & Stoger, E. (2012). Using storage organelles for the accumulation and encapsulation of recombinant proteins. *Biotechnology Journal, 7*(9), 1099–1108.
Kircheis, R., Halanek, N., Koller, I., Jost, W., Schuster, M., Gorr, G., et al. (2012). Correlation of ADCC activity with cytokine release induced by the stably expressed, glyco-engineered

humanized Lewis Y-specific monoclonal antibody MB314. *MAbs, 4*(4), 532–541. Taylor & Francis.

Ko, S., Liu, J. R., Yamakawa, T., & Matsumoto, Y. (2006). Expression of the protective antigen (SpaA) in transgenic hairy roots of tobacco. *Plant Molecular Biology Reporter, 24*(2), 251–251.

Koirala, A., Joo, Y. J., Khatami, A., Chiu, C., & Britton, P. N. (2020). Vaccines for COVID-19: The current state of play. *Paediatric Respiratory Reviews, 35*, 43–49.

Koo, M., Bendahmane, M., Lettieri, G. A., Paoletti, A. D., Lane, T. E., Fitchen, J. H., et al. (1999). Protective immunity against murine hepatitis virus (MHV) induced by intranasal or subcutaneous administration of hybrids of tobacco mosaic virus that carries an MHV epitope. *Proceedings of the National Academy of Sciences, 96*(14), 7774–7779.

Kumar, G. S., Ganapathi, T. R., Srinivas, L., Revathi, C. J., & Bapat, V. A. (2005). Secretion of hepatitis B surface antigen in transformed tobacco cell suspension cultures. *Biotechnology Letters, 27*(13), 927–932.

Kumar, G. S., Ganapathi, T. R., Srinivas, L., Revathi, C. J., & Bapat, V. A. (2006). Expression of hepatitis B surface antigen in potato hairy roots. *Plant Science, 170*(5), 918–925.

Kwon, T. H., Seo, J. E., Kim, J., Lee, J. H., Jang, Y. S., & Yang, M. S. (2003). Expression and secretion of the heterodimeric protein interleukin-12 in plant cell suspension culture. *Biotechnology and Bioengineering, 81*(7), 870–875.

Liu, A., & Li, Y. (2020). Antibody responses against SARS-CoV-2 in COVID-19 patients. *Journal of Medical Virology, 93*, 144–148.

Liu, C., Towler, M. J., Medrano, G., Cramer, C. L., & Weathers, P. J. (2009). Production of mouse interleukin-12 is greater in tobacco hairy roots grown in a mist reactor than in an airlift reactor. *Biotechnology and Bioengineering, 102*(4), 1074–1086.

Long, Q. X., Liu, B. Z., Deng, H. J., Wu, G. C., Deng, K., Chen, Y. K., et al. (2020). Antibody responses to SARS-CoV-2 in patients with COVID-19. *Nature Medicine, 26*(6), 845–848.

Ma, J. K., Drake, P. M., & Christou, P. (2003). The production of recombinant pharmaceutical proteins in plants. *Nature Reviews Genetics, 4*(10), 794–805.

MacDonald, J., Doshi, K., Dussault, M., Hall, J. C., Holbrook, L., Jones, G., et al. (2015). Bringing plant-based veterinary vaccines to market: Managing regulatory and commercial hurdles. *Biotechnology Advances, 33*(8), 1572–1581.

Magy, B., Tollet, J., Laterre, R., Boutry, M., & Navarre, C. (2014). Accumulation of secreted antibodies in plant cell cultures varies according to the isotype, host species and culture conditions. *Plant Biotechnology Journal, 12*(4), 457–467.

Manuell, A. L., Beligni, M. V., Elder, J. H., Siefker, D. T., Tran, M., Weber, A., et al. (2007). Robust expression of a bioactive mammalian protein in *Chlamydomonas* chloroplast. *Plant Biotechnology Journal, 5*(3), 402–412.

Martín-Alonso, J. M., Castañón, S., Alonso, P., Parra, F., & Ordás, R. (2003). Oral immunization using tuber extracts from transgenic potato plants expressing rabbit hemorrhagic disease virus capsid protein. *Transgenic Research, 12*(1), 127–130.

Mason, H. S., Lam, D. M., & Arntzen, C. J. (1992). Expression of hepatitis B surface antigen in transgenic plants. *Proceedings of the National Academy of Sciences, 89*(24), 11,745–11,749.

Mathieu-Rivet, E., Kiefer-Meyer, M. C., Vanier, G., Ovide, C., Burel, C., Lerouge, P., & Bardor, M. (2014). Protein N-glycosylation in eukaryotic microalgae and its impact on the production of nuclear expressed biopharmaceuticals. *Frontiers in Plant Science, 5*, 359.

Medina-Bolivar, F., Wright, R., Funk, V., Sentz, D., Barroso, L., Wilkins, T. D., et al. (2003). A non-toxic lectin for antigen delivery of plant-based mucosal vaccines. *Vaccine, 21*(9–10), 997–1005.

Niederkrüger, H., Dabrowska-Schlepp, P., & Schaaf, A. (2014). Suspension culture of plant cells under phototrophic conditions. In H.-P. Meyer & D. R. Schmidhalter (Eds.), *Industrial scale suspension culture of living cells* (pp. 259–292). Wiley.

Nölke, G., Fischer, R., & Schillberg, S. (2003). Production of therapeutic antibodies in plants. *Expert Opinion on Biological Therapy, 3*(7), 1153–1162.

Ou, J., Guo, Z., Shi, J., Wang, X., Liu, J., Shi, B., et al. (2014). Transgenic rice endosperm as a bioreactor for molecular pharming. *Plant Cell Reports, 33*(4), 585–594.

Paul, M., & Ma, J. K. C. (2011). Plant-made pharmaceuticals: Leading products and production platforms. *Biotechnology and Applied Biochemistry, 58*(1), 58–67.

Peyret, H., & Lomonossoff, G. P. (2015). When plant virology met agrobacterium: The rise of the deconstructed clones. *Plant Biotechnology Journal, 13*(8), 1121–1135.

Phoolcharoen, W., Bhoo, S. H., Lai, H., Ma, J., Arntzen, C. J., Chen, Q., & Mason, H. S. (2011). Expression of an immunogenic bola immune complex in *Nicotiana benthamiana. Plant Biotechnology Journal, 9*(7), 807–816. https://doi.org/10.1111/j.1467-7652.2011.00593.x

PhycoBiologics Inc. (n.d.) Home Page. Retrieved from https://www.phycotransgenics.com.

Popov, S. V., Golovchenko, V. V., Ovodova, R. G., Smirnov, V. V., Khramova, D. S., Popova, G. Y., & Ovodov, Y. S. (2006). Characterisation of the oral adjuvant effect of lemnan, a pectic polysaccharide of *Lemna minor* L. *Vaccine, 24*(26), 5413–5419.

PromoKine Home Page. (2015). Retrieved from https://www.promokine.info

Qiu, X., Wong, G., Audet, J., Bello, A., Fernando, L., Alimonti, J. B., et al. (2014). Reversion of advanced Ebola virus disease in nonhuman primates with ZMapp. *Nature, 514*(7520), 47–53.

Rasala, B. A., & Mayfield, S. P. (2011). The microalga Chlamydomonas reinhardtii as a platform for the production of human protein therapeutics. *Bioengineered Bugs, 2*(1), 50–54.

Raven, N., Rasche, S., Kuehn, C., Anderlei, T., Klöckner, W., Schuster, F., et al. (2015). Scaled-up manufacturing of recombinant antibodies produced by plant cells in a 200-L orbitally-shaken disposable bioreactor. *Biotechnology and Bioengineering, 112*(2), 308–321.

Ren, W., Sun, H., Gao, G. F., Chen, J., Sun, S., Zhao, R., et al. (2020). Recombinant SARS-CoV-2 spike S1-Fc fusion protein induced high levels of neutralizing responses in nonhuman primates. *Vaccine, 38*(35), 5653–5658.

Rensing, S. A., Fritzowsky, D., Lang, D., & Reski, R. (2005). Protein encoding genes in an ancient plant: Analysis of codon usage, retained genes and splice sites in a moss, Physcomitrella patens. *BMC Genomics, 6*(1), 1–13.

Rensing, S. A., Lang, D., Zimmer, A. D., Terry, A., Salamov, A., Shapiro, H., & Boore, J. L. (2008). The Physcomitrella genome reveals evolutionary insights into the conquest of land by plants. *Science, 319*(5859), 64–69.

Reski, R., Parsons, J., & Decker, E. L. (2015). Moss-made pharmaceuticals: From bench to bedside. *Plant Biotechnology Journal, 13*(8), 1191–1198.

Rup, B., Alon, S., Amit-Cohen, B. C., Brill Almon, E., Chertkoff, R., Tekoah, Y., & Rudd, P. M. (2017). Immunogenicity of glycans on biotherapeutic drugs produced in plant expression systems—The taliglucerase alfa story. *PLoS One, 12*(10), e0186211.

Sack, M., Rademacher, T., Spiegel, H., Boes, A., Hellwig, S., Drossard, J., & Fischer, R. (2015). From gene to harvest: Insights into upstream process development for the GMP production of a monoclonal antibody in transgenic tobacco plants. *Plant Biotechnology Journal, 13*(8), 1094–1105.

Sainsbury, F., & Lomonossoff, G. P. (2014). Transient expressions of synthetic biology in plants. *Current Opinion in Plant Biology, 19*, 1–7.

Santi, L. (2009). Plant derived veterinary vaccines. *Veterinary Research Communications, 33*(1), 61–66.

Scheller, J., Leps, M., & Conrad, U. (2006). Forcing single-chain variable fragment production in tobacco seeds by fusion to elastin-like polypeptides. *Plant Biotechnology Journal, 4*(2), 243–249.

Schillberg, S., Raven, N., Fischer, R. M., Twyman, R., & Schiermeyer, A. (2013). Molecular farming of pharmaceutical proteins using plant suspension cell and tissue cultures. *Current Pharmaceutical Design, 19*(31), 5531–5542.

Sijmons, P. C., Dekker, B. M., Schrammeijer, B., Verwoerd, T. C., Van Den Elzen, P. J., & Hoekema, A. (1990). Production of correctly processed human serum albumin in transgenic plants. *Bio/Technology, 8*(3), 217–221.

Smith, M. L., Mason, H. S., & Shuler, M. L. (2002). Hepatitis B surface antigen (HBsAg) expression in plant cell culture: Kinetics of antigen accumulation in batch culture and its intracellular form. *Biotechnology and Bioengineering, 80*(7), 812–822.

Sojikul, P., Buehner, N., & Mason, H. S. (2003). A plant signal peptide–hepatitis B surface antigen fusion protein with enhanced stability and immunogenicity expressed in plant cells. *Proceedings of the National Academy of Sciences, 100*(5), 2209–2214.

Spiegel, H., Stöger, E., Twyman, R. M., & Buyel, J. F. (2018). Current status and perspectives of the molecular farming landscape. In A. R. Kermode & L. Jiang (Eds.), *Molecular pharming: Applications, challenges and emerging areas* (pp. 3–23). Wiley.

Spök, A., Karner, S., Stein, A. J., & Rodríguez-Cerezo, E. (2008). Plant molecular farming. Opportunities and challenges. *JRC Scientific and Technical Reports.*

Stomp, A. M. (2005). The duckweeds: A valuable plant for biomanufacturing. *Biotechnology Annual Review, 11*, 69–99.

Strauss, D. M. (2012). Liability for genetically modified food: Are GMOs a tort waiting to happen? *The SciTech Lawyer, 9*(2), 8–13.

Streatfield, S. J., Jilka, J. M., Hood, E. E., Turner, D. D., Bailey, M. R., Mayor, J. M., et al. (2001). Plant-based vaccines: Unique advantages. *Vaccine, 19*(17–19), 2742–2748.

Sun, M., Qian, K., Su, N., Chang, H., Liu, J., & Shen, G. (2003). Foot-and-mouth disease virus VP1 protein fused with cholera toxin B subunit expressed in *Chlamydomonas reinhardtii* chloroplast. *Biotechnology Letters, 25*(13), 1087–1092.

Tan, Y., Wang, K. Y., Wang, N., Li, G., & Liu, D. (2014). Ectopic expression of human acidic fibroblast growth factor 1 in the medicinal plant, *Salvia miltiorrhiza*, accelerates the healing of burn wounds. *BMC Biotechnology, 14*(1), 1–10.

Teng, C., Qin, S., Liu, J., Yu, D., Liang, C., & Tseng, C. (2002). Transient expression of lacZ in bombarded unicellular green alga *Haematococcus pluvialis*. *Journal of Applied Phycology, 14* (6), 497–500.

Tortorici, M. A., Beltramello, M., Lempp, F. A., Pinto, D., Dang, H. V., Rosen, L. E., et al. (2020). Ultrapotent human antibodies protect against SARS-CoV-2 challenge via multiple mechanisms. *Science, 370*(6519), 950–957.

Twyman, R. M., Schillberg, S., & Fischer, R. (2005). Transgenic plants in the biopharmaceutical market. *Expert Opinion on Emerging Drugs, 10*(1), 185–218.

Twyman, R. M., Stoger, E., Schillberg, S., Christou, P., & Fischer, R. (2003). Molecular farming in plants: Host systems and expression technology. *Trends in Biotechnology, 21*(12), 570–578.

U.S. Department of Health & Human Services. (n.d.). *Partnership advances experimental Ebola drug*. Retrieved from http://www.hhs.gov/about/news/2015/07/20/hhspartnership-advances-experimental-ebola-drug.html

Vasilev, N., Smales, C. M., Schillberg, S., Fischer, R., & Schiermeyer, A. (2016). Developments in the production of mucosal antibodies in plants. *Biotechnology Advances, 34*(2), 77–87.

Vermij, P., & Waltz, E. (2006). USDA approves the first plant-based vaccine. *Nature Biotechnology, 24*(3), 234.

Vunsh, R., Li, J., Hanania, U., Edelman, M., Flaishman, M., Perl, A., et al. (2007). High expression of transgene protein in Spirodela. *Plant Cell Reports, 26*(9), 1511–1519.

Walsh, G. (2018). Biopharmaceutical benchmarks 2018. *Nature Biotechnology, 36*, 1136–1145. https://doi.org/10.1038/nbt.4305

Wang, K. Y., Tull, L., Cooper, E., Wang, N., & Liu, D. (2013). Recombinant protein production of earthworm lumbrokinase for potential antithrombotic application. Evidence-Based Complementary and Alternative Medicine, 2013:783971.

Wang, X., Brandsma, M., Tremblay, R., Maxwell, D., Jevnikar, A. M., Huner, N., & Ma, S. (2008). A novel expression platform for the production of diabetes-associated autoantigen human glutamic acid decarboxylase (hGAD65). *BMC Biotechnology, 8*(1), 1–13.

Xu, J., Tan, L., Goodrum, K. J., & Kieliszewski, M. J. (2007). High-yields and extended serum half-life of human interferon α2b expressed in tobacco cells as arabinogalactan-protein fusions. *Biotechnology and Bioengineering, 97*(5), 997–1008.

Yamamoto, Y. T., Rajbhandari, N., Lin, X., Bergmann, B. A., Nishimura, Y., & Stomp, A. M. (2001). Genetic transformation of duckweed *Lemna gibba* and *Lemna minor*. *In Vitro Cellular & Developmental Biology-Plant, 37*(3), 349–353.

Yusibov, V. (2002). Expression in plants and immunogenicity of plant virus-based experimental rabies vaccine. *Vaccine, 20*(25–26), 3155–3164. https://doi.org/10.1016/S0264-410X(02)00260-8

Zhong, Q., Gu, Z., & Glatz, C. E. (2006). Extraction of recombinant dog gastric lipase from transgenic corn seed. *Journal of Agricultural and Food Chemistry, 54*(21), 8086–8092.

Zhou, J. Y., Cheng, L. Q., Zheng, X. J., Wu, J. X., Shang, S. B., Wang, J. Y., & Chen, J. G. (2004a). Generation of the transgenic potato expressing full-length spike protein of infectious bronchitis virus. *Journal of Biotechnology, 111*(2), 121–130.

Zhou, T., Wang, H., Luo, D., Rowe, T., Wang, Z., Hogan, R. J., et al. (2004b). An exposed domain in the severe acute respiratory syndrome coronavirus spike protein induces neutralizing antibodies. *Journal of Virology, 78*(13), 7217–7226.

Zimran, A., Gonzalez-Rodriguez, D. E., Abrahamov, A., Cooper, P. A., Varughese, S., Giraldo, P., et al. (2018). Long-term safety and efficacy of taliglucerase alfa in pediatric Gaucher disease patients who were treatment-naive or previously treated with imiglucerase. *Blood Cells, Molecules, and Diseases, 68*, 163–172.

Chapter 11
CRISPR/Cas-Mediated Genome Engineering for Abiotic Stress Resilience in Plants

Deepu Pandita

Abstract The world population will terrifically grow up to 9–11 billion by 2050, and hence the food production needs a massive intensification. The abiotic stresses such as water deficiency, salinity, heat, cold, flooding, and heavy metal pollution have considerably limited the worldwide crop yield and sustainable production of agriculture and pose paramount threat to global food security. This warning is advancing in lieu of upcoming global warming and climate changes. The new-generation revolutionary tools of CRISPR/Cas metabolic engineering platforms are prospective tools to mitigate complex machinery of tolerance to abiotic stress in plants and in developing novel climate smart abiotic stress-tolerant crop varieties. CRISPR/Cas facilitated genome editing enables abiotic stress tolerance through knockout, knock in, multiplex editing for broad spectrum tolerance, promoter enhancement, and precise base editing.

Keywords CRISPR/Cas · Genome engineering · Abiotic stress resilience · Food security

Introduction

World population is predicted to grow to almost 9–11 billion by 2050 (Röös et al., 2017). To feed the extra hungry mouths in the upcoming scenario of climate change as well, global agricultural challenge is to more than double agricultural productivity by 2050. Majority of plants face a variety of abiotic stress or adverse environmental conditions detrimental to plants. These include water deficiency, extreme low and high temperatures, salinity, mineral toxicity, oxidative stress, etc. (Chang et al., 2020; Gong et al., 2020). The rates of abiotic stresses will accelerate in the future because of global warming and climate changes (Mahalingam, 2015). Abiotic stress activates changes in cellular gene expression, metabolism, growth yield of crop

---

D. Pandita (✉)
Government Department of School Education, Jammu, Jammu and Kashmir, India

T. Aftab, K. R. Hakeem (eds.), *Metabolic Engineering in Plants*,
https://doi.org/10.1007/978-981-16-7262-0_11

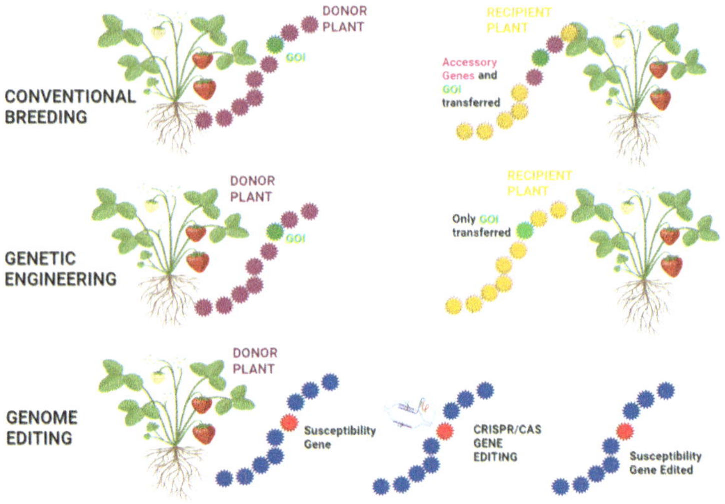

Fig. 11.1 Conventional breeding, genetic engineering, and genome editing

plants, and in turn insufficient food supply and food production (Gupta et al., 2013a, 2013b; Wani & Sah, 2014). Improvement of characteristics for crafting abiotic stress-resistant crops is a persistent sequence at universal level. Conventional cross-breeding and mutation breeding are grueling, laborious, expensive methodology and take 8–10 years for multiple back crossing (Kharkwal et al., 2004; Bado et al., 2015) and transfer undesirable genes besides GENE OF INTEREST (GOI) from donor to progeny generations (Fig. 11.1). Molecular marker technology characterizes stressed plants, but production of stress-tolerant plants because of complex inheritance, genetic, and environment interactions is not possible (Rao et al., 2016; Bhat et al., 2016). Adapting unconventional genome engineering, genetic manipulation, and biotechnological technologies is required for the sustainable agriculture and climate-smart designer crops (Lobell & Gourdji, 2012). Genetic engineering uses price-tagged recombinant DNA technology for transgenic breeding which has social and ethical issues due to negative effects of genetic modification and biosafety sanction limitations (Robinson, 1999; Prado et al., 2014; Raman, 2017). But in genetically modified organisms with desired traits, only external GENE OF INTEREST (GOI) (Anwar & Kim, 2020) (Fig. 11.1) gets transmitted to recipient crop. Novel genome editing technologies enabling targeted gene manipulation have transformed abiotic stressed crops. CRISPR/Cas (Clustered Regularly Interspaced Short Palindromic Repeat)/CRISPR-associated protein editing toolkit is simple, user-friendly, low budget, strong, rapid (takes approximately 4–6 years), highly competent, precise, and site-specific (Zhang et al., 2018). CRISPR/Cas includes a complex of Cas9 endonuclease (creates double-stranded breaks (DSBs) or "molecular scissors" and single guide RNA (sgRNA) with target specificity. DSBs may be monitored by the non-homologous end joining (NHEJ) and homology directed repair (HDR) DNA repair mechanism of plant system and generate mutants (Chen & Gao, 2014; Voytas & Gao, 2014). Quite a lot of chapters shed light on the connotation of genome editing technologies of CRISPR/Cas in plants (Pandita, 2021a, 2021b, 2021c). This chapter focuses on different types of CRISPR/Cas, mode of action of CRISPR/Cas, and distinctive highpoints on the CRISPR/Cas-mediated genome engineering tool in abiotic stress resilience or improvement of plants.

Cataloguing of CRISPR/Cas

CRISPR/Cas systems are currently catalogued into 2 main classes (Class 1 and Class 2) (Fig. 11.2), 6 types (type I–VI), and 33 subtypes on the basis of effector modules with unique signatures and *cas* genes structure, *cas* operon organization, function, and phylogeny of conserved Cas proteins. Class 1 CRISPR/Cas systems have heteromeric multiprotein subunit effector modules with many Cas proteins, viz., Cas3, Cas5–Cas8, Cas10, and Cas11. Class 2 systems involve solitary, bulky, multi-domain effector module, viz., Cas9, Cas12, or Cas13 (Makarova et al., 2015; Makarova et al., 2018; Makarova et al., 2020). Class 1 divides into type I, type III, and type IV and 16 subtypes and is far more abundant bacteria and archaea. The class 2 includes type II, type V, and type VI and 17 subtypes (Makarova et al., 2020). The subtypes have different loci organizations/architecture and encode subtype-specific Cas proteins (Makarova et al., 2020). The types I, II, and V target DNA strands. Type VI targets RNA strands and type III targets DNA as well as RNA (Samai et al., 2015; Koonin et al., 2017). Class 1 and class 2 variants often encoded by mobile

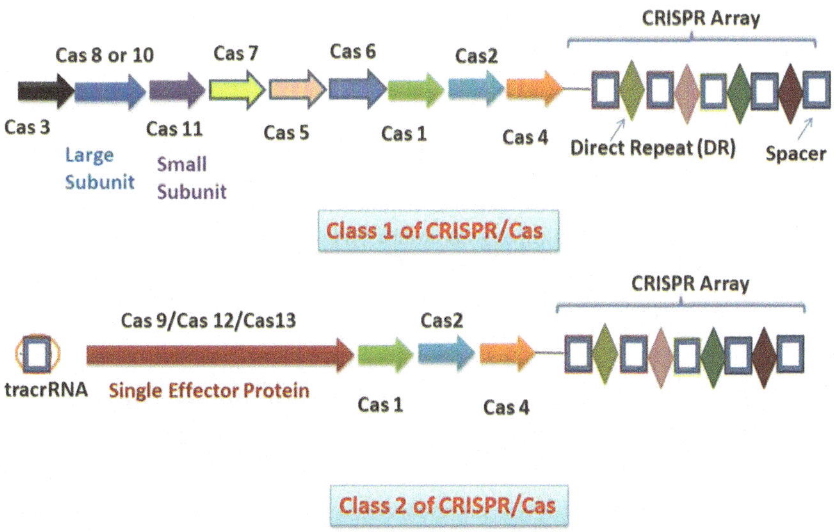

Fig. 11.2 Generic organization of Class 1 and Class 2 of CRISPR/Cas loci

genetic elements show nonexistence of targeted cleavage and do functions different from adaptive immunity (Makarova et al., 2020). Class 2 includes Cas9, Cas12, and Cas13 endonucleases with profound use in genome editing (Wu et al., 2018). There are some exceptions to this complex modular structured and flexible Cas loci-based classification. For example, Cas13d targets RNA instead of DNA (Wu et al., 2018).

Plant Genome Engineering with CRISPR/Cas Systems

CRISPR/Cas editing toolkit has two components of Cas endonuclease and single guide RNA (sgRNA). Cas-sgRNA complex binds to DNA. The sgRNA comprising of CRISPR RNA (crRNA) and trans-activating crRNA (tracrRNA) recognizes and guides Cas to target loci. Cas9 has HNH and RuvC-like endonuclease domains targeting two DNA strands at loci of interest (Hsu et al., 2014; Li et al., 2013; Malzahn et al., 2017). Cas9 endonuclease DNA targeting is based on Protospacer Adjacent Motif (PAM, 5′-NGG-3′) present directly downstream of 20-nt target DNA sequence (Zhang et al., 2018). PAM sequence can vary in Cas9 variants and Cas9 orthologs (Leenay & Beisel, 2017). The Cas9 endonuclease cleaves at 3-nt upstream of PAM within DNA target site, generating blunt ended double-strand breaks (DSBs) (Jiang & Doudna, 2017). CRISPR/Cas12 endonuclease of class 2 type V lacks HNH domain, needs T-rich PAM, and generates staggered cuts for the plant genetic manipulations (Zetsche et al., 2015; Stella et al., 2017). Site-specific double-strand DNA break (DSBs) damage activates inherent mechanisms of DNA repair pathways of cell through non-homologous end joining (NHEJ) or homology-directed repair (HDR) (Manghwar et al., 2019). NHEJ is most dominant, quick, and active error-prone DNA repair pathway. NHEJ repairs DSBs without any homologous template but generates point and frame shift mutations and in turn genome modifications. Classical NHEJ bring about insertions or deletions (indels) (Fig. 11.3), while microhomology based alternative NHEJ generates deletion of DNA resulting in knockout of gene or silencing of gene at target sites (Jiang & Doudna, 2017; Rong & Golic, 2000; Mushtaq et al., 2018; Hasley et al., 2021). Less efficient HDR on which precise genome engineering is based, repairs site through knock-in or integration of homologous donor DNA sequence) (Fig. 11.3) through homologous recombination (Char et al., 2017).

CRISPR/Cas-Mediated Genome Engineering for Abiotic Stress Resilience in Plants

Plant trait upgrading by CRISPR/Cas9-mediated deletion, insertion, and substitution of genes has been very proficient. CRISPR/Cas9 5′-NGG-3′ PAM sequence specificity is bottleneck. Multiple Cas variants with different PAM sequence specificities

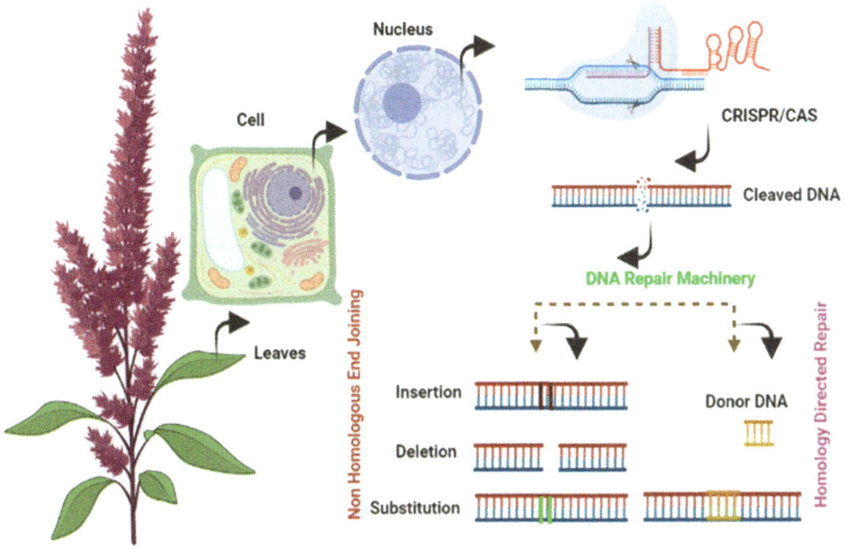

Fig. 11.3 Mode of action of CRISPR/Cas

are available, for instance, SpCas9-NAG and xCas9 used in *Oryza sativa* (Meng et al., 2018; Wrighton, 2018; Hu et al., 2018). CRISPR/Cas12a endonuclease identifies T-rich (5′-TTTN-3′) PAM sequence. CRISPR/Cas12a dominates target constraints of CRISPR/Cas9 and produces 4–5 nucleotide cohesive over hanged DSBs (Zetsche et al., 2015). The Cas12a variants facilitate multiplex gene editing and recognize TYCV PAM sequences (Li et al., 2018a, 2018b). Plants edited by CRISPR/Cas display limitations of off-target mutations and incompetent regeneration capability for crop improvement. Edited pollen grains and immature embryos can overcome these blockades (Kelliher et al., 2019; Nandy et al., 2019).

CRISPR/Cas9-Based Genome Editing (GE)

CRISPR/Cas9-based genome editing framework initially described in *Arabidopsis thaliana* L., *Oryza sativa* L., and *Nicotiana benthamiana*, besides *Triticum aestivum* L., *Solanum lycopersicum* L., *Glycine max* L., *Zea mays* L., *Hordeum vulgare* L., and Physcomitrella patens, assisted in designing of abiotic stress resilient plants (Li et al., 2013; Shan et al., 2013; Nekrasov et al., 2013; Osakabe & Osakabe, 2017; Hussain et al., 2018; Ahmad et al., 2021).

Zea Mays L.

ARGOS8 negatively regulates ethylene response, and CRISPR/Cas9-enhanced ARGOS8 expression increased drought tolerance and yield in *Zea mays* (Shi et al., 2017).

Arabidopsis Thaliana L.

The truncated gRNAs (tru-gRNAs) derived from a promoter of AtEF1 and Cas9 changes in OST2/AHA1 enhanced stomata response in *Arabidopsis thaliana* (Osakabe & Osakabe, 2017).

CRISPR/dCas9HAT genome-editing activated endogenous abscisic acid responsive element binding protein 1/ABRE binding factor 2 (ABRE1/ABF2) promoter and improved tolerance to water deficiency in *Arabidopsis thaliana* L. (Paixão et al., 2019). OST2 (for stomatal movement) overexpression in germ line cells of stressed *Arabidopsis* showed inheritable stress-tolerance (Osakabe et al., 2016).

Oryza Sativa L.

OsRR22, OsSIT1, and OsNAC041 genes increased salt tolerance in *Oryza sativa* L. (Li et al., 2014; Zhang et al., 2019a, 2019b; Bo et al., 2019). MYB30, *OsPIN5b*, *GS3* OsCOLD1, TIFY1a, and TIFY1b genes increased cold resistance in *Oryza sativa* L. (Zeng et al., 2020; Ma et al., 2015; Huang et al., 2017). Sub1A, SK1, and SK2 enhanced resistance to flooding (Fukao et al., 2006; Xu et al., 2006; Hattori et al., 2009). DRO1 enhanced resistance to drought (Uga et al., 2011, 2013). Oryza sativa phosphate transporter 4 (OsPT4) knockout reduced arsenic uptake from soil (Ye et al., 2017). Natural resistance-associated macrophage protein 5 (OsNRAM5) of *Oryza sativa* L. knockout reduces uptake of cadmium from soil (Tang et al., 2017a). *Oryza sativa* PARAQUAT TOLERANCE 3 (OsPQT3) knockout mutants exhibited increased tolerance to salt and oxidative stresses (Alfatih et al., 2020). Abscisic acid receptor gene (PYL1/4/6) knockouts show substantial heat tolerance (Miao et al., 2018). Inactivation of OsHAK1, a K+ transporter, leads to the development of trivial cesium-accumulating rice (Nieves-Cordones et al., 2017). The knockout mutants of OsARM1 showed improved tolerance to arsenic (Wang et al., 2017a, 2017b). Metal transporter OsNramp5 gene knockout exhibited low accumulation of the cadmium (Tang et al., 2017a).

Solanum Lycopersicum L.

Editing of NPRI, SINPR1 (knockout), and mitogen-activated protein kinase 3 (SIMAPK3) improved drought sensitivity in *Solanum lycopersicum*

L. (Erpen-Dalla Corte et al., 2019; Li et al., 2019; Wang et al., 2017c, 2017d). Knockout of *SlMAPK3*, a negative regulator of thermotolerance enhances tolerance to heat stress, and *slmapk3* mutants exhibit more tolerance to high temperature stress (Yu et al., 2019)

Tetraselmis Suecica

Editing of heat shock protein 90 (HSP90) increased tolerance to high temperature in *Tetraselmis suecica* (Xu et al., 2020).

Glycine max L.

Drb2a and Drb2b editing increased salinity and water deficiency tolerance in *Glycine max* L. SAPK1 and SAPK2 enhanced salinity resistance in *Glycine max* L. (Lou et al., 2017; Curtin et al., 2018).

Carica Papaya L.

CpDreb2 editing increased resistance to water deficiency, heat, and cold stress in *Carica papaya* L. (Arroyo-Herrera et al., 2016). CpRap2.4a and CpRap2.4b editing enhanced resistance to heat and cold stress in *Carica papaya* L. (Figueroa-Yañez et al., 2016).

Gossypium Spp.

GhPIN1–3 and GhPIN2 and GhRDL1 editing enhanced resistance to drought in *Gossypium* spp. (Dass et al., 2017; He et al., 2017).

Musa Acuminata Colla

Editing of MaSWEET-4d, MaSWEET-1b, MaSWEET-4b, MaAPS1, and MaAPL3 enhanced resistance to cold and salt in *Musa acuminata* Colla (Miao et al., 2017a; Miao et al., 2017b).

CRISPR/Cas12-Based Genome Editing (GE)

CRISPR/Cas12a has been proficiently used in *Arabidopsis thaliana* L., *Oryza sativa* L., *Nicotiana tabacum*, and *Glycine max* L. (Endo et al., 2016; Tang et al., 2017b;

Tang et al., 2018). CRISPR/Cas12-mediated editing of OsEPFL9 controls density of stomata in leaves of *Oryza sativa* L. (Yin et al., 2019). CRISPR/Cas12-mediated editing of OsDEP1, OsEPFL9, and OsPDS increases tolerance to abiotic stress in *Oryza sativa* L. (Zhong et al., 2018). In *Solanum lycopersicum* L., CRISPR/Cas12 mediated editing of HKT1; 2 HDR enhanced tolerance to multiple stresses (Vu et al., 2020). Some other reports of CRISPR/Cas-based genome editing of plants for improvement and enhancement of abiotic stress resilience are summarized in Table 11.1.

Conclusion

Climatic changes and snowballing population are the biggest threats and new challenges to food security. Plant breeding, tools of genetic engineering, and genome editing mitigate plant abiotic stress. However, only CRISPR/Cas-based genome editing toolkit is a revolutionary technology for crafting the desirable genes for climate smart crops to ensure worldwide nutritional safety and sustainable agriculture and further in probing the plant biology in an efficient and precise mode within short duration. Many CRISPR/Cas9-based success stories on abiotic stress in food crop plants are discussed in this chapter.

Table 11.1 CRISPR/Cas-mediated genome engineering for plant abiotic stress resilience

Plant species	Target gene	Abiotic stress traits improved	Reference/s
CRISPR/Cas9			
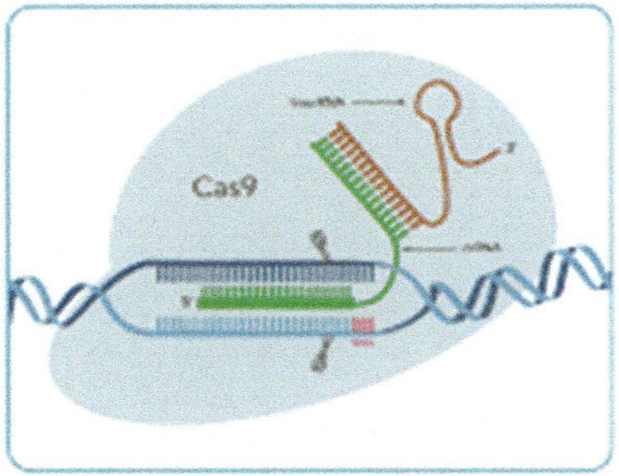			
Arabidopsis thaliana L.	miR169a	Negative factor of drought tolerance	Zhao et al. (2016)
	Vacuolar H+-pyrophosphatase (AVP1)	Transcription factor and drought stress tolerance	Park et al. (2017)
	AREB1	ABA signaling and drought stress tolerance	Paixão et al. (2019)
Carica papaya L.	CpDreb2	Resistance to drought, heat, and cold	Arroyo-Herrera et al. (2016)
	CpRap2.4a and CpRap2.4b	Resistance to heat and cold	Figueroa-Yañez et al. (2016)
Gossypium spp.	GhPIN1–3 and GhPIN2	Resistance to drought	He et al. (2017)
	GhRDL1	Resistance to drought	Dass et al. (2017)
Hordeum vulgare L.	ALMT	High Al3+	Zhou et al. (2014)
Manihot esculenta Crantz	MeKUPs	Resistance to salt, osmosis, cold, drought	Ou et al. (2018)

(continued)

Table 11.1 (continued)

Plant species	Target gene	Abiotic stress traits improved	Reference/s
	MeMAPKKK	Resistance to drought	Ye et al. (2017)
Musa acuminata Colla	MaSWEET-4d, MaSWEET-1b, and MaSWEET-4b	Resistance to cold and salt	Miao et al. (2017a)
	MaAPS1 and MaAPL3	Resistance to cold and salt	Miao et al. (2017b)
Oryza sativa L.	Sub1A, SK1, and SK2	Resistance to flooding	Xu et al. (2006), Fukao et al. (2006), Hattori et al. (2009)
	DRO1	Resistance to drought	Uga et al. (2011, 2013)
	NRAT1	High Al3+	Li et al. (2014)
	PSTOL1 at the Pup1 locus	Low pi	Gamuyao et al. (2012), Chin et al. (2011)
	Oryza sativa phosphate transporter 4 (OsPT4)	Absence of OsPT4 activity reduces arsenic uptake from soil in rice plant	Ye et al. (2017)
	Natural resistance-associated macrophage proteins 5 (OsNRAM5) of *Oryza sativa*	Absence of OsNRAMP5 reduces the uptake of cadmium from soil	Tang et al. (2017a)
	Arsenite-responsive MYB1 of *Oryza sativa* (OsARM1)	Plant lacking the activity of OsARM1 shows tolerance to arsenic stress	Wang et al. (2017a, 2017b)
	OsHAK1	Inactivation of OsHAK1 reduces the uptake of 137Cs + from soil	Nieves-Cordones et al. (2017)
	OsNAC14	OsNAC14 knockout reduces expression of drought tolerance OsRAD51A1, OsFbox341, Piz-t, DR, and OsPAE1	Shim et al. (2018)
	OsMYB30	Cold tolerance	Zeng et al. (2020)

	GTγ-2 transcription factor	Salinity tolerance	Liu et al. (2020)
	PIL14 transcription factor	Salinity tolerance	Mo et al. (2020)
	PQT3 Ubiquitin ligase	Salinity tolerance	Alfatih et al. (2020)
	BGE3 cytokinin transport	Salinity tolerance	Yin et al. (2020)
	DST Zinc finger transcription factor	Salinity tolerance	Santosh Kumar et al. (2020)
	FLN2	Salinity tolerance	Chen et al. (2019)
	RR9, RR10	Signaling of cytokinin and salinity tolerance	Wang et al. (2019)
	DOF15 transcription factor	Salinity tolerance	Qin et al. (2019)
	NCA1a, NCA1b	Salinity tolerance	Liu et al. (2019)
	RR22 transcription factor	Salinity tolerance	Liu et al. (2019)
	NAC041 transcription factor	Salinity tolerance	Bo et al. (2019)
	SAPK1, SAPK2	SAPK1, SAPK2 are part of ABA pathway regulator and improve salinity tolerance	Lou et al. (2018)
	BBS1	Chaperone-mediated signaling and salinity tolerance	Zeng et al. (2018)
	OsMIR528 positive regulator of salt stress	Salinity tolerance	Zhou et al. (2017)
	SAPK2 primary mediator in ABA signaling SAPK2 gene in imparting drought tolerance	SAPK2 knockout mutants were susceptible to drought and oxidative Stress	Lou et al. (2017)
	OsANN3; a gene encoding Ca 2+-dependent phospholipid binding protein	Cold tolerance	Shen et al. (2017)
	Drought and salt tolerance (OsDST) gene	MTU1010 dst mutant showed drought stress tolerance	Ganie et al. (2021), Santosh Kumar et al. (2020)

(continued)

Table 11.1 (continued)

Plant species	Target gene	Abiotic stress traits improved	Reference/s
	Ann3	Cold tolerance	Romero and Gatica-Arias (2019)
	OTS1 Salt stress response regulator	Salt stress response regulator C. Zhang, Srivastava, and Salinity tolerance	Zhang et al. (2019a, 2019b)
Phoenix dactylifera L.	Pdpcs and Pdmt	Cadmium and chromium toxicity	Chaâbene et al. (2018)
Saccharum officinarum L.	ScAPX6	Resistance to copper	Liu et al. (2017)
	ScNsLTP	Resistance to drought and chilling	Chen et al. (2017)
	Mitogen-activated protein kinase 3 (SlMAPK3) in *Solanum lycopersicum*	SlMAPK3 knockout mutants are susceptible to drought stress	Wang et al. (2017d)
	Annexin ANN3	Stress response	Shen et al. (2017)
	NPR1	Drought tolerance	Li et al. (2019)
	SP5G, SP	Regulates day length sensitivity and salinity stress tolerance	Li et al. (2018a, 2018b)
	GGP1	Vitamin C synthesis and salinity stress tolerance	Li et al. (2018a, 2018b)
	WUS	Represses and activates gene transcription in shoot apical meristem and salinity stress tolerance	Li et al. (2018a, 2018b)
	CLV3	Regulates shoot and floral meristem development and salinity stress tolerance	Li et al. (2018a, 2018b), Van Eck et al. (2019)
	HSA1 (heat-stress sensitive albino 1	Sensitivity to heat	Qiu et al. (2018)
Triticum aestivum L. and *Hordeum vulgare* L.	VRN1 at the FR1 locus and CBFs at the FR2 locus	Resistance to low temperature	Dhillon et al. (2012), Stockinger et al. (2007), Knox

			et al. (2010), Francia et al. (2007)
Triticum aestivum L. and *Oryza sativa* L.	TaHKT1;5	Resistance to salinity	Dubcovsky et al. (1996)
Triticum aestivum L.	Dehydration Responsive Element Binding Protein 2 (TaDREB2) and Ethylene Responsive Factor 3 (TaERF3)	Improved tolerance to drought	Kim et al. (2018)
Zea mays L.	MATE1	High Al3+	Maron et al. (2013)
CRISPR/Cas 12 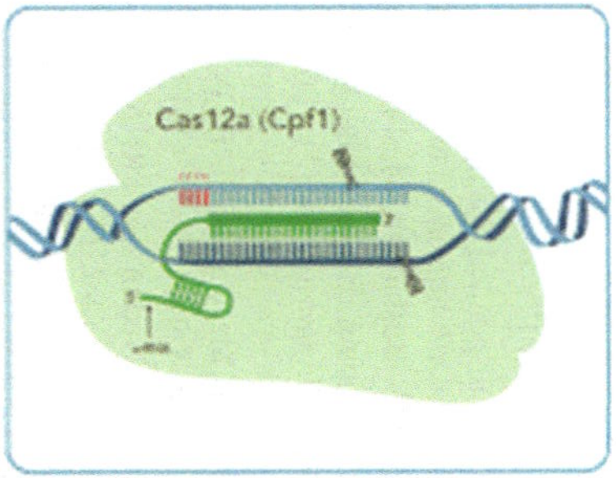			
Oryza sativa L.	OsEPFL9	Regulates stomatal density in leaves	Yin et al. (2019)
	OsDEP1, OsPDS, and OsEPFL9	Abiotic stress tolerance	Zhong et al. (2018)
Solanum lycopersicum L.	HKT1;2 HDR	Multiple-stress tolerance	Vu et al. (2020)

References

Ahmad, S., Sheng, Z., Jalal, R. S., Tabassum, J., Ahmed, F. K., Hu, S., Shao, G., Wei, X., Abd-Elsalam, K. A., Hu, P., et al. (2021). CRISPR–Cas technology towards improvement of abiotic stress tolerance in plants. In *CRISPR and RNAi Systems* (pp. 755–772). Elsevier.

Alfatih, A., Wu, J., Jan, S. U., Zhang, Z. S., Xia, J. Q., & Xiang, C. B. (2020). Loss of rice PARAQUAT TOLERANCE 3 confers enhanced resistance to abiotic stresses and increases grain yield in field. *Plant, Cell and Environment, 43*(11), 2743–2754. https://doi.org/10.1111/pce.13856

Anwar, A., & Kim, J. K. (2020). Transgenic breeding approaches for improving abiotic stress tolerance: Recent progress and future perspectives. *International Journal of Molecular Sciences, 21*(8), 2695. https://doi.org/10.3390/ijms21082695

Arroyo-Herrera, A., Figueroa-Yáñez, L., Castaño, E., Santamaría, J., Pereira-Santana, A., Espadas-Alcocer, J., & Sánchez-Teyer, F. (2016). A novel Dreb2-type gene from Carica papaya confers tolerance under abiotic stress. *Plant Cell, Tissue and Organ Culture, 125*(1), 119–133. https://doi.org/10.1007/s11240-015-0934-9

Bado, S., Forster, B. P., Nielen, S., Ali, A. M., Lagoda, P. J., Till, B. J., & Laimer, M. (2015). Plant mutation breeding: Current progress and future assessment. In J. Janick (Ed.), *Plant breeding reviews* (Vol. 39, pp. 23–88). Wiley-Blackwell. https://doi.org/10.1002/9781119107743.ch02

Bhat, J. A., Ali, S., Salgotra, R. K., Mir, Z. A., Dutta, S., Jadon, V., Tyagi, A., Mushtaq, M., Jain, N., Singh, P. K., & Singh, G. P. (2016). Genomic selection in the era of next generation sequencing for complex traits in plant breeding. *Frontiers in Genetics, 7*, 221.

Bo, W., Zhaohui, Z., Huanhuan, Z., Xia, W., Binglin, L., Lijia, Y., et al. (2019). Targeted mutagenesis of NAC transcription factor gene, OsNAC041, leading to salt sensitivity in rice. *Rice Science, 26*(2), 98–108.

Chaâbene, Z., Rorat, A., Hakim, I., Bernard, F., Douglas, G. C., Elleuch, A., et al. (2018). Insight into the expression variation of metal-responsive genes in the seedling of date palm (Phoenix dactylifera). *Chemosphere, 197*, 123–134. https://doi.org/10.1016/j.chemosphere.2017.12.146

Chang, Y. N., Zhu, C., Jiang, J., Zhang, H., Zhu, J. K., & Duan, C. G. (2020). Epigenetic regulation in plant abiotic stress responses. *Journal of Integrative Plant Biology, 62*(5), 563–580. https://doi.org/10.1111/jipb.12901

Char, S. N., Neelakandan, A. K., Nahampun, H., Frame, B., Main, M., Spalding, M. H., Becraft, P. W., Meyers, B. C., Walbot, V., Wang, K., & Yang, B. (2017). An agrobacterium- delivered CRISPR/Cas9 system for high-frequency targeted mutagenesis in maize. *Plant Biotechnology Journal, 15*, 257–268. https://doi.org/10.1111/pbi.12611

Chen, G., Hu, J., Dong, L., Zeng, D., Guo, L., Zhang, G., et al. (2019). The tolerance of salinity in rice requires the presence of a functional copy of FLN2. *Biomolecules, 10*(1), 17.

Chen, K., & Gao, C. (2014). Targeted genome modification technologies and their applications in crop improvements. *Plant Cell Reports, 33*, 575–583. https://doi.org/10.1007/s00299-013-1539-6

Chen, Y., Ma, J., Zhang, X., et al. (2017). A novel non-specific lipid transfer protein gene from sugarcane (NsLTPs), obviously responded to abiotic stresses and signaling molecules of SA and MeJA. *Sugar Technology, 19*(1), 17. https://doi.org/10.1007/s12355-016-0431-4

Chin, J. H., Gamuyao, R., Dalid, C., Bustamam, M., Prasetiyono, J., Moeljopawiro, S., Wissuwa, M., & Heuer, S. (2011). Developing rice with high yield under phosphorus deficiency: Pup1 sequence to application. *Plant Physiology, 156*(3), 1202–1216. https://doi.org/10.1104/pp.111.175471

Curtin, S. J., Xiong, Y., Michno, J. M., Campbell, B. W., Stec, A. O., Cermák, T., Starker, C., Voytas, D. F., Eamens, A. L., & Stupar, R. M. (2018). CRISPR/Cas9 and TALENs generate heritable mutations for genes involved in small RNA processing of *Glycine max* and *Medicago truncatula*. *Plant Biotechnology Journal, 16*, 1125–1137.

Dass, A., Abdin, M. Z., Reddy, V. S., & Leelavathi, S. (2017). Isolation and characterization of the dehydration stress-inducible GhRDL1 promoter from the cultivated upland cotton (*Gossypium hirsutum*). *Journal of Plant Biochemistry and Biotechnology, 26*(1), 113–119. https://doi.org/10.1007/s13562-016-0369-3

Dhillon, T., Pearce, S. P., Stockinger, E. J., Distelfeld, A., Li, C., Knox, A. K., et al. (2012). Regulation of freezing tolerance and flowering in temperate cereals: The VRN-1 connection. *Plant Physiology, 153*(4), 1846–1858. https://doi.org/10.1104/pp.110.159079

Dubcovsky J, María GS, Epstein E, Luo MC, Dvořák J (1996) Mapping of the K+/Na+ discrimination locus Kna1 in wheat. Theoretical and Applied Genetics 92(3–4): 448–454. doi: https://doi.org/10.1007/BF00223692.

Endo, A., Masafumi, M., Kaya, H., & Toki, S. (2016). Efficient targeted mutagenesis of rice and tobacco genomes using Cpf1 from Francisella novicida. *Scientific Reports, 6*, 38,169.

Erpen-Dalla Corte, L., Mahmoud, L. M., Moraes, T. S., Mou, Z., Grosser, J. W., & Dutt, M. (2019). Development of improved fruit, vegetable, and ornamental crops using the crispr/cas9 genome editing technique. *Plants, 8*, 601.

Figueroa-Yañez, L., Pereira-Santana, A., Arroyo-Herrera, A., Rodriguez-Corona, U., Sanchez-Teyer, F., Espadas Alcocer, J., et al. (2016). RAP2.4a is transported through the phloem to regulate cold and heat tolerance in papaya tree (*Carica papaya* cv. Maradol): Implications for protection against abiotic stress. *PLoS One, 11*(10), 0165030. https://doi.org/10.1371/journal.pone.0165030

Francia, E., Barabaschi, D., Tondelli, A., Laidò, G., Rizza, F., Stanca, A. M., Busconi, M., Fogher, C., Stockinger, E. J., & Pecchioni, N. (2007). Fine mapping of a HvCBF gene cluster at the frost resistance locus Fr-H2 in barley. *Theoretical and Applied Genetics, 115*(8), 1083–1091. https://doi.org/10.1007/s00122-007-0634-x

Fukao, T., Xu, K., Ronald, P. C., & Bailey-Serres, J. (2006). A variable cluster of ethylene response factor–like genes regulates metabolic and developmental acclimation responses to submergence in rice. *The Plant Cell, 18*, 2021–2034. https://doi.org/10.1105/tpc.106.043000

Gamuyao, R., Chin, J. H., Pariasca-Tanaka, J., Pesaresi, P., Catausan, S., Dalid, C., et al. (2012). The protein kinase Pstol1 from traditional rice confers tolerance of phosphorus deficiency. *Nature, 488*(7412), 535–539. https://doi.org/10.1038/nature11346

Ganie, S. A., Wani, S. H., Henry, R., & Hensel, G. (2021). Improving rice salt tolerance by precision breeding in a new era. *Current Opinion in Plant Biology, 60*, 101996.

Gong, Z., Xiong, L., Shi, H., Yang, S., Herrera-Estrella, L. R., Xu, G., Chao, D. Y., Li, J., Wang, P. Y., Qin, F., & Li, J. (2020). Plant abiotic stress response and nutrient use efficiency. *Science China. Life Sciences, 63*(5), 635–674. https://doi.org/10.1007/s11427-020-1683-x

Gupta, K., Dey, A., & Gupta, B. (2013a). Plant polyamines in abiotic stress responses. *Acta Physiologiae Plantarum, 35*(7), 2015–2036. https://doi.org/10.1007/s11738-013-1239-4

Gupta, K., Dey, A., & Gupta, B. (2013b). Polyamines and their role in plant osmotic stress tolerance. In N. Tuteja & S. S. Gill (Eds.), *Climate change and plant abiotic stress tolerance* (pp. 1053–1072). https://doi.org/10.1002/9783527675265.ch40

Hasley, J. A. R., Navet, N., & Tian, M. (2021). CRISPR/Cas9-mediated mutagenesis of sweet basil candidate susceptibility gene ObDMR6 enhances downy mildew resistance. *PLoS One, 16*(6), e0253245. https://doi.org/10.1371/journal.pone.0253245

Hattori, Y., Nagai, K., Furukawa, S., Song, X. J., Kawano, R., Sakakibara, H., Wu, J., Matsumoto, T., Yoshimura, A., Kitano, H., Matsuoka, M., Mori, H., & Ashikari, M. (2009). The ethylene response factors SNORKEL1 and SNORKEL2 allow rice to adapt to deep water. *Nature, 460*(7258), 1026–1030. https://doi.org/10.1038/nature08258

He, P., Zhao, P., Wang, L., Zhang, Y., Wang, X., Xiao, H., et al. (2017). The PIN gene family in cotton (*Gossypium hirsutum*): Genome-wide identification and gene expression analyses during root development and abiotic stress responses. *BMC Genomics, 18*, 507. https://doi.org/10.1186/s12864-017-3901-5

Hsu, P. D., Lander, E. S., & Zhang, F. (2014). Development and applications of CRISPR-Cas9 for genome engineering. *Cell, 157*, 1262–1278. https://doi.org/10.1016/j.cell.2014.05.010

Hu, J. H., Miller, S. M., Geurts, M. H., Tang, W., Chen, L., Sun, N., et al. (2018). Evolved Cas9 variants with broad PAM compatibility and high DNA specificity. *Nature, 556*, 57–63. https://doi.org/10.1038/nature26155

Huang, X. Z., Zeng, X. F., Li, J. R., & Zhao, D. G. (2017). Construction and analysis of tify1a and tify1b mutants in rice (*Oryza sativa*) based on CRISPR/Cas9 technology. *Journal of Agricultural Biotechnology, 25*, 1003–1012. https://doi.org/10.3969/j.issn.1674-7968.2017.06.015

Hussain, B., Lucas, S. J., & Budak, H. (2018). CRISPR/Cas9 in plants: At play in the genome and at work for crop improvement. *Briefings in Functional Genomics, 17*, 319–328.

Jiang, F., & Doudna, J. A. (2017). CRISPR-Cas9 structures and mechanisms. *Annual Review of Biophysics, 46*, 505–529. https://doi.org/10.1146/annurev-biophys-062215-010822

Kelliher, T., Starr, D., Su, X., Tang, G., Chen, Z., Carter, J., et al. (2019). One step genome editing of elite crop germplasm during haploid induction. *Nature Biotechnology, 37*, 287–292. https://doi.org/10.1038/s41587-019-0038-x

Kharkwal, M. C., Pandey, R. N., & Pawar, S. E. (2004). Mutation breeding for crop improvement. In H. K. Jain & M. C. Kharkwal (Eds.), *Plant Breeding* (pp. 601–645). Springer.

Kim, D., Alptekin, B., & Budak, H. (2018). CRISPR/Cas9 genome editing in wheat. *Functional and Integrative Genomics, 18*(1), 31–41.

Knox, A. K., Dhillon, T., Cheng, H., Tondelli, A., Pecchioni, N., & Stockinger, E. J. (2010). CBF gene copy number variation at frost Resistance-2 is associated with levels of freezing tolerance in temperate-climate cereals. *Theoretical and Applied Genetics, 121*(1), 21–35. https://doi.org/10.1007/s00122-010-1288-7

Koonin, E. V., Makarova, K. S., & Zhang, F. (2017). Diversity, classification and evolution of CRISPR-Cas systems. *Current Opinion in Microbiology, 37*, 67–78.

Leenay, R. T., & Beisel, C. L. (2017). Deciphering, communicating, and engineering the CRISPR PAM. *Journal of Molecular Biology, 429*(2), 177–191. https://doi.org/10.1016/j.jmb.2016.11.024

Li, C. H., Wang, G., Zhao, J. L., Zhang, L. Q., Ai, L. F., Han, Y. F., et al. (2014). The receptor-like kinase SIT1 mediates salt sensitivity by activating MAPK3/6 and regulating ethylene homeostasis in rice. *The Plant Cell, 26*(6), 2538–2553. https://doi.org/10.1105/tpc.114.125187

Li, J. F., Norville, J. E., Aach, J., McCormack, M., Zhang, D., Bush, J., Church, G. M., & Sheen, J. (2013). Multiplex and homologous recombination-mediated genome editing in *Arabidopsis* and *Nicotiana benthamiana* using guide RNA and Cas9. *Nature Biotechnology, 31*, 688–691. https://doi.org/10.1038/nbt.2654

Li, L., Wei, K., Zheng, G., Liu, X., Chen, S., Jiang, W., & Lu, Y. (2018a). CRISPRCpf1-assisted multiplex genome editing and transcriptional repression in Streptomyces. *Applied and Environmental Microbiology, 84*, e00827–e0e918. https://doi.org/10.1128/AEM.00827-18

Li, R., Liu, C., Zhao, R., Wang, L., Chen, L., Yu, W., Zhang, S., Shen, J., & Shen, L. (2019). CRISPR/Cas9-mediated SlNPR1 mutagenesis reduces tomato plant drought tolerance. *BMC Plant Biology, 19*, 38. https://doi.org/10.1186/s12870-018-1627-4

Li, T., Yang, X., Yu, Y., Si, X., Zhai, X., Zhang, H., et al. (2018b). Domestication of wild tomato is accelerated by genome editing. *Nature Biotechnology, 36*, 1160–1163.

Liu, F., Huang, N., Wang, L., Ling, H., Sun, T., Ahmad, W., et al. (2017). A novel L-ascorbate peroxidase 6 gene, ScAPX6, plays an important role in the regulation of response to biotic and abiotic stresses in sugarcane. *Frontiers in Plant Science, 8*, 2262. https://doi.org/10.3389/fpls.2017.02262

Liu, J., Cui, L., Xie, Z., Zhang, Z., Liu, E., & Peng, X. (2019). Two NCA1 isoforms interact with catalase in a mutually exclusive manner to redundantly regulate its activity in rice. *BMC Plant Biology, 19*, 105.

Liu, X., Wu, D., Shan, T., Xu, S., Qin, R., Li, H., et al. (2020). The trihelix transcription factor OsGTg-2 is involved adaption to salt stress in rice. *Plant Molecular Biology, 103*, 545–560.

Lobell, D. B., & Gourdji, S. M. (2012). The influence of climate change on global crop productivity. *Plant Physiology, 160*(4), 1686–1697. https://doi.org/10.1104/pp.112.208298

Lou, D., Wang, H., Liang, G., & Yu, D. (2017). OsSAPK2 confers abscisic acid sensitivity and tolerance to drought stress in rice. *Frontiers in Plant Science, 8*, 993.

Lou, D., Wang, H., & Yu, D. (2018). The sucrose non-fermenting-1-related protein kinases SAPK1 and SAPK2 function collaboratively as positive regulators of salt stress tolerance in rice. *BMC Plant Biology, 18*(1), 203.

Ma, Y., Dai, X., Xu, Y., Luo, W., Zheng, X., Zeng, D., Pan, Y., Lin, X., Liu, H., Zhang, D., et al. (2015). COLD1 confers chilling tolerance in rice. *Cell, 160*(6), 1209–1221. https://doi.org/10.1016/j.cell.2015.01.046

Mahalingam, R. (2015). Consideration of combined stress: A crucial paradigm for improving multiple stress tolerance in plants. In R. Mahalingam (Ed.), *Combined stresses in plants*. Springer. https://doi.org/10.1007/978-3-319-07899-1_1

Makarova, K., Wolf, Y., Alkhnbashi, O., et al. (2015). An updated evolutionary classification of CRISPR–Cas systems. *Nature Reviews. Microbiology, 13*, 722–736. https://doi.org/10.1038/nrmicro3569

Makarova, K. S., Wolf, Y. I., Iranzo, J., Shmakov, S. A., Alkhnbashi, O. S., Brouns, S. J. J., Charpentier, E., Cheng, D., Haft, D. H., Horvath, P., et al. (2020). Evolutionary classification of CRISPR-Cas systems: A burst of class 2 and derived variants. *Nature Reviews. Microbiology, 18*, 67–83. https://doi.org/10.1038/s41579-019-0299-x

Makarova, K. S., Wolf, Y. I., & Koonin, E. V. (2018). Classification and nomenclature of CRISPR-Cas systems: Where from here? *CRISPR Journal, 1*, 325–336. https://doi.org/10.1089/crispr.2018.0033

Malzahn, A., Lowder, L., & Qi, Y. (2017). Plant genome editing with TALEN and CRISPR. *Cell & Bioscience, 7*, 21.

Manghwar, H., Lindsey, K., Zhang, X., & Jin, S. (2019). CRISPR/Cas system: Recent advances and future prospects for genome editing. *Trends in Plant Science, 24*, 1102–1125. https://doi.org/10.1016/j.tplants.2019.09.006

Maron, L. G., Guimarães, C. T., Kirst, M., Albert, P. S., Birchler, J. A., Bradbury, P. J., Buckler, E. S., Coluccio, A. E., Danilova, T. V., Kudrna, D., Magalhaes, J. V., et al. (2013). Aluminum tolerance in maize is associated with higher MATE1 gene copy number. *Proceedings of the National Academy of Sciences, 110*(13), 5241–5246. https://doi.org/10.1073/pnas.1220766110

Meng, X., Hu, X., Liu, Q., Song, X., Gao, C., Li, J., et al. (2018). Robust genome editing of CRISPR-Cas9 at NAG PAMs in rice. *Science China. Life Sciences, 61*, 122–125. https://doi.org/10.1007/s11427-017-9247-9

Miao, C., Xiao, L., Hua, K., Zou, C., Zhao, Y., Bressan, R. A., & Zhu, J. K. (2018). Mutations in a subfamily of abscisic acid receptor genes promote rice growth and productivity. *Proceedings of the National Academy of Sciences of the United States of America, 115*, 6058–6063. https://doi.org/10.1073/pnas.1804774115

Miao, H., Sun, P., Liu, Q., Miao, Y., Liu, J., Xu, B., et al. (2017b). The AGPase family proteins in banana: Genome-wide identification, phylogeny, and expression analyses reveal their involvement in the development, ripening, and abiotic/biotic stress responses. *International Journal of Molecular Science, 18*(8), 1581. https://doi.org/10.3390/ijms18081581

Miao, H., Sun, P., Liu, Q., Miao, Y., Liu, J., Zhang, K., et al. (2017a). Genome-wide analyses of SWEET family proteins reveal involvement in fruit development and abiotic/biotic stress responses in banana. *Scientific Reports, 7*, 3536. https://doi.org/10.1038/s41598-017-03872-w

Mo, W., Tang, W., Du, Y., Jing, Y., Bu, Q., & Lin, R. (2020). Phytochromeinteracting factor-LIKE14 and SLENDER RICE1 interaction controls seedling growth under salt stress. *Plant Physiology, 184*, 506–517.

Mushtaq, M., Bhat, J. A., Mir, Z. A., Sakina, A., Ali, S., Singh, A. K., Tyagi, A., Salgotra, R. K., Dar, A. A., & Bhat, R. (2018). CRISPR/Cas approach: A new way of looking at plant-abiotic interactions. *Journal of Plant Physiology, 224-225*, 156–162. https://doi.org/10.1016/j.jplph.2018.04.001

Nandy, S., Pathak, B., Zhao, S., & Srivastava, V. (2019). Heat-shock-inducible CRISPR/Cas9 system generates heritable mutations in rice. *Plant Direct, 3*, e00145. https://doi.org/10.1002/pld3.145

Nekrasov, V., Staskawicz, B., Weigel, D., Jones, J. D., & Kamoun, S. (2013). Targeted mutagenesis in the model plant Nicotiana benthamiana using Cas9 RNA-guided endonuclease. *Nature Biotechnology, 31*, 691.

Nieves-Cordones, M., Mohamed, S., Tanoi, K., Kobayashi, N. I., Takagi, K., Vernet, A., et al. (2017). Production of low-Cs+ rice plants by inactivation of the K+ transporter OsHAK1 with the CRISPR-Cas system. *Plant Journal, 92*(1), 43–56. https://doi.org/10.1111/tpj.13632

Osakabe, Y., & Osakabe, K. (2017). Genome editing to improve abiotic stress responses in plants. In *Progress in molecular biology and translational science* (Vol. 149, pp. 99–109). Elsevier.

Osakabe, Y., Watanabe, T., Sugano, S. S., Ueta, R., & Ishihara, R. (2016). Optimization of CRISPR/Cas9 genome editing to modify abiotic stress responses in plants. *Scientific Reports, 1–10*. https://doi.org/10.1038/srep26685

Ou, W., Mao, X., Huang, C., Tie, W., Yan, Y., Ding, Z., et al. (2018). Genome-wide identification and expression analysis of the KUP family under abiotic stress in cassava (*Manihot esculenta* Crantz). *Frontiers in Physiology, 9*, 17. https://doi.org/10.3389/fphys.2018.00017

Paixão, J. F. R., Gillet, F., Ribeiro, T. P., Bournaud, C., Lourenço-tess, I. T., Noriega, D. D., De Melo, B. P., De Almeida-engler, J., & Grossi-de-sa, M. F. (2019). Improved drought stress tolerance in Arabidopsis by CRISPR/dCas9 fusion with a histone Acetyltransferase. *Scientific Reports, 1–9*. https://doi.org/10.1038/s41598-019-44571-y

Pandita, D. (2021a). Cas9 technology: An innovative approach to enhance phytoremediation. In T. B. Pirzadah, B. Malik, & K. R. Hakeem (Eds.), *Plant-microbe dynamics: Recent advances for sustainable agriculture*. CRC Press, ISBN: 1000388913, 9781000388916.

Pandita, D. (2021b). CRISPR/Cas mediated genome editing for improved stress tolerance in plants. In T. Aftab & K. R. Hakeem (Eds.), *Frontiers in plant–soil interaction: Molecular insights into plant adaptation* (pp. 259–291). Academic Press, Elsevier. https://doi.org/10.1016/B978-0-323-90943-3.00001-8

Pandita, D. (2021c). CRISPR/Cas mediated genome editing technologies in plants. In T. Aftab & K. R. Hakeem (Eds.), *Plant abiotic stress physiology, volume 1: Responses and adaptations*. CRC Press. Hard ISBN: 9781774630167.

Park, J. J., Dempewolf, E., Zhang, W., & Wang, Z. Y. (2017). RNA-guided transcriptional activation via CRISPR/dCas9 mimics overexpression phenotypes in Arabidopsis. *PLoS One, 12*, e0179410.

Prado, J. R., Segers, G., Voelker, T., Carson, D., Dobert, R., Phillips, J., Cook, K., Cornejo, C., Monken, J., Grapes, L., Reynolds, T., & Martino-Catt, R. (2014). Genetically engineered crops: From idea to product. *Annual Review of Plant Biology, 65*, 769–790. https://doi.org/10.1146/annurev-arplant-050213-040039

Qin, H., Wang, J., Chen, X., Wang, F., Peng, P., Zhou, Y., et al. (2019). Rice OsDOF15 contributes to ethylene inhibited primary root elongation under salt stress. *The New Phytologist, 223*, 798–813.

Qiu, Z., Kang, S., He, L., Zhao, J., Zhang, S., Hu, J., et al. (2018). The newly identified heat-stress sensitive albino 1 gene affects chloroplast development in rice. *Plant Science, 267*, 168–179.

Raman, R. (2017). The impact of genetically modified (GM) crops in modern agriculture: A review. *GM Crops & Food, 8*, 195–208.

Rao, G. J. N., Reddy, J. N., Variar, M., & Mahender, A. (2016). Molecular breeding to improve plant resistance to abiotic stresses. In J. Al-Khayri, S. Jain, & D. Johnson (Eds.), *Advances in plant breeding strategies: Agronomic, abiotic and biotic stress traits*. Springer. https://doi.org/10.1007/978-3-319-22518-0_8

Robinson, J. (1999). Ethics and transgenic crops: A review. *Electronic Journal of Biotechnology, 2*(2), 5–6.

Romero, F. M., & Gatica-Arias, A. (2019). CRISPR/Cas9: Development and application in rice breeding. *Rice Science, 26*, 265–281.

Rong, Y. S., & Golic, K. G. (2000). Gene targeting by homologous recombination in drosophila. *Science, 288*, 2013–2018.

Röös, E., Bajželj, B., Smith, P., Patel, M., Little, D., & Garnett, T. (2017). Greedy or needy? Land use and climate impacts of food in 2050 under different livestock futures. *Global Environmental Change, 47*, 1–12.

Samai, P., Pyenson, N., Jiang, W., Goldberg, G. W., Hatoum-Aslan, A., & Marraffini, L. A. (2015). Co-transcriptional DNA and RNA cleavage during type III CRISPR-Cas immunity. *Cell, 161*, 1164–1174.

Santosh Kumar, V. V., Verma, R. K., Yadav, S. K., Yadav, P., Watts, A., Rao, M. V., et al. (2020). CRISPR-Cas9 mediated genome editing of drought and salt tolerance (OsDST) gene in indica mega rice cultivar MTU1010. *Physiology and Molecular Biology of Plants, 26*, 1099–1110.

Shan, Q., Wang, Y., Li, J., Zhang, Y., Chen, K., Liang, Z., Zhang, K., Liu, J., Xi, J. J., & Qiu, J. (2013). Targeted genome modification of crop plants using a CRISPR-Cas system. *Nature Biotechnology, 31*, 686.

Shen, C., Que, Z., Xia, Y., Tang, N., Li, D., He, R., et al. (2017). Knock out of the annexin gene OsAnn3 via CRISPR/Cas9-mediated genome editing decreased cold tolerance in rice. *Journal of Plant Biology, 60*(6), 539–547.

Shi, J., Gao, H., Wang, H., Lafitte, H. R., Archibald, R. L., Yang, M., Hakimi, S. M., Mo, H., & Habben, J. E. (2017). ARGOS8 variants generated by CRISPR-Cas9 improve maize grain yield under field drought stress conditions. *Plant Biotechnology Journal, 15*, 207–216.

Shim, J. S., Oh, N., Chung, P. J., Kim, Y. S., Choi, Y. D., & Kim, J. K. (2018). Overexpression of osnac14 improves drought tolerance in rice. *Frontiers in Plant Science, 9*, 310. https://doi.org/10.3389/fpls.2018.00310

Stella, S., Alcón, P., & Montoya, G. J. N. (2017). Structure of the Cpf1 endonuclease R-loop complex after target DNA cleavage. *Nature, 546*, 559.

Stockinger, E. J., Skinner, J. S., Gardner, K. G., Francia, E., & Pecchioni, N. (2007). Expression levels of barley Cbf genes at the frost resistance-H2 locus are dependent upon alleles at Fr-H1 and Fr-H2. *The Plant Journal, 51*(2), 308–321. https://doi.org/10.1111/j.1365-313X.2007.0141.x

Tang, L., Mao, B., Li, Y., Lv, Q., Zhang, L., Chen, C., He, H., Wang, W., Zeng, X., Shao, Y., et al. (2017a). Knockout of OsNramp5 using the CRISPR/Cas9 system produces low Cd-accumulating indica rice without compromising yield. *Scientific Reports, 7*, 14,438. https://doi.org/10.1038/s41598-017-14832-9

Tang, X., Liu, G., Zhou, J., Ren, Q., You, Q., Tian, L., Xin, X., Zhong, Z., Liu, B., & Zheng, X. (2018). A large-scale whole-genome sequencing analysis reveals highly specific genome editing by both Cas9 and Cpf1 (Cas12a) nucleases in rice. *Genome Biology, 19*, 84.

Tang, X., Lowder, L. G., Zhang, T., Malzahn, A. A., Zheng, X., Voytas, D. F., Zhong, Z., Chen, Y., Ren, Q., Li, Q., Kirkland, E. R., Zhang, Y., & Qi, Y. (2017b). A CRISPR-Cpf1 system for efficient genome editing and transcriptional repression in plants. *Nature Plants, 3*, 17103. https://doi.org/10.1038/nplants.2017.18

Uga, Y., Okuno, K., & Yano, M. (2011). Dro1, a major QTL involved in deep rooting of rice under upland field conditions. *Journal of Experimental Botany, 62*(8), 2485–2494. https://doi.org/10.1093/jxb/erq429

Uga, Y., Sugimoto, K., Ogawa, S., Rane, J., Ishitani, M., Hara, N., Kitomi, Y., Inukai, Y., Ono, K., & Kanno, N. (2013). Control of root system architecture by DEEPER ROOTING 1 increases rice yield under drought conditions. *Nature Genetics, 45*(9), 1097–1102. https://doi.org/10.1038/ng.2725

Van Eck, J., Keen, P., & Tjahjadi, M. (2019). Agrobacterium tumefaciens- mediated transformation of tomato. *Methods in Molecular Biology, 1864*, 225–234.

Voytas, D. F., & Gao, C. (2014). Precision genome engineering and agriculture: Opportunities and regulatory challenges. *PLoS Biology, 12*, e1001877. https://doi.org/10.1371/journal.pbio.1001877

Vu, T. V., Sivankalyani, V., Kim, E. J., Doan, D. T. H., Tran, M. T., Kim, J., Sung, Y. W., Park, M., Kang, Y. J., & Kim, J. Y. (2020). Highly efficient homology-directed repair using CRISPR/Cpf1-geminiviral replicon in tomato. *Plant Biotechnology Journal, 1–11*. https://doi.org/10.1111/pbi.13373

Wang, F. Z., Chen, M. X., Yu, L. J., Xie, L. J., Yuan, L. B., Qi, H., Xiao, M., Guo, W., Chen, Z., Yi, K., Zhang, J., Qiu, R., Shu, W., Xiao, S., & Chen, Q. F. (2017b). OsARM1, an R2R3 MYB transcription factor, is involved in regulation of the response to arsenic stress in rice. *Frontiers in Plant Science, 8*, 1868. https://doi.org/10.3389/fpls.2017.01868

Wang, F. Z., Chen, M. X., Yu, L. J., Xie, L. J., Yuan, L. B., Qi, H., Xiao, M., Guo, W., Chen, Z., Yi, K., et al. (2017a). Osarm1, an R2R3 MYB transcription factor, is involved in regulation of the response to arsenic stress in rice. *Frontiers in Plant Science, 8*, 1868. https://doi.org/10.3389/fpls.2017.01868

Wang, L., Chen, L., Li, R., Zhao, R., Yang, M., Sheng, J., & Shen, L. (2017c). Reduced drought tolerance by CRISPR/Cas9-mediated SlMAPK3 mutagenesis in tomato plants. *Journal of Agricultural and Food Chemistry, 65*, 8674–8682. https://doi.org/10.1021/acs.jafc.7b02745

Wang, L., Chen, L., Li, R., Zhao, R., Yang, M., Sheng, J., & Shen, L. (2017d). Reduced drought tolerance by CRISPR/ Cas9-mediated SlMAPK3 mutagenesis in tomato plants. *Journal of Agricultural and Food Chemistry, 65*(39), 8674–8682. https://doi.org/10.1021/acs.jafc.7b02745

Wang, W. C., Lin, T. C., Kieber, J., & Tsai, Y. C. (2019). Response regulators 9 and 10 negatively regulate salinity tolerance in rice. *Plant and Cell Physiology, 60*, 2549–2563.

Wani, S. H., & Sah, S. K. (2014). Biotechnology and abiotic stress tolerance in rice. *Journal of Rice Research, 2*(2), 1000e105. https://doi.org/10.4172/jrr.1000e105

Wrighton, K. (2018). Expanding the reach of Cas9. *Nature Reviews. Genetics, 19*, 250–251. https://doi.org/10.1038/nrg.2018.15

Wu, W. Y., Lebbink, J. H. G., Kanaar, R., Geijsen, N., & van der Oost, J. (2018). Genome editing by natural and engineered CRISPR-associated nucleases. *Nature Chemical Biology, 14*, 642–651.

Xu, J., Soni, V., Chopra, M., & Chan, O. (2020). Genetic modification of the HSP90 gene using CRISPR-Cas9 to enhance thermotolerance in T. Suecica. *URNCST Journal, 21*, 4(4). https://doi.org/10.26685/urncst.178

Xu, K., Xu, X., Fukao, T., Canlas, P., Maghirang-Rodriguez, R., Heuer, S., et al. (2006). Sub1Ais an ethyleneresponse-factor-like gene that confers submergence tolerance to rice. *Nature, 442*(7103), 705–708. https://doi.org/10.1038/nature04920

Ye, Y., Li, P., Xu, T., Zeng, L., Cheng, D., Yang, M., Luo, J., & Lian, X. (2017). Ospt4 contributes to arsenate uptake and transport in rice. *Frontiers in Plant Science, 8*, 2197. https://doi.org/10.3389/fpls.2017.02197

Yin, W., Xiao, Y., Niu, M., Meng, W., Li, L., Zhang, X., et al. (2020). ARGONAUTE2 enhances grain length and salt tolerance by activating 2292–2306. *The Plant Cell, 32*, 2292–2306.

Yin, X., Anand, A., Quick, P., & Bandyopadhyay, A. (2019). Editing a stomatal developmental gene in rice with CRISPR/Cpf1. In Y. Qi (Ed.), *Plant genome editing with CRISPR systems. Methods in molecular biology* (Vol. 1917). Humana Press.

Yu, W., Wang, L., Zhao, R., et al. (2019). Knockout of *SlMAPK3* enhances tolerance to heat stress involving ROS homeostasis in tomato plants. *BMC Plant Biology, 19*, 354. https://doi.org/10.1186/s12870-019-1939-z

Zeng, D. D., Yang, C. C., Qin, R., Alamin, M., Yue, E. K., Jin, X. L., et al. (2018). A guanine insert in OsBBS1 leads to early leaf senescence and salt stress sensitivity in rice (Oryza sativa L.). *Plant Cell Reports, 37*, 933–946.

Zeng, Y., Wen, J., Zhao, W., Wang, Q., & Huang, W. (2020). Rational improvement of Rice yield and cold tolerance by editing the three genes OsPIN5b, GS3, and OsMYB30 with the CRISPR–Cas9 system. Frontiers. *Plant Science, 10*.

Zetsche, B., Gootenberg, J. S., Abudayyeh, O. O., Slaymaker, I. M., Makarova, K. S., Essletzbichler, P., Volz, S. E., Joung, J., Oost, J. V., Regev, A., Koonin, E. V., & Zhang,

F. (2015). Cpf1 is a single RNA-guided endonuclease of a class 2 CRISPR-Cas system. *Cell, 163*, 759–771. https://doi.org/10.1016/j.cell.2015.09.038

Zhang, A., Liu, Y., Wang, F., Li, T., Chen, Z., Kong, D., Bi, J., Zhang, F., Luo, X., & Wang, J. (2019a). Enhanced rice salinity tolerance via CRISPR/Cas9-targeted mutagenesis of the OsRR22 gene. *Molecular Breeding, 39*, 47.

Zhang, C., Srivastava, A. K., & Sadanandom, A. (2019b). Targeted mutagenesis of the SUMO protease, overly tolerant to Salt1 in rice through CRISPR/Cas9-mediated genome editing reveals a major role of this SUMO protease in salt tolerance. *bioRxiv*, 555706.

Zhang, Y., Massel, K., Godwin, I. D., & Gao, C. (2018). Applications and potential of genome editing in crop improvement. *Genome Biology, 19*, 210. https://doi.org/10.1186/s13059-018-1586-y

Zhao, Y., Zhang, C., Liu, W., Gao, W., Liu, C., Song, G., et al. (2016). An alternative strategy for targeted gene replacement in plants using a dual-sgRNA/Cas9 design. *Scientific Reports, 6*, 23890.

Zhong, Z., Zhang, Y., You, Q., Tang, X., Ren, Q., Liu, S., et al. (2018). Plant genome editing using FnCpf1 and LbCpf1 nucleases at redefined and altered PAM sites. *Molecular Plant, 11*, 999–1002. https://doi.org/10.1016/j.molp.2018.03.008

Zhou, G., Pereira, J. F., Delhaize, E., Zhou, M., Magalhaes, J. V., & Ryan, P. R. (2014). Enhancing the aluminium tolerance of barley by expressing the citrate transporter genes SbMATE and FRD3. *Journal of Experimental Botany, 65*(9), 2381–2390. https://doi.org/10.1093/jxb/eru121

Zhou, J., Deng, K., Cheng, Y., Zhong, Z., Tian, L., Tang, X., et al. (2017). CRISPR-Cas9 based genome editing reveals new insights into Micro-RNA function and regulation in rice. *Frontiers in Plant Science, 8*, 1598.

Chapter 12
Manipulation of Key Genes Involved in Biosynthesis of Terpenoid Compounds in Plants

Mahak Majeed and Reiaz Ul Rehman

Abstract Terpenoids represent the diverse class of plant secondary metabolites exhibiting immense applications in pharmaceutical and other industrial sectors. Commercial exploitation of terpenoid compounds is mostly hampered due to lower quantities of these compounds synthesized in their natural plant sources, difficulty in their isolation, and extreme structural diversity leading to expensive synthetic approaches. To overcome these shortcomings, plant-based systems provide an attractive platforms in manipulating the key genes of terpenoid pathways involved in biosynthesis of target terpenoid compounds. At cellular level, plants show compartmentalization and comprise cofactors which assist in metabolic manipulation of whole functional pathways taken from other plants. In this chapter, we have highlighted various attempts of metabolic engineering in host plants for enhanced production of target terpenoid compound. Besides, we have also discussed some important limitations associated with plant-based expression systems, future directions, and developments in harnessing the maximum potential of metabolic engineered plant systems.

Keywords Pharmaceutical · Synthetic approaches · Terpenoid · Metabolic engineering

Introduction

Plants synthesize diverse range of primary and secondary metabolites that contribute in their growth and development. Among vast array of plant metabolites, terpenes or terpenoids comprise one of the largest structurally diverse group of more than 80,000 compounds reported from all life forms including plants (Christianson, 2017). Terpenoids exhibit additional functional groups and differ from terpenes which are

M. Majeed · R. U. Rehman (✉)
Department of Bioresources, Faculty of Biological Science, University of Kashmir, Hazratbal, Srinagar, Jammu and Kashmir, India

T. Aftab, K. R. Hakeem (eds.), *Metabolic Engineering in Plants*,
https://doi.org/10.1007/978-981-16-7262-0_12

pure hydrocarbons, made up of C_5-isoprene units. On the basis of number of five carbon or isoprene units, terpenes are classified as hemiterpene (C5), monoterpene (C10), sesquiterpene (C15), diterpene (C20), sesterterpene (C25), triterpene (C30), and tetraterpene (C40) (Dudareva et al., 2006; Chen et al., 2011). Plants contain primary terpenoids (carotenoids, sterols, some phytohormones) and secondary or specialized terpenoids. Terpenoids act as key biomolecules in maintaining membrane fluidity, signaling molecules, and hormones (Ashour et al., 2010). The specialized terpenoids represent the largest class that has important function in plant defense against pathogens and herbivores. They also play important role in attracting the pollinators, seed dispersal, and heat regulation (Nagegowda, 2010; Gutensohn et al., 2012). The biosynthesis of terpenoids takes place in specific structures like resin ducts, trichomes, and oil ducts (Hamberger et al., 2011; Wang et al., 2009), while others are produced in general plant organs like fruits and leaves (Drew et al., 2009), and few terpenoids are synthesized throughout the plants (Bohlmann & Keeling, 2008). The specialized metabolism in plants generates diverse families of plant-derived compounds which showed applications against serious disorders such as antimalarial compounds sesquiterpene lactones; anticholinergic tropane alkaloids derived from Solanaceae; anticancerous taxane terpenoids; poppy isoquinoline alkaloids which showed antitussive, antimicrobial, analgesics, and vasodilators; and monoterpene indole alkaloids having antihypertensive and anticancer effects (Newman & Cragg, 2016). The biosynthesis of specialized terpenoids involves the enzyme terpene synthase and cytochrome P450 that act on diphosphate substrate and catalyze the oxidation reactions of aldehydes (–CHO), alcohols (–ROH), ketones (C =O), and carboxylic acids (COOH) on the basic hydrocarbon skeleton of terpenes (Weitzel & Simonsen, 2015; Yang et al., 2015). The structural diversity in terpenoid compounds arises mainly from the diversity of terpene synthase enzyme (Chen et al., 2011) and cytochrome P450 (Boutanaev et al., 2015). Terpenes are biosynthesized in plants through Mevalonate pathway (MVA) which operates in cytosol of cell and its organelles like endoplasmic reticulum and peroxisomes (Carrie et al., 2007; Hemmerlin et al., 2003; Dudareva et al., 2006; Leivar et al., 2005; Merret et al., 2007; Sapir-Mir et al., 2008; Simkin et al., 2011; Lange & Ahkami, 2013). Other pathway of terpenoid biosynthesis known as methylerythritol pathway (MEP) takes place in plastid that initiates with condensation of pyruvate and glceraldehyde-3-phosphate which leads to formation of 1-deoxy-D-xylulose 5-phosphate by the catalytic action of 1-deoxy-D-xylulose 5-phosphate synthase (Sprenger et al., 1997). MVA pathway is responsible for synthesis of sesquiterpenes, while MEP pathway is mainly involved in synthesis of monoterpenes and diterpenes (Muhlemann et al., 2014) (Fig. 12.1). In recent years, many research studies were conducted on identification of terpene synthase genes, elucidation of terpene biosynthetic pathways which has caused a lot of feasibility and huge advancement in genetic engineering in plants for terpenoid synthesis (Dudareva & Pichersky, 2008; Yu & Utsumi, 2009).

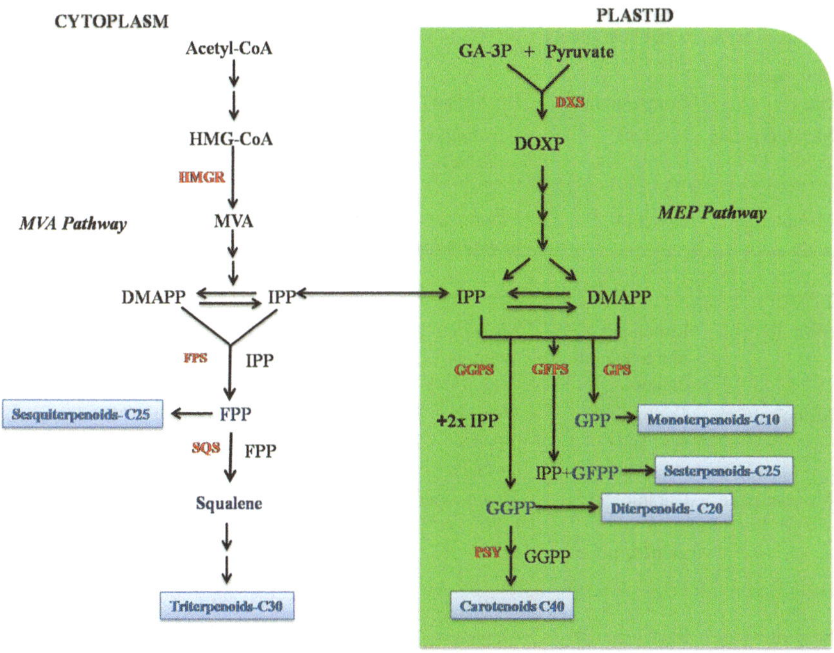

Fig. 12.1 The basic terpenoid metabolic pathways MVA (mevalonic acid pathway) occurring in cytoplasm and MEP (methylerythritol pathway) taking place in plastid. Enzymes are indicated in red text in both pathways. *GA-3P* Glyceraldehyde 3-phosphate, *DXS* 1-Deoxy-D-xylulose 5-phosphate synthase, *DOXP* 1-Deoxy-D-xylulose 5-phosphate, *IPP* Isopentenyl diphosphate, *DMAPP* Dimethylallyl diphosphate, *GGPS* Geranyl geranyl diphosphate synthase, *GFPS* Geranyl farnesyl-diphosphate synthase, *GPS* Geranyl diphosphate synthase, *GPP* Geranyl diphosphate, *GFPP* Geranyl farnesyl siphosphate, *GGPP* Geranyl geranyl diphosphate, *PSY* Phytoene synthase, *Co A* Coenzyme A, *HMG* 3-Hydroxy, 3-methyl glutaryl, *HMGR* HMG Co-A reductase, *FPP* Farnesyl diphosphate, *FPS* FPP synthase, *SQS* Squalene synthase

Terpenoid Metabolic Engineering Approaches

Currently, genetic engineering attempts were mostly performed in model plants which also exhibit thoroughly characterized genome. In manipulating the key genes of terpenoid pathways, co-expression network pathways play an important role in the identification of genes and their expression patterns. The biotechnological interventions used in elevating the level of terpenes in plants involve the amplification of desired genes involved in terpene synthesis from selective plant families. Real-time PCR (RT-PCR) and high-performance liquid chromatography (HPLC) were employed to analyze the biosynthesis of various terpenes at transcriptomic level (Table 12.1, briefly summarizes the biotechnological attempts carried in some model plants for improving the production of different terpenoid compounds).

Table 12.1 Metabolic intervention strategies for enhancing different classes of terpenoid compounds in some model host plants

Terpenoid class	Compound/s	Host plant	Metabolic intervention	References
Monoterpene	Linalool	*Lavandula latifolia* L.	Expression of *LIS* gene	Mendoza-Poudereux et al. (2014)
Monoterpene	Geraniol	*Nicotiana benthamiana* L.	Targeted expression of *GES* and *GPS* genes in plastid, mitochondria, and cytosol	Dong et al. (2016)
Monoterpene	Limonene, myrcene, and linalool	*Nicotiana tabacum* L.	Co-expression of *MpGPS*. *SSU*, and *GPS*	Yin et al. (2017)
Monoterpene	Geranyl acetic acid and cis-geraniol	*Mentha × piperita f. citrata*	Overexpression of *NtLPT1*	Hwang et al. (2020)
Monoterpene	Geraniol and other monoterpenes	*Mentha spicata* L.	RNAi silencing of native monoterpene synthases and independent expression of heterologous *LS*, *MS*, and *GS*	Li et al. (2020)
Sesquiterpene	Santalene and Bergamotene	*Nicotiana tabacum* L.	Targeting co-expression of *HMGR* and *SS* to plastid and cytosol	Yin and Wong (2019)
Sesquiterpene	Patchoulol	*Nicotiana tabacum* L.	Targeting expression of *STS* and *FDS*	Wu et al. (2006), Zhang et al. (2011)
Sesquiterpene	β-Caryophyllene	*Zea mays* L.	Expression of *BCAR*	Degenhardt et al. (2003a, 2003b), Robert et al. (2013)
Sesquiterpene	3S-(E)-nerolidol	*Arabidopsis thaliana* L.	Targeting expression of *NES* in mitochondria	Kappers et al. (2005)
Sesquiterpene	Valencene	*Nicotiana benthamiana* L.	Heterologous expression of *STS* and silencing of native *SQS* and *STS*	Cankar et al. (2015)
Sesquiterpene	Costunolide	*Nicotiana benthamiana* L.	Targeting expression of *STS* in mitochondria	Liu et al. (2011)
Diterpene	Taxadiene	*Nicotiana benthamiana* L.	Silencing the endogenous *PSY*	Hasan et al. (2014)
Diterpene	Toxoid	*Nicotiana benthamiana* L.	Targeted expression of *TS*, *TH* and *CPR* in plastids	Li et al. (2019)
Diterpene	Taxol	*Taxus chinensis* L.	Expression of *NCED* gene	Li et al. (2012)

(continued)

Table 12.1 (continued)

Terpenoid class	Compound/s	Host plant	Metabolic intervention	References
Diterpene	Tanshinones	*Salvia miltiorrhiza* L.	Targeted co-expression of *SmGGPPS* and *SmDXSII* genes in hairy roots	Shi et al. (2016)
Diterpene	Casbene, jolkinol C, and epi-jolkinol C	*Nicotiana benthamiana* L.	Targeting expression of MEP key genes, *GGPPS*, and *JcCAS* genes	Forestier et al. (2021)
Triterpene	Squalene	*Nicotiana tabacum* L.	Targeting expression of *FDS* in chloroplast	Jiang et al. (2016)
Triterpene	Botryococcene	*Nicotiana tabacum* L.	Targeting expression of *BS* in chloroplast	Jiang et al. (2016)
Triterpene	β- amyrin	*Nicotiana benthamiana* L.	Targeting expression of *SQS*	Reed et al. (2017)
Triterpene	Cucurbitacin E	*Nicotiana benthamiana* L.	Targeting co-expression of *CpSEs* and triterpene cyclases	Dong et al. (2018)
Triterpene	Ginsenoside	*Panax ginseng* L.	Targeting expression of *PgMVD* and *PgFPS*	Kim et al. (2014)
Triterpene	Botryococcene	*Brachypodium distachyon* L.	Subcellular targeting of *BS* and *FPS* and amplification of *BDSQE1*	Kempinski et al. (2019)
Triterpene	Dammarenediol-II	*Nicotiana tabacum* L.	Targeted expression of *PgDDS* into genome of cell suspension culture	Han et al. (2014)
Triterpene	Botryococcene and squalene	*Arabidopsis thaliana* L.	Targeting co-expression of *TS*, *FPS* and *DXS* to plastid and cytosol	Kempinski and Chappell (2019)
Triterpene	Protopanaxadiol	*Nicotiana tabacum* L.	Targeting co-expression of *PgDDS* and *CYP716A47* in cell suspension cultures	Chun et al. (2015)
Tetraterpene	Astaxanthin	*Oryza sativa* L.	Targeting co-expression of *sZmPSY1*, *sPaCrtI*, *sCrBKT*, and *sHpBHY* in rice endosperm	Paine et al. (2005), Zhu et al. (2018)

Engineering Monoterpenes

The leaves of *Lavandula latifolia* were genetically engineered for elevated synthesis of monoterpene such as linalool (Fig. 12.2). The linalool synthase gene *(LIS)* transferred into *Lavandula latifolia* was derived from *Clarkia breweri*, a wild flowering plant. It was observed that transgenic plants showed highest synthesis of linalool particularly in their youngest leaves (Mendoza-Poudereux et al., 2014). *LIS* gene encodes enzyme S-linalool synthase that mediates the conversion of geranyl pyrophosphate into linalool in plants (Lücker et al., 2001). Monoterpene synthase

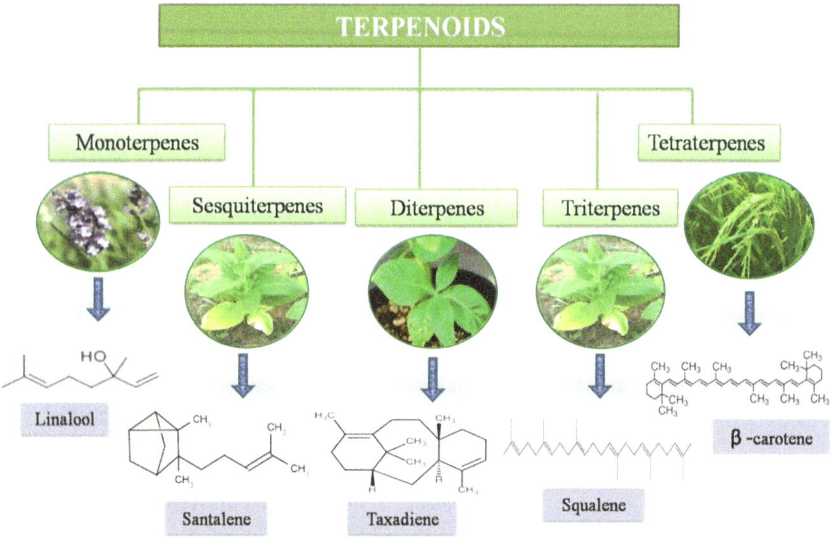

Fig. 12.2 Representative terpenoid compounds from their respective plant sources

has been observed as key gene responsible for synthesis of diverse monoterpenoids. The manipulation of monoterpene synthase gene expressing diverse monoterpene synthases was found to be the primary step that enhances the monoterpenoid profile of essential oils. This genetic metabolic strategy has been exploited in plants or tissues which showed elevated synthesis of monoterpenes as compared to control plants (Aharoni et al., 2005; Degenhardt et al., 2003a, 2003b; Lewinsohn et al., 2001; Wu et al., 2006). Moreover, targeting monoterpene biosynthesizing enzymes such as geraniol synthase (GES) and geraniol diphosphate synthase (GPS) in compartments of cell of *N. benthamiana* such as mitochondria, chloroplast, and cytosol leads to formation of monoterpene compounds such as geraniol. It was also observed that subcellular targeting of *GES* and *GPS* genes caused highest production of geraniol and other monoterpene derivatives in chloroplasts followed by mitochondria and cytosol (Dong et al., 2016). The results of compartmentalized targeting highlighted that transient expression of key terpenoid genes enhanced the production of target metabolites. Modifying the subcellular components for expression of terpene synthase was also found to be an efficient mechanism for elevating the monoterpene synthesis in transgenic plants (Farhi et al., 2011; Wu et al., 2012). The metabolic engineering attempts were carried by incorporation of small subunit of enzyme geranyl diphosphate synthase isolated from *Mentha* × *Piperita* (Mp.GPS. SSU) in enhancing the levels of various kinds of monoterpenes such as (−)-limonene, (−)-linalool, and myrcene in transgenic *Nicotiana benthamiana* L. *and Nicotiana tabacum* L. through overexpression of activity in key gene *geranyl diphosphate synthase (GPS).* Besides, the overexpression of *Mp.GPS.SSU* in *N. tabacum* L. was observed to elevate the levels of diterpenoids (GA_3) by enhancing the transcription levels of MEP pathway genes and geranylgeranyl diphosphate synthases (GGPS3, GGPS4). *N. benthamiana* L. and *N. tabacum* L. were grown in glasshouse. *Agrobacterium*-mediated transient expression of gene *Mp.GPS.SSU* was carried in *N. benthamiana* L. while, in *N. tabacum*, L. *Agrobacterium*-mediated transformation was achieved by following standard protocol (Yin et al., 2017). In other study, silencing of native *limonene synthase (MsLS)* was carried by RNA interference (RNAi) to achieve the heterologous production of monoterpenes in *Mentha spicata* L. (spearmint). The heterologous monoterpene synthase genes namely *linalool synthase (LS)* from *Picea abies* L. and *geraniol synthase (GS)* derived from *Cananga odorata* L. were incorporated in genome of RNAi silenced transgenic lines after successful cloning. The expression of these heterologous monoterpene synthases substantially enhanced the titer of monoterpenes derivatives. However, among all monoterpene, the marked increase was observed in derivatives of geraniol (Li et al., 2020). It was also demonstrated that overexpression of lipid transfer protein gene isolated from tobacco *(NtLTP1)* inserted in transgenic (*Mentha* × *Piperita citrata*) through *Agrobacterium*-mediated transformation. The transformed *Mentha* × *Piperita citrata* lines showed enhanced production of volatile monoterpenoids as compared to control plants. Almost 11 monoterpenoids were identified in volatile emissions of transgenic orange mint by gas chromatography-mass spectroscopy (GC-MS). Besides the monoterpenoid compound, linalyl acetate was predominantly found along with cis geraniol and geranyl acetate. The results from study showed that *LTP* gene played key role in releasing the lipids effectively from glandular trichome heads that results in enhanced synthesis of monoterpenoid emissions from these glandular trichomes with further increase in the diameter of trichome heads (Hwang et al., 2020).

Engineering Sesquiterpenoids

Biosynthesis of some sesquiterpenoid compounds such as santalene and bergamotene was enhanced in transgenic tobacco plant, *Nicotiana tabacum* L., by targeting mevalonate pathways occurring in chloroplast and cytosol. In transgenic tobacco, the co-expression of *3-hydroxy-3-methylglutaryl-CoA reductase (HMGR)* gene or truncated form of *HMGR* with *santalene synthase (SaSSy)* facilitated the elevated synthesis of santalene (Fig. 12.2) and bergamotene (Yin & Wong, 2019). Various metabolic strategies were employed in *N. tabacum* to enhance the production of various kinds of sesquiterpenoids due to ease in cultivation and lesser cost input (Tremblay et al., 2010). Besides, the biosynthetic step catalyzed by *HMGR* has been observed as the major rate-limiting step in mevalonic acid pathway (MVA). The manipulation of key gene, *HMGR,* by its overexpression has led to three to six times more terpenoid production compared to wild-type tobacco plants (Chappell et al., 1995). The overexpression of other genes like *sesquiterpene synthase (STS)* and *farnesyl diphosphate synthase (FDS)* in chloroplast has caused thousand times accumulation of target sesquiterpenoids such as patchoulol (amorpha-4,11-diene) than other transgenic tobacco plants which showed only cytosolic expression of *STS* gene (Wu et al., 2006; Zhang et al., 2011). Other attempts were also made to engineer plants for enhanced sesquiterpenoid production which involved incorporation of *β-caryophyllene (BCAR)* gene from Oregano into wild and cultivated maize. Consequently, genetically modified maize crops showed constitutive expression of volatile sesquiterpenes, (E)-β-caryophyllene. (Degenhardt et al., 2003a, 2003b; Robert et al., 2013). Meanwhile, 3S-(E)-nerolidol, an important sesquiterpene compound exhibiting role in plant protection, showed enhanced synthesis in *Arabidopsis thaliana*. *Nerolidol synthase* gene was introduced from strawberry into mitochondria of *A. thaliana*. Metabolic manipulation of *A. thaliana* involved constitutive promoters that caused overexpression of target metabolites throughout the plant (Kappers et al., 2005). Various cyclized sesquiterpenoids were produced at elevated levels in *N. benthamiana* by targeting co-expression of two key enzymes geranylfarnesyl diphosphate synthases (GFPS) and sesquiterpene synthase from *A. thaliana* (Huang et al., 2017). Further simultaneous silencing of *squalene synthase (SQS)* and *sesquiterpene synthase (STS)* genes in mevalonate pathway of *N. benthamiana* allowed only heterologous sesquiterpene synthase to act upon available substrates farnesyl diphosphate (FPP) that resulted in about 2.8 times increase in sesquiterpene compound (valencene) (Cankar et al., 2015). Subcellular targeting of *sesquiterpene synthase* in mitochondria caused 15 times more expression of end product (costunolide) than its expression observed in control lines of *N. benthamiana* (Liu et al., 2011).

Engineering Diterpenoids

The diterpene, toxoids were synthesized in *Nicotiana benthamiana* by Agrobacterium mediated transfer of selective diterpenoid synthesizing genes derived from Yew tree. The key genes involved in taxol pathway including *taxadiene synthase (TS)*, *taxadiene-5α-hydroxylase (TH)* and *cytochrome P450 reductase (CPR)* were targeted in chloroplasts of *N. benthamiana* along with supply of terpene precursors. The transgenic lines showed considerable enhancement of taxol intermediates (taxadiene and taxadiene-5α-ol) (Fig. 12.2) (Li et al., 2019). The silencing of *Phytoene synthase gene (PSY)* in MEP pathway of *N. benthamiana* was also found to increase the accumulation of taxadiene by inhibiting the enzymatic competition of endogenous phytoene synthase with taxadiene synthase for basic precursor geranylgeranyl diphosphate (GGPP) (Hasan et al., 2014). The key gene *9-cis-epoxycarotenoid dioxygenase (NCED)* involved in abscisic acid pathway (ABA) was overexpressed in cells of *Taxus chinensis* L. The overexpression of *NCED* leads to about 2.7 times increased production of taxol than that of control lines. Moreover, the transformed cells showed 48% increased production of ABA (Li et al., 2012). Hairy roots of *Salvia miltiorrhiza* L. were genetically manipulated by incorporation of key genes involved in production of diterpenoid, tanshinones. It was observed that the overexpression of these two key genes *geranylgeranyl diphosphate synthase (SmGGPPS)* and *1-deoxy-D-xylulose-5-phosphate synthase (SmDXSII)* in hairy roots showed higher biosynthesis of diterpenoids than transgenic lines transformed with single gene and control plants. Also, the transgenic lines which were transformed with both genes *SmGGPPS* and *SmDXSII* showed highest yield of tanshinones. Besides, there were increased synthesis of other significant compounds such as carotenes, chlorophylls, indole acetic acid, and gibberellins (Shi et al., 2016). Recently, *Nicotiana benthamiana* L. was genetically engineered for enhanced production of novel diterpenoids which involved the upregulation of key genes of MEP pathway for the biosynthesis of precursor compounds (GGPP) obtained from *Arabidopsis thaliana* L. Various metabolic engineering attempts were devised to optimize the production of casbene from GGPP. The eight genes from MEP pathway, *GGPP synthase (GGPPS)*, and *Jatropha curcas casbene synthase (JcCAS)* were evaluated in combination for biosynthesis diterpenoids from GGPP precursors. Transient expression of target metabolites was achieved when genes of MEP pathway were used with most appropriate cluster of gene promoters for casbene synthesis. The insertion of selective MEP pathway gene construct in single vector enhanced the production of casbene as compared to single gene transformation of casbene synthase. The ability of vector carrying MEP pathway genes to upregulate the MEP pathway and direct flux of casbene played key role in metabolic engineering of diterpenoids and its derived compounds such as jolkinol C and epi-jolkinol C (Forestier et al., 2021).

Engineering Triterpenoids

Nicotiana tabacum was metabolically engineered by introducing specific reaction steps of triterpene metabolism occurring in green alga *Botryococcus braunii*. Further, the supply of prenyl diphosphate precursors (FPP, IPP) was diverted from MVA and MEP pathways to the chloroplast by targeting overexpression of enzyme coding genes like *farnesyl diphosphate synthase (FDS) and botrycoccene synthases (BS)* to attain high levels of triterpene synthesis including squalene (C30) (Fig. 12.2), botryococcene (C30), and other methyl derivatives (C31–C37) (Jiang et al., 2016). Although the strategy of metabolic engineering of tobacco plants was found successful in elevating the synthesis of various kinds of terpenes, the highest terpene accumulation was revealed in triterpenes reported as 200–1000 μg/g fresh weight of triterpenes (Wu et al., 2006; Kempinski et al., 2015). Further for obtaining higher yields of methyl triterpenes, triterpene methyl transferases (TMTs) were introduced in chloroplast of transgenic tobacco plants to express higher amounts of triterpenes (Jiang et al., 2016). The metabolic engineering attempt in mevalonate pathway of tobacco plant by transient expression of key limiting enzyme SQS derived from *Avena sativa* L. leads to three times increased synthesis of triterpene (β-amyrin) (Reed et al., 2017). Three squalene epoxidases were isolated from *Cucurbita pepo (CpSEs)* which were co-expressed with four triterpene cyclases in *Nicotiana benthamiana* L. The transient expression of *CpSEs* along with triterpene cyclases enhanced the triterpene synthesis. Among all three *CpSEs, CpSE 2* was found to be most effective in triggering the triterpene synthesis. Hairy root lines of *C. pepo* transformed with *CpSE 2* resulted into two times increase in cucurbitacin *Eas* compared to control lines. These results showed the effect of SEs in increasing the substrate channeling to triterpene cyclases which enhanced the production of triterpenes in plants (Dong et al., 2018). *Agrobacterium rhizogenes* mediated genetic transformation of genes responsible for expression of key enzymes involved in terpene biosynthesis such as *mevalonate-5-phosphate decarboxylase (MVD)* and *farnesyl pyrophosphate synthase (FPS)* in hairy roots of *Panax ginseng* L. which revealed the higher expression of these enzyme coding genes *PgMVD* and *PgFPS* as compared to control. The transgenic lines transformed with *PgMVD* showed about four times increase in stigmasterol content. Similarly, transgenic lines expressing *PgFPS* revealed two times enhanced yield of ginsenoside than reported in control lines (Kim et al., 2014). The elevated synthesis of tetracyclic triterpene sapogenin (protopanaxadiol) was reported in transgenic tobacco. The two key gene coding enzymes involved in biosynthesis of protopanaxadiol were reported as *PgDDS* and *CYP716A47. PgDDS* aided in synthesis of primary precursor dammarenediol-II (DD), while *CYP716A47* mediated the synthesis of protopanaxadiol by hydroxylation of DD. The co-overexpression of the genes *PgDDS* and *CYP716A47* in cell suspension cultures of transgenic tobacco lead to increased synthesis of protopanaxadiol (Chun et al., 2015). High levels of triterpene (botryococcene) synthesis were reported in *Brachypodium distachyon* L., a model monocot. Accumulation of triterpene compound was achieved due to subcellular targeting of *botryococcene synthase (BS)* and *FPP synthase (FPS)* in either cytosol or chloroplast. However, the increased titers of botryococcene were observed in cytosolic targeting of BS, FPS enzymes. The genetic engineering of *B. distachyon* L. was attained with amplification of putative gene *BDSQE1* from the cDNA of wild plants using polymerase (PrimeStar) and primers (P42 and P43) (Kempinski et al., 2019). Further, the enhanced synthesis of linear hydrocarbon terpenes like botryococcene and squalene were observed in oil seeds produced from genetically engineered *Arabidopsis thaliana* L. The strategy employed the usage of seed-specific promoters for observing the expression of *terpene synthase (TS)* alone or in combination with *farnesyl diphosphate synthase (FPS)* and *1-deoxyxylulose 5-phosphate synthase (DXS)*. In contrast to cytosolic targeting of MVA pathway, highest yield of squalene and botryococcene was reported when the gene construct comprising *TS, FPS, and DXS* was targeted to plastid. Besides, increased heterologous synthesis of botryococcene was reported in cytosol than natively produced squalene (Kempinski & Chappell, 2019). The cell suspension culture of *Nicotiana tabacum* L. was genetically engineered for enhanced production of tetracyclic triterpenoid (Dammarenediol-II). The biosynthesis of tetracyclic terpene was induced in cell suspension culture of tobacco plants by *Agrobacterium*-mediated insertion of *Dammarenediol-II synthase gene (PgDDS)* obtained from *Panax ginseng* L. into genome of *N. tabacum*. Dammarenediol-II levels were reported highest in roots of transgenic lines. The overexpression of *PgDDS* was mediated by strong constitutive 35S promoters (Han et al., 2014).

Engineering Tetraterpenoids

Research was conducted in bioengineering of astaxanthin (a red-colored ketocarotenoid) biosynthesis in rice endosperm by introducing the key genes *sZmPSY1*, *sPaCrtI*, *sCrBKT*, and *sHpBHY*. These genes encode the enzymes phytoene synthase, phytoene desaturase, beta-carotene ketolase, and β-carotene hydroxylase, respectively. Metabolic engineering of biosynthetic pathways of carotenoids was carried in rice endosperm by transient overexpression of set of four genes, namely, *sZmPSY1*, *sPaCrtI*, *sCrBKT*, and *sHpBHY* genes mediated by specific rice endosperm promoters that enabled the de novo biosynthesis of carotenoids. The genetic manipulation of biosynthetic pathways of endosperm leads to synthesis of various varieties of germplasm. For example, the golden rice that was found to accumulate higher levels of β-carotene, and orange red germplasm showed higher synthesis of canthaxanthin and astaxanthin (Fig. 12.2). Various studies have suggested that endosperm of *Oryza sativa* L. seeds acted as most suitable bioreactors for synthesis of various terpenoids through metabolic engineering that also aids in crop biofortification (Paine et al., 2005; Blancquaert et al., 2015; Zhu et al., 2017). It was also observed that transgenic expression of key genes involved in β-carotene biosynthesis caused production of golden rice 2 variety (Paine et al., 2005; Zhu et al., 2018).

Potential Limitations and Future Prospects

Various studies have clearly suggested that various model plants like *N. benthamiana* L. were feasible for engineering terpenoid metabolic pathways in enhancing the yield of desired metabolite. However, the major drawback was observed in matter of scale from milligrams to gram quantities of target metabolite. Infiltrating the volume of plants was found technically laborious that was mitigated by utilizing the vacuum infiltration carried in *N. benthamiana* L. (Andersen-Ranberg et al., 2016; Huang et al., 2017; Reed et al., 2017). In the process of vacuum infiltration, the aerial parts of *N. benthamiana* L. such as leaves were used to remove air from the interstitial spaces between leaf cells using vacuum (Reed et al., 2017). In *N. benthamiana* L., the commercial production of various pharmaceutical proteins was elevated using vacuum infiltration process (Holtz et al., 2015). In other metabolic engineering studies, the undesirable production of side products is formed by chemical changes (oxidation, glycosylation, and dephosphorylation) of target compound (Brückner & Tissier, 2013; Dong et al., 2013; Khakimov et al., 2015; Liu et al., 2011; Wang et al., 2016). In some cases, formation of new bioactive compound interferes with native metabolism of host plant which causes unfavorable changes in native metabolic profile of plant (Liu et al., 2014). The genome editing strategies were followed in silencing the native *sugar transferases* and *oxidases* which crosstalk with heterologous genes. The genome editing approach was applied in *N. benthamiana* L. by silencing the native glycosyltransferases to prevent the reduction of synthesized precursors that were targeted for glycosylation. This also allowed maximum expression of heterologous genes like *lipid transfer protein 3 (LTP3)* and pleiotropic drug resistance *Transporter 2 (PD2)* isolated from *Artemisia annua* L. incorporated into genome of *N. benthamiana* L. which resulted in enhanced production and targeted transport of desired product, namely, artemisinin precursors to apoplastic compartment (Wang et al., 2016).

In recent times, revolutionary technological advancements in functional genomics, transcriptomics, metabolomics, and proteomics along with noticeable innovations in synthetic and computational biology have substantially eliminated potential constraints and have elucidated unknown metabolic pathways (Jacobowitz & Weng, 2020). This has also helped in thorough understanding of plant metabolism and intermetabolic connections in primary, secondary, and phytohormone biosynthetic pathways (Erb & Kliebenstein, 2020). The metabolic flux analysis techniques have resolved the intricacies of carbon flux distribution among various metabolic pathways, their targeting to subcellular compartments, elucidation of metabolic routes, and inactive metabolic pools (Shih & Morgan, 2020). These technological developments have not only delineated the metabolic target pathways but have enhanced the potential of metabolic modeling. These metabolic models could not be only elusive but can also prove to be predictive to ensure the success of metabolic engineering approaches (Lynch et al., 2021).

Concluding Remarks

Plant terpenoids display remarkable diversity in structure, biological functions, and pharmaceutical applications. Metabolic and biotechnological interventions in plant-based system provide attractive platforms for the exploitation of terpenoid compounds on commercial scale. However, the genetic and metabolic engineering attempts relied upon underlying mechanism and regulation of terpenoid metabolic pathways and functional characterization of key genes encoding terpene biosynthetic enzymes. Recent studies have suggested RNAi technology; inducible and time-specific gene promoters active in specific target tissues have caused significant production of various pharmaceutically important terpenoid compounds. However, lot of research is still needed to be undertaken in various limitations arising from crosstalk between native and heterologous terpene metabolic pathways, genes or gene products, production of undesirable by-products, insignificant yield of target end compounds, and ethical and biosafety issues pertaining to development of transgenic lines.

References

Aharoni, A., Jongsma, M. A., & Bouwmeester, H. J. (2005). Volatile science? Metabolic engineering of terpenoids in plants. *Trends in Plant Science, 10*(12), 594–602.

Andersen-Ranberg, J., Kongstad, K. T., Nielsen, M. T., Jensen, N. B., Pateraki, I., Bach, S. S., et al. (2016). Expanding the landscape of diterpene structural diversity through stereochemically controlled combinatorial biosynthesis. *Angewandte Chemie International Edition, 55*(6), 2142–2146.

Ashour, M., Wink, M., & Gershenzon, J. (2010). Biochemistry of terpenoids: Monoterpenes, sesquiterpenes and diterpenes. *Annual Plant Reviews, 40*, 258–303. Biochemistry of plant secondary metabolism.

Blancquaert, D., Van Daele, J., Strobbe, S., Kiekens, F., Storozhenko, S., De Steur, H., et al. (2015). Improving folate (vitamin B 9) stability in biofortified rice through metabolic engineering. *Nature Biotechnology, 33*(10), 1076–1078.

Bohlmann, J., & Keeling, C. I. (2008). Terpenoid biomaterials. *The Plant Journal, 54*(4), 656–669.

Boutanaev, A. M., Moses, T., Zi, J., Nelson, D. R., Mugford, S. T., Peters, R. J., & Osbourn, A. (2015). Investigation of terpene diversification across multiple sequenced plant genomes. *Proceedings of the National Academy of Sciences, 112*(1), E81–E88.

Brückner, K., & Tissier, A. (2013). High-level diterpene production by transient expression in *Nicotiana benthamiana*. *Plant Methods, 9*(1), 1–10.

Cankar, K., Jongedijk, E., Klompmaker, M., Majdic, T., Mumm, R., Bouwmeester, H., et al. (2015). (+)-Valencene production in *Nicotiana benthamiana* is increased by down-regulation of competing pathways. *Biotechnology Journal, 10*(1), 180–189.

Carrie, C., Murcha, M. W., Millar, A. H., Smith, S. M., & Whelan, J. (2007). Nine 3-ketoacyl-CoA thiolases (KATs) and acetoacetyl-CoA thiolases (ACATs) encoded by five genes in *Arabidopsis thaliana* are targeted either to peroxisomes or cytosol but not to mitochondria. *Plant Molecular Biology, 63*(1), 97–108.

Chappell, J., Wolf, F., Proulx, J., Cuellar, R., & Saunders, C. (1995). Is the reaction catalyzed by 3-hydroxy-3-methylglutaryl coenzyme A reductase a rate-limiting step for isoprenoid biosynthesis in plants? *Plant Physiology, 109*(4), 1337–1343.

Chen, F., Tholl, D., Bohlmann, J., & Pichersky, E. (2011). The family of terpene synthases in plants: A mid-size family of genes for specialized metabolism that is highly diversified throughout the kingdom. *The Plant Journal, 66*(1), 212–229.

Christianson, D. W. (2017). Structural and chemical biology of terpenoid cyclases. *Chemical Reviews, 117*(17), 11,570–11,648.

Chun, J. H., Adhikari, P. B., Park, S. B., Han, J. Y., & Choi, Y. E. (2015). Production of the dammarene sapogenin (protopanaxadiol) in transgenic tobacco plants and cultured cells by heterologous expression of *PgDDS* and *CYP716A47*. *Plant Cell Reports, 34*(9), 1551–1560.

Degenhardt, J., Gershenzon, J., Baldwin, I. T., & Kessler, A. (2003a). Attracting friends to feast on foes: Engineering terpene emission to make crop plants more attractive to herbivore enemies. *Current Opinion in Biotechnology, 14*(2), 169–176.

Degenhardt, L., Hall, W., & Lynskey, M. (2003b). Exploring the association between cannabis use and depression. *Addiction, 98*(11), 1493–1504.

Dong, L., Miettinen, K., Goedbloed, M., Verstappen, F. W., Voster, A., Jongsma, M. A., & Bouwmeester, H. J. (2013). Characterization of two geraniol synthases from Valeriana officinalis and Lippia dulcis: Similar activity but difference in subcellular localization. *Metabolic Engineering, 20*, 198–211.

Dong, L., Jongedijk, E., Bouwmeester, H., & Van Der Krol, A. (2016). Monoterpene biosynthesis potential of plant subcellular compartments. *New Phytologist, 209*(2), 679–690.

Dong, L., Pollier, J., Bassard, J. E., Ntallas, G., Almeida, A., Lazaridi, E., et al. (2018). Co-expression of *squalene epoxidases* with *triterpene cyclases* boosts production of triterpenoids in plants and yeast. *Metabolic Engineering, 49*, 1–12.

Drew, D. P., Krichau, N., Reichwald, K., & Simonsen, H. T. (2009). Guaianolides in Apiaceae: Perspectives on pharmacology and biosynthesis. *Phytochemistry Reviews, 8*(3), 581–599.

Dudareva, N., Negre, F., Nagegowda, D. A., & Orlova, I. (2006). Plant volatiles: Recent advances and future perspectives. *Critical Reviews in Plant Sciences, 25*(5), 417–440.
Dudareva, N., & Pichersky, E. (2008). Metabolic engineering of plant volatiles. *Current Opinion in Biotechnology, 19*(2), 181–189.
Erb, M., & Kliebenstein, D. J. (2020). Plant secondary metabolites as defenses, regulators, and primary metabolites: The blurred functional trichotomy. *Plant Physiology, 184*(1), 39–52.
Farhi, M., Marhevka, E., Ben-Ari, J., Algamas-Dimantov, A., Liang, Z., Zeevi, V., et al. (2011). Generation of the potent anti-malarial drug artemisinin in tobacco. *Nature Biotechnology, 29*(12), 1072–1074.
Forestier, E. C., Czechowski, T., Cording, A. C., Gilday, A. D., King, A. J., Brown, G. D., & Graham, I. A. (2021). Developing a *Nicotiana benthamiana* transgenic platform for high-value diterpene production and candidate gene evaluation. *Plant Biotechnology Journal.*
Gutensohn, M., Nagegowda, D. A., & Dudareva, N. (2012). Involvement of compartmentalization in monoterpene and sesquiterpene biosynthesis in plants. In *Isoprenoid synthesis in plants and microorganisms* (pp. 155–169). Springer.
Hamberger, B., Ohnishi, T., Hamberger, B., Séguin, A., & Bohlmann, J. (2011). Evolution of diterpene metabolism: Sitka spruce CYP720B4 catalyzes multiple oxidations in resin acid biosynthesis of conifer defense against insects. *Plant Physiology, 157*(4), 1677–1695.
Han, J. Y., Wang, H. Y., & Choi, Y. E. (2014). Production of dammarenediol-II triterpene in a cell suspension culture of transgenic tobacco. *Plant Cell Reports, 33*(2), 225–233.
Hasan, M. M., Kim, H. S., Jeon, J. H., Kim, S. H., Moon, B., Song, J. Y., et al. (2014). Metabolic engineering of *Nicotiana benthamiana* for the increased production of taxadiene. *Plant Cell Reports, 33*(6), 895–904.
Hemmerlin, A., Hoeffler, J. F., Meyer, O., Tritsch, D., Kagan, I. A., Grosdemange-Billiard, C., et al. (2003). Cross-talk between the cytosolic mevalonate and the plastidial methyl erythritol phosphate pathways in tobacco bright yellow-2 cells. *Journal of Biological Chemistry, 278*(29), 26,666–26,676.
Holtz, B. R., Berquist, B. R., Bennett, L. D., Kommineni, V. J., Munigunti, R. K., White, E. L., et al. (2015). Commercial-scale biotherapeutics manufacturing facility for plant-made pharmaceuticals. *Plant Biotechnology Journal, 13*(8), 1180–1190.
Huang, A. C., Kautsar, S. A., Hong, Y. J., Medema, M. H., Bond, A. D., Tantillo, D. J., & Osbourn, A. (2017). Unearthing a sesterterpene biosynthetic repertoire in the Brassicaceae through genome mining reveals convergent evolution. *Proceedings of the National Academy of Sciences, 114*(29), E6005–E6014.
Hwang, H. S., Adhikari, P. B., Jo, H. J., Han, J. Y., & Choi, Y. E. (2020). Enhanced monoterpene emission in transgenic orange mint *(Mentha× piperita f. citrata)* overexpressing a tobacco *lipid transfer protein (NtLTP1). Planta, 252*(3), 1–12.
Jacobowitz, J. R., & Weng, J. K. (2020). Exploring uncharted territories of plant specialized metabolism in the post genomic era. *Annual Review of Plant Biology, 71*, 631–658.
Jiang, Z., Kempinski, C., Bush, C. J., Nybo, S. E., & Chappell, J. (2016). Engineering triterpene and methylated triterpene production in plants provides biochemical and physiological insights into terpene metabolism. *Plant Physiology, 170*(2), 702–716.
Kappers, I. F., Aharoni, A., Van Herpen, T. W., Luckerhoff, L. L., Dicke, M., & Bouwmeester, H. J. (2005). Genetic engineering of terpenoid metabolism attracts bodyguards to Arabidopsis. *Science, 309*(5743), 2070–2072.
Kempinski, C., & Chappell, J. (2019). Engineering triterpene metabolism in the oilseed of *Arabidopsis thaliana. Plant Biotechnology Journal, 17*(2), 386–396.
Kempinski, C., Jiang, Z., Bell, S., & Chappell, J. (2015). Metabolic engineering of higher plants and algae for isoprenoid production. *Biotechnology of Isoprenoids*, 161–199.
Kempinski, C., Jiang, Z., Zinck, G., Sato, S. J., Ge, Z., Clemente, T. E., & Chappell, J. (2019). Engineering linear, branched-chain triterpene metabolism in monocots. *Plant Biotechnology Journal, 17*(2), 373–385.
Khakimov, B., Kuzina, V., Erthmann, P. Ø., Fukushima, E. O., Augustin, J. M., Olsen, C. E., et al. (2015). Identification and genome organization of saponin pathway genes from a wild crucifer,

and their use for transient production of saponins in *Nicotiana benthamiana*. *The Plant Journal, 84*(3), 478–490.

Kim, Y. K., Kim, Y. B., Uddin, M. R., Lee, S., Kim, S. U., & Park, S. U. (2014). Enhanced triterpene accumulation in *Panax ginseng* hairy roots overexpressing mevalonate-5-pyrophosphate decarboxylase and farnesyl pyrophosphate synthase. *ACS Synthetic Biology, 3*(10), 773–779.

Lange, B. M., & Ahkami, A. (2013). Metabolic engineering of plant monoterpenes, sesquiterpenes and diterpenes—Current status and future opportunities. *Plant Biotechnology Journal, 11*(2), 169–196.

Leivar, P., González, V. M., Castel, S., Trelease, R. N., López-Iglesias, C., Arró, M., et al. (2005). Subcellular localization of Arabidopsis 3-hydroxy-3-methylglutaryl-coenzyme A reductase. *Plant Physiology, 137*(1), 57–69.

Lewinsohn, E., Schalechet, F., Wilkinson, J., Matsui, K., Tadmor, Y., Nam, K. H., et al. (2001). Enhanced levels of the aroma and flavor compound S-linalool by metabolic engineering of the terpenoid pathway in tomato fruits. *Plant Physiology, 127*(3), 1256–1265.

Li, C., Sarangapani, S., Wang, Q., Nadimuthu, K., & Sarojam, R. (2020). Metabolic engineering of the native monoterpene pathway in spearmint for production of heterologous monoterpenes reveals complex metabolism and pathway interactions. *International Journal of Molecular Sciences, 21*(17), 6164.

Li, J., Mutanda, I., Wang, K., Yang, L., Wang, J., & Wang, Y. (2019). Chloroplastic metabolic engineering coupled with isoprenoid pool enhancement for committed taxanes biosynthesis in *Nicotiana benthamiana*. *Nature Communications, 10*(1), 1–12.

Li, S. T., Fu, C. H., Zhang, M., Zhang, Y., Xie, S., & Yu, L. J. (2012). Enhancing taxol biosynthesis by overexpressing a 9-cis-epoxycarotenoid dioxygenase gene in transgenic cell lines of *Taxus chinensis*. *Plant Molecular Biology Reporter, 30*(5), 1125–1130.

Liu, Q., Majdi, M., Cankar, K., Goedbloed, M., Charnikhova, T., Verstappen, F. W., et al. (2011). Reconstitution of the costunolide biosynthetic pathway in yeast and Nicotiana benthamiana. *PLoS One, 6*(8), e23255.

Liu, Q., Manzano, D., Tanić, N., Pesic, M., Bankovic, J., Pateraki, I., Ricard, L., Ferrer, A., de Vos, R., van de Krol, S., & Bouwmeester, H. (2014). Elucidation and in planta reconstitution of the parthenolide biosynthetic pathway. *Metabolic Engineering, 23*, 145–153.

Lücker, J., Bouwmeester, H. J., Schwab, W., Blaas, J., Van Der Plas, L. H., & Verhoeven, H. A. (2001). Expression of clarkia S-linalool synthase in transgenic petunia plants results in the accumulation of S-linalyl-β-d-glucopyranoside. *The Plant Journal, 27*(4), 315–324.

Lynch, J. H., Huang, X. Q., & Dudareva, N. (2021). Silent constraints: The hidden challenges faced in plant metabolic engineering. *Current Opinion in Biotechnology, 69*, 112–117.

Mendoza-Poudereux, I., Muñoz-Bertomeu, J., Navarro, A., Arrillaga, I., & Segura, J. (2014). Enhanced levels of S-linalool by metabolic engineering of the terpenoid pathway in spike lavender leaves. *Metabolic Engineering, 23*, 136 144.

Merret, R., Cirioni, J. R., Bach, T. J., & Hemmerlin, A. (2007). A serine involved in actin-dependent subcellular localization of a stress-induced tobacco BY-2 hydroxymethylglutaryl-CoA reductase isoform. *FEBS Letters, 581*(27), 5295–5299.

Muhlemann, J. K., Klempien, A., & Dudareva, N. (2014). Floral volatiles: From biosynthesis to function. *Plant, Cell & Environment, 37*(8), 1936–1949.

Nagegowda, D. A. (2010). Plant volatile terpenoid metabolism: Biosynthetic genes, transcriptional regulation and subcellular compartmentation. *FEBS Letters, 584*(14), 2965–2973.

Newman, D. J., & Cragg, G. M. (2016). Natural products as sources of new drugs from 1981 to 2014. *Journal of Natural Products, 79*(3), 629–661.

Paine, J. A., Shipton, C. A., Chaggar, S., Howells, R. M., Kennedy, M. J., Vernon, G., et al. (2005). Improving the nutritional value of Golden Rice through increased pro-vitamin A content. *Nature Biotechnology, 23*(4), 482–487.

Reed, J., Stephenson, M. J., Miettinen, K., Brouwer, B., Leveau, A., Brett, P., et al. (2017). A translational synthetic biology platform for rapid access to gram-scale quantities of novel drug-like molecules. *Metabolic Engineering, 42*, 185–193.

Robert, C. A. M., Erb, M., Hiltpold, I., Hibbard, B. E., Gaillard, M. D. P., Bilat, J., et al. (2013). Genetically engineered maize plants reveal distinct costs and benefits of constitutive volatile emissions in the field. *Plant Biotechnology Journal, 11*(5), 628–639.

Sapir-Mir, M., Mett, A., Belausov, E., Tal-Meshulam, S., Frydman, A., Gidoni, D., & Eyal, Y. (2008). Peroxisomal localization of Arabidopsis isopentenyl diphosphate isomerases suggests that part of the plant isoprenoid mevalonic acid pathway is compartmentalized to peroxisomes. *Plant Physiology, 148*(3), 1219–1228.

Shi, M., Luo, X., Ju, G., Li, L., Huang, S., Zhang, T., et al. (2016). Enhanced diterpene tanshinone accumulation and bioactivity of transgenic *Salvia miltiorrhiza* hairy roots by pathway engineering. *Journal of Agricultural and Food Chemistry, 64*(12), 2523–2530.

Shih, M. L., & Morgan, J. A. (2020). Metabolic flux analysis of secondary metabolism in plants. *Metabolic Engineering Communications, 10*, e00123.

Simkin, A. J., Guirimand, G., Papon, N., Courdavault, V., Thabet, I., Ginis, O., et al. (2011). Peroxisomal localisation of the final steps of the mevalonic acid pathway in planta. *Planta, 234*(5), 903–914.

Sprenger, G. A., Schörken, U., Wiegert, T., Grolle, S., De Graaf, A. A., Taylor, S. V., et al. (1997). Identification of a thiamin-dependent synthase in Escherichia coli required for the formation of the 1-deoxy-D-xylulose 5-phosphate precursor to isoprenoids, thiamin, and pyridoxol. *Proceedings of the National Academy of Sciences, 94*(24), 12,857–12,862.

Tremblay, R., Wang, D., Jevnikar, A. M., & Ma, S. (2010). Tobacco, a highly efficient green bioreactor for production of therapeutic proteins. *Biotechnology Advances, 28*(2), 214–221.

Wang, B., Kashkooli, A. B., Sallets, A., Ting, H. M., de Ruijter, N. C., Olofsson, L., et al. (2016). Transient production of artemisinin in *Nicotiana benthamiana* is boosted by a specific lipid transfer protein from A. annua. *Metabolic Engineering, 38*, 159–169.

Wang, W., Wang, Y., Zhang, Q., Qi, Y., & Guo, D. (2009). Global characterization of Artemisia annua glandular trichome transcriptome using 454 pyrosequencing. *BMC Genomics, 10*(1), 1–10.

Weitzel, C., & Simonsen, H. T. (2015). Cytochrome P450-enzymes involved in the biosynthesis of mono-and sesquiterpenes. *Phytochemistry Reviews, 14*(1), 7–24.

Wu, S., Jiang, Z., Kempinski, C., Nybo, S. E., Husodo, S., Williams, R., & Chappell, J. (2012). Engineering triterpene metabolism in tobacco. *Planta, 236*(3), 867–877.

Wu, S., Schalk, M., Clark, A., Miles, R. B., Coates, R., & Chappell, J. (2006). Redirection of cytosolic or plastidic isoprenoid precursors elevates terpene production in plants. *Nature Biotechnology, 24*(11), 1441–1447.

Yang, K., Monafared, R. S., Wang, H., Lundgren, A., & Brodelius, P. E. (2015). The activity of the artemisinic aldehyde Δ11 (13) reductase promoter is important for artemisinin yield in different chemotypes of Artemisia annua L. *Plant Molecular Biology, 88*(4), 325–340.

Yin, J. L., & Wong, W. S. (2019). Production of santalenes and bergamotene in *Nicotiana tabacum* plants. *PLoS One, 14*(1), e0203249.

Yin, J. L., Wong, W. S., Jang, I. C., & Chua, N. H. (2017). Co-expression of peppermint geranyl diphosphate synthase small subunit enhances monoterpene production in transgenic tobacco plants. *New Phytologist, 213*(3), 1133–1144.

Yu, F., & Utsumi, R. (2009). Diversity, regulation, and genetic manipulation of plant mono-and sesquiterpenoid biosynthesis. *Cellular and Molecular Life Sciences, 66*(18), 3043–3052.

Zhang, Y., Nowak, G., Reed, D. W., & Covello, P. S. (2011). The production of artemisinin precursors in tobacco. *Plant Biotechnology Journal, 9*(4), 445–454.

Zhu, Q., Yu, S., Zeng, D., Liu, H., Wang, H., Yang, Z., & Liu, Y. G. (2017). Development of "purple endosperm rice" by engineering anthocyanin biosynthesis in the endosperm with a high-efficiency transgene stacking system. *Molecular Plant, 10*(7), 918–929.

Zhu, Q., Zeng, D., Yu, S., Cui, C., Li, J., Li, H., & Liu, Y. G. (2018). From golden rice to aSTARice: Bioengineering astaxanthin biosynthesis in rice endosperm. *Molecular Plant, 11* (12), 1440–1448.

Chapter 13
Use of Metabolic Engineering/Biotechnology in Crops Breeding and Development of New Crops

Junaid A. Magray, Shabir A. Zargar, and Tajamul Islam

Abstract In the current era of population explosion, food security is one of the major challenges to the world. In fact, agricultural activity is considered most vulnerable to the climatic variations particularly extreme temperature and drought conditions and biotic stresses (pathogens and pests); same time, it increases the nutritional value of different plants. New approaches are therefore required for development of more productive and more resilient varieties of crop plants. Genetic engineering offers a promising tool to combat the negative consequences of climate variation on crop production. Genetic engineering systems enable precise manipulation of targeted organism's genome for development of genetically modified crops with improved agricultural traits. Various genome-editing tools are used to modify crop plant, which possess tolerance and resistance against one or more useful traits. These technologies are now becoming user-friendly tools for development of modified crops with desired agro-based traits. This chapter summarizes the use of different biotechnological tools for development of new crop with the main motive to benefit the humankind.

Keywords Agricultural traits · Genetically modified crops · Population explosion · Genetic engineering conventional

Introduction

Agriculture could not prosper and survive in a civilized world without new crop varieties, and from this perspective, it is apparent that plant breeding is among the key pillars of civilization. Plant breeding is a co-evolutionary mechanism between humans and edible plants (Ladizinsky, 1998). Humans changed the plants that were used in agriculture, resulting in the creation of new varieties (Jones et al., 2021). The human population is expected to hit 9 billion people in the coming decades which is

J. A. Magray · S. A. Zargar (✉) · T. Islam
Department of Botany, University of Kashmir, Srinagar, J&K, India

T. Aftab, K. R. Hakeem (eds.), *Metabolic Engineering in Plants*,
https://doi.org/10.1007/978-981-16-7262-0_13

a major concern as ten children die of hunger in every minute (Pinstrup-Andersen & Watson II, 2011; Kendall & Pimentel, 1994). One billion people worldwide are hungry, and two billion people have one or more micronutrient deficiencies, particularly vitamin A, iodine, and iron which are often grouped together as "hidden hunger" (Burchi et al., 2011). Agriculture is the primary source of food, but modern, intensive agriculture has a negative impact on the climate, contaminating drinking water, causing soil depletion and deforestation, and reducing biodiversity (Zargar et al., 2021). As a result of the increased demand for food, humans are confronted with major challenges. To ensure future agricultural food production, we must protect biodiversity and genetic diversity to ensure ecosystem resilience. Agriculture's long-term viability is no longer a choice; it is now a necessity. Changes in the growing season of different crops, for example, would necessitate increased crop breeding efforts (Jacobsen et al., 2013). Such breeding programmers need not only knowledge of current practices but also access to a large pool of genetic resources from established crops and breeds, as well as their wild relatives, in order to provide the genes needed to adapt to changes in agricultural production. As a result, agrobiodiversity should be a central component of long-term agricultural production, rather than just a source of traits for current breeding program. The philosophy of sustainability is based on the idea that current needs must be met without jeopardizing future generations demands (Lichtfouse et al., 2009). Sustainable agriculture is a method of resolving fundamental and basic problems of food production in an eco-friendly manner (Frison et al., 2011). If growing cropland area does not result in a substantial increase in cereal production, yield improvements would be necessary to satisfy demand. Due to low price of cereals, some farmers are vanishing the cultivation of cereals in favor of more lucrative crops, resulting in a decrease in ongoing investment in science, set-up, and irrigation facilities (Hazell, 2010). Furthermore, in some areas where water and fertilizer inputs are already high and the maximum profitable economic return has been achieved, sustaining the same rate of yield growth will be difficult (Pinstrup-Andersen et al., 1999).

Methods of Plant Breeding

Conventional Methods

Selection and hybridization of desirable trait have been used since the dawn of agronomy; the reason behind trait transfer retention of characters was not revealed till the mid-nineteenth century. In the eighteenth century, Gregor Mendel validated heredity by crossing different pea (*Pisum sativum*) varieties and observing flower and seed color, seed and pod form, flower position, and height of the plant in successive generations (Halford, 2006). At the roots of plant breeding is the selection of improved traits in terms of yield and quality of edible bits, convenience of production, harvest and processing, tolerance to stress conditions, and pest resistance (Rubatzky & Yamaguchi, 2012). Manipulation of a one trait while ignoring all others is simple; however, this will unlikely to produce useful variety. As a result, a large number of plants have been breeded to carry desirable traits like pest resistance (Russell, 2013). Resistant traits against blight from a Mexican potato species (*Solanum demissum*) have been concentrate into 50% of all potato cultivars is an example of conventional plant breeding (Colton et al., 2006). Other resistant crops include rust-resistant wheat (*Triticum aestivum*), blight-resistant corn (*Zea mays*), and aphid-resistant alfalfa (*Medicago sativa*). Advances in traditional plant breeding and pest-resistant plants have contributed to significant increases in crop yields of Corn yields (Jauhar, 2006).

Genetic Engineering Methods

Plants that have their genomes altered using genetic modification methods to enhance existing traits or introduce a new character or trait not found in the plant population naturally are known as genetically modified (GM) crops (Conko et al., 2016). Transgenic plants are those that have foreign segments of DNA inserted into their genome using methods such as *Agrobacterium*-mediated transformation or direct gene transfer (Kumar et al., 2020). The transgene or inserted gene may be from a different plant, virus, fungus, bacteria, or animal species (Rivera et al., 2012). As a result, genetic engineering methods resolve one of the main limitations of traditional plant breeding methods which only crosses sexually compatible organisms (Hansen, 2000). *Agrobacterium tumefaciens'* natural ability to stably insert Ti plasmid DNA (T-DNA) into host plant cell genome was discovered in 1977 (Schell et al., 1979), and Ti plasmid was suggested as a vector to introduce foreign genes into plant cells. This marks a significant step forward in the growth of transgenic plants. Following that, it was first stated that a particular gene sequence could be transferred to a plant cell using recombinant DNA and a transformation technique (Puchta & Fauser, 2014). In the same year, the first transgenic plants, tobacco and petunia, were created that are antibiotic-resistant. Kumar et al. (2020)investigated the expression of the bean's "phaseolin" gene in sunflower, paving the way for genes from distantly related plants species to be transferred and expressed. The Food and Drug Administration (FDA) approved the transgenic tomato "Flavr Savr" produced by Calgene (Monsanto) for sale in the United States in 1994. It has the property of delayed ripening or longer shelf life than normal tomatoes (Diretto et al., 2020). Scientists have focused their efforts over the last two decades on expanding genetic engineering techniques to include recombinant DNA (rDNA) techniques. rDNA allows development of new varieties in short span of time (Chawla, 2011). DNA segments from remotely interrelated species or even different biological kingdoms may be inserted using rDNA methods. Despite the fact that the location of a gene's insertion will affect its expression for the production of large number of plants, the plants with suitable level of gene expression will be selected. After a trait has been inserted using transgenic technique, the resulting plant can be crossed with those crop varieties that are obtained through conventional breeding (Bhat & Srinivasan, 2002).

Genetic engineering played a vital role in crop enhancement by inserting useful external gene(s) or suppressing the expression of existing gene(s) in plants. Abiotic stress tolerance, insect resistance, herbicide tolerance, nutritional enhancement, and disease resistance are all characteristics found in genetically modified plants. Approximately, 525 GMO are developed till now, and for cultivation around the world, only 32 crops have been cleared (Kumar et al., 2020). This technology has been proven to boost crop yields, reduction in the use of pesticide and insecticide, reduction in CO_2 emissions, and reduction in crop production costs. However, concerns about human toxicity and allergenicity as well as possible environmental risks, such as gene transfer, adverse effects on other organisms, and weed and insect resistance evolution, are preventing worldwide acceptance of GMO crops. In response to these issues' other technologies such as intragenesis and cisgenesis, techniques of genome editing have been recently introduced. Some of these substitute technologies can be used to grow cropland plant varieties that are free from any external genes; as a result, such crops are expected to have higher market acceptance and receive faster regulatory approvals than other transgenic crops (Lucht, 2015).

Approaches Used for the Development of Transgenic Crops

The development of transgenic crops both in monocots and dicots, has been widely grown using two main methods. The first method is biolistic bombardment, and the second method is transformation induced by *Agrobacterium*. The BioRad PDS 1000/He helium-powered gun and related designs, as well as the particle inflow gun, are the primary delivering systems in the biolistic protocol.

Various Application of Genetically Modified Crops

Nutritional Value

The world's most widely consumed grain is rice (*Oryza sativa* L.) that has no C_{40} carotenoid or carotene (provitamin A) precursors in its endosperm (Paine et al., 2005).. Children in developing countries, such as Sub-Saharan Africa (48 percent) and South Asia, have the highest prevalence of VAD 44 percent. β-carotene is a precursor molecule for vitamin A biosynthesis that does not occur naturally in edible sections of staple food crops like rice. Transgenic rice enriched with provitamin A in its endosperm was created to overcome vitamin A deficiency by engineering a pathway for β-carotene biosynthesis (Ye et al., 2000). Because of its yellow color, this genetically modified rice was dubbed "Golden rice." Two foreign genes were introduced into the japonica rice cultivar Taipei309 to recreate the carotenoid biosynthetic pathway within the rice endosperm (Ye et al., 2000).

Omega-3 Oils

This begins in the late 1990s and involves modifying the plant genome to accumulate omega-3 long-chain polyunsaturated fatty acids (LC-PUFAs) (Lakra et al., 2019). This is due to the absence of the omega-3 LC-PUFAs eicosapentaenoic acid (EPA) and docosa-hexanoic acid (DHA) in higher plants, which are considered to play an important role in human development and health. In terms of economics, the price of fish oil is very high than that of the vegetable oil making it a desirable trait to incorporate into plants. EPA and DHA biosynthesis unlike Golden Rice and beta-carotene requires a large number of enzymes and is not found in higher plants. Microalgae which form the foundation of aquatic food webs in which these fatty acids accumulate at any trophic stage are responsible for the primary biosynthesis of these omega-3 LC-PUFAs. Plant seed oils now contain amounts of EPA and DHA that are comparable to or greater than those contained in genuine fish oils, thanks to advances from many different research teams (Haslam et al., 2016). Field trials of GM camelina and canola that accumulate EPA and DHA have recently been performed in the United Kingdom, the United States, and Australia (Colombo et al., 2018).

Essential Amino Acids

Animals and humans do not synthesize amino acids like methionine (Met) and tryptophan (Trp) so they must be acquired through the diet. Lysine (Lys), tryptophan (Trp), and methionine (Met) are the three essential amino acids that are scarce in legumes (Met) and cereals (Lys and Trp) (Galili & Amir, 2013). Several transgenic approaches have been utilized in the last decade to change the composition of amino acids in plant proteins by integrating key amino acid metabolic pathways, hence improving the quality of usable amino acid(s) in plants. Transgenic wheat and rice has been created via incongruent expression of lysine-rich pea legumin protein in the endosperm of wheat and rice (Ufaz & Galili, 2008). Another accomplishment was the production of a seed storage protein from *Amaranthus hypochondriacus*. This protein includes all of the necessary amino acids needed by humans. Transgenic maize seeds expressing the AH protein were developed (Rascón-Cruz et al., 2004). Essential amino acids like lysine, tryptophan, and isoleucine, as well as up to 32 percent more protein, were higher in these seeds than in wild-type seeds. Bicar et al. (2008) produced transgenic maize by heterologously expressing lysine-rich animal proteins such as porcine-lactalbumin, resulting in a 47 percent increase in lysine content. To present, only two transgenic maize events with altered amino acid characteristics have been commercialized. These experiments used the *cordapA* gene from *Corynebacterium glutamicum* to increase free lysine content in maize kernels by expressing the transgene embryo-specifically. *Cordap*A codes for an enzyme dihydrodipicolinate synthase (DHDPS) that catalyzes the first committed step in the biosynthesis of lysine in plants and bacteria (Azevedo & Lea, 2001). In plants, lysine inhibits DHDPS, showing feedback inhibition making it the rate-limiting stage in the lysine production process (Galili & Amir, 2013). The responsiveness of the enzyme DHDPS sequestered from bacteria is >50 times lower than that of the plant enzyme making possible synthesis of lysine under high lysine concentration (Dewaele et al., 2002). One of the two commercially available maize events is a stacked case with increased production of lysine and insect resistance (cry1Ab gene). So many quality traits can be targeted to improve crop yield nutritional status. Carbohydrates, fats, oils, vitamins, iron, and amino acids are among them. End consumers, manufacturers, and the agro-based industry all have an effect on target trait selection. This research exemplifies the shift in focus from single gene agronomic traits like herbicide and insect resistance to more nuanced traits that directly benefit the customer like seed quality modification (Bicar et al., 2008). For example transgenic rice with the ability to develop beta-carotene which helps treat vitamin A deficiency. Similarly, in transgenic rice, genes involved in the development of an iron-binding protein that promotes iron availability in the human diet have been used to create transgenic rice with increased iron levels (Meng et al., 2005). Protein concentrations, fatty acid composition, vitamins, and amino acid composition are all being targeted for value enhancement. It is now possible to change the fatty acid composition so that polyunsaturated (e.g., linoleic acid) content decreases while monounsaturated (e.g., oleic acid) content rises allowing processing without the use of hydrogenation and thereby eliminating trans-fatty acids (Usher et al., 2015). To improve the nutritional quality of cereal grains, important amino acids such as lysine, methionine, threonine, and tryptophan can be added. The ratio of amylose to amylopectin in starch has also been altered through transgenic alterations (Schwall et al., 2000). Reduced oligosaccharides in the diet (such as raffinose and stachyose) improves digestion and reduces flatulence. Anti-nutritional factors can also be suppressed via transgenic technology (Valentine et al., 2017).

Photosynthetic Efficiency and Improved Yield

Changing plant biochemistry components in order to incorporate the C4 photosynthesis process into C3 plants such as *Arabidopsis* is an exciting experimental approach for significantly increasing crop production as well as potato. The oxygenase reaction of ribulose 1, 5-biophosphate carboxylase/oxygenase (Rubisco) can hinder photosynthesis in plants resulting in CO2 loss through photorespiration. C4 plants on the other hand such as maize have developed a mechanism to overcome this inhibition (Sharma et al., 2002). The activity of the enzyme phosphoenolpyruvate carboxylase (PEPC) is the key feature of this process (Arias-Baldrich et al.,

2017) that is responsible for the Co_2 fixation in the cytoplasm of mesophyll cells. PEPC was recently transmitted from intact maize to C3 plants via an *Agrobacterium*-mediated transformation technique (Miyao, 2003). These plants had least O_2 photosynthesis inhibition and photosynthetic concentrations that were equivalent to non-transformed control plants physiologically. In the C4 dicotyledonous species *Flaveria bidentis*, studies into the manipulation of Rubisco, pyruvate phosphate kinase, and other primary photosynthetic enzymes (PPDK) and NADP malate dehydrogenase (NADPMDH) have also been reported (Ruan et al., 2012).

Genes responsible for plant height in *Arabidopsis* are orthologous (similar) to cereal dwarf genes used in conventional plant breeding during the "Green Revolution" [11]. These genes (NORIN 10) were first integrated into western wheat varieties in the 1950s and have since been isolated, with similar phenotypes being recreated in other crops through genetic transformation (Waines & Ehdaie, 2007). In a number of crop species, these dwarfing genes can now be used to increase crop productivity. Manipulation of fructose-1, 6-bisphosphate aldolase (FDA), an enzyme that catalyzes the conversion of triosephosphate to fructose-1, 6-bisphosphate in a reversible manner, can also boost yield. FDA from *E. coli* expressed in the chloroplast of transgenic plants shows substantially increased starch accumulation lower sucrose concentration and higher root mass. Modifying plastid number as well as the expression of a hybrid protein composed of a yeast gene encoding 5-amino levulinic acid synthase and an N-terminal transit sequence for the small subunit of carboxydismutase may be a more generic method for changing plant results. Manipulation of the chlorophyll a/b binding genes has also affected chlorophyll levels. Another non-photosynthetic method for enhancing shoot and root yield is to overexpress a cyclin gene, such as the cycla gene from *Arabidopsis thaliana* L. (Makandar & O'Kennedy, 2002).

Delayed Ripening/Increased Shelf Life

In tomato, the first commercial varieties were developed to slow down the ripening process, and this leads to increase in the shelf life of tomato and was the first variety approved for human consumption in the United States in 1994 (Wang et al., 2019). Consumers prefer mature fruit, but only fully ripened fruit is mature which occurs only after the degradation and decay of cell wall which was a challenging problem for the fruit industry. The unstiffening of cell walls, sweetening, and growth of color, flavor, and aroma compounds are all part of the ripening process. Delaying the synthesis of the plant hormone ethylene by interfering either with ethylene development or with the biosynthesis pathway has been used to delay ripening or increase the shelf life of ripe fruit (Gupta et al., 2013a).

Antisense and co-suppression affected the expression of the polygalacturonase (PG) gene in GM tomato varieties, an enzyme that plays a key role in cell wall weakening during ripening. Around the same time, two competing groups created these kinds. In the United States, Calgene used an antisense strategy, while Zeneca joined with Grierson's group to use co-suppression. Calgene's product was a fresh fruit variety called "Flavr Savr" (Redenbaugh, 1992). It was first widely grown in 1996, but it was not a commercial success, and it was removed from the market within a year. Zeneca opted to use the trait in processing tomatoes, which proved to be a much more effective technique. These tomato varieties had hard cell wall than conventional varieties which results in less wastage lower processing costs and a thicker paste consistency during the paste production process. From its inception in 1996 until 1999, when most merchants dropped it due to environmental and ethical concerns, this product was accessible in several countries and was quite popular in the United Kingdom (Moran et al., 2001).

Fungal Resistance

Plant fungal diseases result in significant crop losses. The Irish famine in the nineteenth century, i.e., the causative agent was fungus *Phytophthora infestans* responsible for the late blight disease, is a severe example. Resistance genes, also known as R genes, confer infection resistance on the plant. R genes code for proteins that function as pathogen receptors, triggering a hypersensitive response (HR) in which cells die quickly near the fungus' entry point, halting disease progression. R genes (*Avr* genes) identify pathogen proteins encoded by avirulence genes (Brogue et al., 1991). An Avr gene facilitates infecting plants that lack an R gene to recognize it, but it must be discarded by the pathogen in order for resistance provided by an R gene to be resolved. Many years of co-evolution have resulted in this complex relationship between plant genetics and fungal pathogens. The effectiveness of R genes is based on the fact that only a small number of people carry each R gene (Punja, 2001).

A pathogen that retains an *Avr* gene has a selective advantage in that it has enhanced ability to contaminate plants that lack the R gene. This selective advantage does not occur in a field of crop crops that are virtually all the same (Century et al., 1995). As a result, plant breeders' attempts to incorporate specific R genes into crop types have been foiled by the emergence of new disease strains that are unaffected (Lorito et al., 1998). Biotechnologists believe that by understanding how the R gene functions, they will be able to build fungus resistance into crop plants, possibly by stacking many R genes in a single crop variety. Inserting genes that produce fungicidal proteins into crop plants is a different approach to engineering fungal resistance in crop plants. There are a large number of plants that are resistant to fungal diseases; examples of plants that are resistant to fungal attack are in shown in Table 13.1 (Ceasar & Ignacimuthu, 2012).

Table 13.1 Some fungal disease-resistant crops/varieties

Transgenic crop	Description	Gene/trait	Donor(s)
Grape, raspberry, tomato	Inhibitor of polygalacturonase	Polygalacturonase inhibitor protein	Bean, pear
Soybean	Resistance gene	Protein kinase	Soybean
Barley, festuca, potato, soybean	Resistance gene	R-gene	Barley (Rpg1), rice (Pi9), Solanum bulbocastanum (RB2), soybean (Rps1-k)
Wheat	Cell death regulator	Cell death regulator	Baculovirus, chicken, nematode
Barley, wheat	Fusarium toxin detoxifier	Toxin detoxifier	Fusarium sporotrichioides (Deoxynivalenol acetyltransferase, 3-hydroxyl trichoecene acetyltransferase)
Cotton, barley, grape, peanut, potato, rice, sweet potato, sorghum, tobacco, wheat	Pathogenesis-related proteins	PR proteins	Alfalfa (PR-2), Arabidopsis (PR-2), grape (PR-5), pea (PR-2), rice (PR-5), tobacco (PR-1)
Alfalfa, apple, carrot, cotton, melon, onion, papaya, peanut, rice, squash, tobacco, tomato, wheat	Chitin degradation	Chitinase	Alfalfa, barley, bean, petunia, rice, tobacco
Cowpea, bean, lettuce, peanut, potato, soybean, sunflower, tobacco	Reactive oxygen production	Oxalate oxidase	Barley, wheat
Barley, potato, rice	Plant defensin	Thionin	Barley, tobacco
Cotton, grape, plum, poplar, tobacco, wheat	Antimicrobial proteins	Antimicrobial peptide	African clawed frog (*Xenopus laevis*) (magainin), cow (lactoferrin), *Gastrodia elata* (mannose-binding lectin, gastrodianin), *Ustilago maydis* (KP4), wheat (PGL)
Cotton, maize, papaya	Antimicrobial proteins	Cecropin	Giant silk moths (*Hyalophora cecropia*)
Potato, tobacco	Polyphenol	Stilbene synthase	Grape

Herbicide Tolerance

Herbicide resistance is already a prominent characteristic of cultivated GM crops, and it will continue to be so in the near future. The first commercially grown genetically modified crops resistant to the broad-spectrum herbicides glyphosate and glufosinate were developed in the 1990s (Nandula, 2010).

Glyphosate

Glyphosate chemical formula $C_3H_8NO_5P$ IUPAC name N-phosphonomethyl glycine is a polar water-soluble organic acid with a molecular weight of 169 (Duke & Powles, 2008). Glyphosate is a wide-ranging herbicide that is effective against both monocotyledonous and dicotyledonous plants. Due to the widespread cultivation of glyphosate-resistant (GR) crops (such as soybean, corn, cotton, oilseed rape, and others) on millions of hectares around the world, glyphosate-based herbicides (GBHs) are the most commonly utilized herbicides in the world (Mertens et al., 2018). Glyphosate inhibits 5-enolpyruvylshikimate-3-phosphate synthase (EPSPS), a shikimic acid pathway enzyme that helps make aromatic acids including phenylalanine, tryptophan, and tyrosine, as well as a host of other compounds. Glyphosate was once thought to be safe for humans and animals since this mechanism is only present in plants and microorganisms. On the other side, a recent report has raised concerns that glyphosate could be more harmful to animals and humans than previously believed (Myers et al., 2016).

Yields of HR Crops

Differing to popular belief, HR crops do not reliably produce higher yields than traditional crops. Farmers choose HR crops for a variety of reasons not the least of which is increased yield (Ficiciyan et al., 2018). The yield difference between HR and traditional variety may be due to other factors that including increasing scale and area location and size of the form soil, environment, tillage method, weed abundance, genetic background/varieties, crop management, weed control practice, farmer skills, and farm operator education.

Edible Vaccines

Every year, 12 billion vaccines are performed in the world. 35% of these are not carried out in sterile environments. Moreover, hundreds of millions of people decease each year due to diseases that could be avoided with vaccination. People will also travel for days to get their children and themselves vaccinated at a clinic, they are not appropriately covered, and they are unable to repeat the journey with a subsequent booster vaccine (Langridge, 2000).

Hepatitis B is a virus that causes acute and chronic liver disease, as well as liver failure and cancer. It is a major killer in developing countries (Vahdat et al., 2021). The first hepatitis B vaccine was developed in the 1970s using a protein (the surface antigen) obtained from the blood of hepatitis B patients. Unfortunately, it is much too expensive for developing countries to afford, and it is prohibitively expensive even in the United States (Kurup & Thomas, 2020; Judge et al., 2004; Bureau & Directive, 2001).

Genes in Transgenic Plants that Confer Dehydration Stress Resistance

Transgenic plants with improved dehydration stress tolerance have been generated utilizing a number of genes (Table 13.2). This list is not intended to be exhaustive; rather, it is intended to demonstrate the breadth of research that has been done.

Enzymes that Synthesize Osmoprotectants

In response to water stress, many plants collect suitable solutes or osmolytes, which are organic molecules with a low molecular weight. As a result, increasing osmolyte amount in transgenic plants is an important approach for preventing waster stress in plants (Riadh et al., 2010). Dehydration stress is reduced slightly in transgenic plants having genes encoding enzymes involved in the synthesis of osmolytes such as mannitol, proline, and others (Suprasanna et al., 2016).

Mannitol

Plants of tobacco with the mtlD gene which encodes a mannitol-1-phosphate dehydrogenase accumulated mannitol. In comparison to control plants, these plants showed improved tolerance to high salinity (Shaw et al., 2002). Karakas et al. (1997) found that tobacco plants transformed with the mtlD gene displayed a small rise in dry weight when exposed to salt, but no difference in growth was observed when exposed to drought. They also discovered that mannitol accumulation had a marginal effect on osmotic change in tobacco with mtlD gene (Stoop et al., 1996). Mannitol buildup up to 10 mM in transgenic tobacco chloroplasts results in improved resistance to methyl-viologen (MV)-induced oxidative stress, as indicated by greater than before chlorophyll retention in transgenic tissues after MV treatment (Shen et al., 1997).

Table 13.2 Stress responses of transgenic plants overexpressing various genes involved in stress tolerance

Transgenic plant	Gene product (and function)	Genes encoding enzymes that synthesize osmoprotectants	Performance of transgenic plant
Tobacco	Choline dehydrogenase (glycine betaine synthesis)	Beta	Increased tolerance to salt
Arabidopsis thaliana, rice	Choline oxidase (glycine betaine synthesis)	Coda	Seedlings were more salt tolerant, and germination was improved under cold conditions
Tobacco	Myo-inositol O-methyltransferase (D-ononitol synthesis)	IMT1	Under drought and salt, a higher photosynthetic rate means higher yield
Tobacco	Mannitol-1-phosphate dehydrogenase (mannitol synthesis)	mtlD	In terms of % change in height and fresh weight, 6-week-old plants grew quicker under high salinity
Arabidopsis thaliana		mtlD	Enhanced seed germination under high salinity
Tobacco		mtlD	Dry weight increased slightly under salt stress, but there was no difference in growth under drought stress
Tobacco		mtlD	Increased resistance to methylviologen-induced oxidative stress, as evidenced by higher chlorophyll retention in transgenic leaves under stress
Tobacco	Trehalose-6-phosphate synthase	otsA	Drought stress results in increased dry weight and more effective photosynthesis
Tobacco	Trehalose-6-phosphate phosphatase (trehalose synthesis)	otsB	
Tobacco	Δ1-pyrroline-5-carboxylate synthetase (proline synthesis)	p5cs	Salt stress results in increased biomass and flower growth
Rice		p5cs	Transgenic seedlings exposed to 100 mM NaCl for 5 days or 8-week-old plants stressed by water produced greater biomass than control plants
Tobacco	Fructosyl transferase (fructan synthesis)	sacB	Under PEG-induced osmotic tension, 3-week-old plants grew faster
Rice	Arginine decarboxylase (putrescine synthesis)	Adc	Drought stress causes minimal chlorophyll loss

(continued)

Table 13.2 (continued)

Transgenic plant	Gene product (and function)	Genes encoding enzymes that synthesize osmoprotectants	Performance of transgenic plant
Carrot (cell line)	Ornithine decarboxylase (putrescine synthesis)	Odc	Over a short amount of time, transgenic cell lines could resist in high salt levels
LEA or LEA-related genes			
Rice	Group 3 LEA protein	HVA1	Drought resistance improved after 15 days of water deprivation in a 3-month research, as shown by better leaf survival. Transgenic seedlings grew better in the presence of 100 mM NaCl or 200 mM mannitol, demonstrating greater tolerance to water shortage and salt stress; 3-week-old seedlings in solid transgenic plants performed better under water stress conditions
Arabidopsis thaliana	Cold-induced gene	COR15a	Increased freezing tolerance of chloroplasts and protoplasts

Proline

The enzyme 11-pyrroline-5-carboxylate synthetase (P5CS) converts glutamate to 11-pyrroline-5-carboxylate, which is then reduced to proline (Kishor et al., 2005). In transgenic tobacco plants, overexpression of a gene encoding for moth bean P5CS resulted in a 10- to 18-fold increase in proline accumulation and better growth under dehydration stress compared to control plants (Kishor et al., 1995). Transgenic plants also showed enhanced biomass and development of flower in salt stress, as measured by improvement in length of root, dry weight of shoot, dry weight of number of capsules, and number of seeds in each capsule. Under the regulation of an ABA/stress-inducible promoter, the same gene was introduced into rice (Liu & Zhu, 1997). Under tension, transgenic rice plants accumulated 2.5 times more proline than that of control. Stress-inducible expression of the P5CS transgene boosted biomass in second-generation transgenic rice plants when compared to untransformed control plants, as shown by higher fresh shoot and root weight under salt and water stress. Salt tolerance is connected to the amount of proline stored in the body (Hmida-Sayari et al., 2005).

Glycine Betaine

As an adaptation reaction to saline or water-stress circumstances, glycine betaine is found in the number cells of halophytes and bacteria. The introduction of coli betA gene into the tabacco plant which encodes for choline dehydrogenase was employed latter (Nuccio et al., 1998). When transgenic and wild-type plants were weighed at 300 mM NaCl, the transgenic plants were found to be more salt tolerant. The bacterial choline oxidase (*codA*) gene, which converts choline to glycine betaine and was isolated from *Arthrobacter globiformis*, was introduced to *Arabidopsis*. Glycine betaine was accumulated in the transgenic plants, and they were more resistant to salt and cold stress (Hayashi et al., 1997). The *Cyanobacterium Synechococcus* sp. PCC7942 was also found to have improved salt tolerance after being transformed with the codA gene (Deshnium et al., 1997). Recently confirmed that overexpression of the *coda in* transgenic rice plants shows more resistant to salt and low temperatures. The levels of glycine betaine in transgenic plants were as high as 1–5 mol/gm fresh weight of leaves in two types of transgenic plants. In one form, choline oxidase was directed to the chloroplasts (ChlCOD), while in the other it was directed to the cytosol (CytCOD).

Polyamines

Polyamines are nitrogenous cellular compounds that are small and abundant in plants and have been linked to a number of stress responses. Salt and drought are two abiotic stressors that cause polyamine accumulation (Gupta et al., 2013b). Polyamine levels were higher in cultivars with a higher degree of salt tolerance. Furthermore, exogenous polyamine application protected oat leaves from osmotic stress (Besford et al., 1993).

Detoxification Enzymes or Oxidative Stress-Related Genes

Protecting sensitive metabolic reactions by stabilizing protein complexes or membrane structures and developing hydroxyl radical scavenging ability may be an effective strategy for engineering water stress tolerance (Bohnert & Jensen, 1996). Drought conditions caused oxidative stress in pea plants, according to Moran et al. (1994). Under stress, they found significant reductions in photosynthesis and transpiration. Discovered that oxidative stress-related genes were induced in four drought-tolerant tobacco varieties. They discovered increased glutathione reductase, superoxide dismutase, and ascorbate peroxidase activities, as well as catalase activity to a lesser degree. They also linked these behaviors to the photosynthetic pigments' integrity.

Conclusion

Advanced approaches, such as bio-fortification, improve the nutritious quality of cereals and crops, as addressed in this chapter. Many of the genetic modifications in crops that happened during domestication were discovered in cis-regulatory elements (CREs) of genes, and typically CREs of transcription factors (TFs), presumably affecting the time and pattern of gene expression without modifying the gene output. Although there has been great progress in employing synthetic metabolic engineering to bio-fortify crops, there are still certain obstacles to overcome. Genomic, transcriptomic, proteomic, and metabolomics analysis together will improve our understanding of metabolic pathways and their major components. Although biofortification by bioengineering is not the only way to address nutritional issues, it does provide an alternate and complementary strategy to other interventions. In this regard, if possible, traditional breeding and metabolic engineering should work together to generate multi-biofortified crops with health benefits. Metabolic engineering will allow scientists to refine and fine-tune their biofortification tactics as our understanding of micronutrient metabolic pathways and regulation grows.

Acknowledgments We acknowledge all the colleagues who made contribution in this book chapter.

Conflict of Interest The authors declare that there is no conflict of interest.

References

Arias-Baldrich, C., de la Osa, C., Bosch, N., Ruiz-Ballesta, I., Monreal, J. A., & García-Mauriño, S. (2017). Enzymatic activity, gene expression and posttranslational modifications of photosynthetic and non-photosynthetic phosphoenolpyruvate carboxylase in ammonium-stressed sorghum plants. *Journal of Plant Physiology, 214*, 39–47.

Azevedo, R. A., & Lea, P. J. (2001). Lysine metabolism in higher plants. *Amino Acids, 20*(3), 261–279.

Besford, R. T., Richardson, C. M., Campos, J. L., & Tiburcio, A. F. (1993). Effect of polyamines on stabilization of molecular complexes in thylakoid membranes of osmotically stressed oat leaves. *Planta, 189*(2), 201–206.

Bhat, S. R., & Srinivasan, S. (2002). Molecular and genetic analyses of transgenic plants: Considerations and approaches. *Plant Science, 163*(4), 673–681.

Bicar, E. H., Woodman-Clikeman, W., Sangtong, V., Peterson, J. M., Yang, S. S., Lee, M., & Scott, M. P. (2008). Transgenic maize endosperm containing a milk protein has improved amino acid balance. *Transgenic Research, 17*(1), 59–71.

Bohnert, H. J., & Jensen, R. G. (1996). Strategies for engineering water-stress tolerance in plants. *Trends in Biotechnology, 14*(3), 89–97.

Brogue, K., Chet, I., Holliday, M., Cressman, R., Biddle, P., Knowlton, S., et al. (1991). Transgenic plants with enhanced resistance to the fungal pathogen Rhizoctonia solani. *Science, 254*(5035), 1194–1197.

Burchi, F., Fanzo, J., & Frison, E. (2011). The role of food and nutrition system approaches in tackling hidden hunger. *International Journal of Environmental Research and Public Health, 8*(2), 358–373.

Bureau, F., & Directive, EU. (2001). Molecular farming of edible vaccines {1998–2001}.
Ceasar, S. A., & Ignacimuthu, S. (2012). Genetic engineering of crop plants for fungal resistance: Role of antifungal genes. *Biotechnology Letters, 34*(6), 995–1002.
Century, K. S., Holub, E. B., & Staskawicz, B. J. (1995). NDR1, a locus of Arabidopsis thaliana that is required for disease resistance to both a bacterial and a fungal pathogen. *Proceedings of the National Academy of Sciences, 92*(14), 6597–6601.
Chawla, H. (2011). *Introduction to plant biotechnology (3/e)*. CRC Press.
Colombo, S. M., Campbell, L. G., Murphy, E. J., Martin, S. L., & Arts, M. T. (2018). Potential for novel production of omega-3 long-chain fatty acids by genetically engineered oilseed plants to alter terrestrial ecosystem dynamics. *Agricultural Systems, 164*, 31–37.
Colton, L. M., Groza, H. I., Wielgus, S. M., & Jiang, J. (2006). Marker-assisted selection for the broad-spectrum potato late blight resistance conferred by gene RB derived from a wild potato species. *Crop Science, 46*(2), 589–594.
Conko, G., Kershen, D. L., Miller, H., & Parrott, W. A. (2016). A risk-based approach to the regulation of genetically engineered organisms. *Nature Biotechnology, 34*(5), 493–503.
Deshnium, P., Gombos, Z., Nishiyama, Y., & Murata, N. (1997). The action in vivo of glycine betaine in enhancement of tolerance of Synechococcus sp. strain PCC 7942 to low temperature. *Journal of Bacteriology, 179*(2), 339–344.
Dewaele, E., Craciun, A., Vauterin, M., Frankard, V., Suharyanto, E., Tadesse, J., & Jacobs, M. (2002). Metabolic engineering of a complex biochemical pathway: The lysine and threonine biosynthesis as an example. *Phytochemistry Reviews, 1*(1), 125–133.
Diretto, G., Frusciante, S., Fabbri, C., Schauer, N., Busta, L., Wang, Z., et al. (2020). Manipulation of β-carotene levels in tomato fruits results in increased ABA content and extended shelf life. *Plant Biotechnology Journal, 18*(5), 1185–1199.
Duke, S. O., & Powles, S. B. (2008). Glyphosate: A once-in-a-century herbicide. *Pest Management Science: Formerly Pesticide Science, 64*(4), 319–325.
Ficiciyan, A., Loos, J., Sievers-Glotzbach, S., & Tscharntke, T. (2018). More than yield: Ecosystem services of traditional versus modern crop varieties revisited. *Sustainability, 10*(8), 2834.
Frison, E. A., Cherfas, J., & Hodgkin, T. (2011). Agricultural biodiversity is essential for a sustainable improvement in food and nutrition security. *Sustainability, 3*(1), 238–253.
Galili, G., & Amir, R. (2013). Fortifying plants with the essential amino acids lysine and methionine to improve nutritional quality. *Plant Biotechnology Journal, 11*(2), 211–222.
Gupta, A., Pal, R. K., & Rajam, M. V. (2013a). Delayed ripening and improved fruit processing quality in tomato by RNAi-mediated silencing of three homologes of 1-aminopropane-1-carboxylate synthase gene. *Journal of Plant Physiology, 170*(11), 987–995.
Gupta, K., Dey, A., & Gupta, B. (2013b). Plant polyamines in abiotic stress responses. *Acta Physiologiae Plantarum, 35*(7), 2015–2036.
Halford, N. G. (2006). Risk assessment, regulation and labeling. In N. G. Halford (Ed.), *Current and future applications of genetically modified crops*. Wiley, 280 pp.
Hansen, M. (2000). *Genetic engineering is not an extension of conventional plant breeding: How genetic engineering differs from conventional breeding, hybridization, wide crosses and horizontal gene transfer*. Consumer Policy Institute/Consumer's Union.
Haslam, R. P., Sayanova, O., Kim, H. J., Cahoon, E. B., & Napier, J. A. (2016). Synthetic redesign of plant lipid metabolism. *The Plant Journal, 87*(1), 76–86.
Hayashi, H., Mustardy, L., Deshnium, P., Ida, M., & Murata, N. (1997). Transformation of *Arabidopsis thaliana* with the coda gene for choline oxidase; accumulation of glycinebetaine and enhanced tolerance to salt and cold stress. *The Plant Journal, 12*(1), 133–142.
Hazell, P. B. (2010). The Asian green revolution. In *Proven successes in agricultural development: A technical compendium to millions fed* (p. 67) (www.ifpri.org/millionsfed)
Hmida-Sayari, A., Gargouri-Bouzid, R., Bidani, A., Jaoua, L., Savouré, A., & Jaoua, S. (2005). Overexpression of Δ1-pyrroline-5-carboxylate synthetase increases proline production and confers salt tolerance in transgenic potato plants. *Plant Science, 169*(4), 746–752.

Jacobsen, S. E., Sørensen, M., Pedersen, S. M., & Weiner, J. (2013). Feeding the world: Genetically modified crops versus agricultural biodiversity. *Agronomy for Sustainable Development, 33*(4), 651–662.

Jauhar, P. P. (2006). Modern biotechnology as an integral supplement to conventional plant breeding: The prospects and challenges. *Crop Science, 46*(5), 1841–1859.

Jones, G., Kluyver, T., Preece, C., Swarbrick, J., Forster, E., Wallace, M., et al. (2021). The origins of agriculture: Intentions and consequences. *Journal of Archaeological Science, 125*, 105,290.

Judge, N. A., Mason, H. S., & O'Brien, A. D. (2004). Plant cell-based intimin vaccine given orally to mice primed with intimin reduces time of *Escherichia coli* O157: H7 shedding in feces. *Infection and Immunity, 72*(1), 168–175.

Karakas, B., Ozias-Akins, P., Stushnoff, C., Suefferheld, M., & Rieger, M. (1997). Salinity and drought tolerance of mannitol-accumulating transgenic tobacco. *Plant, Cell & Environment, 20*(5), 609–616.

Kendall, H. W., & Pimentel, D. (1994). Constraints on the expansion of the global food supply. *Ambio*, 198–205.

Kishor, P. K., Hong, Z., Miao, G. H., Hu, C. A. A., & Verma, D. P. S. (1995). Overexpression of [delta]-pyrroline-5-carboxylate synthetase increases proline production and confers osmotolerance in transgenic plants. *Plant Physiology, 108*(4), 1387–1394.

Kikkert, J. R., Vidal, J. R., & Reisch, B. I. (2005). Stable transformation of plant cells by particle bombardment/biolistics. In *Transgenic plants: Methods and protocols* (pp. 61–78). Humana Press.

Kishor, P. K., Sangam, S., Amrutha, R. N., Laxmi, P. S., Naidu, K. R., Rao, K. S., et al. (2005). Regulation of proline biosynthesis, degradation, uptake and transport in higher plants: Its implications in plant growth and abiotic stress tolerance. *Current Science*, 424–438.

Kumar, K., Gambhir, G., Dass, A., Tripathi, A. K., Singh, A., Jha, A. K., et al. (2020). Genetically modified crops: Current status and future prospects. *Planta, 251*(4), 1–27.

Kurup, V. M., & Thomas, J. (2020). Edible vaccines: Promises and challenges. *Molecular Biotechnology, 62*(2), 79–90.

Ladizinsky, G. (1998). Origin of agriculture. In *Plant evolution under domestication* (pp. 1–60). Springer.

Lakra, N., Mahmood, S., Marwal, A., Sudheep, N. M., & Anwar, K. (2019). Bioengineered plants can be an alternative source of omega-3 fatty acids for human health. In *Plant and human health* (Vol. 2, pp. 361–382). Springer.

Langridge, W. H. (2000). Edible vaccines. *Scientific American, 283*(3), 66–71.

Lichtfouse, E., Navarrete, M., Debaeke, P., Souchère, V., Alberola, C., & Ménassieu, J. (2009). Agronomy for sustainable agriculture: A review. *Sustainable Agriculture*, 1–7.

Liu, J., & Zhu, J. K. (1997). Proline accumulation and salt-stress-induced gene expression in a salt-hypersensitive mutant of Arabidopsis. *Plant Physiology, 114*(2), 591–596.

Lorito, M., Woo, S. L., Fernandez, I. G., Colucci, G., Harman, G. E., Pintor-Toro, J. A., et al. (1998). Genes from mycoparasitic fungi as a source for improving plant resistance to fungal pathogens. *Proceedings of the National Academy of Sciences, 95*(14), 7860–7865.

Lucht, J. M. (2015). Public acceptance of plant biotechnology and GM crops. *Viruses, 7*(8), 4254–4281.

Makandar, S. A., & O'Kennedy, M. (2002). Applications of biotechnology. *Plant Science, 163*, 381–395.

Meng, F., Wei, Y., & Yang, X. (2005). Iron content and bioavailability in rice. *Journal of Trace Elements in Medicine and Biology, 18*(4), 333–338.

Mertens, M., Höss, S., Neumann, G., Afzal, J., & Reichenbecher, W. (2018). Glyphosate, a chelating agent—Relevant for ecological risk assessment? *Environmental Science and Pollution Research, 25*(6), 5298–5317.

Miyao, M. (2003). Molecular evolution and genetic engineering of C4 photosynthetic enzymes. *Journal of Experimental Botany, 54*(381), 179–189.

Moran, J. F., Becana, M., Iturbe-Ormaetxe, I., Frechilla, S., Klucas, R. V., & Aparicio-Tejo, P. (1994). Drought induces oxidative stress in pea plants. *Planta, 194*(3), 346–352.
Moran, K., King, S. R., & Carlson, T. J. (2001). Biodiversity prospecting: Lessons and prospects. *Annual Review of Anthropology, 30*(1), 505–526.
Myers, J. P., Antoniou, M. N., Blumberg, B., Carroll, L., Colborn, T., Everett, L. G., et al. (2016). Concerns over use of glyphosate-based herbicides and risks associated with exposures: A consensus statement. *Environmental Health, 15*(1), 1–13.
Nandula, V. K. (2010). Herbicide resistance: Definitions and concepts. In V. K. Nandula (Ed.), *Glyphosate resistance in crops and weeds: History, development, and management* (pp. 35–43). Wiley.
Nuccio, M. L., Russell, B. L., Nolte, K. D., Rathinasabapathi, B., Gage, D. A., & Hanson, A. D. (1998). The endogenous choline supply limits glycine betaine synthesis in transgenic tobacco expressing choline monooxygenase. *The Plant Journal, 16*(4), 487–496.
Paine, J. A., Shipton, C. A., Chaggar, S., Howells, R. M., Kennedy, M. J., Vernon, G., et al. (2005). Improving the nutritional value of Golden Rice through increased pro-vitamin A content. *Nature Biotechnology, 23*(4), 482–487.
Pinstrup-Andersen, P., Pandya-Lorch, R., & Rosegrant, M. W. (1999). *World food prospects: Critical issues for the early twenty-first century*. doi: https://doi.org/10.22004/ag.econ.42572
Pinstrup-Andersen, P., & Watson, D. D., II. (2011). *Food policy for developing countries*. Cornell University Press.
Puchta, H., & Fauser, F. (2014). Synthetic nucleases for genome engineering in plants: Prospects for a bright future. *The Plant Journal, 78*(5), 727–741.
Punja, Z. K. (2001). Genetic engineering of plants to enhance resistance to fungal pathogens—A review of progress and future prospects. *Canadian Journal of Plant Pathology, 23*(3), 216–235.
Rascón-Cruz, Q., Sinagawa-García, S., Osuna-Castro, J. A., Bohorova, N., & Paredes-López, O. (2004). Accumulation, assembly, and digestibility of amarantin expressed in transgenic tropical maize. *Theoretical and Applied Genetics, 108*(2), 335–342.
Redenbaugh, K. (1992). *Safety assessment of genetically engineered fruits and vegetables: A case study of the Flavr Savr tomato*. CRC Press.
Riadh, K., Wided, M., Hans-Werner, K., & Chedly, A. (2010). Responses of halophytes to environmental stresses with special emphasis to salinity. *Advances in Botanical Research, 53*, 117–145.
Rivera, A. L., Gómez-Lim, M., Fernández, F., & Loske, A. M. (2012). Physical methods for genetic plant transformation. *Physics of Life Reviews, 9*(3), 308–345.
Ruan, C. J., Shao, H. B., & Teixeira da Silva, J. A. (2012). A critical review on the improvement of photosynthetic carbon assimilation in C3 plants using genetic engineering. *Critical Reviews in Biotechnology, 32*(1), 1–21.
Rubatzky, V. E., & Yamaguchi, M. (2012). *World vegetables: Principles, production, and nutritive values*. Springer Science & Business Media.
Russell, G. E. (2013). *Plant breeding for pest and disease resistance: Studies in the agricultural and food sciences*. Butterworths.
Schell, J., Montagu, M. V., Beuckeleer, M. D., Block, M. D., Depicker, A., Wilde, M. D., et al. (1979). Interactions and DNA transfer between agrobacterium tumefaciens, the Ti-plasmid and the plant host. *Proceedings of the Royal Society of London Series B. Biological Sciences, 204*(1155), 251–266.
Schwall, G. P., Safford, R., Westcott, R. J., Jeffcoat, R., Tayal, A., Shi, Y. C., et al. (2000). Production of very-high-amylose potato starch by inhibition of SBE A and B. *Nature Biotechnology, 18*(5), 551–554.
Sharma, H. C., Crouch, J. H., Sharma, K. K., Seetharama, N., & Hash, C. T. (2002). Applications of biotechnology for crop improvement: Prospects and constraints. *Plant Science, 163*(3), 381–395.
Shaw, B., Thomas, T. H., & Cooke, D. T. (2002). Responses of sugar beet (*Beta vulgaris* L.) to drought and nutrient deficiency stress. *Plant Growth Regulation, 37*(1), 77–83.

Shen, B., Jensen, R. G., & Bohnert, H. J. (1997). Mannitol protects against oxidation by hydroxyl radicals. *Plant Physiology, 115*(2), 527–532.

Stoop, J. M., Williamson, J. D., & Pharr, D. M. (1996). Mannitol metabolism in plants: A method for coping with stress. *Trends in Plant Science, 1*(5), 139–144.

Suprasanna, P., Nikalje, G. C., & Rai, A. N. (2016). Osmolyte accumulation and implications in plant abiotic stress tolerance. In N. Iqbal, N. Nazar, & N. A. Khan (Eds.), *Osmolytes and plants acclimation to changing environment: Emerging omics technologies* (pp. 1–12). Springer.

Ufaz, S., & Galili, G. (2008). Improving the content of essential amino acids in crop plants: Goals and opportunities. *Plant Physiology, 147*(3), 954–961.

Usher, S., Haslam, R. P., Ruiz-Lopez, N., Sayanova, O., & Napier, J. A. (2015). Field trial evaluation of the accumulation of omega-3 long chain polyunsaturated fatty acids in transgenic Camelina sativa: Making fish oil substitutes in plants. *Metabolic Engineering Communications, 2*, 93–98.

Vahdat, M. M., Hemati, F., Ghorbani, A., Rutkowska, D., Afsharifar, A., Eskandari, M. H., et al. (2021). Hepatitis B core-based virus-like particles: A platform for vaccine development in plants. *Biotechnology Reports*, e00605.

Valentine, M. F., De Tar, J. R., Mookkan, M., Firman, J. D., & Zhang, Z. J. (2017). Silencing of soybean raffinose synthase gene reduced raffinose family oligosaccharides and increased true metabolizable energy of poultry feed. *Frontiers in Plant Science, 8*, 692.

Waines, J. G., & Ehdaie, B. (2007). Domestication and crop physiology: Roots of green-revolution wheat. *Annals of Botany, 100*(5), 991–998.

Wang, T., Zhang, H., & Zhu, H. (2019). CRISPR technology is revolutionizing the improvement of tomato and other fruit crops. *Horticulture Research, 6*(1), 1–13.

Ye, X., Al-Babili, S., Klöti, A., Zhang, J., Lucca, P., Beyer, P., & Potrykus, I. (2000). Engineering the provitamin A (β-carotene) biosynthetic pathway into (carotenoid-free) rice endosperm. *Science, 287*(5451), 303–305.

Zargar, S. A., Islam, T., & Margay, J. A. (2021). Agricultural diversity of Kashmir Valley. doi: https://doi.org/10.3390/bdee2021-09396

Chapter 14
Improving the Quality of Medicinal and Aromatic Plants Through Metabolic Engineering

Mubeen Fatima, Muhammad Zafar Saleem, Sidqua Zafar, Syed Muhammad Hassan Askri, Muhammad Ashar Ayub, Rabia Saleem, and Kabeer Haneef

Abstract Secondary metabolites in plants having aromatic and medicinal properties are commercially usable and are being used for health protection in different forms such as in drugs, pheromones, flavors, antioxidants, fragrances, insecticides, and dyes. About 25% of the legal drugs have been synthesized directly from natural substances obtained from plants, and 40% involve chemically modified natural substances. Owing to immense demand of these secondary metabolites, the deficient production capacity of plants must be overcomed. Plant breeding is a traditional, time-consuming, and limited way of improving the secondary metabolites production

Mubeen Fatima and Kabeer Haneef have equally contributed to this chapter.

M. Fatima
Center for Applied Molecular Biology and Forensic Sciences, University of Punjab, Lahore, Punjab, Pakistan

Forman Christian University, School of Life Science, Lahore, Pakistan

M. Z. Saleem
Center for Applied Molecular Biology and Forensic Sciences, University of Punjab, Lahore, Punjab, Pakistan

S. Zafar · R. Saleem
Forman Christian University, School of Life Science, Lahore, Pakistan

S. M. H. Askri
College of Agriculture and Biotechnology, Provincial Key Laboratory of Crop Gene Resources, Zhejiang University, Hangzhou, China

M. A. Ayub (✉)
Institute of Soil and Environmental Sciences, University of Agriculture, Faisalabad, Punjab, Pakistan

Indian River Research and Education Centre, Institute of Food and Agricultural Sciences, University of Florida (UF-IFAS), Gainesville, FL, USA

K. Haneef
School of Life Science (SLS), Institute of Immunology (IITU), Tsinghua University, Beijing, China

T. Aftab, K. R. Hakeem (eds.), *Metabolic Engineering in Plants*,
https://doi.org/10.1007/978-981-16-7262-0_14

capacity of medicinal plants. Metabolic engineering in this regard can be a helpful and economical method which involves insertion of genes of interest into other cells and modify target pathways, allowing for improved processing of usable natural substances as well as the development of novel compounds. This review illustrates the importance of medicinal plants in producing secondary metabolites. Here we discussed the pathways involved in producing secondary metabolites and the role of metabolic engineering in manipulating these pathways to get the required results. Additionally, we explained how RNAi-mediated gene silencing helps to regulate different genes involved in synthesizing useful natural substances.

Keywords Secondary metabolites · Medicinal plants · Metabolic engineering · Novel compounds · RNAi-mediated gene silencing

Introduction

In the pharmaceutical industry, a significant role is being played by medicinal plants and other natural products. Most of the plant's natural products revealed a complex and advanced chemical structure, which cannot be synthesized easily. Another major challenge is to synthesize the stereochemistry of natural plant products (Pank, 2007). Enantioselective synthesis of these natural compounds of plants is not easy, so it is preferred to extract these compounds or their biosynthetic precursors. So, for their production and manufacturing, different procedures have to be found. It is necessary to improve the productivity of medicinal plants, but this improvement in its quality traits plays a significant part in future plant breeding. No doubt, the use of classical breeding strategies is essential. Still, for optimizing future medicinal plants, metabolic engineering is considered an effective tool due to considerable developments in genetic techniques (Pank, 2007). The goals of metabolic engineering in medicinal and aromatic plants are to increase the production of natural bioproducts, create new biosimilar compounds with greater biological activity, formation of new colors in flowers and food, impart new taste and flavor in food, improvements in nutrition and nutraceuticals effects of food, decrease the production of toxic and allergic compound's, and provision of resistance against pests and diseases (Verpoorte et al., 1999; Verpoorte & Memelink, 2002). The introduction of genetic techniques can reduce obstruction in the biosynthesis of natural products, and genetically modified plants can be used for the extraction purposes of different compounds (Kayser, 2009).

Several organic compounds perform physiological functions that are associated with plant protection and defense mechanisms. These organic compounds are secondary metabolites because they do not directly play a crucial role in primary metabolism. These compounds serve an assortment of operations because they protect plants from UV radiations, attract insects and further act as a signaling molecules during nitrogen fixation, and harbor a role in the formation of bark and wood. However, several strategies are involved in the production of these secondary metabolites (Razdan, 2018). Many secondary metabolites are present in plants that have medicinal and aromatic properties. It is estimated that in 50,000 plant species, almost 100,000 secondary metabolites are present, and each year from different plant species, nearly 4000 new secondary metabolites are discovered (Verpoorte et al., 1999; Gómez-Galera et al., 2007). Natural products of plants have been utilized in many ways for human healthcare in antioxidants, fragrances, insecticides, drugs, dyes, flavors, and pheromones for thousands of years.

During the last century, synthetic drugs started to replace the use of natural plant compounds but when sideffects of synthetics drugs started to report, consumers are again moving towards natural and plant drived medicines and drugs and the market of these products is flourishing (Joshi et al., 2004). The world market of herbal products and their raw material is growing at a yearly rate of 5–15% owing to rapid adoption of natural products. As the demand of plant drived products (having medicinal and metabolic properties) is increasing many researchers are trying to linearly increase the production potential of these products from medicinal plants. In the past, conventional method of breeding has been used for potential improvement in quality and yield of secondary metabolites of those plants having medicinal and aromatic value, but due to the involvement of complex biosynthetic pathways, long generation time and sterility the desired results are not always achievable (Rhodes et al., 1988). Alternatively, in vitro plant cell culture has been preferred in some cases but rejected (Haq, 2000; Roberts, 2007). In more advanced methods, genomics and post-genomic approaches have been used to enhance the quality and yield of secondary plant metabolites. Due to advancements in genomics and post-genomics approaches, biosynthesis of complex plant secondary metabolites has been possible. Engineering metabolic pathway is a powerful tool for careful economic and scalable production of secondary plant metabolites. Plant biosynthetic pathways are engineered by expressing or silencing the transcriptional factors or by manipulating the main biosynthetic pathway genes (Razdan, 2018).

The changes of endogenous pathways in plants to increase the flux toward the specific required molecule is termed metabolic engineering. The primary aim is to increase the yield of natural products, whereas in some cases, the purpose is to synthesize the novel compound or macromolecule (Barone et al., 2020; Pickens et al., 2011). Modifications in metabolic pathways are essential for exploring cell physiology (Farmer & Liao, 2001). It is estimated that almost 200,000 secondary metabolites are present in the plant kingdom, which are mainly divided into three major groups: phenolic compounds, alkaloids, and terpenoids (Dixon, 2001; Dixon, 2005). Biosynthesis of alkaloids can effectively be done by the use of metabolic engineering (Facchini, 2001). From plants, secondary metabolites as well as terpenoids are the largest group harboring pharmaceutical activity to enhance flavor and color. Metabolic engineering has been applied for mass production of these terpenoids in plants which work as attractants for natural enemies to herbivores' pests (Roberts, 2007; Kappers et al., 2005). The metabolic engineering of the carotenoid pathway both qualitatively and quantitatively can trigger the nutritional quality (Ye et al., 2000). Comprehensive understanding of biosynthetic pathways is less, which is the major obstacle in the metabolic engineering of plant secondary metabolites. Similarly, limited knowledge in the complex metabolic networks for regulating various pathways and metabolic balance of plant cellular systems is unpredictably affected by single gene insertions (Dixon & Steele, 1999; Forkmann & Martens, 2001; Dixon, 2005).

Secondary plant metabolites (Birchfield & McIntosh, 2020) via metabolic engineering, enzymatic reactions (DellaPenna, 2001). Application of recombinant DNA transformations in which the metabolic network is reconstructed for improvements in metabolite production by altering metabolic pathway rates and distribution is metabolic also part of engineering (Kayser, 2009).

Goals and Strategies of Metabolic Engineering

In plants, suitable heterologous genes are introduced to yield novel compounds and precursors (Gómez-Galera et al., 2007). New traits like taste, smell, and color are imparted to flowers, food, and other ornamental plants (Verpoorte et al., 2002). Specific metabolites are expressed to improve the Agronomic traits, and the yield of aromatic and medicinal plant species increases (Verpoorte et al., 2002; Kinney, 2006). In Infeed crops, the level of harmful or anti-nutritional factors is decreased by metabolic engineering (Gómez-Galera et al., 2007).

Metabolic engineering approach have been successfully implemented to enhance yield aspects of various crop plants. Moreover, new opportunities are required to direct use of metabolic engineering in different fields (agriculture, chemical production, environmental applications, and medicines) (Verpoorte et al., 2002; Lessard et al., 2002). Owing to metabolic engineering practices, the precursor compound flux can be increased in plants, competative pathways can be blocked, metabolic activities can be enhanced, genes of interests can be overexpressed, the number of special cells can be enhanced triggering production of product of interest (Gómez-Galera et al., 2007; Koffas et al., 1999).

Metabolic Pathways of Interest

The use of biotechnology and metabolic engineering on aromatic plants is well reported to manipulate secondary metabolic pathways (Kumar & Gupta, 2008). There is a list of metabolic pathways which can be explored in metabolic engineering out of which shikimate, polyketide and terpenoid pathways are most prominent as discussed in our draft (Bentley & Haslam, 1990; Herrmann & Weaver, 1999). The role of these pathways is important in production of quinoline flavonoids, anthocyanins, indole alkaloids, lignin and lignans etc (Verpoorte & Alfermann, 2000).

Recent Advancements in Metabolic Engineering of Plants

In modern era, hidden hunger due to difficiency of certain nutrients have become an issue and metabolic engineering is currently used for multi-biofortification of staple crops to ensure required enrichment of crops with nutrients (Blancquaert et al., 2017). Among various plant derived biomolecules, vitamins are an important class and biosynthesis of these vitamins have many issues out of which stability of these compounds is important. Metabolic engineering is an effective and safe way for efficacious bioproduction of vitamin rich foods (Blancquaert et al., 2017).

The benificial role of metabolites like terpenoids is important for efficacious growth and development of plant and these compounds have widespread uses in industry as well. Owing to their dual purposes (for plant health and industrial use) use of genetic engineering is imminent to enhance the production (Nagegowda & Gupta, 2020). Essential features of biosynthesis and controlling these molecules open up the path for plant trait enhancement and help in the overproduction of required molecules by homologous and heterologous engineering. Recent progress in analytical techniques and functional genomics has directed towards different aspects of regulating and synthesizing these specialized target terpenoids by metabolic engineering in homo- or heterologous host systems with specific pathways (Nagegowda & Gupta, 2020).

Secondary Metabolic Pathways

The series of chemical reactions catalyzed by enzymes in an organism is known as the metabolic pathway. After detailed research, it was concluded that primary and secondary metabolic pathways exist in an organism. Plants use direct metabolic pathways to produce essential metabolites for their survival (Chakraborty, 2018). The secondary metabolic pathway is the pathway that uses primary metabolites as a precursor. For example, the phenylpropanoid pathway originates from amino acid phenylalanine, and phenolic compounds are produced due to this pathway.

Secondary metabolism is a specialized metabolism that consists of a series of reactions (Chakraborty, 2018). Plant secondary metabolism plays various roles in plant life cycles. Some of its roles are, e.g., plant-plant, plant-microorganisms, and plant insects interactions (Verpoorte et al., 2002). Secondary metabolites like terpenoids, phenylpropanoids, and terpenoids have been found necessary in drug development. The shikimic acid pathway and mevalonic acid are the principal pathways for the biosynthesis of these secondary metabolites (Ramawat et al., 2009).

Shikimic Acid Pathway

The shikimate pathway is a source of various precursors for aromatic molecules in apicomplexan, fungus, plants, and bacteria but not in animals (Singh et al., 2020). The aromatic amino acids, e.g., tyrosine, tryptophan, phenylalanine, and other compounds like phenylpropanoids, alkaloids, and phenolics, are synthesized by this pathway. The basic unit of various aromatic compounds, the benzene ring, is produced through the shikimate pathway in plants and fungi (Borah, 2015). The shikimate pathway consists of a sequence of seven enzymatic steps. The seven enzymes that catalyze the shikimate pathway are given in Fig. 14.1.

Phosphoenolpyruvic acid (PEP) and carbohydrate D-erythrose-4-phosphate are two phosphorylated active compounds from the glycolytic and pentose phosphate cycles. In the first step, the indole condensation of these two compounds occurs, and as an end product, 3-deoxy-D-arabinose-heptulosonate acid 7-phosphate (DAHP) is formed. DAHP loses Pi (Phosphate) in the second step. The 3-dehydroquinic acid (DHQ) is produced because of the cyclization of the enolic-type product to another aldol-type reaction. This cyclization is catalyzed by 3-dehydroquinate synthase (Averesch & Krömer, 2018; Bilal et al., 2018; Wilson & Roberts, 2014; Santos-Sánchez et al., 2019).

In the third step, the 3-dehydroshikimic acid (DHS) is further produced due to dehydration of DHQ. A conjugated double carbon-carbon is present in the 3-dehydrogenate, the branch point reactions from DHS give rise to protocatechuic acid (C6-C1) and gallic acid. In the fourth step, the reduction reaction of DHS occurs in the pathway with reduced nicotinamide adenine dinucleotide phosphate (NADPH).

The 3-dehydrogenate dehydratase is a hydro-lyase kind of enzyme, while shikimate dehydrogenase is an oxidoreductase enzyme. In the third step of the pathway, the DHQ is converted into the 3-dehydroshikimic acid (DHS) by DHQ dehydratase by eliminating water. This reaction step is reversible. The carbonyl group at the C5 is reduced due to the catalytic activity of SDH with NADPH, which converts DHS into shikimic acid. The production of DHS is a branch point to the catabolic quinate pathway, and the shikimic acid pathway. If the DHS's dehydration occurs, the protocatechuic acid (C6-C1) and the gallic acid are produced. Gallic acid is a hydroxybenzoic acid that is a part of tannins. There are two types of 3-dehydrogenate dehydratase (DHQ hydratase). Type I is heat-sensitive, while type II is heat stable. Bacteria and higher plants contain type I, while fungi include both types of enzymes (Averesch & Krömer, 2018; Bilal et al., 2018; Wilson & Roberts, 2014; Santos-Sánchez et al., 2019).

In the fifth step, the shikimic acid is activated with ATP (adenosine triphosphate) in the presence of shikimate kinase. The end product is shikimic acid 3 phosphate (S3P). In the sixth step, PEP is added to S3P for the production of 5-enolpyruvyl shikimic acid 3-phosphate. 5-enolpyruvylshikimate 3-phosphate synthase (ESPS) is the enzyme that catalyzes this step (Averesch & Krömer, 2018, Bilal et al., 2018, Wilson & Roberts, 2014, Santos-Sánchez et al., 2019).

Phosphoenolpyruvic acid D-erythrose-4-phosphate

↓ 3-deoxy-D-arabino- heptulosonate acid 7-phosphate synthase

3-deoxy-D-arabino- heptulosonate acid 7-phosphate (DAHP)

↓ 3-dehydroquinate synthase (DHQS)

3-dehydroquinic acid (DHQ)

↓ 3-dehydroquinate dehydratase (DHQ hydratase)

3-dehydroshikimic acid (DHS)

↓ shikimate dehydrogenase (SDH)

Shikimate

↓ Shikimate kinase

Shikimate 3-Phosphate

↓ 5-enolpyruvylshikimate 3-phosphate synthase (EPSPS)

5-enolpyruvylshikimate 3-phosphate synthase (ESPS)

↓ Chorismate synthase

Chorismate

Fig. 14.1 Diagrammatic representation of shikimic acid pathway

Mevalonic Acid (MVA) Pathway

2-C-methyl-D-erythritol 4-phosphate (MEP) and mevalonic acid (MVA) pathways coexist in plants. MEP pathway is a source of precursors for the synthesis of carotenoids, apocarotenoids, monoterpenes, side-chain tocopherols, prenylquinones, and side chains tocopherols including prenyl quinones, and chlorophylls, while the MVA pathway is a source of precursors for synthesizing sesquiterpenes, triterpenes, ubiquinone, phytosterols, vitamin D, and primary metabolites crucial for cell integrity. The end of these two pathways is the formation of IPP. IPP and its allylic isomer DMAPP are C5 compounds frequently used as a precursor for synthesizing isoprenoids (Zhu et al., 2014; Simkin et al., 2011; Yang et al., 2016).

The MVA pathway consists of six enzymatic reactions to produce IPP. In the first reaction, two molecules of Acetyl-CoA are condensed to produce acetoacetyl-CoA. This reaction is further catalyzed by acetoacetyl-CoA thiolase (AACT). The hydroxy-3-methylglutaryl (HMG)-CoA synthase (HMGS) catalyzes an additional condensation of acetyl-CoA. As a result, HMG-CoA is produced that is further reduced to make mevalonate. This reduction is further catalyzed by the HMG-CoA reductase (HMGR) (Zhu et al., 2014; Simkin et al., 2011; Wilson & Roberts, 2014).

The mevalonate kinase (MVK) and 5-phosphomevalonate kinase (PMK) catalyze the next step in which two successive phosphorylations of mevalonate take place, and mevalonate diphosphate is produced. In the last and final stage, the decarboxylation of mevalonate diphosphate is catalyzed by mevalonate 5-diphosphate decarboxylase (MVD). IPP is produced as an end product of aforementioned reaction. Then, the IPP isomerase (IDI) catalyzes the reversible isomerization of IPP, leading to the formation of DMAPP thus, allowing the condensation of IPP/DMAPP for the biogenesis of terpenoids compounds (Zhu et al., 2014, Simkin et al., 2011, Wilson & Roberts, 2014).

2-C-Methyl-D-Erythritol 4-Phosphate (MEP) Pathway

The MEP pathway consists of seven sequential steps catalyzed by enzymes. In the first step, pyruvate with D-glyceraldehyde-3-phosphate undergoes condensation to synthesize 1-deoxy-D'Xylulose-5-phosphate (DXP) which subsequently catalyzes the first reaction of 1-deoxy-D-xylulose-5-phosphate synthase (DXS). The second step involves converting DXP into 2-C-methyl-D-erythritol 4-phosphate (MEP) in the presence of 1-deoxy-D-xylulose-5-phosphate reductoisomerase (DXR) enzyme. MEP is further converted into 4-diphosphocytidyl-2-C-methyl-D-erythritol (CDP-ME) in the third step by 4-diphosphocytidyl-2-C-methyl-D-erythritol synthase (CMS). Then, a CDP-ME kinase (CMK) phosphorylates CDP-ME at its 2-hydroxy group (Zhu et al., 2014; Wilson & Roberts, 2014).

In the fifth step, the 4-diphosphocytidyl-2-C-methyl-D-erythritol 2-phosphate is converted into 2-C Methyl-D-erythritol 2,4-cyclodiphosphate (MECDP) by 2-C-methyl-D-erythritol 2,4- cyclodiphosphate synthase (MECS) enzyme. The sixth step involves the conversion of MECDP into 1-hydroxy-2-methyl-2-butenyl 4-diphosphate (HMBPP).

The sixth step is catalyzed by HMBPP synthase (HDS). In the final step, HMBPP has converted into IPP or dimethylallyl diphosphate (DMAPP) in a 5:1 mixture in the presence of HMBPP reductase (Zhu et al., 2014, Wilson & Roberts, 2014).

Utilization of Metabolic Engineering in the Manipulation of Secondary Metabolism

Plants produce many economically essential compounds used in various sectors and the pharmaceutical industry (Verpoorte et al., 1999). Humans have identified many valuable metabolites naturally produced by plants used in treating diseases like cancer, illness, infections, cardiovascular diseases, and neuromuscular disorders (Courdavault et al., 2021). These active metabolites are particularly synthesized in plants by different metabolic pathways (Sharma et al., 2019). The active metabolites are accumulated in low quality in natural plants. So, when many metabolites, e.g., in case of a pandemic or average condition, are required, these natural resources are overexploited. In the era of 2000, when synthetic biology emerged, various biotechnological approaches for manufacturing these metabolites began to appear (Courdavault et al., 2021). Metabolic engineering is an alternative approach to improving and synthesizing secondary plant metabolites (Sharma et al., 2019).

Metabolic engineering is a very complicated and diverse process in which there is a need to understand the core pathways and genes and transcription factors, flux, metabolic bottleneck, and the effect of metabolic channeling metabolon formation on overall systems (Nabavi et al., 2020; Barone et al., 2020). This understanding is crucial for regulating the biosynthesis of a desired phytochemical and its possible exploitation in nutraceutical industries, food, and medicine (Nabavi et al., 2020). However, when the microbial hosts are utilized for synthetic expression of metabolic pathways several other complexities emerged. There is a possibility of producing unexpected end products and intermediates and the interaction of ways with the host metabolism with potential possibility of producing unexpected end products and intermediates (Barone et al., 2020).

The most crucial step in producing a considerable number of bioactive compounds in different plant cell compartments is gene expression regulation. This regulation of gene expression can be made at any stage from transcription initiation to RNA processing and during the posttranscriptional modifications of the final protein product (Sharma et al., 2019). The modification of protein is one of the main stages of regulation. A conserved protein modification step exists in the biosynthesis of three prominent natural product families named nonribosomal peptides (NRP), fatty acids (FA), and polyketides (PK). In this particular step, the phosphopantetheinylation of encoded carrier proteins (apo form) of their synthases takes place by phosphopantetheinyl transferases (PPtase), and they are converted into their holo form (active form) (Zhang et al., 2017). It is interesting that PPtase can also potentially catalyze the acyl-phosphopantetheinylation of encoded carrier proteins by accepting acyl-CoAs rather than HSCoAs. However, this acyl-phosphopantetheinylation is not suitable for metabolic biosynthesis that is why additional editing enzymes remove this acyl group. This potential of PPtase to accept free or acyl-CoA is a cause of activation or deactivation of metabolic biosynthesis. So, it plays a pivotal role as a regulator in protein modification and can turn on or off the biosynthetic process (Zhang et al., 2017).

Either upregulation or downregulation modifications further carries out the gene regulation. The upregulation results in excessive gene product (required protein), while downregulation mitigate genetic expression. For upregulation or downregulation, the key genes involved in a particular pathway are targeted, e.g., bioactive metabolic pathways in plants (Sharma et al., 2019). Various mechanisms carry out the upregulation of many genes, e.g., inserting highly expressive genes from other plants to enhance the production of a specific bioactive compound in plants, using solid promoters, and modulating transcriptional factors. While in the downregulation, the expression of genes involved in competitive pathways is knocked down to further enhance the production capacity of several metabolic compounds in plant species (Sharma et al., 2019).

RNAi-Mediated Gene Silencing

In 1928, the first paper on RNA silencing was published, but this topic got attention in the past few years. According to genetic and molecular analysis, plants have three pathways for RNA silencing. It is very unusual that green plants compared to other organisms, possess all three pathways of RNA silencing. These pathways specified enzymatic Dicer is involved for double-stranded RNA cleavage into short nucleotides of size 21–26. The first pathway, which is cytoplasmic siRNA silencing, is crucial for virus-infected plant cells and has been widely declared as posttranscriptional gene silencing (PTGS) (Hamilton & Baulcombe, 1999). In the second pathway, miRNAs are involved in silencing of endogenous mRNAs by negatively regulating the gene expression through base pairing to mRNAs (Bartel, 2004). While in the third pathway, DNA methylation and transcription suppression are involved for RNA silencing in plants (Wassenegger et al., 1994; Mette et al., 2000; Jones et al., 2001). Gene expression is concurrently suppressed by a homology-based process of RNA silencing activated by double-stranded RNA (Denli & Hannon, 2003). This process was firstly identified in plants and plays a crucial role against viruses in defense mechanisms. Still, later on, it was found that this phenomenon exists in other organisms too, including plants, animals, and protozoa (Hannon, 2002). Similar speculations have showed that dsRNS is particularly stringent in mediating the prescribed mechanism through concurrent involvement of sense and antisense strands (Fire et al., 1998). Activation of dsRNA is a significant function of RNA silencing in controlling gene expression. However, intensive investigation is still required in gene expression, RNA silencing, host defense, and for the development of special therapeutics. Similar speculations have revealed that genetic engineering of plants is done by targeting genes or their promoters using RNAi (Saurabh et al., 2014).

Numerous products like fiber, oils, food, dyes, wood, and pharmaceuticals are provided by the metabolic engineering of plants using RNAi. Similarly, valuable secondary metabolites are obtained in minimal quantities due to the minimal production of these secondary metabolites. Potential limitations have been overcame by the use of RNAi in metabolic engineering. Likewise, hpRNA-mediated silencing in rice genes have been reported (Kusaba et al., 2003). The major seed storage protein in rice is glutelin. Rice lines that are low in glutelin content are helpful for patients with kidney diseases in which less protein intake is required. The cause of the reduction in glutelin content in rice grain is a dominant mutation *Lgc*1 in which 3.5 kb deletion occurred between two similar glutelin genes that form dsRNA. Therefore, gene silencing is induced (Kusaba et al., 2003). The stable trait demonstrates the stability of dsRNA in the transgenic plant for 20 generations. *ghSAD-1* and *ghFAD2-1* are two enzymes of cotton which were silenced by the fatty acid biosynthesis pathway. Silencing of *ghSAD-1* in cotton increased the seed stearic acid content, and RNAi-mediated mechanism downregulation of *ghFAD2-1* which in turn elevated the oleic acid content (Liu et al., 2002). In producing hypoallergenic grasses, which cause asthma and hay fever, by which 25% of the temperate region population is affected, RNAi shows excellent potential. Pollen proteins *Lolp*1 and *Lol p*2 are the primary allergens. 905 of the population of allergy sufferers are sensitive towards these proteins. Under the control of the maize-specific promoter, *Lolp1* and *Lol p2* levels are downregulated by the antisense cDNA sequence expression (Petrovska et al., 2004). hpRNAi constructs that have been expressed in transgenic Coffea spp. show an accumulation of theobromine and caffeine in the range of 30–50% (Ogita et al., 2004). It shows the involvement of coffee theobromine biosynthesis so engineering decaffeinated coffee plants can be synthesized hpRNAi-mediated gene silencing. *Brassica napus* produces bright yellow flower canopy which absorbs active radiation up to 60%; as a result, yield is reduced; hp. construct is prepared which targets *BPI* gene family under the control of petal specific promoter from *Arabidopsis* for achieving reduced or no leaves (Byzova et al., 2004). As a result, fertile male flowers are produced in which sepaloid petals are formed. Metabolic engineering of plants by using RNAi-mediated gene silencing produces many products like dyes, wood, food, fiber, oil, and pharmaceuticals. As the production of secondary metabolites is deficient so small quantities of valuable secondary metabolites are obtained. So, RNAi in metabolic engineering is used to overcome all these limitations. hpRNA-mediated gene silencing is used in the metabolic engineering of rice genes. Glutelin is a major seed storage protein in rice.

RNAi-mediated metabolic engineering is firstly reported in opium poppy for enzyme silencing in codeine reductase gene (Allen et al., 2004). Increase in carotenoid and flavonoid content is observed in tomato in which improvements are made by the use of fruit-specific promoter for endogenous photomorphogenesis regulatory gene *DET1* suppression. (Davuluri et al., 2005). The ACR2 gene in *Arabidopsis*, which encodes for arsenic reductase, is silenced using hp. construct for bioremediation of heavy metal contaminated soil. The transgenic plant accumulates more arsenic in shoots than the roots in wild type (Dhankher et al., 2006).

RNAi for Disease Resistance in Plants

In 1998, it was first demonstrated that RNAi technology could generate virus resistance in plants (Waterhouse et al., 1998). To overcome the resistance, counter-silencing strategies have been evolved by encoding proteins in plant viruses. Distinct viruses can be targeted in a single construct by designing a hairpin structure for concurrent silencing of diverse plant species. Efforts for controlling single-stranded DNA viruses Gemini viruses by using RNAi have been reported (Mansoor et al., 2006).

Case Studies

Alkaloids

Plants have developed defensive mechanisms like physical barriers and bioactive metabolites or anti-digestive proteins to combat herbivores and microbial pathogens. The transcriptional activation of genes involved in the biosynthesis of metabolic pathways regulates the accumulation of bioactive metabolites through the phytohormone jasmonate-isoleucine. Schweizer et al. (2018) have utilized their newly developed flower petal transformation method to explain the complicated regulatory mechanisms navigating the jasmonate-modulated monoterpenoid indole alkaloid (MIA) biosynthesis in *Catharanthus roseus* (medicinal plant). The anticancer metabolites vincristine and vinblastine also belong to monoterpenoid indole alkaloids (MIAs).

Schweizer and his coworkers in 2018 have overexpressed transcriptional activators like *ORCA3*, *MYC2a*, and *BIS1* in combinations to understand the modular transcriptional control of MIA biosynthesis. They have indicated that an engineered de-repressed *MYC2a* expression works as a trigger for a massive reprogramming of the MIA biosynthesis pathway resulting in tremendous increase in at least 23 MIAs accumulation (Schweizer et al., 2018).

Dehghan and his coworkers et al. (2017) studied the outcomes of metabolic engineering in the context of polyploidy by overexpressing the *h6h* gene in tetraploid *H. muticus* hairy root cultures. According to flow cytometry analysis, every clone's genetic stability was shown except a few of them suggesting the highest scopolamine accumulation and expression of the *h6h* gene compared to diploid clones. Interestingly, metabolic engineering of the pathway of tropane biosynthesis is considered to be a potential system in polyploids to enhance the production of tropane alkaloids (Dehghan et al., 2017).

They also established *Hyoscyamus senecionis* transgenic hairy root cultures through the overexpression of *pmt* with increased scopolamine accumulation contents. At the same time, the clones in which *h6h* was overexpressed were not able to have high scopolamine accumulation compared to intact leaves plants. It was observed that methyl jasmonate application alternatively enhances the pmt. expression and at the same time, it reduces the tropinone reductase II *(trII)* that results in more hyoscyamine and total alkaloid biosynthesis in *H. Senecionis* (Dehghan et al., 2017).

Scopolia lurida is one of the largest producers of Tropane alkaloids, and it is natively found in Tibet as a herbal species. However, the molecular, biotechnological, and biochemical studies of tropane alkaloid biosynthesis have not yet been studied in this particular species. A putative short-chain dehydrogenase *(SDR)* gene's isolation and characterization was reported by Dehghan et al. (2017). According to sequence analysis, *SlTRI* is related to the SDR family. The phylogenetic analysis further showed that SlTRI was clustered with tropine-forming reductases. *SlTRI* and other TA-biosynthesis genes, including hyoscyamine 6β-hydroxylase *(SlH6H)* and putrescine N-methyltransferase *(SlPMT)*, were expressed in the roots of *S. lurida*, preferably or exclusively. The tissue profile of *SlTRI* further showed the involvement of this gene in the biosynthesis of tropane alkaloid (Zhao et al., 2017).

The use of GC-MS has shown the involvement of *SlTRI* in catalyzing the tropinone contents that results in tropine. Tropine is an essential intermediate product of tropane alkaloids. The enzymatic assay is further carried out with a purified recombinant *SlTRI*, obtained from *Escherichia coli* exhibiting *SlTRI* as a tropine-forming reductase. Finally, metabolic engineering in *S. lurida* confirmed that *SITRI* promotes tropane alkaloid biosynthesis (Zhao et al., 2017).

So, the effect of overexpression of *SITRI* on tropane alkaloid accumulation has been observed by establishing the *S. lurida* hairy root cultures. The root cultures with *SITRI* overexpression were found to have hyoscyamine content 1.7- to 2.9-fold higher than the controlled ones. The content of the corresponding scopolamine was also elevated. So, A gene has been functionally identified that can be used to increase the tropane alkaloid biosynthesis in hairy root cultures of *S. lurida* via metabolic engineering (Zhao et al., 2017).

Terpenoids

Terpenoids are natural compounds based on the C5 Isoprene subunits (Ma et al., 2019). Terpenoids have great potential for medicinal and industrial (food and cosmetic fields) applications (Abdallah et al., 2019, Ma et al., 2019). They harbor significant activity against allergies and cancer, with pleasant aromas as well (Ma et al., 2019). Metabolic engineering of microbial hosts has gained rapid attention for producing valuable compounds such as taxol and artemisinin in the last few decades. *Bacillus subtilis* 168 is considered an exciting host harboring metabolic potential. Abdallah has expressed the plant-derived taxadiene synthase (TXS) enzyme, to engineer *Bacillus subtilis* to be a cell factory for the production of chemotherapeutic taxol. The conversion of the precursor geranylgeranyl pyrophosphate (GGPP) into taxa-4,11-diene takes place with the help of TXS enzyme. The GGPP is the first intermediate in the biosynthesis of taxol (Abdallah et al., 2019).

Moreover, Abdallah and his coworkers in 2019 overexpressed the eight enzymes involved in *taxol* biosynthesis to enhance the flux of GGPP. A synthetic operon has been ellucidated harboring *Bacillus subtilis* genes that encode the 2-C-methyl-D-erythritol-4-phosphate (MEP) pathway along with *ispA* (encoding geranyl and farnesyl pyrophosphate synthases), that is stringently responsible for farnesyl pyrophosphate (FPP) production. In addition, a vector was also introduced, harboring the *crtE* gene that encodes *geranylgeranyl pyrophosphate synthase*, GGPPS, of *Pantoea ananatis* for increased supply of GGPP. Correspondingly, an 83-fold increase in production of taxadiene was observed due to the overexpression of the genes mentioned above compared to the strains in which only TXS was overexpressed, and they were dependent on the innate pathway of *Bacillus subtilis*. The overall production of taxadiene was 17.8 mg/l in that strain.

Tanshinones are the diterpenoid compounds involved in the treatment of cardiovascular diseases. There is high demand for tanshinones, but the extraction methods for tanshinones are not sufficient. Wei and his coworkers provided a practical approach to increase the tanshinone content and other natural products by using metabolic engineering. They reported that *SmMDS* (2-c-methyl-d-erythritol 2,4-cyclodiphosphate synthase) is a tanshinone biosynthesis gene, and its overexpression in transgenic *Salvia miltiorrhiza* hairy roots remarkably enhance the tanshinone yield (Wei et al., 2019). Meanwhile, elicitor treatment have significantly increased the tanshinone contents in overexpressing cell lines, suggesting as promising strategy to trigger their production. The total enhanced amount of tanshinones by Ag+, MJ (Methyl Jasmonate), and YE (Yeast Extract) treatment was 2.5, 3.2, and 2.3 mg/g DW (Dry weight), respectively, in comparison with non-induced transgenic lines where tanshinone increase was reported 1.7 mg/g DW (Wei et al., 2019).

Flavonoids

Baicalein, wogonin, and baicalin are important natural flavonoid compounds that are produced by *Scutellaria baicalensis*. Park et al. (2021) have investigated that these three flavonoid productions can be increased by maize transcription factor Lc in *S. baicalensis* hairy root cultures by upregulating baicalein 7-O-glucuronosyltransferase *(UBGAT)*, and flavonoid biosynthesis pathway genes (*SbPAL1*, *Sb4CL*, and *SbC4H*). As a result, up to 80.5 ± 6.15 mg g − 1, dry weight flavonoid content in total is 322% greater than the average contents produced by GUS-overexpressing lines (Park et al., 2021).

Moreover, it was found that *Arabidopsis* transcription factor *PAP1* can also trigger flavonoid accumulation by upregulation of *SbPAL1*, *SbPAL2*, *SbPAL3*, *SbC4H*, *Sb4CL*, *SbCHI*, and *UBGAT* genes. This upregulation enhances the total flavonoid contents up to 133 ± 7.66 mg g^{-1} dry weight, 532% greater than the average flavonoid content produced by three GUS-overexpressing lines (Park et al., 2021).

These findings indicate that the *Agrobacterium* rhizogenes-mediated transformation can be used to achieve metabolic engineering in *S. baicalensis*. Baicalein, wogonin, and baicalin can be increased by overexpression of *AtPAP1* and *ZmLc* in hairy root cultures. According to the results, *AtPAP1* and *ZmLc* can be employed as positive regulators of the flavonoid biosynthesis pathway of hairy root cultures of *S. baicalensis* (Park et al., 2021).

Future Prospects

Plants are a significant part of the medicinal industry as they are a source of beneficial essential compounds. These natural compounds are enhanced and obtained from plants through various methods (Hendrawati et al., 2012). Metabolic engineering is a promising technology to decrease the production costs of commercial products and make food crops more nutritious. However, the major limitation in metabolic engineering is the lack of resources and availability of data. Moreover, genes involved in biosynthetic pathways are not known accurately. The level of expression of various genes and enzymes encoded by them must be understood in a biosynthetic pathway to manipulate the path accordingly (Ghassemi et al., 2020; Wilson & Roberts, 2014; Hendrawati et al., 2012). The generation of omics datasets has been massive in the previous few years. It is essential to successfully implement metabolic processes (Wilson & Roberts, 2014; Rai et al., 2017).

Moreover, the complex interaction of different biosynthetic pathways further complicates the application of metabolic engineering strategies. Therefore, it is essential to develop methods to manipulate multiple genes involved in interacting pathways to obtain the desired end products (Wilson & Roberts, 2014). The triggering of many harmful effects after cold stress in aromatic and medicinal plants influences various biochemical, physiological, and morphological processes. There is a high need to explore the role of multiple metabolites of medicinal plants in detoxification reaction after facing the low temperature. Moreover, the role of non-enzymatic compounds in detoxification reactions is also not known yet (Ghassemi et al., 2020). Proper metabolic models and detailed knowledge of the systems require the implementation of prescribed strategy to explore more medicinal and aromatic plants through metabolic engineering (Wilson & Roberts, 2014).

References

Abdallah, I. I., Pramastya, H., Van Merkerk, R., & Quax, W. J. (2019). Metabolic engineering of *Bacillus subtilis* toward taxadiene biosynthesis as the first committed step for Taxol production. *Frontiers in Microbiology, 10*, 218.

Allen, R. S., Millgate, A. G., Chitty, J. A., Thisleton, J., Miller, J. A., Fist, A. J., et al. (2004). RNAi-mediated replacement of morphine with the nonnarcotic alkaloid reticuline in opium poppy. *Nature Biotechnology, 22*(12), 1559–1566.

Averesch, N. J., & Krömer, J. O. (2018). Metabolic engineering of the shikimate pathway for production of aromatics and derived compounds—Present and future strain construction strategies. *Frontiers in Bioengineering and Biotechnology, 6*, 32.

Barone, R. P., Knittel, D. K., Ooka, J. K., Porter, L. N., Smith, N. T., & Owens, D. K. (2020). The production of plant natural products beneficial to humanity by metabolic engineering. *Current Plant Biology, 24*, 100,121.

Bartel, D. P. (2004). MicroRNAs: Genomics, biogenesis, mechanism, and function. *Cell, 116*(2), 281–297.

Bentley, R., & Haslam, E. (1990). The shikimate pathway—A metabolic tree with many branches. *Critical Reviews in Biochemistry and Molecular Biology, 25*(5), 307–384.

Bilal, M., Wang, S., Iqbal, H. M., Zhao, Y., Hu, H., Wang, W., & Zhang, X. (2018). Metabolic engineering strategies for enhanced shikimate biosynthesis: Current scenario and future developments. *Applied Microbiology and Biotechnology, 102*(18), 7759–7773.

Birchfield, A. S., & McIntosh, C. A. (2020). Metabolic engineering and synthetic biology of plant natural products—A mini review. *Current Plant Biology, 24*, 100,163.

Blancquaert, D., De Steur, H., Gellynck, X., & Van Der Straeten, D. (2017). Metabolic engineering of micronutrients in crop plants. *Annals of the New York Academy of Sciences, 1390*(1), 59–73.

Borah, J. C. (2015). Shikimic acid: A highly prospective molecule in pharmaceutical industry. *Current Science, 109*(9), 1672–1679.

Byzova, M., Verduyn, C., De Brouwer, D., & De Block, M. (2004). Transforming petals into sepaloid organs in Arabidopsis and oilseed rape: Implementation of the hairpin RNA-mediated gene silencing technology in an organ-specific manner. *Planta, 218*(3), 379–387.

Chakraborty, P. (2018). Herbal genomics as tools for dissecting new metabolic pathways of unexplored medicinal plants and drug discovery. *Biochimie open, 6*, 9–16.

Courdavault, V., O'Connor, S. E., Jensen, M. K., & Papon, N. (2021). Metabolic engineering for plant natural products biosynthesis: New procedures, concrete achievements and remaining limits. *Natural Product Reports*.

Davuluri, G. R., Van Tuinen, A., Fraser, P. D., Manfredonia, A., Newman, R., Burgess, D., et al. (2005). Fruit-specific RNAi-mediated suppression of DET1 enhances carotenoid and flavonoid content in tomatoes. *Nature Biotechnology, 23*(7), 890–895.

Dehghan, E., Reed, D. W., Covello, P. S., Hasanpour, Z., Palazon, J., Oksman-Caldentey, K. M., & Ahmadi, F. S. (2017). Genetically engineered hairy root cultures of *Hyoscyamus senecionis* and *H. muticus*: Ploidy as a promising parameter in the metabolic engineering of tropane alkaloids. *Plant Cell Reports, 36*(10), 1615–1626.

DellaPenna, D. (2001). Plant metabolic engineering. *Plant Physiology, 125*(1), 160–163.

Denli, A. M., & Hannon, G. J. (2003). RNAi: An ever-growing puzzle. *Trends in Biochemical Sciences, 28*(4), 196–201.

Dhankher, O. P., Rosen, B. P., McKinney, E. C., & Meagher, R. B. (2006). Hyperaccumulation of arsenic in the shoots of Arabidopsis silenced for *arsenate reductase* (*ACR2*). *Proceedings of the National Academy of Sciences, 103*(14), 5413–5418.

Dixon, R. A., & Steele, C. L. (1999). Flavonoids and isoflavonoids—A gold mine for metabolic engineering. *Trends in Plant Science, 4*(10), 394–400.

Dixon, R. A. (2001). Natural products and plant disease resistance. *Nature, 411*(6839), 843–847.

Dixon, R. A. (2005). Engineering of plant natural product pathways. *Current Opinion in Plant Biology, 8*(3), 329–336.

Facchini, P. J. (2001). Alkaloid biosynthesis in plants: Biochemistry, cell biology, molecular regulation, and metabolic engineering applications. *Annual Review of Plant Biology, 52*(1), 29–66.

Farmer, W. R., & Liao, J. C. (2001). Precursor balancing for metabolic engineering of lycopene production in *Escherichia coli. Biotechnology progress, 17*(1), 57–61.

Fire, A., Xu, S., Montgomery, M. K., Kostas, S. A., Driver, S. E., & Mello, C. C. (1998). Potent and specific genetic interference by double-stranded RNA in *Caenorhabditis elegans. Nature, 391*(6669), 806–811.

Forkmann, G., & Martens, S. (2001). Metabolic engineering and applications of flavonoids. *Current Opinion in Biotechnology, 12*(2), 155–160.

Ghassemi, S., Delangiz, N., Lajayer, B. A., Saghafi, D., & Maggi, F. (2020). Review and future prospects on the mechanisms related to cold stress resistance and tolerance in medicinal plants. *Acta Ecologica Sinica, 41*, 120–129.

Gómez-Galera, S., Pelacho, A. M., Gené, A., Capell, T., & Christou, P. (2007). The genetic manipulation of medicinal and aromatic plants. *Plant Cell Reports, 26*(10), 1689–1715.

Hamilton, A. J., & Baulcombe, D. C. (1999). A species of small antisense RNA in posttranscriptional gene silencing in plants. *Science, 286*(5441), 950–952.

Hannon, G. J. (2002). RNA interference. *Nature, 418*(6894), 244–251.

Haq, J. (2000). Production of bioactive compounds: In vitro production of bioactive compounds from medicinal and aromatic plants. Retrieved from http://www.telmedpak.com/ngonews.Asp

Hendrawati, O., Woerdenbag, H. J., Hille, J., & Kayser, O. (2012). Metabolic engineering of medicinal plants and microorganisms for the production of natural products. In O. Kayser & H. Warzecha (Eds.), *Pharmaceutical biotechnology: Drug discovery and clinical applications* (2nd ed., pp. 491–526).

Herrmann, K. M., & Weaver, L. M. (1999). The shikimate pathway. *Annual Review of Plant Biology, 50*(1), 473–503.

Jones, L., Ratcliff, F., & Baulcombe, D. C. (2001). RNA-directed transcriptional gene silencing in plants can be inherited independently of the RNA trigger and requires Met1 for maintenance. *Current Biology, 11*(10), 747–757.

Joshi, K., Chavan, P., Warude, D., & Patwardhan, B. (2004). Molecular markers in herbal drug technology. *Current Science*, 159–165.

Kappers, I. F., Aharoni, A., Van Herpen, T. W., Luckerhoff, L. L., Dicke, M., & Bouwmeester, H. J. (2005). Genetic engineering of terpenoid metabolism attracts bodyguards to Arabidopsis. *Science, 309*(5743), 2070–2072.

Kayser, O. (2009). Metabolic engineering strategies for the optimization of medicinal and aromatic plants: Expectations and realities. In *IV international symposium on breeding research on medicinal and aromatic plants ISBMAP2009 860* (pp. 199–204).

Kinney, A. J. (2006). Metabolic engineering in plants for human health and nutrition. *Curren Opinion in Biotechnology, 17*(2), 130–138.

Koffas, M., Roberge, C., Lee, K., & Stephanopoulos, G. (1999). Metabolic engineering. *Annual Review of Biomedical Engineering, 1*(1), 535–557.

Kumar, J., & Gupta, P. K. (2008). Molecular approaches for improvement of medicinal and aromatic plants. *Plant Biotechnology Reports, 2*(2), 93–112.

Kusaba, M., Miyahara, K., Iida, S., Fukuoka, H., Takano, T., Sassa, H., et al. (2003). Low glutelin content1: A dominant mutation that suppresses the glutelin multigene family via RNA silencing in rice. *The Plant Cell, 15*(6), 1455–1467.

Lessard, P. A., Kulaveerasingam, H., York, G. M., Strong, A., & Sinskey, A. J. (2002). Manipulating gene expression for the metabolic engineering of plants. *Metabolic Engineering, 4*(1), 67–79.

Liu, Q., Singh, S. P., & Green, A. G. (2002). High-stearic and high-oleic cottonseed oils produced by hairpin RNA-mediated post-transcriptional gene silencing. *Plant Physiology, 129*(4), 1732–1743.

Ma, Y. R., Wang, K. F., Wang, W. J., Ding, Y., Shi, T. Q., Huang, H., & Ji, X. J. (2019). Advances in the metabolic engineering of *Yarrowia lipolytica* for the production of terpenoids. *Bioresource Technology, 281*, 449–456.

Mansoor, S., Amin, I., Hussain, M., Zafar, Y., & Briddon, R. W. (2006). Engineering novel traits in plants through RNA interference. *Trends in Plant Science, 11*(11), 559–565.

Mette, M. F., Aufsatz, W., Van der Winden, J., Matzke, M. A., & Matzke, A. J. M. (2000). Transcriptional silencing and promoter methylation triggered by double-stranded RNA. *The EMBO Journal, 19*(19), 5194–5201.

Nabavi, S. M., Šamec, D., Tomczyk, M., Milella, L., Russo, D., Habtemariam, S., et al. (2020). Flavonoid biosynthetic pathways in plants: Versatile targets for metabolic engineering. *Biotechnology Advances, 38*, 107,316.

Nagegowda, D. A., & Gupta, P. (2020). Advances in biosynthesis, regulation, and metabolic engineering of plant specialized terpenoids. *Plant Science, 294*, 110,457.

Ogita, S., Uefuji, H., Morimoto, M., & Sano, H. (2004). Application of RNAi to confirm theobromine as the major intermediate for caffeine biosynthesis in coffee plants with potential for construction of decaffeinated varieties. *Plant Molecular Biology, 54*(6), 931–941.

Pank, F. (2007). Breeding of medicinal plants. 1807–2007 knowledge for generations.

Park, C. H., Park, Y. E., Yeo, H. J., Yoon, J. S., Park, S. Y., Kim, J. K., & Park, S. U. (2021). Comparative analysis of secondary metabolites and metabolic profiling between Diploid and Tetraploid *Morus alba* L. *Journal of Agricultural and Food Chemistry, 69*(4), 1300–1307.

Petrovska, N., Wu, X., Donato, R., Wang, Z., Ong, E. K., Jones, E., et al. (2004). Transgenic ryegrasses (*Lolium spp.*) with down egulation of main pollen allergens. *Molecular Breeding, 14*(4), 489–501.

Pickens, L. B., Tang, Y., & Chooi, Y. H. (2011). Metabolic engineering for the production of natural products. *Annual Review of Chemical and Biomolecular Engineering, 2*, 211–236.

Rai, A., Saito, K., & Yamazaki, M. (2017). Integrated omics analysis of specialized metabolism in medicinal plants. *The Plant Journal, 90*, 764–787.

Ramawat, K. G., Dass, S., & Mathur, M. (2009). The chemical diversity of bioactive molecules and therapeutic potential of medicinal plants. *Herbal Drugs: Ethnomedicine to Modern Medicine*, 7–32.

Razdan, S. (2018). An update towards the production of plant secondary metabolites. In *Recent trends and techniques in plant metabolic engineering* (pp. 1–17). Springer.

Rhodes, M. J. C., Hamill, J., Parr, A. J., Robins, R. J., & Walton, N. J. (1988). *Strain improvement by screening and selection techniques. Manipulating secondary metabolism in culture* (pp. 83–93). Cambridge University Press.

Roberts, S. C. (2007). Production and engineering of terpenoids in plant cell culture. *Nature Chemical Biology, 3*(7), 387–395.

Santos-Sánchez, N. F., Salas-Coronado, R., Hernández-Carlos, B., & Villanueva Cañongo, C. (2019). Shikimic acid pathway in biosynthesis of phenolic compounds. In M. Soto-Hernández, R. García-Mateos, & M. Palma-Tenango (Eds.), *Plant physiological aspects of phenolic compounds* (pp. 1–15). IntechOpen.

Saurabh, S., Vidyarthi, A. S., & Prasad, D. (2014). RNA interference: Concept to reality in crop improvement. *Planta, 239*(3), 543–564.

Schweizer, F., Colinas, M., Pollier, J., Van Moerkercke, A., Bossche, R. V., De Clercq, R., & Goossens, A. (2018). An engineered combinatorial module of transcription factors boosts production of monoterpenoid indole alkaloids in *Catharanthus roseus*. *Metabolic Engineering, 48*, 150–162.

Sharma, M., Koul, A., Sharma, D., Kaul, S., Swamy, M. K., & Dhar, M. K. (2019). Metabolic engineering strategies for enhancing the production of bio-active compounds from medicinal plants. In *Natural bio-active compounds* (pp. 287–316). Springer.

Simkin, A. J., Guirimand, G., Papon, N., Courdavault, V., Thabet, I., Ginis, O., et al. (2011). Peroxisomal localisation of the final steps of the mevalonic acid pathway in planta. *Planta, 234*(5), 903–914.

Singh, P., Gupta, E., Mishra, N., & Mishra, P. (2020). Shikimic acid as intermediary model for the production of drugs effective against influenza virus. In *Phytochemicals as Lead compounds for new drug discovery* (pp. 245–256). Elsevier.

Verpoorte, R., van der Heijden, R. T. J. M., Ten Hoopen, H. J. G., & Memelink, J. (1999). Metabolic engineering of plant secondary metabolite pathways for the production of fine chemicals. *Biotechnology Letters, 21*(6), 467–479.

Verpoorte, R., & Alfermann, A. W. (Eds.). (2000). *Metabolic engineering of plant secondary metabolism*. Springer Science & Business Media.

Verpoorte, R., Contin, A., & Memelink, J. (2002). Biotechnology for the production of plant secondary metabolites. *Phytochemistry Reviews, 1*(1), 13–25.

Verpoorte, R., & Memelink, J. (2002). Engineering secondary metabolite production in plants. *Current Opinion in Biotechnology, 13*(2), 181–187.

Wassenegger, M., Heimes, S., Riedel, L., & Sänger, H. L. (1994). RNA-directed de novo methylation of genomic sequences in plants. *Cell, 76*(3), 567–576.

Waterhouse, P. M., Graham, M. W., & Wang, M. B. (1998). Virus resistance and gene silencing in plants can be induced by simultaneous expression of sense and antisense RNA. *Proceedings of the National Academy of Sciences, 95*(23), 13,959–13,964.

Wei, Z., Li, C., Gao, T., Zhang, Z., Liang, B., Lv, Z., et al. (2019). Melatonin increases the performance of *Malus hupehensis* after UV-B exposure. *Plant Physiology and Biochemistry, 139*, 630–641.

Wilson, S. A., & Roberts, S. C. (2014). Metabolic engineering approaches for production of biochemicals in food and medicinal plants. *Current Opinion in Biotechnology, 26*, 174,182.

Yang, L., Yang, C., Li, C., Zhao, Q., Liu, L., Fang, X., & Chen, X. Y. (2016). Recent advances in biosynthesis of bioactive compounds in traditional Chinese medicinal plants. *Science Bulletin, 61*(1), 3–17.

Ye, X., Al-Babili, S., Klöti, A., Zhang, J., Lucca, P., Beyer, P., & Potrykus, I. (2000). Engineering the provitamin A (β-carotene) biosynthetic pathway into (carotenoid-free) rice endosperm. *Science, 287*(5451), 303–305.

Zhang, B., Tian, W., Wang, S., Yan, X., Jia, X., Pierens, G. K., et al. (2017). Activation of natural products biosynthetic pathways via a protein modification level regulation. *ACS Chemical Biology, 12*(7), 1732–1736.

Zhu, X., Zeng, X., Sun, C., & Chen, S. (2014). Biosynthetic pathway of terpenoid indole alkaloids in *Catharanthus roseus*. *Frontiers of Medicine, 8*(3), 285–293.

Zhao, K., Zeng, J., Zhao, T., Zhang, H., Qiu, F., Yang, C., et al. (2017). Enhancing tropane alkaloid production based on the functional identification of tropine-forming reductase in *Scopolia lurida*, a tibetan medicinal plant. *Frontiers in Plant Science, 8*, 1745.

Chapter 15
Polymeric Composites: A Promising Tool for Enhancing Photosynthetic Efficiency of Crops

Irsad, Neetu Talreja, Divya Chauhan, R. V. Mangalaraja, Parvez Qamar Rizvi, and Mohammad Ashfaq

Abstract The demand for food continuously increased with increasing the population globally. Significant effort has been made to improve the productivity of the crops using genetic engineering, breeding, and improving agriculture practices. Usually, crops can absorb and use light to provide nutrition to crops. However, the negative impact of climate changes on agriculture remains a concern. In this context, emerging nanomaterials might be a strategic tool that increases the use of the light source. Carbon-based nanomaterials (CBNMs), mainly carbon nanotubes (CNTs), carbon nanofibers (CNFs), graphene, graphene oxide (GO), and fullerenes, are extensively used for improving the yield of crops (plant growth and development) by increasing photosynthetic efficiency. Moreover, incorporating metal and polymers within the CBNMs might improve photosynthetic efficiency, thereby developing and growing the crops. It is considered a unique tool in improving the nutrient uptake and translocation to achieve proper growth of plants in agroecosystem and potential application in crop improvement.

Keywords Nanotechnology · CNTs · Crop production · Plant health · Food security

Irsad · P. Q. Rizvi
Department of Plant Protection, Faculty of Agricultural Sciences, Aligarh Muslim University, Aligarh, India

N. Talreja · R. V. Mangalaraja
Advanced Ceramics and Nanotechnology Laboratory, Department of Materials Engineering, Faculty of Engineering, University of Concepción, Concepción, Chile

D. Chauhan
Department of Chemical and Biomedical Engineering, University of South Florida, Tampa, FL, USA

M. Ashfaq (✉)
Advanced Ceramics and Nanotechnology Laboratory, Department of Materials Engineering, Faculty of Engineering, University of Concepción, Concepción, Chile

School of Life Science, BS Abdur Rahman Crescent Institute of Science and Technology, Chennai, India

T. Aftab, K. R. Hakeem (eds.), *Metabolic Engineering in Plants*,
https://doi.org/10.1007/978-981-16-7262-0_15

Introduction

Agriculture is the prime foundation for developing nations and offers food to sustain the population. The demand for food increased endlessly with the increasing population. To fulfil the requirements, agrochemicals, phytohormones, and fertilizers were used to meet their needs. Agriculture faces numerous challenges, including drought stress, salinity and water deficit, environmental contamination, and soil toxicity due to the accumulation of pesticides and fertilizers (Mittal et al., 2020; Shrivastava & Kumar, 2015; Ashfaq & Khan, 2017; Sultana et al., 2021). In this aspect, nanomaterial-based fertilizers/agrochemicals might be beneficial to resolve such associated issues.

Nanotechnology is one of the most powerful and promising tools in sustainable crop production for human consumption and cattle feeding. Nanomaterials (NMs) can promote the delivery mechanism of nutrients/micronutrients to enhance the development process, including photosynthesis, thereby enhanced crop productivity that overcomes the possibilities of pesticide/fertilizer accumulation during application. Numerous metals (Cu, Zn, Fe, Mn, and Ce) and their metal-oxides (CuO, ZnO, FeO, MnO, and CeO) have been extensively used in various applications like water treatments, antibacterial agents, antifungal agents, photocatalytic, supercapacitors, sensor, and agricultures (Omar et al., 2019a; Omar et al., 2019b; Mustafa et al., 2011; Talreja et al., 2021a; Talreja et al., 2021b; Ashfaq et al., 2021; Chauhan et al., 2021; Talreja & Kumar, 2018). Interestingly, these metals and their metal-oxide-based NMs augment the crops by increasing the photosynthetic efficiency, chlorophyll, protein, water-uptake ability, and translocation ability of micronutrients. However, these metals and their metal-oxide-based NMs show phytotoxicity at higher doses due to the accumulation of metal ions on the plant root surface (Irsad et al., 2020). In this aspect, carbon-based nanomaterials (CBNMs) might resolve issues related to metals and their metal-oxide-based NMs.

CBNMs such as carbon nanotubes (CNTs), carbon nanofibers (CNFs), carbon dots (CDs), graphene, graphene oxide (GO), and fullerenes have been synthesized by chemical vapor deposition (CVD), etching process, liquid exfoliation process, and chemical oxidation process. These CBNMs are comprehensively used for various applications, mainly environmental remediation, drug delivery, sensors, nanomedicine, and agriculture (Afreen et al., 2020; Chauhan et al., 2020; Afreen et al., 2018; Ashfaq et al., 2019; Sasidharan et al., 2021; Kumar et al., 2011; Talreja et al., 2014; Saraswat et al., 2012; Khare et al., 2013; Hermes et al., 2020; Magnabosco et al., 2020). The CNFs are relatively newer NMs and insignificant toxicity compared with the CNTs (Ashfaq et al., 2013), thereby enormous use of CNFs decontamination of water (removal of heavy metal ions, dyes, and pharmaceuticals), antibiotics materials, antibiotics, drug delivery, sensors (chemical and gas), wound dressing materials, nanomedicine, and agriculture (Ashfaq et al., 2018; Bhadauriya et al., 2018; Ashfaq et al., 2017a; Ashfaq et al., 2017b; Ashfaq et al., 2016; Tripathi et al., 2016; Ashfaq et al., 2014; Singh et al., 2013; Talreja et al., 2020; Kumar & Talreja, 2019; Talreja et al., 2016). Remarkably, CNTs, CNFs, and graphene have translocation ability within the plants to penetrate seed coats. The translocation ability of the CNTs, CNFs, and graphene was due to the graphitic nature and negatively charged surface. Interestingly, CNFs hold metal-nanoparticles that release metal ions in a controlled manner. Due to repulsion force, the highly negatively charged CNFs easily translocate from root to shoot to leaf within the plants. The CNFs increase photosynthetic efficiency by increasing the protein content, protein content, and water uptake ability, thereby improving crop yield (Ashfaq et al., 2017a; Gupta et al., 2019). However, the direct application of CNFs remains a concern. In this aspect, polymeric composite or CBNM-based hybrid materials might be beneficial for delivering agricultural or fertilizers or micronutrients.

Numerous polymers such as chitosan, polyvinyl alcohol (PVA), starch, cellulose, etc. have been used for the nanomedicine drug delivery and micronutrients/fertilizers/agrochemical delivery system (Alimardani et al., 2021; Chen et al., 2019a; Ashfaq & Ahmad, 2021). These polymers are biocompatible and biodegradable. Those can release micronutrients/fertilizers/agrochemicals in a controlled manner. The encapsulation of nanomaterials might improve the biocompatibility of the nanofertilizers, translocation ability, and less accumulation, thereby resulting in high photosynthesis efficiency and high yield of crops (Kumar et al., 2018; Mohammad Ashfaq et al., 2019). The present book chapter focuses on the CBNM-based hybrid materials or polymeric composite and their effects on the photosynthetic efficiency of the crops. Also, we discuss the high chlorophyll content has improved the photosynthetic activity of plants, leading to higher production. This book chapter provides newer insight into the safer use of functional nanomaterial or CBNM-based hybrid materials or polymeric composite and their application to crops.

Role of Photosynthesis for Improving the Growth of the Plants

Plants are the major producer of food, thereby known as the primary producer, and we all are completely dependent upon them for food intake. Photosynthesis is the biological process that can use energy derived from light to produce chemical energy to synthesize carbon compounds. Chloroplasts are the principal resource of chemical energy in nourishment supplies to the plants. These plant organelles have excellent potential to convert light energy into three major sugars, (1) maltose, (2) triosephosphate, and (3) glucose that promote plant growth by collecting atmospheric carbon. The chloroplasts have components to self-repair photodamaged proteins and double-stranded circular-DNA associated with the photosynthesis process. Improving the photosynthetic efficiency might be required to enhance the adsorption of solar energy that can penetrate more deeply into living beings. However, photosystem interfaces with nanomaterials are broadly considered (Neales

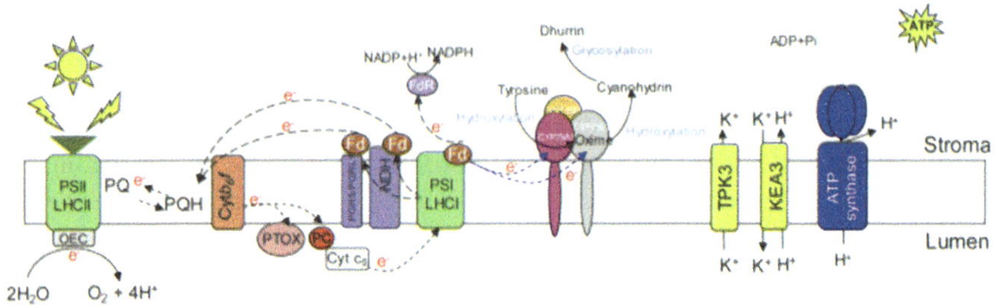

Fig. 15.1 Improvement strategies for photosynthesis. This image was taken with permission (Batista-Silva et al., 2020)

& Incoll, 1968; Baslam et al., 2020; Flügge et al., 2016; Batista-Silva et al., 2020). Figure 15.1 shows the improvement strategies for photosynthesis. The light-dependent process or reactions of photosynthesis consists of two photochemical systems, (1) photosystems I and (2) (PSI and PSII). The PSI and PSII work in conjunction with light-absorbing antenna systems, light-harvesting complex I, and light-harvesting complex II, connected by an electron-transport chain. The light-absorbing molecules (chlorophyll) are conscientious for collecting the light and transferring it into energy (Batista-Silva et al., 2020; Wilhelm & Selmar, 2011; Li et al., 2019; Swift et al., 2019).

Moreover, the activity of the photosynthetic enzymes, mesophyll cells, integrity of chloroplasts, and carbon dioxide aggregation also influence photosynthesis. Nanoengineered or NM-based chloroplast photosynthesis for improving solar energy, thereby improving the yield of crops. The NMs or CBNMs, or polymeric composite, have the potential ability to empower chloroplast-based photocatalytic complexes. The high photosynthesis efficiency might improve the crops' yield.

Effect of Metal-/Ceramic-Based Nanomaterials (MC-NMs) on the Photosynthesis

Numerous MC-NMs such as Cu, Zn, Ce, Fe, Si, Ti, and its oxide have been used to improve photosynthetic efficiency. These NMs show the potential ability to enhance photosynthetic efficacy by improving chlorophyll content, thereby improving the plant's growth. However, some NMs affect the plants' cellular processes, developmental processes, and photosynthetic efficiency.

Usually, NMs interfere with the plant's light-harvesting molecules (chlorophyll), ETC, PEP, carbonic anhydrase, and RuBisCo. With the help of these NMs, we can increase the yield of crops ultimately resolving the issue associated with the food demand. Photosynthesis in PS II was also higher using nano mesoporous silica due to higher oxygen transport and light accumulation. TiO_2 might improve the acquisition of light that promotes the functioning of RuBisCO enzyme, thereby

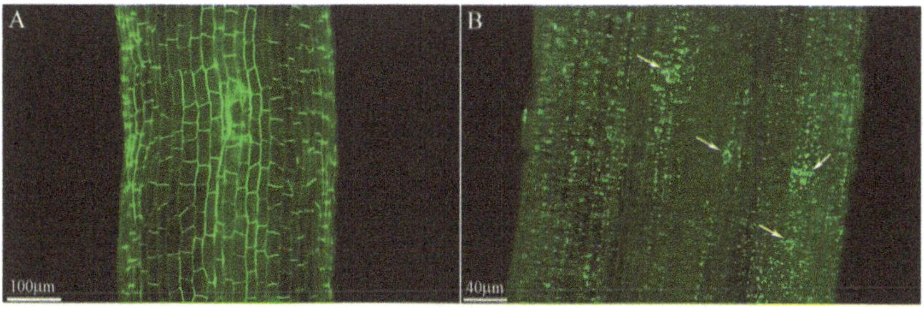

Fig. 15.2 Fluorescence microscopy images of the D. Moldavica roots (**a**) Control and (**b**) 200 mg/L TiO_2. The image was taken with permission (Gohari et al., 2020)

assimilating carbon (Poddar et al., 2020). The nano-TiO_2 can perform an advanced oxidation process in the presence of light that promotes the photosynthesis and growth of spinach plants. The electron transfer rate, oxygen evolution, and photophosphorylation of chloroplast (Chl) are increased under solar irradiation in nano-TiO_2-exposed spinach. The data suggested that the nano-TiO_2 affects the whole chain electron transport, photoreduction activity of PS-II, O_2-evolving, and photophosphorylation activity of spinach (Lei et al., 2007). Another study with similar materials shows that TiO_2 considerably changes the PS-II microenvironment, increases solar light utilization, improves energy transfer, and fastens chlorophyll's energy transport. The photochemical activity of PSII (fluorescence quantum yield) and its oxygen-evolving rate were enhanced by TiO_2. The data indicate that the TiO_2 NMs can promote energy transfer and oxygen evolution in PSII of spinach plants (Mingyu et al., 2007). Another study with similar materials suggested that the TiO_2 NPs can induce the light-harvesting complex II (LHCII) b gene expression and LHCII II content on the thylakoid membrane of *A. thaliana*. The distribution of light energy from PS-I to PS-II increased by increasing the LHCII, light energy to electronic energy, photolysis, and oxygen evolution (Gao et al., 2008). Gohari et al. synthesized TiO_2 NPs and tested them against *Moldavian balm* under salt stress conditions. The data suggested that the TiO_2 NPs increased the agronomic traits like antioxidant enzyme activity under salt stress conditions (Gohari et al., 2020). Figure 15.2 shows the fluorescence microscopy images of the roots of *Moldavian balm*. The higher concentration of TiO_2 NP-treated plants shows the aggregation; the arrow indicates the fluorescent light spot that confirms the plants actively took the TiO_2 NPs. The lower concentration of the TiO_2 NPs might increase the enzymatic activity that leads to photosynthetic efficiency, thereby improving yield of crops.

SiO_2-based NMs increase the phytotoxic effect of Al that decreases the growth and photosynthesis in maize grown in acidic soils. The photosynthetic efficiency decreased due to metal detoxification of Al-stressed maize (de Sousa et al., 2019). CeO at 70% moisture on-field capacity shows a positive photosynthesis outcome with a 20% increase in chlorophyll in soybean (Cao et al., 2018). Silicon oxide (SO) NPs can also increase the drought resistance in hawthorn seedlings by improving the photosynthetic rate and stomatal conductance and decreasing xylem water potential and malondialdehyde (MDA). The SO-NPs show the increase in chlorophyll content of basil grown under salinity stress. Another study suggested that the foliar spray of nano SO-NPs improved photosynthesis rate ~23.6%, PS II (8.8%) activity, stomatal conductance (20.4%), and electron transport (34.1%) at 300 mg/L concentration in *Indocalamus barbatus* (Cao et al., 2018).

In general, these MC-NMs show positive effects or increase the plant's photosynthetic efficiency, thereby increasing crop yield. However, a high dose of NMs might decrease the photosynthetic efficiency.

Effect of Carbon-Based Nanomaterials (CBNMs) on the Photosynthesis

Numerous CBNMs, mainly CNTs, CNFs, CDs, graphene, GO, and fullerenes, have been extensively used for the increasing yield of crops. The effects of these CBNMs vary with the plants, species, and types of the materials (size, shape, and composition). For example, Gao et al. synthesized polyhydroxylated fullerenes (PHF) and tested against the growth and life span of the different biological plant models. The PHF with 20 mg/L increased the lifespan and stimulated the reproduction of *Daphnia* by 38%. The PHF also enhanced fungus, *Aspergillus niger* at 10 mg/L, and *Arabidopsis thaliana* seedlings, resulting in longer hypocotyls on 100 and 200 mg/L concentrations (Gao et al., 2011). Kole et al. synthesized fullerol and tested it against the bitter guard. The data suggested that the fullerol increases the yield around 54% yield with 24% water content in bitter guard. Interestingly, two anticancer phytomedicines, cucurbitacin-B and lycopene, were enhanced up to 74% and 82%, respectively, and contents of two antidiabetic phytomedicines, charantin and insulin, were augmented up to 20% and 91%, respectively. The data suggested that the fullerol increased photosynthesis efficiency by increasing chlorophyll content, thereby high crop yield (Kole et al., 2013).

Tripathy et al. synthesized water-soluble CNTs (Ws-CNTs) and tested them against *Cicer arietinum*. The Ws-CNTs, less than 10 nm diameter, were observed to be an excellent promoter of plant growth in 10 nm diameter in *Cicer arietinum* (Tripathi et al., 2011). Another study shows the oxidized multi-walled CNTs (OMW-CNTs) with a 50–630 nm size range. The OMW-CNTs promote root growth and higher vegetative biomass in *Triticum aestivum* at 10–160 μg/mL (Wang et al., 2012). Khodakovskaya et al. synthesized CNTs and tested them against *Arabidopsis*. The data suggested that the overexpression of *Arabidopsis* aquaporin in tobacco increases plant growth and photosynthetic efficiency. The CNTs can stimulate growth and activate gene and protein expression of aquaporin in tobacco cells (Khodakovskaya et al., 2012). Khodakovskaya et al. synthesized CNTs and tested them against tomato plants. CNT-treated tomato plants produced almost double flowers and fruits at the concentrations (50 and 200 μg/mL). The data suggested

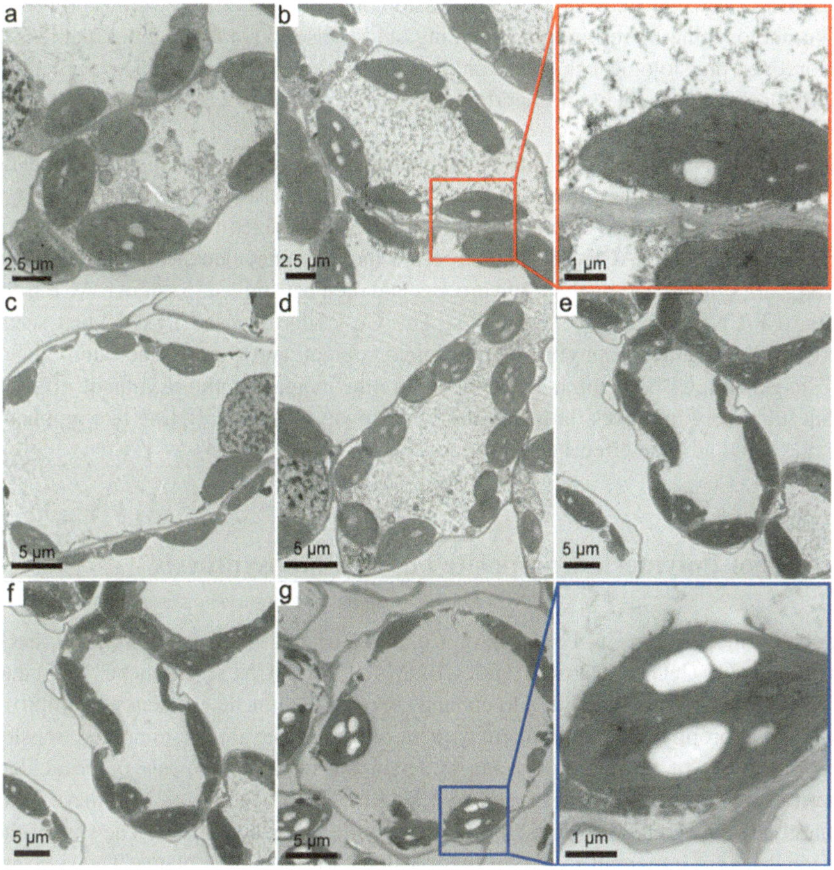

Fig. 15.3 Exposure of graphene with peas leaf. The image was taken with permission (Chen et al., 2019b)

the upregulation of involved genes in cell division and water transport. CNTs can easily penetrate the thick seed coat and enhance the uptake of water inside seeds which is further responsible for faster germination and higher biomass production (Khodakovskaya et al., 2013).

Like CNTs, graphene is also effectively used for improving photosynthetic efficiency. Chen et al. synthesized the GO and reduced-GO (RGO) and tested them against pea plants. The data suggested that the graphene efficiently translocates within the pea plants from root to leaf, inhibiting PS-II activity by damaging the oxygen evolve complex. Figure 15.3 shows the leaf of pea plants with the treatment of GO and RGO. The ultrastructure images indicate GO damage of the membrane, whereas RGO did not show any damage even at high concentrations (Chen et al., 2019b). Another study of a similar group shows the synthesis of graphene and is tested against alfalfa plants under stress conditions. The data suggested that graphene, salt, and alkali show the expression of genes involved in various processes like photosynthesis, respiration, signaling, and transcriptional regulation pathways. Moreover, physiological changes confirmed that the antioxidant defence mechanism and photosynthesis play an important role in reacting with stress conditions (Chen et al., 2021).

Other CBNMs like CNFs make suitable candidates for improving photosynthetic efficiency due to their translocation ability and release of micronutrients in a controlled manner. For example, Ashfaq et al. synthesized Cu NP-dispersed activated carbon fiber (ACFs)/CNFs and tested them against chickpea seed. Cu-ACF/CNFs are synthesized by the CVD process and then ball-milled to remove ACF from Cu-ACF/CNFs to produce Cu-CNFs. The Cu-CNFs exposed to chickpea seeds positively affect chlorophyll content, protein content, and germination rate.

In general, CBNMs might be beneficial for the increasing photosynthetic efficacy that ultimately improves the yield of crops. However, some CBNMs like graphene oxide show adverse effects at higher concentrations.

Effect of Polymeric Composite on the Photosynthesis

Usually, metals (Cu, Zn, Fe, Mn, Ag, Ce, and Au) and its oxide (CuO, ZnO, FeO, MnO, AgO, CeO, and AuO) and CBNMs (CNTs, CNFs, graphene, GO, and fullerens) have potential ability to enhance the photosynthetic efficiency by improving the chlorophyll content, protein content, water uptake ability, improved translocation ability of the micronutrients or fertilizers or agrochemicals or pesticides, germination rate, shoot length, and root length. The mode of action of these NMs mainly depends only on two factors, (1) size and (2) surface charge. The size of the NMs is a very important factor for the translocation within the plants. The size less than 50 nm translocates within the plants through plasmodesmata, whereas size more than 50 nm translocate through apoplastic and symplastic pathways. Another important factor is the surface charge of the NMs; the negatively charged NMs easily translocate within the plants through root to shoot to leaf, as plant cells itself negatively charged. The negative-negative charged (NM-plant cells) gives strong repulsion force, thereby relatively minor accumulation and translocation ability. The positively charged NMs with negatively charged plant cells show strong attraction force, thereby high accumulation and insignificant translocation ability leads to phytotoxicity within the plants (Pérez-de-Luque, 2017; Zhang et al., 2021; Wu & Li, 2021; Cañas et al., 2008; Su et al., 2009).

Based on the above discussion regarding translocation ability, we can say that NMs with around 500 nm size and the negatively charged surface might be beneficial for the translocation ability that improved photosynthetic ability and the growth of the plants. In this aspect, metals and metal-oxide and CBNMs need to modify with surface functional groups or polymers that improved the translocation ability and release behavior of the NMs. Although some of the CBNMs, like CNFs, have strong negatively charged that easily translocate within the plants through root

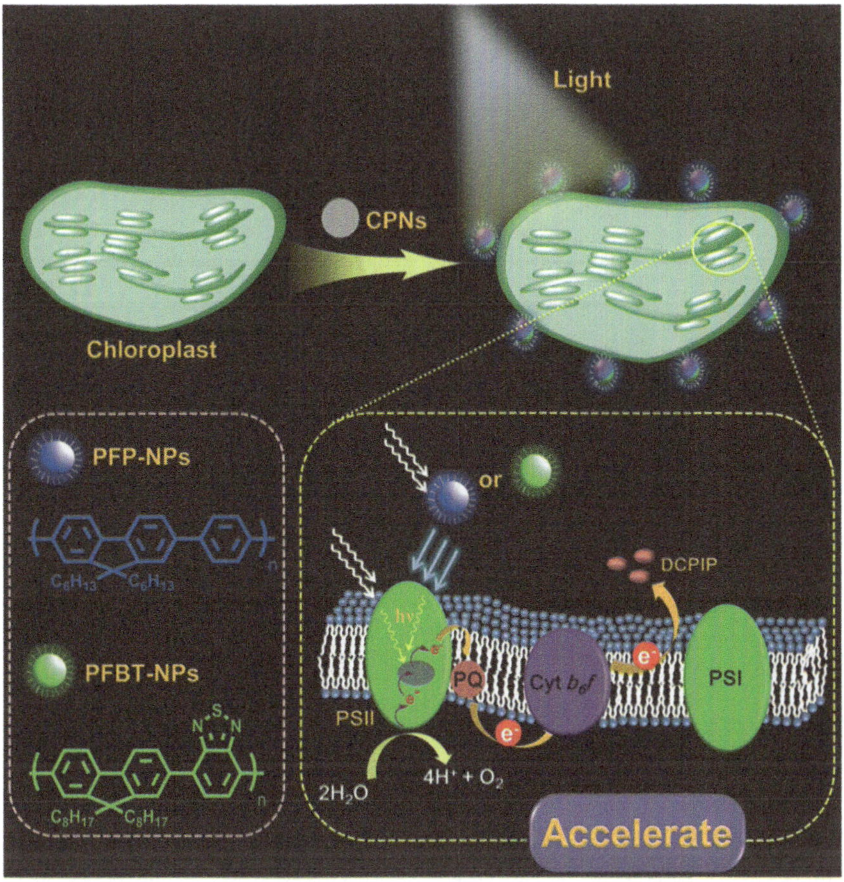

Fig. 15.4 Improvement strategies for light reactions by using CCP-NPs. The image was taken with permission (Wang et al., 2017)

to shoot to leaf. However, the direct application of the NMs remains a concern. In this aspect, the polymeric delivery system might be beneficial to overcome such associated issues. Several polymers have been used to encapsulate NMs, including metal and metal-oxides and CBNMs that control the release behavior of the NMs, reduce phytotoxicity, and improve biocompatibility and biodegradability of the materials. For example, Wang et al. synthesized chloroplast-coated conjugated polymeric NPs (CCP-NPS) to improve photosynthesis in the chloroplast. The data suggested that the CCP-NPs can capture the wide range of solar irradiation that enhances the electron transport rates in the PS-II system, thereby improving photosynthesis (Wang et al., 2017). Figure 15.4 shows the improvement strategies using CCP-NPs. The two types of conjugated polymeric NPs with fluorene unit like poly [2,7-(9,9-dihexylfluorene)-co-alt-p-phenylene] (PFP) andpoly [(9,9-dioctylfluorenyl-2,7-diyl)-co-(1,4-benzo-123-thiadiazole)] (PFBT) are used to

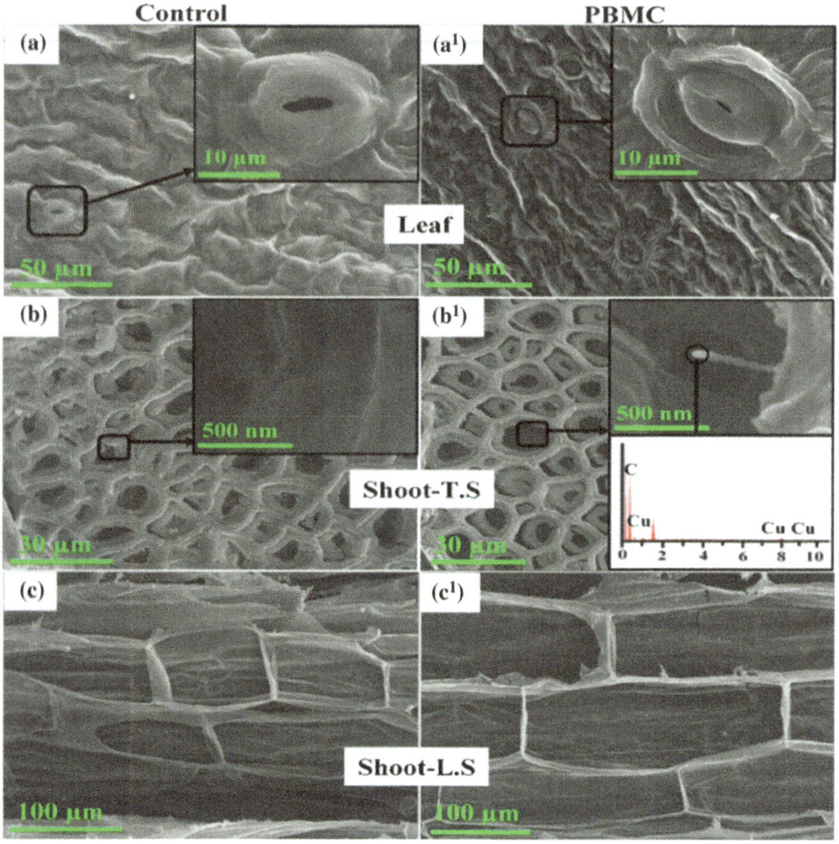

Fig. 15.5 SEM images of the chickpea plant before and after exposure to the PBMC. The image was taken with permission (Kumar et al., 2018)

synthesize PFP-NPs and PFBT-NPs. The protein complex in chloroplasts captures the light under solar irradiation to break down water into electrons, oxygen, and protons. The CCP-NPs adsorbed solar irradiation, especially ultraviolet and emission of visible light for the adsorption of the chloroplast. The PS-II system adsorbed more light after exposure of CCP-NPs that generate more electrons in the electron transport chain, thereby enhancing photosynthetic efficiency.

Kumar et al. synthesized PVA-starch incorporated Cu-Zn-CNFs (PBMC) based on composite materials and tested them against chickpea. The data suggested that the incorporation of polymeric composite within the Cu-Zn-CNFs might be advantageous for the delivery of micronutrients (Cu-Zn-CNFs) in a controlled manner that augments the chlorophyll content, protein content, germination rate, and growth of the plant by improving photosynthetic efficiency (Kumar et al., 2018). Figure 15.5 shows the SEM images of the chickpea plant before and after exposure to the PBMC sample. The leaf SEM images show the stomata; after exposure to PBMC,

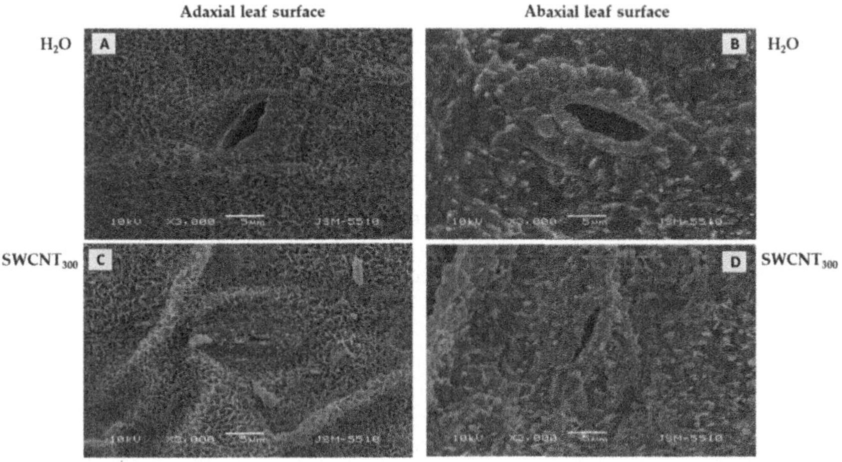

Fig. 15.6 Effect of TBC-SWCNTs onto the morphology of pea leaf. The image was taken with permission (Velikova et al., 2021)

significant changes in the stomata were observed. The exposed leaf samples show intact stomata or healthy stomata that indicate more photosynthetic efficiency, which conformed from the chlorophyll content and growth of the plants.

Leonardi et al. synthesized chitosan/alginate loaded with CuO-NPs (CAC-NPs) and tested against *Fortunella margarita* Swingle seeds. The data suggested that the synthesized CAC-NPs release CuO-NPs in a controlled manner and improve the plant's growth by increasing chlorophyll content, resulting in high photosynthetic efficiency (Leonardi et al., 2021). Velikova et al. synthesized tri-block copolymer (poly(ethylene oxide)-poly(propylene oxide)-poly(ethylene oxide)), incorporated SW-CNTs (TBC-SWCNTs), and tested against peas plants. The data suggested that the low concentrations (10 mg/L) did not show any adverse effects against pea plants, whereas at higher concentrations shows adverse effects in term of weakening photosynthesis system (Velikova et al., 2021). Figure 15.6 shows the SEM images of pea plant leaves before and after exposure to the TBC-SWCNTs. The SEM images of pea leaf without any exposure of the TBC-SWCNTs show the elastic edges of stomata. A significant change was observed, like more closure of stomata and changes in the structure after exposure of TBC-SWCNTs, indicating impaired photosynthesis at higher concentrations.

Table 15.1 summarized the comparative data of the different NMs or polymeric composites augmenting the photosynthesis process and subsequently improving the growth of the plants. The studies suggested that the effects of NMs or polymeric composite are plant or species-dependent. These NMs have the potential ability to increase photosynthesis efficiency. The incorporation of polymeric composite within the NMs leads to photosynthetic efficiency and decrease in toxicity.

Table 15.1 Effect of different NMs or polymeric composite for improving photosynthetic efficiency

S. No.	Materials	Plants/targeted molecules	Photosynthetic efficiency	References
1.	CNTs	*Arabidopsis* leaves	Improved electron transport system	Cañas et al. (2008)
2.	TiO_2	Spinach	Improved the photosynthesis	Su et al. (2009)
3.	Ws-CNTs	Chick pea	Improved chlorophyll content	Tripathi et al. (2011)
4.	CCP-NPs	Chloroplast	Improved photosynthesis	Wang et al. (2017)
5.	PBMCs	Chick pea	Improved chlorophyll content	Kumar et al. (2018)
6.	CAC-NPs	*Fortunella margarita* Swingle seeds	Increased chlorophyll	Leonardi et al. (2021)
7.	TBC-SWCNTs	Peas plant	Low dose (10 mg/L) is effective for increasing efficiency, whereas high dose decrease efficiency.	Velikova et al. (2021)

Conclusion

The CBNM-based hybrid materials or polymeric composite has extraordinary materials that can potentially improve the photosynthetic efficiency of the crops. Usually, these CBNM-based hybrid materials or polymeric composite enhances the photosynthetic efficiency by improving the chlorophyll content, protein content, water uptake ability, improved translocation ability of the micronutrients or fertilizers or agrochemicals or pesticides, germination rate, shoot length, and root length. The interaction of NMs with plants might be increased by increasing the surface functional groups or polymers that improved the translocation ability and release behavior of the NMs. With the help of NMs, we can easily increase photosynthetic efficiency. This book chapter provides newer insight into the safer use of functional nanomaterial or CBNM-based hybrid materials or polymeric composite and their application to crops.

Acknowledgments The authors acknowledge the financial support by ANID through FONDECYT project No. 3190515 and 3190581.

References

Afreen, S., Omar, R. A., Talreja, N., Chauhan, D., & Ashfaq, M. (2018). Carbon-based nanostructured materials for energy and environmental remediation applications. In R. Prasad & E. Aranda (Eds.), *Approaches in bioremediation: The new era of environmental microbiology and nanobiotechnology*. Springer International Publishing.

Afreen, S., Talreja, N., Chauhan, D., & Ashfaq, M. (2020). Chapter 15—Polymer/metal/carbon-based hybrid materials for the detection of heavy metal ions. In K. A. Abd-Elsalam (Ed.), *Multifunctional hybrid nanomaterials for sustainable agri-food and ecosystems*. Elsevier.

Alimardani, V., Abolmaali, S. S., Tamaddon, A. M., & Ashfaq, M. (2021). Recent advances on microneedle arrays-mediated technology in cancer diagnosis and therapy. *Drug Delivery and Translational Research, 11*(3), 788–816.

Ashfaq, M., & Ahmad, A. (2021). 3-Chitosan and its derivatives-based dimensional frameworks as carrier for gene delivery. In S. A. Bhawani, Z. Karim, & M. Jawaid (Eds.), *Polysaccharide-based nanocomposites for gene delivery and tissue engineering*. Woodhead Publishing.

Ashfaq, M., & Khan, S. (2017). Role of phytohormones in improving the yield of oilseed crops. *Oilseed Crops*. https://doi.org/10.1002/9781119048800.CH9

Ashfaq, M., Khan, S., & Verma, N. (2014). Synthesis of PVA-CAP-based biomaterial in situ dispersed with Cu nanoparticles and carbon micro-nanofibers for antibiotic drug delivery applications. *Biochemical Engineering Journal, 90*, 79–89.

Ashfaq, M., Singh, S., Sharma, A., & Verma, N. (2013). Cytotoxic evaluation of the hierarchical web of carbon micronanofibers. *Industrial & Engineering Chemistry Research, 52*(12), 4672–4682.

Ashfaq, M., Talreja, N., Chauhan, D., Rodríguez, C. A., Mera, A. C., & Mangalaraja, R. V. (2021). A novel bimetallic (Fe/Bi)-povidone-iodine micro-flowers composite for photocatalytic and antibacterial applications. *Journal of Photochemistry and Photobiology B: Biology, 219*, 112,204.

Ashfaq, M., Talreja, N., Chuahan, D., & Srituravanich, W. (2019). Carbon nanostructure-based materials: A novel tool for detection of Alzheimer's disease. In G. M. Ashraf & A. Alexiou (Eds.), *Biological, diagnostic and therapeutic advances in Alzheimer's disease: Non-pharmacological therapies for Alzheimer's disease*. Springer.

Ashfaq, M., Verma, N., & Khan, S. (2016). Copper/zinc bimetal nanoparticles-dispersed carbon nanofibers: A novel potential antibiotic material. *Materials Science and Engineering C, 59*, 938–947.

Ashfaq, M., Verma, N., & Khan, S. (2017a). Carbon nanofibers as a micronutrient carrier in plants: Efficient translocation and controlled release of Cu nanoparticles. *Environmental Science: Nano, 4*(1), 138–148.

Ashfaq, M., Verma, N., & Khan, S. (2017b). Highly effective Cu/Zn-carbon micro/nanofiber-polymer nanocomposite-based wound dressing biomaterial against the P. aeruginosa multi- and extensively drug-resistant strains. *Materials Science and Engineering C, 77*, 630–641.

Ashfaq, M., Verma, N., & Khan, S. (2018). Novel polymeric composite grafted with metal nanoparticle-dispersed CNFs as a chemiresistive non-destructive fruit sensor material. *Materials Chemistry and Physics, 217*, 216–227.

Baslam, M., Mitsui, T., Hodges, M., Priesack, E., Herritt, M. T., Aranjuelo, I., & Sanz-Sáez, Á. (2020). Photosynthesis in a changing global climate: Scaling up and scaling down in crops. *Frontiers in Plant Science, 11*, 882.

Batista-Silva, W., da Fonseca-Pereira, P., Martins, A. O., Zsögön, A., Nunes-Nesi, A., & Araújo, W. L. (2020). Engineering improved photosynthesis in the era of synthetic biology. *Plant Communications, 1*(2), 100,032.

Bhadauriya, P., Mamtani, H., Ashfaq, M., Raghav, A., Teotia, A. K., Kumar, A., & Verma, N. (2018). Synthesis of yeast-immobilized and copper nanoparticle-dispersed carbon nanofiber-based diabetic wound dressing material: Simultaneous control of glucose and bacterial infections. *ACS Applied Bio Materials, 1*(2), 246–258.

Cañas, J. E., Long, M., Nations, S., Vadan, R., Dai, L., Luo, M., Ambikapathi, R., Lee, E. H., & Olszyk, D. (2008). Effects of functionalized and nonfunctionalized single-walled carbon nanotubes on root elongation of select crop species. *Environmental Toxicology and Chemistry, 27*(9), 1922–1931.

Cao, Z., Rossi, L., Stowers, C., Zhang, W., Lombardini, L., & Ma, X. (2018). The impact of cerium oxide nanoparticles on the physiology of soybean (Glycine max (L.) Merr.) under different soil moisture conditions. *Environmental Science and Pollution Research, 25*(1), 930–939.

Chauhan, D., Afreen, S., Talreja, N., & Ashfaq, M. (2020). Chapter 8—Multifunctional copper polymer-based nanocomposite for environmental and agricultural applications. In K. A. Abd-Elsalam (Ed.), *Multifunctional hybrid nanomaterials for sustainable Agri-food and ecosystems*. Elsevier.

Chauhan, D., Talreja, N., & Ashfaq, M. (2021). Chapter 13—Nanoadsorbents for wastewater remediation. In K. A. Abd-Elsalam & M. Zahid (Eds.), *Aquananotechnology*. Elsevier.

Chen, B. Z., Yang, Y., Wang, B. B., Ashfaq, M., & Guo, X. D. (2019a). Self-implanted tiny needles as alternative to traditional parenteral administrations for controlled transdermal drug delivery. *International Journal of Pharmaceutics, 556*, 338–348.

Chen, L., Wang, C., Yang, S., Guan, X., Zhang, Q., Shi, M., Yang, S.-T., Chen, C., & Chang, X.-L. (2019b). Chemical reduction of graphene enhances in vivo translocation and photosynthetic inhibition in pea plants. *Environmental Science: Nano, 6*(4), 1077–1088.

Chen, Z., Niu, J., Guo, Z., Sui, X., Xu, N., Kareem, H. A., Hassan, M. U., Yan, M., Zhang, Q., Cui, J., Kang, J., Wang, Z., Mi, F., Karagić, Đ., & Wang, Q. (2021). Graphene enhances photosynthesis and the antioxidative defense system and alleviates salinity and alkalinity stresses in alfalfa (Medicago sativa L.) by regulating gene expression. *Environmental Science: Nano, 8*, 2731–2748.

de Sousa, A., Saleh, A. M., Habeeb, T. H., Hassan, Y. M., Zrieq, R., Wadaan, M. A. M., Hozzein, W. N., Selim, S., Matos, M., & AbdElgawad, H. (2019). Silicon dioxide nanoparticles ameliorate the phytotoxic hazards of aluminum in maize grown on acidic soil. *Science of the Total Environment, 693*, 133,636.

Flügge, U.-I., Westhoff, P., & Leister, D. (2016). Recent advances in understanding photosynthesis. *F1000Research, 5*, 2890.

Gao, F., Liu, C., Qu, C., Zheng, L., Yang, F., Su, M., & Hong, F. (2008). Was improvement of spinach growth by nano-TiO2 treatment related to the changes of Rubisco activase? *Biometals, 21*(2), 211–217.

Gao, J., Wang, Y., Folta, K. M., Krishna, V., Bai, W., Indeglia, P., Georgieva, A., Nakamura, H., Koopman, B., & Moudgil, B. (2011). Polyhydroxy fullerenes (Fullerols or Fullerenols): Beneficial effects on growth and lifespan in diverse biological models. *PLoS One, 6*(5), e19976.

Gohari, G., Mohammadi, A., Akbari, A., Panahirad, S., Dadpour, M. R., Fotopoulos, V., & Kimura, S. (2020). Titanium dioxide nanoparticles (TiO2 NPs) promote growth and ameliorate salinity stress effects on essential oil profile and biochemical attributes of *Dracocephalum moldavica*. *Scientific Reports, 10*(1), 912.

Gupta, G. S., Kumar, A., & Verma, N. (2019). Bacterial homoserine lactones as a nanocomposite fertilizer and defense regulator for chickpeas. *Environmental Science: Nano, 6*(4), 1246–1258.

Hermes, P.-H., Gabriela, M.-P., Ileana, V.-R., Fusaro, C., Fernando, L.-V., Mariana, M.-A., Padilla-Rodríguez, C., & Fabián, F.-L. (2020). Carbon nanotubes as plant growth regulators: Prospects. In J. K. Patra, L. F. Fraceto, G. Das, & E. V. R. Campos (Eds.), *Green nanoparticles: Synthesis and biomedical applications*. Springer International Publishing.

Irsad, Talreja, N., Chauhan, D., Rodríguez, C. A., Mera, A. C., & Ashfaq, M. (2020). Nanocarriers: An emerging tool for micronutrient delivery in plants. In T. Aftab & K. R. Hakeem (Eds.), *Plant micronutrients: Deficiency and toxicity management*. Springer International Publishing.

Khare, P., Talreja, N., Deva, D., Sharma, A., & Verma, N. (2013). Carbon nanofibers containing metal-doped porous carbon beads for environmental remediation applications. *Chemical Engineering Journal, 229*, 72–81.

Khodakovskaya, M. V., de Silva, K., Biris, A. S., Dervishi, E., & Villagarcia, H. (2012). Carbon nanotubes induce growth enhancement of tobacco cells. *ACS Nano, 6*(3), 2128–2135.

Khodakovskaya, M. V., Kim, B.-S., Kim, J. N., Alimohammadi, M., Dervishi, E., Mustafa, T., & Cernigla, C. E. (2013). Carbon nanotubes as plant growth regulators: Effects on tomato growth, reproductive system, and soil microbial community. *Small, 9*(1), 115–123.

Kole, C., Kole, P., Randunu, K. M., Choudhary, P., Podila, R., Ke, P. C., Rao, A. M., & Marcus, R. K. (2013). Nanobiotechnology can boost crop production and quality: First evidence from increased plant biomass, fruit yield and phytomedicine content in bitter melon (Momordica charantia). *BMC Biotechnology, 13*(1), 37.

Kumar, D., & Talreja, N. (2019). Nickel nanoparticles-doped rhodamine grafted carbon nanofibers as colorimetric probe: Naked eye detection and highly sensitive measurement of aqueous Cr3+ and Pb2+. *Korean Journal of Chemical Engineering, 36*(1), 126–135.

Kumar, R., Ashfaq, M., & Verma, N. (2018). Synthesis of novel PVA–starch formulation-supported Cu–Zn nanoparticle carrying carbon nanofibers as a nanofertilizer: Controlled release of micronutrients. *Journal of Materials Science, 53*(10), 7150–7164.

Kumar, V., Talreja, N., Deva, D., Sankararamakrishnan, N., Sharma, A., & Verma, N. (2011). Development of bi-metal doped micro- and nano multi-functional polymeric adsorbents for the removal of fluoride and arsenic(V) from wastewater. *Desalination, 282*, 27–38.

Lei, Z., Mingyu, S., Chao, L., Liang, C., Hao, H., Xiao, W., Xiaoqing, L., Fan, Y., Fengqing, G., & Fashui, H. (2007). Effects of nanoanatase TiO2 on photosynthesis of spinach chloroplasts under different light illumination. *Biological Trace Element Research, 119*(1), 68–76.

Leonardi, M., Caruso, G. M., Carroccio, S. C., Boninelli, S., Curcuruto, G., Zimbone, M., Allegra, M., Torrisi, B., Ferlito, F., & Miritello, M. (2021). Smart nanocomposites of chitosan/alginate nanoparticles loaded with copper oxide as alternative nanofertilizers. *Environmental Science: Nano, 8*(1), 174–187.

Li, Y., Kong, T., & Shen, S. (2019). Artificial photosynthesis with polymeric carbon nitride: When meeting metal nanoparticles, single atoms, and molecular complexes. *Small, 15*(32), 1900772.

Magnabosco, G., Pantano, M. F., Rapino, S., Di Giosia, M., Valle, F., Taxis, L., Sparla, F., Falini, G., Pugno, N. M., & Calvaresi, M. (2020). *A plant bioreactor for the synthesis of carbon nanotube bionic nanocomposites* (Vol. 8, p. 1287). Frontiers in Bioengineering and Biotechnology.

Mingyu, S., Xiao, W., Chao, L., Chunxiang, Q., Xiaoqing, L., Liang, C., Hao, H., & Fashui, H. (2007). Promotion of energy transfer and oxygen evolution in spinach photosystem II by nano-anatase TiO2. *Biological Trace Element Research, 119*(2), 183–192.

Mittal, D., Kaur, G., Singh, P., Yadav, K., & Ali, S. A. (2020). Nanoparticle-based sustainable agriculture and food science: Recent advances and future outlook. *Frontiers in Nanotechnology, 2*, 10.

Mohammad Ashfaq, N. T., Chuahan, D., & Srituravanich, W. (2019). *Polymeric nanocomposite-based agriculture delivery system: emerging technology for agriculture*. IntechOPen.

Mustafa, S., Khan, H. M., Shukla, I., Shujatullah, F., Shahid, M., Ashfaq, M., & Azam, A. (2011). Effect of ZnO nanoparticles on ESBL producing Escherichia coli & Klebsiella spp. *Eastern Journal of Medicine, 16*(4), 253–257.

Neales, T. F., & Incoll, L. D. (1968). The control of leaf photosynthesis rate by the level of assimilate concentration in the leaf: A review of the hypothesis. *The Botanical Review, 34*(2), 107.

Omar, R. A., Afreen, S., Talreja, N., Chauhan, D., & Ashfaq, M. (2019a). Impact of nanomaterials in plant systems. In R. Prasad (Ed.), *Plant nanobionics: Volume 1, advances in the understanding of nanomaterials research and applications*. Springer International Publishing.

Omar, R. A., Afreen, S., Talreja, N., Chauhan, D., Ashfaq, M., & Srituravanich, W. (2019b). Impact of nanomaterials on the microbial system. In R. Prasad (Ed.), *Microbial nanobionics: Volume 1, state-of-the-art*. Springer International Publishing.

Pérez-de-Luque, A. (2017). Interaction of nanomaterials with plants: What do we need for real applications in agriculture? *Frontiers in Environmental Science, 5*(12). https://doi.org/10.3389/fenvs.2017.00012

Poddar, K., Sarkar, D., & Sarkar, A. (2020). Nanoparticles on photosynthesis of plants: Effects and role. In J. K. Patra, L. F. Fraceto, G. Das, & E. V. R. Campos (Eds.), *Green nanoparticles: Synthesis and biomedical applications*. Springer International Publishing.

Saraswat, R., Talreja, N., Deva, D., Sankararamakrishnan, N., Sharma, A., & Verma, N. (2012). Development of novel in situ nickel-doped, phenolic resin-based micro–nano-activated carbon adsorbents for the removal of vitamin B-12. *Chemical Engineering Journal, 197*, 250–260.

Sasidharan, V., Sachan, D., Chauhan, D., Talreja, N., & Ashfaq, M. (2021). Three-dimensional (3D) polymer—Metal–carbon framework for efficient removal of chemical and biological contaminants. *Scientific Reports, 11*(1), 7708.

Shrivastava, P., & Kumar, R. (2015). Soil salinity: A serious environmental issue and plant growth promoting bacteria as one of the tools for its alleviation. *Saudi Journal of Biological Sciences, 22*(2), 123–131.

Singh, S., Ashfaq, M., Singh, R. K., Joshi, H. C., Srivastava, A., Sharma, A., & Verma, N. (2013). Preparation of surfactant-mediated silver and copper nanoparticles dispersed in hierarchical carbon micro-nanofibers for antibacterial applications. *New Biotechnology, 30*(6), 656–665.

Su, M., Liu, H., Liu, C., Qu, C., Zheng, L., & Hong, F. (2009). Promotion of nano-anatase TiO2 on the spectral responses and photochemical activities of D1/D2/Cyt b559 complex of spinach. *Spectrochimica Acta Part A: Molecular and Biomolecular Spectroscopy, 72*(5), 1112–1116.

Sultana, A., Talreja, N., Chauhan, D., & Ashfaq, M. (2021). Chapter 4—Nanotechnology-based biofortification: A plant–soil interaction modulator/enhancer. In T. Aftab & K. R. Hakeem (Eds.), *Frontiers in plant-soil interaction*. Academic Press.

Swift, T. A., Oliver, T. A. A., Galan, M. C., & Whitney, H. M. (2019). Functional nanomaterials to augment photosynthesis: Evidence and considerations for their responsible use in agricultural applications. *Interface Focus, 9*(1), 20180048.

Talreja, N., Afreen, S., Ashfaq, M., Chauhan, D., Mera, A. C., Rodríguez, C. A., & Mangalaraja, R. V. (2021a). Bimetal (Fe/Zn) doped BiOI photocatalyst: An effective photodegradation of tetracycline and bacteria. *Chemosphere, 280*, 130,803.

Talreja, N., Ashfaq, M., Chauhan, D., Mera, A. C., & Rodríguez, C. A. (2021b). Strategic doping approach of the Fe–BiOI microstructure: An improved photodegradation efficiency of tetracycline. *ACS Omega, 6*(2), 1575–1583.

Talreja, N., Jung, S., Yen, L. T. H., & Kim, T. (2020). Phenol-formaldehyde-resin-based activated carbons with controlled pore size distribution for high-performance supercapacitors. *Chemical Engineering Journal, 379*, 122,332.

Talreja, N., & Kumar, D. (2018). Engineered nanoparticles' toxicity: Environmental aspects. *Nanotechnology in Environmental Science*.

Talreja, N., Kumar, D., & Verma, N. (2014). Removal of hexavalent chromium from water using Fe-grown carbon nanofibers containing porous carbon microbeads. *Journal of Water Process Engineering, 3*, 34–45.

Talreja, N., Verma, N., & Kumar, D. (2016). Carbon bead-supported ethylene diamine-functionalized carbon nanofibers: An efficient adsorbent for salicylic acid. *Clean—Soil, Air, Water, 44*(11), 1461–1470.

Tripathi, K. M., Tyagi, A., Ashfaq, M., & Gupta, R. K. (2016). Temperature dependent, shape variant synthesis of photoluminescent and biocompatible carbon nanostructures from almond husk for applications in dye removal. *RSC Advances, 6*(35), 29,545–29,553.

Tripathi, S., Sonkar, S. K., & Sarkar, S. (2011). Growth stimulation of gram (Cicer arietinum) plant by water soluble carbon nanotubes. *Nanoscale, 3*(3), 1176–1181.

Velikova, V., Petrova, N., Kovács, L., Petrova, A., Koleva, D., & Tsonev, T. (2021). Single-walled carbon nanotubes modify leaf micromorphology, chloroplast ultrastructure and photosynthetic activity of pea plants. *International Journal of Molecular Sciences, 22*(9), 4878.

Wang, X., Han, H., Liu, X., Gu, X., Chen, K., & Lu, D. (2012). Multi-walled carbon nanotubes can enhance root elongation of wheat (Triticum aestivum) plants. *Journal of Nanoparticle Research, 14*(6), 841.

Wang, Y., Li, S., Liu, L., Lv, F., & Wang, S. (2017). Conjugated polymer nanoparticles to augment photosynthesis of chloroplasts. *Angewandte Chemie International Edition, 56*(19), 5308–5311.

Wilhelm, C., & Selmar, D. (2011). Energy dissipation is an essential mechanism to sustain the viability of plants: The physiological limits of improved photosynthesis. *Journal of Plant Physiology, 168*(2), 79–87.

Wu, H., & Li, Z. (2021). Recent advances in nano-enabled agriculture for improving plant performance. *The Crop Journal.* https://doi.org/10.1016/j.cj.2021.06.002

Zhang, P., Guo, Z., Ullah, S., Melagraki, G., Afantitis, A., & Lynch, I. (2021). Nanotechnology and artificial intelligence to enable sustainable and precision agriculture. *Nature Plants, 7*(7), 864–876.

Chapter 16
Metabolic Engineering Approaches to Produce Compounds of Interest in Plants

Lubna Azmi, Reena Bharti, and Siva Reddy

Abstract Plants are an important source of nutrition as well as medicines. Recent developments in metabolic engineering have allowed scientists not just to raise the overall quantity of targeted chemicals, and also to incorporate alternative biological functions into a range of species, leading to increased nutritious or commercial value. The development of understanding models has been developed to enable individually targeted suppression tactics or layering of several genes inside a chromosomal region to increase metabolic engineering capabilities. To boost performance, a variety of analytical techniques are used, including reasonable metabolic engineering, evolutionary engineering, inverse biotechnological applications, secretary route manufacturing, and performance improvement. The aim to reduce premiums, enhance yields, reduce prices, improve customer acceptance, as well as employ sustainable feedstocks to create a more environmentally sustainable procedure than conventional techniques drives mitochondrial application possibilities.

Keywords Metabolic engineering · Nutrient deficiency · Multigene transformation · Modern biotechnology

Introduction

Nutrient deficiency and increased cases of different diseases in humans demand the production of rich sources of nutrients and new drug molecules to combat the diseases. Natural compounds from plants, fungi, animals, and microorganisms are rich sources for the production of bioactive compounds like nutrient supplements,

L. Azmi (✉)
Hygia Institute of Pharmaceutical Education and Research, Lucknow, Uttar Pradesh, India

R. Bharti
Jawaharlal Nehru University, New Delhi, India

S. Reddy
Academy of Scientific and Innovative Research, Ghaziabad, Uttar Pradesh, India

T. Aftab, K. R. Hakeem (eds.), *Metabolic Engineering in Plants*,
https://doi.org/10.1007/978-981-16-7262-0_16

drugs, or precursors of drug molecules for human welfare, but the extraction and synthesis of bioactive compounds from these are difficult because of long breeding cycles and limited nutrient content variability. Modern biotechnology methods like the metabolic engineering approach to produce pharmaceutical compounds are vital (Datta et al., 2002). Developments in metabolic engineering had allowed scientists not just to boost the synthesis of natural antioxidants from natural origin and to also introduce novel biological functions to improve the nutritional and economical value of foods. Metabolic engineering entails tinkering with one or even more critical genes or frequency proteins in the synthesis or metabolic engineering. The bioactive compounds from plant sources like flavonoids, alkaloids, betalains, and glucosinolates are important for the production of drug molecules (Vega et al., 2008).

Flavonoids

Flavonoids belong to the plant's secondary metabolites. They protect the plants from UV radiation, and herbivores also help in assisting the symbiotic relationship with *rhizobia* species in fixing nitrogen. Many flavonoids have a health benefit to humans as well such as inhibiting angiogenesis and upregulate apoptosis in cancerous cells. The extensive benefit of flavonoids and their application is one of the reasons for its need to produce these compounds through recombinant means (Zhu et al., 2008). Thereby, the pathways for the synthesis of flavonoids were elucidated and reviewed. Most terrestrial plants produce flavonoids, but the final product and the amounts differ significantly. Flavonoid derivatives which are not found in nature have been created and are studied for their potential as pharmaceuticals (Møldrup et al., 2011). The biosynthesis of flavonoids involves several steps and requires enzymes. Flavonoids are biologically active when glycosylated. The synthesis steps can be followed in *E. coli* and *S. cerevisiae*. A mixture of *E. coli* variants is used for manufacturing, and each is responsible for a different biosynthetic phase. The method started with plasmids that overexpress phenylpropanoid acids, especially p-coumaric acids like caffeic, in *E. coli* cultures (Naqvi et al., 2009).

The other method involves the use of *C. glutamicum* to express naringenin and resveratrol to improve engineering strategies that increase the malonyl CoA availability. Because of increased malonyl CoA production, comparable efforts were made using *E. coli* and CRISPER editing to enhance naringenin manufacturing.

Other method includes using the yeast *Y. lipolytica* for engineer increased synthesis with hydroxylated flavonoids. The microorganism creates a lipid-soluble atmosphere that promotes the aggressive metabolite of flavonoid-producing enzymes including chalcone synthesis and cytochrome P40 reductase. The malonyl CoA flow in *Y. lipolytica* is generally rich (Kohli et al., 2006). The engineered strain can produce naringenin at a concentration at a titer of 250 mg/L and 140 mg/L, respectively (Chen et al., 1998).

Alkaloids

Alkaloids are used by plants in defense of UV radiation and herbivores, and they also impact plant-insect interactions. Plants also use alkaloids as nitrogen sinks for supplying important chemicals needed for wound healing (Chong et al., 1997). Several alkaloids have been found to benefit human and hence used by a human from centuries. Alkaloids like taxol and opiates are used as chemotherapeutics to treat pain. These are having fundamental high medicinal values, but extraction of them in large amounts may affect the environment and also require large space to grow the plants. Therefore, it is necessary to search for alternatives for the extraction of alkaloids important for medicinal perspectives. Metabolic engineering towards the synthesis of important alkaloids has found interest in research interest (Farhi et al., 2011).

Chemical synthesis of alkaloid is insufficient to fulfill the high requirement for existence medicines. Semi-synthesis of alkaloids utilizing microbial biosynthesis produces a large amount of precursor alkaloids for medication production. This technique has driven down the cost of manufacturing the drug. Despite applying semisynthetic and engineering techniques, the production of some alkaloids needs to be produced only by plants. Significant research is focused on the enhanced and optimized expression of alkaloids. The techniques like CRISPR/Cas9 gene-editing tools have been employed for the upregulation of the expression of metabolites to regulate alkaloids biosynthesis. The method was designed to regulate the production of Baine, noscapine, as well as morphine in *P. Somniferum* by knocking down gene 4′OMT2. The techniques also produced the transgenic poppy plant to regulate the biosynthesis of alkaloids. Anti-cancer chemicals including vinblastine and vincristine were already found in plants like *C. roseus* (Rischer et al., 2013).

Betalains

Plants of the *Caryophyllales* family, including such cacti and root beets, generate nitrogen-containing chemicals called betalains. The two main subclasses of betalains are betacyanins and betaxathins. The significance of these plants in improving human health has taken attention in the production of the compounds from these plants. They have a positive effect on cancer, hypertension, and many more. They are likewise utilized in the food industry as food colorants as well as colors. Application of metabolic engineering has been applied in biosynthesis, metabolism, and amino acid participation. Metabolic engineering also helps in identifying the key genes and transcription factors. It involves the engineering of betanins in rice endosperm as a health benefit and providing raw material for commercial supplement production (Zhu et al., 2008).

Glucosinolates

Brassicas and other similar plants contain glucosinolates, which are naturally occurring secondary metabolites produced from both aromatic and aliphatic acids. There are around 150 naturally occurring glucosinolates. Glucosinolates have an impact on human health besides having a positive impact on animal health and nutrition. Health benefit of glucosinolates on human includes prevention of cancer and helps in inhibition of proliferation, heart diseases, and activation of the immune system. Because of all the available benefits of glucosinolates, there is a need for the large-scale production of glucosinolates. Metabolic engineering using glucosinolates includes introducing genes within plants or systems which does not usually produce glucosinolates, as well as increasing the nutritious value of agricultural plant. The use of *E. coli* has shown the significance of advancement in technological approach in the production of glucosinolates through metabolic engineering. *E. coli* has been engineered to produce benzyl glucosinolates. The process involves series of stages from evaluating host strains, optimizing culture conditions, and also monitoring protein expression to increase the production fivefold. The efforts of Liou et al. have identified the operon used by human gut symbiont for the transformation of glucosinolates into the useful products of health-promoting isothiocyanates. It involves the engineering of operon into non-metabolizing strain and thus enhancing the gain of function (Perkins et al., 1999).

Because most staple crops lack appropriate nutritional benefits is the reason why metabolic engineering of vitamins is beneficial. Vitamin A, for example, is necessary for the prevention of blindness and the activation of the immune system. Vitamin is also important because of its antioxidant properties. However, none of these compounds are free in large amounts in these plants, and this is the reason for deaths for those who have a deficiency of nutrition and mostly rely on a cereal-based diet. Multigene metabolic engineering in the plant has been introduced to increase the synthesis of carotenoid (vitamin A), tocochromanol (vitamin E), and folate synthesis.

Gene Transfer and Expression Modification Method

Agronomic traits are quantitative and are controlled polygenetically in the vast majority of agronomic traits. The genetic engineering from single-gene trait transfer (e.g., resistance to insects, diseases, and herbicides) to multigene trait transfer, coding for bacterial operon, metabolic pathway, or biopharmaceuticals requiring the assembly of complex subunit proteins. In early plant biotechnology, gene transfer experiments involve two strategies:

1. Selectable marker under the control of a constitutive promoter to facilitate the selective propagation of transformed cells

2. Any promoter-controlled gene that is responsible for altering the plant's phenotypic in a certain way

Multigene transformation (MGT) is the advanced approach to generate plants with multiphenotype and function with the advancement of metabolic engineering. The application of MGT involves the following:

1. Enhancing the activity of the enzyme at multiple rate-limiting steps in target pathways
2. Enhancing flux across the target pathways, increasing the availability of upstream precursors 3) Preventing the loss of flux
3. Promoting the development of sink compartment to store target compounds

Genetic recombination into plants can take two forms: linked genes (many genes on the same plasmid) and unlinked genes (single genes on distinct plasmids). *Agrobacterium*-mediated transformation is the method to transfer multiple linked genes using a standard binary vector containing multiple genes within a single TDNA or multiple TDNA in a single gene. It is designed to handle up to 15 distinct transgenic organisms in a single lifetime (Naqvi et al., 2009).

Multigene Transformation Methods

Stacking and Retransformation

Crosses between transgenic lines or retransformation of transformed plants with the addition of transgenes can be used to stack several transgenes in plats. Both procedures are time-consuming, and the stacking process necessitates several breeding generations. However, the method helps increase the number of transgenes with the help of selectable markers.

High-Capacity TDNA Vectors

Due to the limitation of carrying small gene capacity in *Agrobacterium tumefaciens*, to integrate big DNA pieces, high-capacity artificial chromosome vectors must be introduced. Bacterial artificial chromosome (BAC) vectors may transport huge DNA pieces together, and when combined with binary vector components, chimeric binary vectors like BIBAC and TAC arise. Initially, they lack the unique restriction sites, but the Gateway site-specific recombination technology is used to combine the vector and can integrate several genes into the genome of plants (Vega et al., 2008).

TDNA and Bombardment Vectors

To enable subcloning, both the *Agrobacterium*-mediated transformation and the direct DNA transfer have been tuned to replicate in *E. coli.* To clone large genes, it becomes difficult for *Agrobacterium* to clone due to the small size. Vector becomes unstable and thereby prone to eject DNA. With conventional vectors, direct DNA transfer may be done effectively. The employment of two bacterial strains, each harboring TDNAs having two or more transgenes, can address the difficulty of cloning big DNA. Up to 15 unlinked transgenes can be transferred directly from DNA (Zhu et al., 2008).

Split Reading Frames

By maintaining multiple polypeptides in a single open reading frame (ORF) regulated by a single promoter, linker peptides can aid MGT. The foot-and-mouth disease virus 2A polyprotein system, for example, permits up to four polypeptides to be co-expressed in tobacco (*Nicotiana tabacum*) plants. The 2A linker is fewer than 20 amino acids long and cleaves at the C-terminus, allowing downstream polypeptides to be released after synthesis. To produce new ketocarotenoid by metabolic engineering, the *Paracoccus* crtW and crtZ genes were expressed as a polyprotein with an intervening 2A linker in transgenic tobacco and tomato plants.

Controlling the Expression of Multiple Transgenes

Promoters have been proven in the literature to drive numerous transgenes with no deleterious consequences. The barley (*Hordeum Vulgare*) D-hordein promoter, for example, is used to induce high endosperm specific expression of three transgenes in maize. Transgene silencing can also be caused by the same promoter. The presence of secondary structures encourages de novo methylations of mRNA, which saturate the cell's polyadenylation machinery and result in hairpin RNA formation. The effects like integration sites and juxtaposition of transgene copies encourage the formation of double-stranded RNA (Kohli et al., 2006).

Combinatorial Transformation

The combinatorial transformation has been developed. The method entails the construction of metabolic libraries, which are made up of plants that have been altered with a randomly selected of certain transgenes showed (Fig. 16.1).

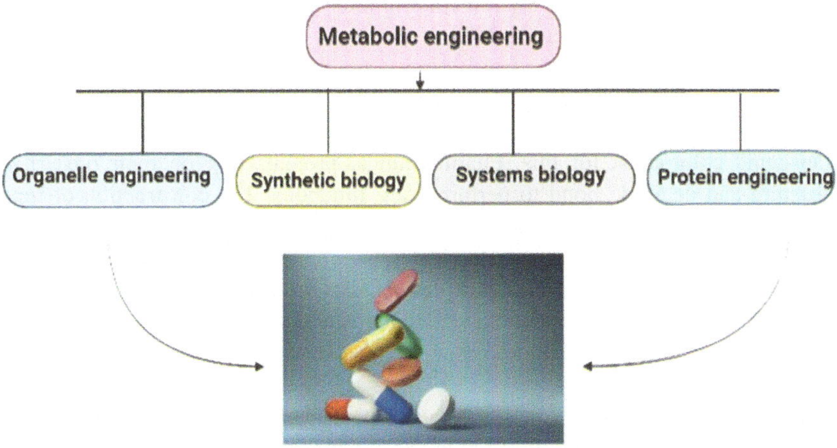

Fig. 16.1 Metabolic engineering, organelle engineering (mitochondrial engineering), synthetic biology (biosensors and genetic circuits), systems biology, structural biology, and protein engineering are all topics of research in the lab

A detailed look at the production of glucosinolates is presented. It covers aliphatic and aromatic amino acid synthesis, as well as critical research to find and understand genes. The transformation of genes into non-glucosinolate-producing plants or into the model systems helps in the identification of key genes and potential regulatory factors. It helps in identifying the function of the cytochrome $P4_{50}$ enzyme as well as the hydrolytic enzymes. The identification of potential key genes in *Arabidopsis* helps in the identification of key genes in other species through comparative genetics (Naqvi et al., 2009).

The development of rice varieties having enriched provitamin A is one of the best examples of gene transfer and gene modification systems in the early plant biotechnology era. The transgenic approach in the chloroplast is another example of the facilitated expression of bacterial operon and biopharmaceuticals. Chen et al. (1998) reported that particle bombardment with 13 distinct plasmids containing various marker genes successfully transformed rice (*Oryza sativa*) plants, resulting in plant regeneration harboring all input genes at one locus (Chen et al., 1998).

Plant biotechnology promoters are typically unidirectional; however, bidirectional promoters are becoming more common in sophisticated metabolic engineering techniques since they allow for simultaneous production of two gene products. To improve the nutritional value of transgenic potatoes, auxin-inducible, bidirectional mannose synthase (mas) promoters have been used to generate *luciferase* from human-casein gene and a bacterial marker gene producing *luciferase.* In transgenic chloroplasts, human serum albumin (HSA) expression under the control of the optimum chloroplast ribosome binding site (GGAGG) could not be clearly identified (0.02 percent tsp). However, the identical regulatory region has resulted in significant amounts of numerous additional foreign proteins accumulating up to 21 percent tsp. The study documents the highest concentration of medicinal proteins ever found in transgenic plants (Vega et al., 2008).

There is a need to develop vaccines on a wide scale for illnesses like cholera, which can reach epidemic proportions and pose a significant threat as a bioterrorism weapon. The cholera toxin component (CTB) was expressed as a cistron in the transgenic chloroplast for this. Pharmaceutical proteins rely on their quaternary structure and disulfide bond to perform their functions. The construction of functional proteins in transgenic chloroplasts has been demonstrated using CTB. Native subunit gene (ctxB) expression was 410 times greater than in nuclear transgenic plants. Engineering CTB in transgenic chloroplasts has resulted in generating edible vaccines in transgenic crops in a cost-effective way, thanks to the advent of chloroplast transformation in edible crops and the availability of plant-derived selectable markers (Datta et al., 2002).

Artemisinin is a drug molecule against Malaria caused by *Plasmodium falciparum* and has been produced in transgenic tobacco by multigene engineering. A vector containing the gene for *Artemisia* like cytochrome P450 reductase (CPR), Amorpha-4,11-diene synthase, artemisinin aldehyde reductase, and the yeast 3-hydroxy-3 methyl glutaryl coenzyme. A reductase, each under the control of the different promoter. In another vector, the ADS sequence was fused with COX 4 signal peptide for import into the mitochondria, to increase the synthesis of terpenoid. The vector was introduced through *Agrobacterium*-mediated transformation in the tobacco plant and resulted in transgenic produce amorpha 4,11-diene at 26–72 ng/g (normal ADS) and 137–827 ng/g fresh weight (mitochondrial ADS) (Kohli et al., 2006).

Opium poppy (*Papaver somniferum*) is one of the most medicinal plants having importance in cancer therapeutics drug noscapine, analgesic, and narcotic drugs such as morphine and codeine. It is also a muscle relaxant drug, papaverine. Morphine-type alkaloids are produced in plant cell culture since the 1970s. MGT has the application in silencing some of the genes to gain insight into the steps which are majorly important for the synthesis of morphine.

Production of Pharmaceutically Relevant Molecules and Precursors

Vitamins

Bacterial metabolic engineering has aided the synthesis of vitamins and vitamin-like substances. Misawa and colleagues in Japan, for example, have cloned DNA from the external bacteria *Erwinia uredova* and *Agrobacterium aurantiacum* into food-grade yeast *Candida utilis* to synthesize lycopene, b-carotene, and astaxanthin (Miura et al., 1998a, 1998b). They showed that just by upregulating the HMG CoA reductase genes and interrupting the squalene synthase gene, lycopene synthesis can be boosted sevenfold (to 7.8 mg/g dry cell weight) [(2)]. They also devised a method that requires altering the early non-mevalonate path in *E. coli*, which can produce either b-carotene or zeaxanthin when combined with allogeneic genes (Albrecht et al., 1999). Using an alternative method, the Liao group developed an *E. coli* variant that produced 50 times more astaxanthin (1.25 mg/g cell dry weight) than that formerly observed in the literature (Wang et al., 1999). A team from Omni Gene Bioproducts (Cambridge, MA, USA) and BioTechnica International (Cambridge, MA, USA) showed greater synthesis rates of riboflavin, another vital vitamin. Upregulation of homologous synthesizing genes combined with multiple modifications targeted at increasing the carbon flux towards the molecule of focus resulted in the manufacture of riboflavin at 14 g/L in a *Bacillus subtilis* variant (Perkins et al., 1999). These demonstrations show that the commercial synthesis of numerous vitamins by bacterial fermentation might be revived and increased with the help of metabolic engineering.

Metabolic Synthesis of Vitamin A

Microbes, algae, fungus, and flora all produce carotenoids, which are secondary metabolites. They are prolifically incorporated in poultry, food, beauty products, nutritional supplements, and, because of their antioxidant qualities and vitamin A actions, medicines (Saini and Keum, 2017). To accomplish product buildup, transgenic plants generate provitamin A by increasing metabolic flux upstream of the desired product. Increased levels of one or more biogenesis enzymes are an effective approach to increase metabolic flux around the target molecule. For example, Welsch et al. found that *Cassava* plants overexpressed a phytoene synthase transgenic gene, resulting in yellow-fleshed roots with high carotenoid content. According to the findings, breeding and genetic modification can increase the content of provitamin A in cassava roots by increasing the expression of the phytoene synthase (PSY) gene (Welsch et al., 2010). Paul et al. increased the Fe′s banana-derived phytoene synthase 2a (MtPsy2a) genotype in bananas, resulting in a 55 g/g provitamin A production (Paul et al., 2017).

Commercially pertinent carotenoids are produced in large quantities by microalgae species (Christaki et al., 2013). β-Carotene is a valuable pigment found in a variety of microalgal species; yet, the low productivity of β-carotene in microalgae is a major stumbling block.

Microbes, in conjunction with plants and algae, can create huge quantities of β-carotene in the natural world. Filamentous fungi such as *Blakeslea trispora* and yeast-like *Phaffia* rhodozyma and *Torulopsis candida* are used to produce β-carotene. In the presence of β-carotene synthesis genes and the complete MVA pathway, the transgenic *E. coli* DH5 increased β-carotene production to 465 mg/L at a 2% (w/v) glycerol concentration. It utilized ergosterol biosynthesis inhibitors to increase the flow of farnesyl diphosphate (FPP) into the carotenoid cascade by overexpressing 3-hydroxy-3-methylglutaryl coenzyme A (HMG-CoA) reductase recombinant *S. cerevisiae* may generate up to 6.29 mg/g of β-carotene (Yan et al., 2012; Yoon et al., 2009)

Metabolic Production of Vitamin D

Under ultraviolet light, ergosterol is transformed to the vitamin D2 precursor, which is produced via thermal isomerization and by the shedding of two hydrogen atoms. Vitamin D2 is frequently utilized in medicine, food, and animal feed and has a promising future. Active vitamin D3 is a lipid-soluble vitamin that regulates calcium and phosphorus metabolism (Liu et al., 2007) as well as manages the proliferation and development of skeletal muscles in the human body (Li et al., 2019). As of yet, there are two main methods for producing active vitamin D3: the first is by thermalization method of research, and the second is the bacterial reformation process using P450 enzymes. Site-directed mutagenesis of CYP105A1 based on its crystal structure and controlled evolution of CYP107 resulted in a significant increase in vitamin D3 hydroxylase activity (Eniyan et al., 2018). Moreover, the actinomycete *Pseudonocardia autotrophica* may transform vitamin D3 to its bioactive forms: 1,25-dihydroxyvitamin D3 and 1,25-dihydroxyvitamin D3 are two types of vitamin D3. Fujii et al. isolated and transformed vitamin D3 hydroxylase (Vdh) from *Pseudonocardia autotrophica* NBRC 12743, an actinobacterium with Vdh coexpression, which can hydroxylate vitamin D3 via *Rhodococcus erythropolis* (Fujii et al., 2009).

Vitamin K Is Produced in the Body Through Metabolic Processes

Vitamin K is divided into two types: vitamin K1 and vitamin K2. Vitamin K1, also known as phylloquinone, is produced naturally in plants and is found in abundance in leafy green vegetables (Widhalm et al., 2012). Vitamin K2, also known as menaquinones, is produced in both mammalian and microbial cells (Mahdinia et al., 2019). Unlike other forms of vitamin K, MK-7 can still be generated by a wide range of microorganisms (Geleijnse et al., 2004). Investigators have been looking into metabolic engineering approaches for the commercial synthesis of MK-7 for the past decade. In *Bacillus subtilis natto*, the development of pellicle was important for MK-7 synthesis. In contrary to the static condition, increased agitation and aeration rates greatly improve MK7 synthesis. In addition, agitating can help with heat transmission in commercial MK-7 synthesis (Berenjian et al., 2015).

Steroids

The most notable case is a French group that engineered *Saccharomyces cerevisiae* to synthesize pregnenolone and progesterone. Till now, these crucial steroids were synthesized using a blend of synthetic and metabolic procedures, usually using a steroid precursor derived from cholesterol or phytosterols. In conjunction with Roussel and Transgene, Pompon and colleagues at the CNRS genetically engineered and co-expressed multiple heterogeneous genes in *S. cerevisiae*. They were successful to create pregnenolone and progesterone effectively from a basic source of carbon such as galactose by combining *S. cerevisiae's* intrinsic capacity to generate sterols, which are employed as endogenous precursors, with genetically inserted metabolic capabilities. Despite the low documented production quantities (60 mg/L), this experiment is a great example of how bacterial metabolic engineering can be used to generate a useful medicine in a single step (Duport et al., 1998).

Unnatural Amino Acids

An even more fascinating case is the creation of so-called "unnatural" amino acids. The utilization of D-amino acids or artificial L-amino acids in the production of peptidomimetic substances that serve as blockers for diverse enzyme actions like thrombin, HIV protease, and ACE is common in structure-based drug development (Taylor et al., 1998). They were successful in developing D-phenylalanine, D-tyrosine, D-tryptophan, and L-leucine from their corresponding L-isomers by cloning three enzymes from various microorganisms into *E. coli*. In bioreactors, NSC has shown conversion yields of around 80%, which they anticipate would enable cost-effective bulk synthesis of these critical D-amino acids.

Metabolites of Pharmacological Significance

Another pertinent pharmaceutical illustration is the collaboration between the Sinskey and Stephanopoulos groups at the Massachusetts Institute of Technology (MIT) and Merck scientists. They have started working on a *Rhodococcus* variant that can make cis-1S,2R-indandiol and trans-1R,2R indandiol, two probable precursors to CRIXIVANR, a well-known HIV protease inhibitor (Chartrain et al., 1998). The necessity of carotenoids including lutein and zeaxanthin is essential for muscular strength, and certain secondary metabolites have clear effectiveness in healthcare that is redundant to standard vitamin utility (Bai et al., 2011; Roberts et al., 2009).

Strategies of Metabolic Engineering Used for Enhancement of Bioactive Compounds

Plants have a variety of metabolic processes that are necessary for the formation of physiologically functional compounds. Metabolic engineering is a new method for identifying and enhancing important metabolites in plants. Numerous bioengineering strategies were incorporated in plant species to stimulate the production of bioactive substances, to screen and select the largest yielding cell line, to standardize culture media, to elicit, to commercialize creation using bioreactors, to immobilize cells, to feed metabolic precursors, as well as to biotransform (Parsaeimehr et al., 2011). The most recent advancements in genetic modification strategies have greatly aided in utilizing plant cells' ability to improve the synthesis of bioactive chemicals (Woo et al., 2015).

Production of Bioactive Compounds from Cell Culture

Herbal medicines harvested from wild populations lead to genetic biodiversity depletion as well as the degradation of plant habitats and environments. As a result, adequate schemes for quick growth and enhancement of therapeutically significant plants must be developed for increased output. Tissue culture approaches provide benefits over traditional production strategies in satisfying the ever-increasing need for therapeutic plants (Rout et al., 2000). Cell and organ cultivation in diverse plants, also serves as an alternate method for the production of medicinally/industrially valuable metabolites (Vanisree et al., 2004; Vongpaseuth and Roberts, 2007). Geraniol, a significant monoterpenoid utilized in the flavor and perfume sectors, was made utilizing a cell culture technique (Chen and Viljoen, 2010). Cell culture has shown to be an economically effective and easy-to-manage technology for the synthesis of various natural products (Vasilev et al., 2014). Plant cell culture is a well-demonstrated technical base for constantly producing plant therapeutics. This method permits cells to multiply at faster speeds in a contained and regulated environment, regardless of ecological or climatic factors (Niraula et al., 2010).

Regulation of Biosynthesis Pathway Genes

The greatest important phase in plant developmental stages is the modulation of gene expression, which leads to the formation of a huge variety of bioactive compounds in various regions of a plant cell. Gene transcription can be changed at any point throughout the transcriptional process, from initiation to RNA processing to protein post-translational alteration (Petrillo et al., 2014). Expression of genes that encode TFs, such as MYB, which govern jasmonic acid (JA), culminated in increased terpene build-up in *Arabidopsis*, according to Fits and Memelin (van der Fits and Memelink, 2000). Watanabe et al. investigated the formation of several carotenoids in the petals of *Ipomoea nil*, a Japanese morning glory that only deposits minimal levels of carotenoids (Watanabe et al., 2017). The use of RNA silencing to downregulate the major enzyme 9-cis-epoxycarotenoid dioxygenase (NCED) of the abscisic acid biosynthesis cascade has resulted in enhanced accumulation of upstream byproducts, primarily carotene and lycopene (Guo et al., 2016).

Current Trends in Metabolic Engineering

The genes expressing diverse enzymes engaged in certain metabolic pathways have become easier to find, characterize, and change as a result of the advancement of genomic sequencing, genetic modification technologies, and bioinformatics tools (Olivoto et al., 2017). This technique has proven to be effective for isolating and improving high-value plant-based bioactive chemicals by reestablishing plant mechanisms in heterologous hosts or indigenous producers to bestow new features to the chosen plant species (Tatsis and O'Connor, 2016). The introduction of high-throughput or next-generation sequencing technology (NGS) in the last decade has enabled genome-enabled technologies to change the identification of novel compounds in plants, including biosynthetic patterns of genes, their control, regulating pathways, and so on. The method is based on a basic forward genetic premise, wherein newer genes (DNA sequencing) and their transcription trends (RNA-Seq) are first identified using NGS technologies that are subsequently employed to correlate novel proteins and intermediates using standard metabolomics and proteogenomics techniques. Millions of single nucleotide polymorphisms (SNPs) can be identified in a genome using NGS technologies, with several of them belonging to secondary metabolic processes. When both phenotypical data and SNP mappings are available in a genome, users can uncover the connection on a genome-wide scale. NGS has improved the selection procedure of cultivars with favorable agronomic features for generating larger quantities of desirable pharmacological metabolites, resulting in improved natural biomedicines (Unamba et al., 2015).

References

Albrecht, M., Misawa, N., & Sandmann, G. (1999). Metabolic engineering of the terpenoid biosynthetic pathway of *Escherichia coli* for production of the carotenoids β-carotene and zeaxanthin. *Biotechnology Letters, 21*, 791–795.

Bai, C., Twyman, R. M., Farré, G., Sanahuja, G., Christou, P., Capell, T., & Zhu, C. (2011). A golden era—pro-vitamin A enhancement in diverse crops. *In Vitro Cellular & Developmental Biology—Plant, 47*, 205–221.

Berenjian, A., Mahanama, R., Kavanagh, J., & Dehghani, F. (2015). Vitamin K series: Current status and future prospects. *Critical Reviews in Biotechnology, 35*, 199–208.

Chartrain, M., Jackey, B., Taylor, C., Sandford, V., Gbewonyo, K., Lister, L., Dimichele, L., Hirsch, C., Heimbuch, B., Maxwell, C., et al. (1998). Bioconversion of indene to cis (1S, 2R) indandiol and trans (1R, 2R) indandiol by *Rhodococcus* species. *Journal of Fermentation and Bioengineering, 86*, 550–558.

Chen, L., Marmey, P., Taylor, N. J., Brizard, J.-P., Espinoza, C., D'Cruz, P., Huet, H., Zhang, S., de Kochko, A., Beachy, R. N., et al. (1998). Expression and inheritance of multiple transgenes in rice plants. *Nature Biotechnology, 16*, 1060–1064.

Chen, W., & Viljoen, A. M. (2010). Geraniol—A review of a commercially important fragrance material. *South African Journal of Botany, 76*, 643–651.

Chong, D. K. X., Roberts, W., Arakawa, T., Illes, K., Bagi, G., Slattery, C. W., & Langridge, W. H. R. (1997). Expression of the human milk protein β-casein in transgenic potato plants. *Transgenic Research, 6*, 289–296.

Datta, K., Baisakh, N., Thet, K. M., Tu, J., & Datta, S. (2002). Pyramiding transgenes for multiple resistance in rice against bacterial blight, yellow stem borer and sheath blight. *Theoretical and Applied Genetics, 106*, 1–8.

Duport, C., Spagnoli, R., Degryse, E., & Pompon, D. (1998). Self-sufficient biosynthesis of pregnenolone and progesterone in engineered yeast. *Nature Biotechnology, 16*, 186–189.

Eniyan, K., Dharavath, S., Vijayan, R., Bajpai, U., & Gourinath, S. (2018). Crystal structure of UDP-N-acetylglucosamineenolpyruvate reductase (MurB) from *Mycobacterium tuberculosis*. *Biochimica et Biophysica Acta (BBA)—Proteins and Proteomics, 1866*, 397–406.

Farhi, M., Marhevka, E., Ben-Ari, J., Algamas-Dimantov, A., Liang, Z., Zeevi, V., Edelbaum, O., Spitzer-Rimon, B., Abeliovich, H., Schwartz, B., et al. (2011). Generation of the potent anti-malarial drug artemisinin in tobacco. *Nature Biotechnology, 29*, 1072–1074.

Fujii, Y., Kabumoto, H., Nishimura, K., Fujii, T., Yanai, S., Takeda, K., Tamura, N., Arisawa, A., & Tamura, T. (2009). Purification, characterization, and directed evolution study of a vitamin D3 hydroxylase from *Pseudonocardia autotrophica*. *Biochemical and Biophysical Research Communications, 385*, 170–175.

Geleijnse, J. M., Vermeer, C., Grobbee, D. E., Schurgers, L. J., Knapen, M. H. J., Van Der Meer, I. M., Hofman, A., & Witteman, J. C. M. (2004). Dietary intake of menaquinone is associated with a reduced risk of coronary heart disease: The Rotterdam study. *The Journal of Nutrition, 134*, 3100–3105.

Christaki, E., Bonos, E., Giannenas, I., & Florou-Paneri, P. (2013). Functional properties of carotenoids originating from algae. *Journal of the Science of Food and Agriculture, 93*, 5–11. https://doi.org/10.1002/jsfa.5902

Guo, Q., Liu, Q. A., Smith, N., Liang, G., & Wang, M.-B. (2016). RNA silencing in plants: Mechanisms, technologies and applications in horticultural crops. *Current Genomics, 17*, 476–489.

Kohli, A., González-Melendi, P., Abranches, R., Capell, T., Stoger, E., & Christou, P. (2006). The quest to understand the basis and mechanisms that control expression of introduced transgenes in crop plants. *Plant Signaling & Behavior, 1*, 185–195.

Li, L., Chen, J., & Jiang, Y. (2019). The association between vitamin D level and Sjögren's syndrome: A meta-analysis. *International Journal of Rheumatic Diseases, 22*, 532–533.

Liu, L., Li, Y., Zhu, Y., Du, G., & Chen, J. (2007). Redistribution of carbon flux in Torulopsis glabrata by altering vitamin and calcium level. *Metabolic Engineering, 9*, 21–29.

Mahdinia, E., Demirci, A., & Berenjian, A. (2019). Biofilm reactors as a promising method for vitamin K (menaquinone-7) production. *Applied Microbiology and Biotechnology, 103*, 5583–5592.

Miura, Y., Kondo, K., Saito, T., Shimada, H., Fraser, P. D., & Misawa, N. (1998a). Production of the carotenoids lycopene, β-carotene, and astaxanthin in the food yeast Candida utilis. *Applied and Environmental Microbiology, 64*, 1226–1229.

Miura, Y., Kondo, K., Shimada, H., Saito, T., Nakamura, K., & Misawa, N. (1998b). Production of lycopene by the food yeast, Candida utilis that does not naturally synthesize carotenoid. *Biotechnology and Bioengineering, 58*, 306–308.

Møldrup, M. E., Geu-Flores, F., Olsen, C. E., & Halkier, B. A. (2011). Modulation of sulfur metabolism enables efficient glucosinolate engineering. *BMC Biotechnology, 11*, 1–8.

Naqvi, S., Zhu, C., Farre, G., Ramessar, K., Bassie, L., Breitenbach, J., Conesa, D. P., Ros, G., Sandmann, G., Capell, T., et al. (2009). Transgenic multivitamin corn through biofortification of endosperm with three vitamins representing three distinct metabolic pathways. *Proceedings of the National Academy of Sciences, 106*, 7762–7767.

Niraula, N. P., Kim, S.-H., Sohng, J. K., & Kim, E.-S. (2010). Biotechnological doxorubicin production: Pathway and regulation engineering of strains for enhanced production. *Applied Microbiology and Biotechnology, 87*, 1187–1194.

Olivoto, T., Nardino, M., Carvalho, I. R., Follmann, D. N., Szareski, V. J., Ferrari, M., de Pelegrin, A. J., & de Souza, V. Q. (2017). Plant secondary metabolites and its dynamical systems of induction in response to environmental factors: A review. *African Journal of Agricultural Research, 12*, 71–84.

Parsaeimehr, A., Sargsyan, E., & Vardanyan, A. (2011). Expression of secondary metabolites in plants and their useful perspective in animal health, *3*, 115–124.

Paul, J.-Y., Khanna, H., Kleidon, J., Hoang, P., Geijskes, J., Daniells, J., Zaplin, E., Rosenberg, Y., James, A., Mlalazi, B., et al. (2017). Golden bananas in the field: Elevated fruit pro-vitamin A from the expression of a single banana transgene. *Plant Biotechnology Journal, 15*, 520–532.

Perkins, J. B., Sloma, A., Hermann, T., Theriault, K., Zachgo, E., Erdenberger, T., Hannett, N., Chatterjee, N. P., Williams, V., II, Rufo, G. A., Jr., et al. (1999). Genetic engineering of *Bacillus subtilis* for the commercial production of riboflavin. *Journal of Industrial Microbiology & Biotechnology, 22*, 8–18.

Petrillo, E., Godoy Herz, M. A., Barta, A., Kalyna, M., & Kornblihtt, A. R. (2014). Let there be light: Regulation of gene expression in plants. *RNA Biology, 11*, 1215–1220.

Rischer, H. T., Hakkinen, S., Ritala, A., Seppanen-Laakso, T., Miralpeix, B., Capell, T., Christou, P., & Oksman-Caldentey, K.-M. (2013). Plant cells as pharmaceutical factories. *Current Pharmaceutical Design, 19*, 5640–5660.

Roberts, R. L., Green, J., & Lewis, B. (2009). Lutein and zeaxanthin in eye and skin health. *Clinics in Dermatology, 27*, 195–201.

Rout, G. R., Samantaray, S., & Das, P. (2000). In vitro manipulation and propagation of medicinal plants. *Biotechnology Advances, 18*, 91–120.

Saini, R. K., & Keum, Y.-S. (2017). Progress in microbial carotenoids production. *Indian Journal of Microbiology, 57*, 129–130.

Tatsis, E. C., & O'Connor, S. E. (2016). New developments in engineering plant metabolic pathways. *Current Opinion in Biotechnology, 42*, 126–132.

Taylor, P. P., Pantaleone, D. P., Senkpeil, R. F., & Fotheringham, I. G. (1998). Novel biosynthetic approaches to the production of unnatural amino acids using transaminases. *Trends in Biotechnology, 16*, 412–418.

Unamba, C. I. N., Nag, A., & Sharma, R. K. (2015). Next generation sequencing technologies: The doorway to the unexplored genomics of non-model plants. *Frontiers in Plant Science, 6*, 1074.

van der Fits, L., & Memelink, J. (2000). ORCA3, a jasmonate-responsive transcriptional regulator of plant primary and secondary metabolism. *Science (80-.), 289*, 295–297.

Vanisree, M., Lee, C.-Y., Lo, S.-F., Nalawade, S. M., Lin, C. Y., & Tsay, H.-S. (2004). Studies on the production of some important secondary metabolites from medicinal plants by plant tissue cultures. *Botanical Bulletin of Academia Sinica, 45*, 1–22.

Vasilev, N., Schmitz, C., Grömping, U., Fischer, R., & Schillberg, S. (2014). Assessment of cultivation factors that affect biomass and geraniol production in transgenic tobacco cell suspension cultures. *PLoS One, 9*, e104620.

Vega, J. M., Yu, W., Han, F., Kato, A., Peters, E. M., Zhang, Z. J., & Birchler, J. A. (2008). *Agrobacterium-mediated* transformation of maize (*Zea mays*) with Cre-lox site specific recombination cassettes in BIBAC vectors. *Plant Molecular Biology, 66*, 587–598.

Vongpaseuth, K., & Roberts, S. C. (2007). Advancements in the understanding of paclitaxel metabolism in tissue culture. *Current Pharmaceutical Biotechnology, 8*, 219–236.

Wang, C.-W., Oh, M.-K., & Liao, J. C. (1999). Engineered isoprenoid pathway enhances astaxanthin production in *Escherichia coli*. *Biotechnology and Bioengineering, 62*, 235–241.

Watanabe, K., Kobayashi, A., Endo, M., Sage-Ono, K., Toki, S., & Ono, M. (2017). CRISPR/Cas9-mediated mutagenesis of the dihydroflavonol-4-reductase-B (DFR-B) locus in the Japanese morning glory ipomoea (Pharbitis) nil. *Scientific Reports, 7*, 1–9.

Welsch, R., Arango, J., Bär, C., Salazar, B., Al-Babili, S., Beltrán, J., Chavarriaga, P., Ceballos, H., Tohme, J., & Beyer, P. (2010). Provitamin A accumulation in cassava (*Manihot esculenta*) roots driven by a single nucleotide polymorphism in a phytoene synthase gene. *Plant Cell, 22*, 3348–3356.

Widhalm, J. R., Ducluzeau, A.-L., Buller, N. E., Elowsky, C. G., Olsen, L. J., & Basset, G. J. C. (2012). Phylloquinone (vitamin K1) biosynthesis in plants: Two peroxisomal thioesterases of *lactobacillales* origin hydrolyze 1, 4-dihydroxy-2-naphthoyl-coa. *The Plant Journal, 71*, 205–215.

Woo, J. W., Kim, J., Kwon, S. I., Corvalán, C., Cho, S. W., Kim, H., Kim, S.-G., Kim, S.-T., Choe, S., & Kim, J.-S. (2015). DNA-free genome editing in plants with preassembled CRISPR-Cas9 ribonucleoproteins. *Nature Biotechnology, 33*, 1162–1164.

Yan, G., Wen, K., & Duan, C. (2012). Enhancement of \$β\$-carotene production by over-expression of HMG-CoA reductase coupled with addition of ergosterol biosynthesis inhibitors in recombinant *Saccharomyces cerevisiae*. *Current Microbiology, 64*, 159–163.

Yoon, S.-H., Lee, S.-H., Das, A., Ryu, H.-K., Jang, H.-J., Kim, J.-Y., Oh, D.-K., Keasling, J. D., & Kim, S.-W. (2009). Combinatorial expression of bacterial whole mevalonate pathway for the production of \$β\$-carotene in E. coli. *Journal of Biotechnology, 140*, 218–226.

Zhu, C., Naqvi, S., Breitenbach, J., Sandmann, G., Christou, P., & Capell, T. (2008). Combinatorial genetic transformation generates a library of metabolic phenotypes for the carotenoid pathway in maize. *Proceedings of the National Academy of Sciences, 105*, 18,232–18,237.

Chapter 17
Carbon-Based Nanomaterials: An Efficient Tool for Improving the Nutritional Quality of Crops

V. Sasidharan, Fouad Damiri, Neetu Talreja, Divya Chauhan, R. V. Mangalaraja, Mohammed Berrada, and Mohammad Ashfaq

Abstract One of the essential tasks while practicing agricultural sustainability is maintaining nutrition quality while improving crop growth and production, as nutritional deficiencies within the plants are serious threats that affect human health. In this context, carbon-based nanomaterials (CNMs), including carbon nanotubes (CNTs), carbon nanofibers (CNFs), graphene, etc., and their metallic, non-metallic composites have great potential to improve crop production as well as nutritional quality due to their significant properties such as high surface polarity, large pore size distribution, translocation, low toxicity, high biocompatibility, and binding at the target site due to opposite polarity. Herein, we discuss the CNMs and their role in improving the nutrition quality of the plant. The participation of CNMs and their composites in several nutrition cycles inside the plant and their working direction have also been discussed in detail. This chapter also summarizes the toxic effects of

V. Sasidharan
Department of Biotechnology, SASTRA University, Tanjore, Tamil Nadu, India

F. Damiri · M. Berrada
Department of Chemistry, Laboratory of Biomolecules and Organic Synthesis (BioSynthO), Faculty of Sciences Ben M'Sick, University Hassan II of Casablanca, Casablanca, Morocco

N. Talreja · R. V. Mangalaraja
Advanced Ceramics and Nanotechnology Laboratory, Department of Materials Engineering, Faculty of Engineering, University of Concepción, Concepción, Chile

D. Chauhan
Department of Chemical and Biomedical Engineering, University of South Florida, Tampa, FL, USA

M. Ashfaq (✉)
Advanced Ceramics and Nanotechnology Laboratory, Department of Materials Engineering, Faculty of Engineering, University of Concepción, Concepción, Chile

Department of Chemical and Biomedical Engineering, University of South Florida, Tampa, FL, USA

School of Life Science, BS Abdur Rahman Crescent Institute of Science and Technology, Chennai, India

T. Aftab, K. R. Hakeem (eds.), *Metabolic Engineering in Plants*,
https://doi.org/10.1007/978-981-16-7262-0_17

engineered CNMs in plants and their ultimate development in the nutritional quality of the crop. Comparison of a wide range of CNMs applied to improve the quality of crops production and their nutritional values.

Keywords Carbon-based nanomaterials · Plant growth · Nutrition · Composite · Translocation

Introduction

Metabolic engineering involves changes in cell metabolism by changing its pathway enzyme(s) or regulatory protein(s), thereby increasing or improving the production of plant products. Metabolic function in plants is mainly performed and controlled by enzymes and elements, respectively, based on the phenotypic distribution of species (Yadav et al., 2012; Pickens et al., 2011; Cravens et al., 2019; Yuan & Grotewold, 2015). High nutritional foods or a balanced diet decrease the prevalence of chronic diseases because of high antioxidant and health-promoting abilities. Therefore, the requirement of the high demand for nutritious, quality food increased day by day. However, sodium (Na) and chloride (Cl) changes alter the nutritional imbalance due to more minor or insignificant uptake like Ca, Mg, K, nitrates, carbohydrates, proteins, and phosphate, thereby showing a decrease in the growth of the plants as well as a decrease in the nutritional quality of the food. The uses of chemicals or agrochemicals or fertilizers regulate phytohormones that significantly enhanced the yield of crops. These practices (chemicals or agrochemicals, or fertilizers) remain incompetent to fulfill foods' requirements (Sultana et al., 2021; Shang et al., 2019; Ashfaq & Khan, 2017). The use of nitrogen and phosphorus-based fertilizers significantly enhanced the developing countries in hybrid crops. The loss of fertilizers from agricultural land leads to environmental consequences. However, the lack of other micronutrients like Fe, Cu, and Zn for the uptake of plants that decrease the agricultural biodiversity leading to imbalance of nutritional quality of crops is still a significant challenge. The imbalance of the crops' dietary values might increase consumer products or foods (Tilman, 1999; Mitter et al., 2021; Yousaf et al., 2017). Therefore, such materials or agrochemicals or fertilizers that improve the crops' nutritional quality are needed. In this regard, several metal- and metal-oxide nanoparticles (M-NPs) and CNMs have the potential ability to increase the crops' nutritional value and yield.

M-NPs and CNMs can participate in several metabolic functions during plant growth to maintain nutrition quality. M-NPs and their oxides such as Cu, Zn, Fe, Ce, Mn, and Mg are extensively used in various applications, including agriculture. These M-NPs augment the plant's growth and increase nutritional value. These M-NPs also have an antibacterial and antifungal activity that protects the crops against pathogens. Interestingly, M-NPs, mainly Fe, Cu, and Zn, are essential micronutrients required to maintain the nutritional value of the crops. However, ineffective dispersion, agglomeration, accumulation, and relatively more minor translocation ability lead to phytotoxicity (Mustafa et al., 2011; Omar et al., 2019a, 2019b; Irsad et al., 2020; Mittal et al., 2020; Faizan et al., 2021; Skiba et al., 2020; Sultana et al., 2021; Ashfaq et al., 2021; Talreja et al., 2021a, 2021b; Talreja & Kumar, 2018; Chauhan et al., 2021). Therefore, there is need to develop newer NMs or modify existing NMs that effectively translocate within the plants without any adverse effects. In this aspect, CNMs have the potential ability that might be next-generation tools for improving the yield of crops and the nutritional quality of food.

CNMs such as carbon nanotubes (CNTs), carbon nanofibers (CNFs), graphene, graphene oxide (GO), and fullerenes consist of unique characteristics such as high surface area, electrical, optical, mechanical, large pore size, and ease to functionalize. High biocompatibility makes them an excellent candidate for numerous applications, including removal of heavy metal ions, pharmaceuticals compounds (antibiotics and vitamins), dyes, sensors (chemical and gas), drug delivery, antibiotic materials, energy, and agriculture (Sasidharan et al., 2021; Ashfaq et al., 2019; Kumar & Talreja, 2019; Talreja et al., 2020; Talreja et al., 2016; Talreja et al., 2014; Kumar et al., 2011; Saraswat et al., 2012; Khare et al., 2013; Afreen et al., 2018; Singh et al., 2013; Ashfaq et al., 2016, 2014, 2017b, 2018; Tripathi et al., 2016; Bhadauriya et al., 2018). Remarkably, CNMs, especially CNTs, CNFs, and graphene, easily penetrate seed coats and translocate within the plants through root to shoot to leaf due to its graphitic nature. These CNMs like CNTs, CNFs, and graphene acted as a carrier for the delivery of agrochemicals/fertilizers/ micronutrients within the plants, thereby increasing the nutritional quality as well as plant growth (Das et al., 2018; Chen et al., 2015; Mathew et al., 2021). Among all of them, CNFs are relatively more minor toxicity or insignificant toxic to human and plant cells, thereby promising tools for next-generation biological application (Kumar et al., 2018; Ashfaq et al., 2013, 2017a).

Usually, a lack of micronutrients within the soil decreases the nutritional value of the crops. The CNMs manage equilibrium among carbon metabolism by increasing mineral uptake, water uptake ability, and nutrient efficiency in plants. In this chapter, we focus on the different CNMs and their effect on the growth of the plants. This chapter also summarizes the toxic results of engineered CNMs in plants and their ultimate development in the nutritional quality of the crop.

Role of Metabolic Engineering to Improve the Nutrition Quality of Plants

Considerable advancement has been done to improve the nutritional quality of food with the help of metabolic engineering through managing various metabolic pathways, accumulation, and translocation of micronutrients. With the use of genomics and proteomics technologies, these metabolic pathways easily regulate, improving the nutritional quality of foods or crops (Kumar, 2015). Usually, metabolic engineering is the process that enhances the production of the targeted compounds or

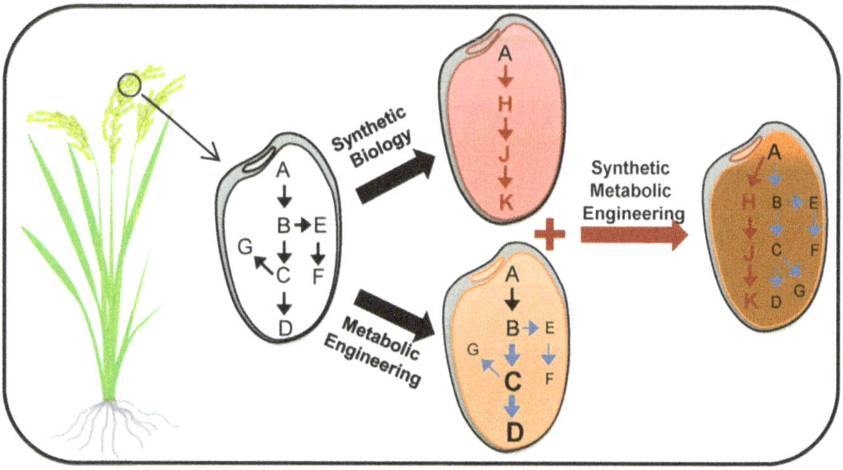

Fig. 17.1 Schematic representation of metabolic engineering in plants. The image was taken with permission (Zhu et al., 2020)

molecules through the modulation of metabolic pathways. The metabolic pathways are the biochemical chain reaction, where numerous enzymes catalyzed that convert substrate to targeted product. The production of the targeted molecules or compounds can increase by multiple strategies: (1) improving gene expression, (2) inhibiting the expression of the gene, and (3) overexpressing transcription factors (Zhu et al., 2020). Figure 17.1 shows the schematic representation of the metabolic engineering in the plant. In general, metabolic engineering or metabolic pathways are the critical processes for the development and growth of plants. The nutritional quality of the foods or crops efficiently improves by altering or changing the metabolic pathways using various strategies.

Role of NMs in Metabolic Engineering

NMs interact with the biomolecules, and plant cells play a significant role in any cellular process, thereby enhancing the nutritional value of crops. Numerous studies show the metal-NPs have been used for the improvement of nutritional quality (Nair & Chung, 2015; Patra et al., 2013; Hong et al., 2015; Arora et al., 2012; Bandyopadhyay et al., 2015; Abd-Alla et al., 2016; Raliya et al., 2015; Rico et al., 2013; Cai et al., 2020; Chauhan et al., 2020). For example, Rico et al. synthesized nano-CeO_2 and tested its effects on plant growth, the yield of crops, and nutritional quality in wheat. The data suggested that the nano-CeO_2 increases the plant biomass, plant growth, and yield of crops. Interestingly, nano-CeO_2 alters S and Mn storage and amino acid composition and increased linolenic acid compared with control. The data indicated that nano-CeO_2 improved plant growth, the yield of grains, and the

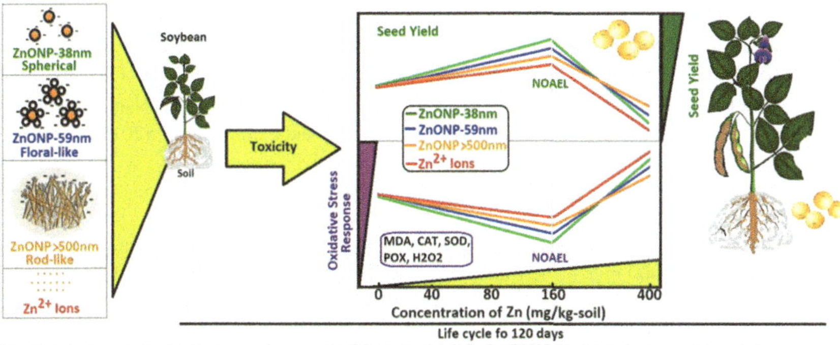

Fig. 17.2 Schematic illustration of the ZnO nanoparticles and their effects on soybean plants. The image was taken with permission (Yusefi-Tanha et al., 2020)

quality of food by modifying the physiology of the crops/plants (Rico et al., 2014). Another study of the same group focuses on the nano-CeO_2 and understands its effects on the barley's plant physiology, productivity, and macromolecular composition. The data suggested that the nano-CeO_2 increases the development of the plant, shoot biomass, the micronutrient contents (Ca, Mg, S, P, K, Fe, Zn, Cu, and Al), and the amino acid contents. However, some other types of nano-CeO_2 show toxic effects. Based on the results, we can say that nano-CeO_2 shows synergetic effects to increase the grains' plant development, growth, and nutritional quality. Moreover, different nano-CeO_2 shows toxic effects on the plants (Rico et al., 2015). Sadak and Bakry et al. synthesized ZnO nanoparticles and ZnO and tested their effects on the flax plant's growth rate, photosynthetic pigment, and chemical composition or nutritional quality. The data suggested that the ZnO nanoparticles increase the root and shoot length, plant biomass, chlorophyll content, amino acid, carbohydrate, and yield of crops. The data revealed that the ZnO nanoparticles increase the growth of the plants and the nutritional quality of the food (Sadak & Bakry, 2020). Another study focused on the synthesis of urea-coated ZnO nanoparticles and tested their effects on the nutritional quality of the wheat under drought stress conditions. The data suggested that the bulk and ZnO nanoparticles show an insignificant increase in the yield. The urea-coated ZnO shows high uptake of the Zn in comparison with that of the uncoated ZnO. Moreover, urea-coated ZnO did not affect the N and P uptake. The urea-coated ZnO nanoparticles show the high yield of crops and nutritional quality of wheat (Dimkpa et al., 2020). Yusefi-Tanha et al. synthesized ZnO nanoparticles and tested their effects on antioxidant defense and soybean seed yield. The data suggested that the growth and nutritional quality of the soybean, thereby ZnO nanoparticles effectively used even at Zn-deficient soil (Yusefi-Tanha et al., 2020). Figure 17.2 shows the schematic representation of the ZnO-based nanofertilizers and their effects on the soybean plant. Yang et al. synthesized ZnO nanoparticles and tested their effect on rice's biofortification and nutritional quality. The data suggested that both ZnO nanoparticles and Zn salt increased grain yield, NPK level, and growth of the plants (Yang et al., 2021).

Table 17.1 Metal-NP and its effects on plant

S. No.	Material	Experimental condition	Plant	Effect over plant	References
1.	CuO-NP	0–500 mg/L 14 days exposure	*Brassica juncea L.*	Decrease in total chlorophyll and carotenoids contents. Increase in H_2O_2 content	Nair and Chung (2015)
2.	ZnO-NP	500–4000 mg/L 10 days exposure	*Vigna radiata*	Increase in total chlorophyll content and root and shoot length	Patra et al. (2013)
3.	nCu, nCuO, Cu $(OH)_2$	0–20 mg/L 15 days exposure	*Lettuce, alfalfa*	Fe and P content decrease	Hong et al. (2015)
4.	AuNPs	0–100 mg/kg 50–70 days exposure	*Mustard*	Increase in sugar content with redox activity	Arora et al. (2012)
5.	ZnO-NP	250–750 mg/kg 30 days exposure	*Alfalfa*	Decrease in biomass	Bandyopadhyay et al. (2015)
6.	ZnO-NP	100–1000 mg/L 20 days exposure	*Garden pea*	No significant effect was observed on plant germination rate	Abd-Alla et al. (2016)
7.	Ag-NP	800 mg/kg 37 days exposure	*Faba bean*	Root and shoot length and germination rate was declined	Raliya et al. (2015)
8.	CeO_2-NP	0–500 mg/L 4 days exposure	*Rice*	Variations in fatty acids and starch were observed	Rico et al. (2013)
9.	Fe_3O_4-NP	100 µg/mL 12 days exposure	*Nicotiana benthamiana*	Improve the activity of antioxidants, upregulated SA synthesis and the expression of SA-responsive PR genes	Cai et al. (2020)

Table 17.1 shows the comparative data table of Metal-NP and its effects on the plant. The data suggested that the metal-NPs enhance the plant's development and the yield of grains. Moreover, a higher concentration of metal-NP might be toxic that decreases the growth of the plant.

In general, based on the literature above studied, metal-NPs, especially nano-CeO_2 and ZnO-NP, enhance chlorophyll content, protein content, plant biomass, root length, and shoot length, the yield of grains, and nutritional value of the food. However, these metal-NPs show toxic effect at a higher dose. The toxic effect might vary with different metal-NPs, plants, species, and concentrations.

Role of Carbon-Based Nanomaterials (CNMs) in Metabolic Engineering

The CNMs, mainly CNTs, CNFs, graphene, GO, carbon dots, and fullerene, are extensively used in various end applications, including agriculture. These CNMs increase the development, growth of the plant, yield of crops, and nutritional value of the crops. Several literature reports use carbon-based materials to enhance plant growth, simultaneously increasing the nutrition content in the plant. For example, Ashfaq et al. synthesized Cu-NP-dispersed CNFs (Cu-CNFs) using the chemical vapor deposition technique (CVD) process. The Cu-CNFs were grown onto the activated carbon fiber (ACF) fabric. The synthesized Cu-CNFs were ball-milled to detached Cu-CNFs from the ACFs. The produced Cu-CNFs were tested against chickpea. The data suggested that the synthesized Cu-CNFs efficiently translocate within the plants through roots to shoot to leaf.

Moreover, Cu-CNFs release Cu ions within the plant in a controlled manner. The Cu-CNFs increase the chlorophyll contents, protein contents, germination rate, and water uptake ability, thereby increasing the development and growth of the plants and resulting in high nutritional value (Ashfaq et al., 2017a). Another study of a similar group focuses on synthesizing the Cu-Zn-CNFs and encapsulating it with polyvinyl alcohol and starch-based polymeric composite. The synthesized Cu-Zn-CNF-based polymeric composite is efficiently used for the plants' growth (Kumar et al., 2018). Shekhawat et al. studied the application of carbon nanoparticles (CNPs) on the development of Vigna radiate. In this study, different doses of CNPs were applied, and their effect on the growth was observed after 96 h; results showed that those seedlings are subjected under CNP dose having high chlorophyll content, high protein content, and higher plant biomass. Also, a decrease in ROS level was observed (Shekhawat et al., 2021). The effect of MW-CNTs, functionalized-CNTs, and fullerenes was studied on broccoli, MR219 paddy variety, and bitter melon by Martínez-Ballesta et al., Yatim et al. and Azamal Husen et al. (Martínez-Ballesta et al., 2016; Yatim et al., 2018; Husen & Siddiqi, 2014). All these studies show a positive impact of MW-CNTs on plant growth and nutrition quality as Ballesta et al. observed improvement in lipid composition, rigidity, and permeability in root membrane, which subsequently increases water uptake and improves the negative effects of salt stress. Yatim et al. prove the increase in the yield of paddy plant, and Husen et al. show the increase in yield of bitter melon plant and phytomedicine contents within plants. The impact of graphene and GO-based carbon material also shows significant improvement in nutrition quality and plant growth. Jiao et al. show graphene oxide-treated plant samples with shorter seminal root length. However, it also indicates an increase in the numbers of adventitious roots (Jiao et al., 2016). Another study by Chakravarty et al. synthesized graphene quantum dots and observed their effect on coriander and garlic plant growth. Results showed improvements in leaves, roots, shoots, flowers, and fruits (Chakravarty et al., 2015).

Similarly, Zhang et al. study the positive impact of graphene and graphene oxide on the growth and development of apple, rice, and aloe vera plants as graphene oxide can induce the generation of oxidative stress enzymes, including catalase (CAT), peroxidase (POD), and superoxide dismutase (SOD) in apple plant, while in rice plant graphene oxide reduces abscisic acid (ABA), indole-3-acetic acid (IAA), and malondialdehyde (MDA) and the activity of superoxide dismutase (SOD) and catalase (CAT) enzymes. Also, in the aloe vera plant, graphene oxide increases leaves' photosynthesis properties and increases nutrient content (Zhang et al., 2016; Li et al., 2018; Shen et al., 2019). In contrast, P. Zheng studied the toxic impact of graphene on the nutritional content of the wheat plant. According to the author, graphene exposure induces several adverse effects like reducing the shoot biomass, chlorophyll content, PSII activity, and levels of several nutrient elements (N, K, Ca, Mg, Fe, Zn, and Cu), which can inhibit plant growth by inhibiting the photosynthesis and by creating the imbalance of nutrient homeostasis (Zhang et al., 2021). Table 17.2 summarized the different CNMs for improving nutritional value of the crops. All these studies demonstrates the CNMs have several advantage and disadvantage in increasing the crop production and improving the nutrient quality by interfering the basic mechanism of plant growth. The positive and adverse effects of the CNMs are observed to mainly depend on the doses, plants, and different materials. The CNMs could regulate expression of genes that leads the nutritional value of the crops as well as growth of the plants. However, research still needs to focus on mechanistic pathways of action of CNMs in growth and development of agricultural.

Interaction of CNMs with Plant

The understanding of CNMs and plant interaction is necessary to determine cellular responses, translocation, accumulation, and detoxification pathways. Usually, CNMs, mainly CNTs and CNFs, efficiently penetrate the seed coat and translocate within the plants via root to shoot to leaf. The negatively charged surface CNMs like CNFs might be advantageous for the translocation, whereas positively charged surface leads to accumulation on the root surface. The negatively charged CNMs do not interact with negatively charged plant cells due to the strong repulsion force, thereby showing high translocation ability. On the other hand, positively charged CNMs interact with negatively charged plant surfaces due to the strong attraction forces, thereby showing high accumulation onto the root surface. Based on the surface charged concept, we can say that negatively charged CNMs show insignificant toxicity or relatively more minor toxicity than positively charged surface CNMs. On the other hand, negatively charged surface CNMs do not show phytotoxic effect even at higher doses or concentrations, whereas positively charged surface CNMs show toxic effect even at a lower amount (Lin et al., 2009; Chichiriccò & Poma, 2015; Deng et al., 2017; Lv et al., 2019; Ashfaq et al.,

Table 17.2 Different CNMs for improving the nutritional quality of the plant

S. No.	Carbon material	Plant	Nutrition quality of plant	Reference
1.	C-NPs	*Vigna radiata*	Increase nutrition quality (chlorophyll content (1.9-fold), protein content (1.14-fold), and plant biomass)	Shekhawat et al. (2021)
2.	MW-CNTs	*Broccoli*	Enhanced net assimilation of CO_2. Increased lipid composition, rigidity, and permeability of the root plasma membranes relative to salt-stressed plants. Also, enhanced aquaporin transduction occurred, which improved water uptake and transport, alleviating the adverse effects of salt stress	Martínez-Ballesta et al. (2016)
3.	f-MWNTs	MR219 paddy variety	Crop growth of plants significantly increased by 22.6% and 38.5% and 35% more grain yield	Yatim et al. (2018)
4.	Fullerene, C_{60} and CNTs	Bitter melon	Increase the phytomedicine contents such as cucurbitacin-B (74%), lycopene (82%), charantin (20%), and insulin (91%)	Husen and Siddiqi (2014)
5.	GO	Tobacco Root	Increase transcript levels of IAA3, IAA4, IAA7, ARF2, and ARF8. Also, higher SOD, POD, and CAT activities and also lower MDA content	Jiao et al. (2016)
6.	GQDs	Coriander and garlic plants	Growth rates of the plants increase in the presence of GQD	Chakravarty et al. (2015))
7.	Graphene	Wheat (*Triticum aestivum* L.)	After 30 days of graphene exposure, shoot biomass, chlorophyll content, PSII activity, and levels of several nutrient elements (N, K, Ca, Mg, Fe, Zn and Cu) were reduced	Zhang et al. (2016)
8.	GO	Rice plant	Reduced the root length, fresh weight, and dry weight for rice. Reduces the levels of abscisic acid (ABA), indole-3-acetic acid (IAA), and malondialdehyde (MDA) and the activity of superoxide dismutase (SOD) and catalase (CAT) enzymes	Shen et al. (2019)
9.	GO	*Aloe vera* L.	Increases the photosynthetic capacity of leaves, increase the yield and morphological characters of root and leaf, improve the nutrient (protein and amino acid) contents of leaf, without reducing the content of the main bioactive compound aloin	Zhang et al. (2021)

2017a). Figure 17.3 shows the experimental scheme and microscopic images of the CNM interaction with plants.

In general, CNMs, especially negatively charged surface CNMs like CNFs, efficiently translocate within the plant via root to shoot to leaf without any toxic effect even at higher concentrations. The negatively charged surface CNMs might be

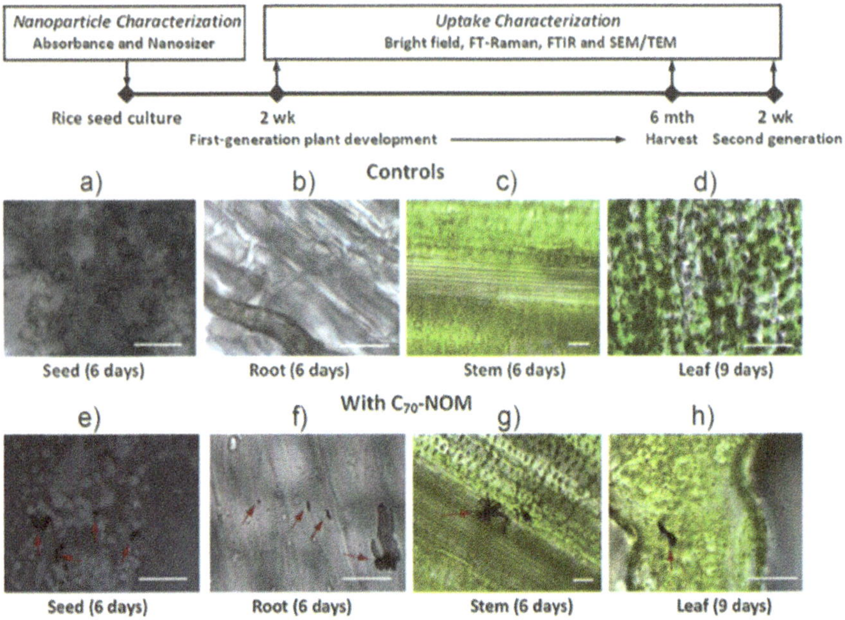

Fig. 17.3 Experimental scheme and microscopic images of the CNM interaction with plants. The picture was taken with permission (Lin et al., 2009)

acted as carriers for agrochemicals or fertilizers, or micronutrients that delivered these compounds in a controlled manner. Moreover, these CNMs significantly improved the development and growth of the plants that improved the nutritional quality of the crops.

Conclusion

This book chapter focuses on the CNM role in metabolic engineering. The metal-NPs especially nano-CeO_2 and ZnO-NP enhance chlorophyll content, protein content, plant biomass, root length, shoot length, the yield of grains, and nutritional value of the food. However, these metal-NPs show toxic effect at a higher dose. The poisonous impact might vary with different metal-NPs, plants, species, and concentrations. CNMs like CNTs and CNFs have penetration ability that increases water uptake, germination rate, chlorophyll, protein, root, and shoot length. Interestingly, negatively charged CNMs might be advantageous for improving the growth of the plants and nutritional value. However, the direct application of the NMs and CNMs is complex. Therefore, polymeric composite might be beneficial for the delivery of these NMs and CNMs.

Acknowledgment The authors acknowledge the financial support by ANID through FONDECYT projectNo. 3190515 and 3190581.

References

Abd-Alla, M. H., Nafady, N. A., & Khalaf, D. M. (2016). Assessment of silver nanoparticles contamination on faba bean-Rhizobium leguminosarum bv. viciae-Glomus aggregatum symbiosis: Implications for induction of autophagy process in root nodule. *Agriculture, Ecosystems & Environment, 218*, 163–177.

Afreen, S., Omar, R. A., Talreja, N., Chauhan, D., & Ashfaq, M. (2018). Carbon-based nanostructured materials for energy and environmental remediation applications. In R. Prasad & E. Aranda (Eds.), *Approaches in bioremediation: The new era of environmental microbiology and nanobiotechnology*. Springer International Publishing.

Arora, S., Sharma, P., Kumar, S., Nayan, R., Khanna, P. K., & Zaidi, M. G. H. (2012). Gold-nanoparticle induced enhancement in growth and seed yield of Brassica juncea. *Plant Growth Regulation, 66*(3), 303–310.

Ashfaq, M., & Khan, S. (2017). Role of phytohormones in improving the yield of oilseed crops. In P. Ahmad (Ed.), *Oilseed crops*. Wiley.

Ashfaq, M., Khan, S., & Verma, N. (2014). Synthesis of PVA-CAP-based biomaterial in situ dispersed with Cu nanoparticles and carbon micro-nanofibers for antibiotic drug delivery applications. *Biochemical Engineering Journal, 90*, 79–89.

Ashfaq, M., Singh, S., Sharma, A., & Verma, N. (2013). Cytotoxic evaluation of the hierarchical web of carbon micronanofibers. *Industrial & Engineering Chemistry Research, 52*(12), 4672–4682.

Ashfaq, M., Talreja, N., Chauhan, D., Rodríguez, C. A., Mera, A. C., & Mangalaraja, R. V. (2021). A novel bimetallic (Fe/Bi)-povidone-iodine micro-flowers composite for photocatalytic and antibacterial applications. *Journal of Photochemistry and Photobiology B: Biology, 219*, 112,204.

Ashfaq, M., Talreja, N., Chuahan, D., & Srituravanich, W. (2019). Carbon nanostructure-based materials: A novel tool for detection of Alzheimer's disease. In G. M. Ashraf & A. Alexiou (Eds.), *Biological, diagnostic and therapeutic advances in Alzheimer's disease: Non-pharmacological therapies for Alzheimer's disease*. Singapore.

Ashfaq, M., Verma, N., & Khan, S. (2016). Copper/zinc bimetal nanoparticles-dispersed carbon nanofibers: A novel potential antibiotic material. *Materials Science and Engineering C, 59*, 938–947.

Ashfaq, M., Verma, N., & Khan, S. (2017a). Carbon nanofibers as a micronutrient carrier in plants: Efficient translocation and controlled release of Cu nanoparticles. *Environmental Science: Nano, 4*(1), 138–148.

Ashfaq, M., Verma, N., & Khan, S. (2017b). Highly effective Cu/Zn-carbon micro/nanofiber-polymer nanocomposite-based wound dressing biomaterial against the P. aeruginosa multi- and extensively drug-resistant strains. *Materials Science and Engineering: C, 77*, 630–641.

Ashfaq, M., Verma, N., & Khan, S. (2018). Novel polymeric composite grafted with metal nanoparticle-dispersed CNFs as a chemiresistive non-destructive fruit sensor material. *Materials Chemistry and Physics, 217*, 216–227.

Bandyopadhyay, S., Plascencia-Villa, G., Mukherjee, A., Rico, C. M., José-Yacamán, M., Peralta-Videa, J. R., & Gardea-Torresdey, J. L. (2015). Comparative phytotoxicity of ZnO NPs, bulk ZnO, and ionic zinc onto the alfalfa plants symbiotically associated with Sinorhizobium meliloti in soil. *Science of the Total Environment, 515–516*, 60–69.

Bhadauriya, P., Mamtani, H., Ashfaq, M., Raghav, A., Teotia, A. K., Kumar, A., & Verma, N. (2018). Synthesis of yeast-immobilized and copper nanoparticle-dispersed carbon nanofiber-based diabetic wound dressing material: Simultaneous control of glucose and bacterial infections. *ACS Applied Bio Materials, 1*(2), 246–258.

Cai, L., Cai, L., Jia, H., Liu, C., Wang, D., & Sun, X. (2020). Foliar exposure of Fe3O4 nanoparticles on Nicotiana benthamiana: Evidence for nanoparticles uptake, plant growth promoter and defense response elicitor against plant virus. *Journal of Hazardous Materials, 393*, 122,415.

Chakravarty, D., Erande, M. B., & Late, D. J. (2015). Graphene quantum dots as enhanced plant growth regulators: effects on coriander and garlic plants. *Journal of the Science of Food and Agriculture, 95*(13), 2772–2778.

Chauhan, D., Afreen, S., Talreja, N., & Ashfaq, M. (2020). Chapter 8—Multifunctional copper polymer-based nanocomposite for environmental and agricultural applications. In K. A. Abd-Elsalam (Ed.), *Multifunctional hybrid nanomaterials for sustainable agri-food and ecosystems*. Elsevier.

Chauhan, D., Talreja, N., & Ashfaq, M. (2021). Chapter 13—Nanoadsorbents for wastewater remediation. In K. A. Abd-Elsalam & M. Zahid (Eds.), *Aquananotechnology*. Elsevier.

Chen, G., Qiu, J., Liu, Y., Jiang, R., Cai, S., Liu, Y., Zhu, F., Zeng, F., Luan, T., & Ouyang, G. (2015). Carbon nanotubes act as contaminant carriers and translocate within plants. Scientific Reports, 5(1), pp 15682. *Scientific Reports, 5*(1), 15,682.

Chichiriccò, G., & Poma, A. (2015). Penetration and toxicity of nanomaterials in higher plants. *Nanomaterials, 5*(2), 851–873.

Cravens, A., Payne, J., & Smolke, C. D. (2019). Synthetic biology strategies for microbial biosynthesis of plant natural products. *Nature Communications, 10*(1), 2142.

Das, K. K., You, Y., Torres, M., Barrios-Masias, F., Wang, X., Tao, S., Xing, B., & Yang, Y. (2018). Development and application of a digestion-Raman analysis approach for studying multiwall carbon nanotube uptake in lettuce. *Environmental Science: Nano, 5*(3), 659–668.

Deng, Y., Eitzer, B., White, J. C., & Xing, B. (2017). Impact of multiwall carbon nanotubes on the accumulation and distribution of carbamazepine in collard greens (Brassica oleracea). *Environmental Science: Nano, 4*(1), 149–159.

Dimkpa, C. O., Andrews, J., Fugice, J., Singh, U., Bindraban, P. S., Elmer, W. H., Gardea-Torresdey, J. L., & White, J. C. (2020). Facile coating of urea with low-dose ZnO nanoparticles promotes wheat performance and enhances zn uptake under drought stress. *Frontiers in Plant Science, 11*, 168.

Faizan, M., Bhat, J. A., Noureldeen, A., Ahmad, P., & Yu, F. (2021). Zinc oxide nanoparticles and 24-epibrassinolide alleviates Cu toxicity in tomato by regulating ROS scavenging, stomatal movement and photosynthesis. *Ecotoxicology and Environmental Safety, 218*, 112,293.

Hong, J., Rico, C. M., Zhao, L., Adeleye, A. S., Keller, A. A., Peralta-Videa, J. R., & Gardea-Torresdey, J. L. (2015). Toxic effects of copper-based nanoparticles or compounds to lettuce (Lactuca sativa) and alfalfa (Medicago sativa). *Environmental Science. Processes & Impacts, 17* (1), 177–185.

Husen, A., & Siddiqi, K. S. (2014). Carbon and fullerene nanomaterials in plant system. *Journal of Nanobiotechnology, 12*(1), 16.

Irsad, Talreja, N., Chauhan, D., Rodríguez, C. A., Mera, A. C., & Ashfaq, M. (2020). Nanocarriers: An emerging tool for micronutrient delivery in plants. In T. Aftab & K. R. Hakeem (Eds.), *Plant micronutrients: deficiency and toxicity management*. Springer International Publishing.

Jiao, J., Yuan, C., Wang, J., Xia, Z., Xie, L., Chen, F., Li, Z., & Xu, B. (2016). The role of graphene oxide on tobacco root growth and its preliminary mechanism. *Journal of Nanoscience and Nanotechnology, 16*(12), 12,449–12,454.

Khare, P., Talreja, N., Deva, D., Sharma, A., & Verma, N. (2013). Carbon nanofibers containing metal-doped porous carbon beads for environmental remediation applications. *Chemical Engineering Journal, 229*, 72–81.

Kumar, A. (2015). Metabolic engineering in plants. In B. Bahadur, M. Venkat Rajam, L. Sahijram, & K. V. Krishnamurthy (Eds.), *Plant biology and biotechnology: volume II: Plant genomics and biotechnology*. New Delhi.

Kumar, D., & Talreja, N. (2019). Nickel nanoparticles-doped rhodamine grafted carbon nanofibers as colorimetric probe: Naked eye detection and highly sensitive measurement of aqueous Cr3+ and Pb2+. *Korean Journal of Chemical Engineering, 36*(1), 126–135.

Kumar, R., Ashfaq, M., & Verma, N. (2018). Synthesis of novel PVA–starch formulation-supported Cu–Zn nanoparticle carrying carbon nanofibers as a nanofertilizer: controlled release of micronutrients. *Journal of Materials Science, 53*(10), 7150–7164.

Kumar, V., Talreja, N., Deva, D., Sankararamakrishnan, N., Sharma, A., & Verma, N. (2011). Development of bi-metal doped micro- and nano multi-functional polymeric adsorbents for the removal of fluoride and arsenic (V) from wastewater. *Desalination, 282*, 27–38.

Li, F., Sun, C., Li, X., Yu, X., Luo, C., Shen, Y., & Qu, S. (2018). The effect of graphene oxide on adventitious root formation and growth in apple. *Plant Physiology and Biochemistry, 129*, 122–129.

Lin, S., Reppert, J., Hu, Q., Hudson, J. S., Reid, M. L., Ratnikova, T. A., Rao, A. M., Luo, H., & Ke, P. C. (2009). Uptake, translocation, and transmission of carbon nanomaterials in rice plants. *Small, 5*(10), 1128–1132.

Lv, J., Christie, P., & Zhang, S. (2019). Uptake, translocation, and transformation of metal-based nanoparticles in plants: recent advances and methodological challenges. *Environmental Science: Nano, 6*(1), 41–59.

Martínez-Ballesta, M. C., Zapata, L., Chalbi, N., & Carvajal, M. (2016). Multiwalled carbon nanotubes enter broccoli cells enhancing growth and water uptake of plants exposed to salinity. *Journal of Nanobiotechnology, 14*(1), 42.

Mathew, S., Tiwari, D. K., & Tripathi, D. (2021). Interaction of carbon nanotubes with plant system: A review. *Carbon Letters, 31*(2), 167–176.

Mittal, D., Kaur, G., Singh, P., Yadav, K., & Ali, S. A. (2020). Nanoparticle-based sustainable agriculture and food science: Recent advances and future outlook. *Frontiers in Nanotechnology, 2*(10). https://doi.org/10.3389/fnano.2020.579954

Mitter, E. K., Tosi, M., Obregón, D., Dunfield, K. E., & Germida, J. J. (2021). Rethinking crop nutrition in times of modern microbiology: Innovative biofertilizer technologies. *Frontiers in Sustainable Food Systems, 5*(29). https://doi.org/10.3389/fsufs.2021.606815

Mustafa, S., Khan, H. M., Shukla, I., Shujatullah, F., Shahid, M., Ashfaq, M., & Azam, A. (2011). Effect of ZnO nanoparticles on ESBL producing Escherichia coli & Klebsiella spp. *Eastern Journal of Medicine, 16*(4), 253–257.

Nair, P. M., & Chung, I. M. (2015). Study on the correlation between copper oxide nanoparticles induced growth suppression and enhanced lignification in Indian mustard (Brassica juncea L.). *Ecotoxicology and Environmental Safety, 113*, 302–313.

Omar, R. A., Afreen, S., Talreja, N., Chauhan, D., & Ashfaq, M. (2019a). Impact of nanomaterials in plant systems. In R. Prasad (Ed.), *Plant nanobionics: Volume 1, advances in the understanding of nanomaterials research and applications*. Springer International Publishing.

Omar, R. A., Afreen, S., Talreja, N., Chauhan, D., Ashfaq, M., & Srituravanich, W. (2019b). Impact of nanomaterials on the microbial system. In R. Prasad (Ed.), *Microbial nanobionics: Volume 1, state-of-the-art*. Springer International Publishing.

Patra, P., Choudhury, S. R., Mandal, S., Basu, A., Goswami, A., Gogoi, R., Srivastava, C., Kumar, R., & Gopal, M. (2013). Effect sulfur and ZnO nanoparticles on stress physiology and plant (Vigna radiata) nutrition. In P. K. Giri, D. K. Goswami, & A. Perumal (Eds.), *Advanced nanomaterials and nanotechnology* (pp. 301–309). Springer.

Pickens, L. B., Tang, Y., & Chooi, Y.-H. (2011). Metabolic engineering for the production of natural products. *Annual Review of Chemical and Biomolecular Engineering, 2*, 211–236.

Raliya, R., Nair, R., Chavalmane, S., Wang, W.-N., & Biswas, P. (2015). Mechanistic evaluation of translocation and physiological impact of titanium dioxide and zinc oxide nanoparticles on the tomato (Solanum lycopersicum L.) plant. *Metallomics, 7*(12), 1584–1594.

Rico, C. M., Barrios, A. C., Tan, W., Rubenecia, R., Lee, S. C., Varela-Ramirez, A., Peralta-Videa, J. R., & Gardea-Torresdey, J. L. (2015). Physiological and biochemical response of soil-grown barley (Hordeum vulgare L.) to cerium oxide nanoparticles. *Environmental Science and Pollution Research, 22*(14), 10,551–10,558.

Rico, C. M., Hong, J., Morales, M. I., Zhao, L., Barrios, A. C., Zhang, J. Y., Peralta-Videa, J. R., & Gardea-Torresdey, J. L. (2013). Effect of cerium oxide nanoparticles on rice: a study involving the antioxidant defense system and in vivo fluorescence imaging. *Environmental Science & Technology, 47*(11), 5635–5642.

Rico, C. M., Lee, S. C., Rubenecia, R., Mukherjee, A., Hong, J., Peralta-Videa, J. R., & Gardea-Torresdey, J. L. (2014). Cerium oxide nanoparticles impact yield and modify nutritional parameters in wheat (Triticum aestivum L.). *Journal of Agricultural and Food Chemistry, 62* (40), 9669–9675.

Sadak, M. S., & Bakry, B. A. (2020). Zinc-oxide and nano ZnO oxide effects on growth, some biochemical aspects, yield quantity, and quality of flax (Linum uitatissimum L.) in absence and presence of compost under sandy soil. *Bulletin of the National Research Centre, 44*(1), 98.

Saraswat, R., Talreja, N., Deva, D., Sankararamakrishnan, N., Sharma, A., & Verma, N. (2012). Development of novel in situ nickel-doped, phenolic resin-based micro–nano-activated carbon adsorbents for the removal of vitamin B-12. *Chemical Engineering Journal, 197*, 250–260.

Sasidharan, V., Sachan, D., Chauhan, D., Talreja, N., & Ashfaq, M. (2021). Three-dimensional (3D) polymer—metal–carbon framework for efficient removal of chemical and biological contaminants. *Scientific Reports, 11*(1), 7708.

Shang, Y., Hasan, M. K., Ahammed, G. J., Li, M., Yin, H., & Zhou, J. (2019). Applications of nanotechnology in plant growth and crop protection: A review. *Molecules (Basel, Switzerland), 24*(14), 2558.

Shekhawat, G. S., Mahawar, L., Rajput, P., Rajput, V. D., Minkina, T., & Singh, R. K. (2021). Role of engineered carbon nanoparticles (CNPs) in promoting growth and metabolism of Vigna radiata (L.) Wilczek: Insights into the biochemical and physiological responses. *Plants, 10*(7), 1317.

Shen, S., Liu, Y., Wang, F., Yao, G., Xie, L., & Xu, B. (2019). Graphene oxide regulates root development and influences IAA concentration in rice. *Journal of Plant Growth Regulation, 38* (1), 241–248.

Singh, S., Ashfaq, M., Singh, R. K., Joshi, H. C., Srivastava, A., Sharma, A., & Verma, N. (2013). Preparation of surfactant-mediated silver and copper nanoparticles dispersed in hierarchical carbon micro-nanofibers for antibacterial applications. *New Biotechnology, 30*(6), 656–665.

Skiba, E., Michlewska, S., Pietrzak, M., & Wolf, W. M. (2020). Additive interactions of nanoparticulate ZnO with copper, manganese and iron in Pisum sativum L., a hydroponic study. *Scientific Reports, 10*(1), 13,574.

Sultana, A., Talreja, N., Chauhan, D., & Ashfaq, M. (2021). Chapter 4—Nanotechnology-based biofortification: a plant–soil interaction modulator/enhancer. In T. Aftab & K. R. Hakeem (Eds.), *Frontiers in plant-soil interaction*. Academic Press.

Talreja, N., Afreen, S., Ashfaq, M., Chauhan, D., Mera, A. C., Rodríguez, C. A., & Mangalaraja, R. V. (2021a). Bimetal (Fe/Zn) doped BiOI photocatalyst: An effective photodegradation of tetracycline and bacteria. *Chemosphere, 280*, 130,803.

Talreja, N., Ashfaq, M., Chauhan, D., Mera, A. C., & Rodríguez, C. A. (2021b). Strategic doping approach of the Fe–BiOI microstructure: An improved photodegradation efficiency of tetracycline. *ACS Omega, 6*(2), 1575–1583.

Talreja, N., Jung, S., Yen, L. T. H., & Kim, T. (2020). Phenol-formaldehyde-resin-based activated carbons with controlled pore size distribution for high-performance supercapacitors. *Chemical Engineering Journal, 379*, 122,332.

Talreja, N., & Kumar, D. (2018). Engineered nanoparticles' toxicity: Environmental aspects. In C. M. Hussain & A. K. Mishra (Eds.), *Nanotechnology in environmental science*. Wiley.

Talreja, N., Kumar, D., & Verma, N. (2014). Removal of hexavalent chromium from water using Fe-grown carbon nanofibers containing porous carbon microbeads. *Journal of Water Process Engineering, 3*, 34–45.

Talreja, N., Verma, N., & Kumar, D. (2016). Carbon bead-supported ethylene diamine-functionalized carbon nanofibers: An efficient adsorbent for salicylic acid. *CLEAN—Soil, Air, Water, 44*(11), 1461–1470.

Tilman, D. (1999). Global environmental impacts of agricultural expansion: The need for sustainable and efficient practices. *Proceedings of the National Academy of Sciences of the United States of America, 96*(11), 5995–6000.

Tripathi, K. M., Tyagi, A., Ashfaq, M., & Gupta, R. K. (2016). Temperature dependent, shape variant synthesis of photoluminescent and biocompatible carbon nanostructures from almond husk for applications in dye removal. *RSC Advances, 6*(35), 29545–29553.

Yadav, V. G., De Mey, M., Lim, C. G., Ajikumar, P. K., & Stephanopoulos, G. (2012). The future of metabolic engineering and synthetic biology: Towards a systematic practice. *Metabolic Engineering, 14*(3), 233–241.

Yang, G., Yuan, H., Ji, H., Liu, H., Zhang, Y., Wang, G., Chen, L., & Guo, Z. (2021). Effect of ZnO nanoparticles on the productivity, Zn biofortification, and nutritional quality of rice in a life cycle study. *Plant Physiology and Biochemistry, 163*, 87–94.

Yatim, N. M., Shaaban, A., Dimin, M. F., Yusof, F., & Razak, J. A. (2018). Effect of functionalised and non-functionalised carbon nanotubes-urea fertilizer on the growth of paddy. *Tropical life Sciences Research, 29*(1), 17–35.

Yousaf, M., Li, J., Lu, J., Ren, T., Cong, R., Fahad, S., & Li, X. (2017). Effects of fertilization on crop production and nutrient-supplying capacity under rice-oilseed rape rotation system. *Scientific Reports, 7*(1), 1270.

Yuan, L., & Grotewold, E. (2015). Metabolic engineering to enhance the value of plants as green factories. *Metabolic Engineering, 27*, 83–91.

Yusefi-Tanha, E., Fallah, S., Rostamnejadi, A., & Pokhrel, L. R. (2020). Zinc oxide nanoparticles (ZnONPs) as a novel nanofertilizer: Influence on seed yield and antioxidant defense system in soil grown soybean (Glycine max cv. Kowsar). *Science of the Total Environment, 738*, 140,240.

Zhang, P., Zhang, R., Fang, X., Song, T., Cai, X., Liu, H., & Du, S. (2016). Toxic effects of graphene on the growth and nutritional levels of Wheat (Triticum aestivum L.): Short- and long-term exposure studies. *Journal of Hazardous Materials, 317*, 543–551.

Zhang, X., Cao, H., & Zhao, J. (2021). Graphene oxide exhibited positive effects on the growth of Aloe vera L. *Physiology and Molecular Biology of Plants, 27*(4), 815–824.

Zhu, Q., Wang, B., Tan, J., Liu, T., Li, L., & Liu, Y.-G. (2020). Plant synthetic metabolic engineering for enhancing crop nutritional quality. *Plant Communications, 1*(1), 100,017.

Chapter 18
Plant Metabolic Engineering for a Futuristic Economy

Ishani Chakrabartty

Abstract Metabolic engineering rearranges or redirects more than one enzymatic reaction to produce new variety of compounds for the overall improvement of the metabolic reactions occurring in an organism. Introduction and utilization of such processes in plants have witnessed tremendous technological advancements in plant molecular biology, breeding, transformation, protein targeting, etc. Starting from photosynthesis to pheromones, plants are involved in the synthesis of many beneficial metabolites. Dissection of different plant pathways for engineering of plant metabolism has yielded several successful reactions and has even led to a number of unanticipated results. It is indeed an exciting endeavor to decipher and alter the metabolic network of plants that involve more than 10,000 unique compounds! As such, the use of sustainable organisms like plants, may help to meet the challenges of food security and health of the growing human population. Successful engineering of plants, together with the use of systems and synthetic biology, results in the high yield of primary food sources and biofuel feedstocks, pharmaceuticals, and platform chemicals. Further, engineering of lesser-known biosynthesis pathways in plants and diversified process developments for production of platform chemicals are essential to overcome the hurdles for sustainable production of value-added biomolecules from plants. The present chapter focuses on the use of plant metabolic engineering as an effort to enhance the production of different biomolecules—starting from nutrients to pharmaceuticals—for a sustainable living. Such an attempt may hold the key for a paradigm shift towards a biobased economy.

Keywords Biomolecules · Economy · Metabolic engineering · Pharmaceuticals · Plants · Sustainable

I. Chakrabartty (✉)
Department of Applied Biology, School of Biological Sciences, University of Science and Technology Megahalaya (USTM), Ri-Bhoi, Baridua, Meghalaya, India

T. Aftab, K. R. Hakeem (eds.), *Metabolic Engineering in Plants*,
https://doi.org/10.1007/978-981-16-7262-0_18

Introduction

From the time a living organism is born or comes into existence, different biochemical reactions operate in its body. Such reactions are crucial for an organism to survive and maintain itself in an equilibrium state. The myriad of reactions includes important processes like generation of energy and production of fundamental building blocks that are required for structural and functional organization of an organism, besides having other specialized functions. Metabolism—the process of generating important chemicals and biomolecules, many of which are required by mankind for various applications, is a vital and crucial phenomenon of a living body. Like all living organisms, plants are also involved in different metabolic reactions. A number of physical and chemical processes work together as a network to synthesize an array of substances and transform those into energy that is used for the functioning of the entire plant machinery/system (Bowsher, 2019). By the process of metabolism, plants synthesize and produce variety of important chemical compounds. These so-called value-added chemicals include diverse range of bioactive secondary metabolites such as important drugs and drug-like molecules (e.g., artemisinin), chemicals required as the raw material for the synthesis of other molecules such as lactic acid, chemicals imparting flavor, color, and fragrance to food and plant material such as terpenes, biofuels and associated chemicals such as ethanol and butanol, and others (Kulkarni, 2016). From time immemorial, nature has been exploited as the source of medicines. Natural products, including plants, animals, bacteria, and minerals, have been used by man for treatment of diseases (Patwardhan et al., 2004; Lahlou, 2007). In 1985, the World Health Organization (WHO) reported that 65% of the world population relies on plant-derived medicines for their healthcare and treatment of ailments, while the remaining population relies on chemical or synthesized medicines inspired from plant drugs. A survey in the countries hosting WHO-Traditional Medicinal Centers reports that out of 122 compounds identified, 80% were used for ethnomedicinal purposes and derived only from 94 species of plants alone (Farnsworth et al., 1985). Plants and plant parts are bestowed with a wealth of chemicals like flavonoids, alkaloids, saponins, tannins, etc. (Ghosh & Rangan, 2013; Singh et al., 2016; Basak et al., 2017; Chakrabartty et al., 2019); these secondary metabolites are first synthesized and then produced by the plants to adapt themselves to the ever-changing environment or as a mechanism for defense against insects, pests, hungry animals, etc. (Dewick, 2002; Colegate & Molyneux, 2008). Secondary metabolites, such as terpene derivatives, obtained from plant sources, have a wide range of biological activities like effects against pathogenic bacteria, diabetes, soreness, rheumatism, etc. (Landau et al., 1994; Fortuna et al., 2001; Cabral et al., 2008b; Steinbach et al., 2008); the sesquiterpene, cnicin, can induce death of myeloma cells even in presence of cytokines (Jöhrer et al., 2012).

The traditional way of utilization of such chemicals is to cultivate the host organisms producing these chemicals followed by harvesting the desired biochemical. However, this method fails to achieve the required quantities of useful chemicals in the cells, thus demanding cultivation of the organisms on a very large scale. To address this problem, organic chemists devised strategies to chemically synthesize such value-added chemicals from the petroleum crude material. However, rapidly depleting petroleum resources and generation of harmful by-products have forced scientists to find out other alternative ways to generate these value-added biochemicals. Since living organisms already produce these chemicals, copying their biosynthesis mechanisms for their semi-natural production appears to be an excellent alternative to their isolation from natural resources and the chemical synthesis. Such an approach for the production of desired chemicals in living organisms is called metabolic engineering (Kulkarni, 2016). However, this phenomenon is not something very new in science; organisms have been altering metabolic pathways to improve their cell properties and enhancing their survival since millions of years—a strategy that is very commonly adopted by microorganisms, especially extremophiles (Koffas et al., 1999). Humans have been involved in the intentional manipulation of different metabolic pathways of organisms to improve the properties and productivity of microorganisms for their benefit using different tools and techniques of gene manipulation.

In the mid-1980s, technological advancements in gene manipulation led to the dissection of several metabolic pathways in plants and, thus, engineer the plant metabolism. Approximately 100,000 unique compounds are produced in the entire plant kingdom, and elucidating their metabolic network is a challenging, yet exciting endeavor (DellaPenna, 2001). The metabolites produced by plants have occupied some major sectors of the market, viz., food, fiber, animal feed, fuel, and pharmaceuticals. With the growing population, the demand for energy, medicinal chemicals, and other resources has been increasing at an exponential rate. Plants, which have taken the center stage to meet human demands, can be used efficiently as sustainable sources—metabolic engineering of different biosynthetic plant pathways may fulfill the growing human needs, by utilizing microorganisms as suitable hosts (Yoon et al., 2013).

The present chapter focuses on the efforts of using plant metabolic engineering to enhance production of different important chemicals, ranging from nutrients to industrial applications. Different genetic modifications are required for the redirection of carbon flux to desired biomass component production. It describes the alteration of these metabolic pathways that can help to achieve high yield of natural medicinal terpenes not only from a plant itself but also from gene expression in plants and microbial hosts. Finally, the challenges and limitations associated with the shift to a biobased platform are discussed, along with its impact on the global economy.

Strategies and Tools to Engineer Plant Pathways

A metabolic pathway is defined as a sequence/series or pathway of biochemical reactions that generates a specific group of metabolites as products from a group of inputs. Plants may be made to produce a variety of desired products/substances by engineering or playing with their metabolic pathways in any one of the following ways (Alper et al., 2018):

1. Enhancing the inherent metabolic pathways
2. Introducing new and modified pathways by de novo pathway designing or introducing similar pathways from other organisms

In addition, detailed knowledge and clear description of the cellular processes involved are an utmost requirement. For these, a number of computational tools, programming, and modeling are required in order to understand the plant host, discover, design, and remodel its metabolic pathways, combined with strategies of synthetic biology tools to understand and increase the capabilities and expedite genetic interventions (Fu et al., 2018). Recently, two innovative methods to engineer plant metabolic pathways have been developed by researchers at Northwestern University which are fast and extremely efficient. This approach utilizes cell-free protein synthesis and self-assembled monolayer desorption ionization or SAMDI-combined mass spectrometry to help metabolic engineers understand biochemical pathways and generate valuable and complex biomolecules (O'Kane et al., 2019; Northwestern University, 2019). Using genome-scale models for the understanding of signal transduction, coordinated metabolism, regulation, and its networking on the gene as well as protein level and significant interactions among proteins have key roles in developing strategies for plant metabolic engineering; basic tools of genetic engineering like cloning, expression, and gene editing (deletion and insertion) have led to the significant. However, these basic techniques and strategies need to be advanced and upgraded at a fairly quick pace to understand the cellular properties on a molecular level, gene-protein relationships, cell dynamics, enzyme kinetics etc., particularly in response to different stress and disturbances in the cell (Dasgupta et al., 2020) in order to rearrange or manipulate them according to human requirements. Figure 18.1 shows some of the major strategies applied in plant metabolic engineering from application point of view.

Inverse or reverse engineering is a new concept that brings in a fresh perspective to the concept of metabolic engineering in plants. In a lot of research areas, the metabolic pathways are difficult to comprehend, and hence, the specific target genes are not recognized. Inverse engineering, coupled with metabolic control analysis (MCA) can help in designing efficient and strategic models in plants, particularly in the field of plant on natural product-based drug discovery (Fig. 18.2). Such techniques may help to identify specific DNA/RNA sequences or exact genes for the production of a desired secondary metabolite and also decipher the pattern of its production in the plant by tracking the generation pathway (Harvey et al., 2015; Dasgupta et al., 2020). This strategy simply involves genetic analysis of the selected

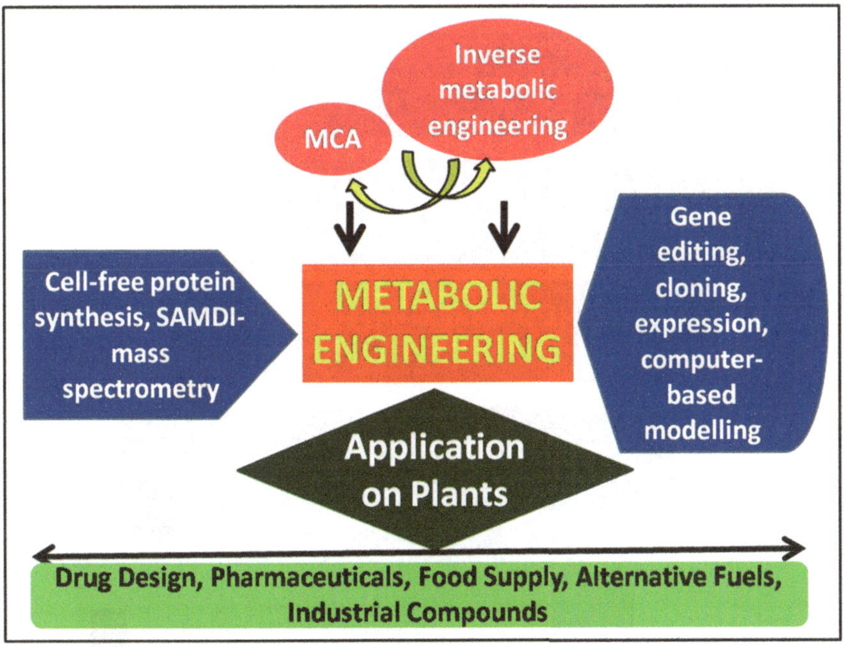

Fig. 18.1 Major strategies applied for plant metabolic engineering

particular phenotype, which reveals the underlying genetic mechanism responsible for the desired cell properties. A library of desired phenotypes, with its associated genotypes is created—more diverse the library, more successful the identification of the required genotype and the cell pathway (Skretas & Kolisis, 2012).

Metabolic Engineering for Production of Terpenoids

Pharmaceutical Terpenoids

Nature has been an important source of drugs since time immemorial. Numerous organisms, harbored in the heart of nature, have contributed immensely to the world of modern medicine for centuries together. In 1763, Reverend Edward Stone of Oxfordshire gave the "scientific use" of willow to treat feverish patients, followed by the synthesis of acetylsalicylic acid (ASA) by Felix Hoffman in 1897, which began to be marketed under the brand name of ***aspirin*** (Chamberlain, 2015). The scientific community has come a long way since then—in 2015, Tu Youyou of the Chinese National Academy of Sciences was awarded the Nobel Prize in Physiology or Medicine for her pioneering work of discovering artemisinin, a novel therapeutic compound against malaria. She had combed through age-old documents of the

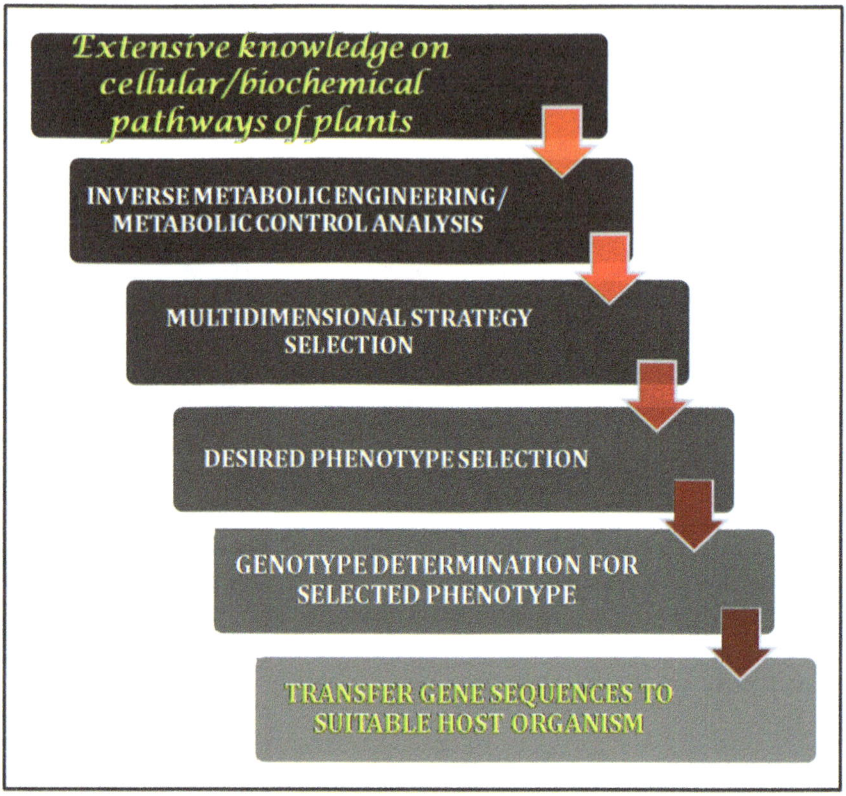

Fig. 18.2 Schematic representation on inverse metabolic engineering and MCA process

Chinese traditional medicine for her groundbreaking work and thereby further solidifying the importance of traditional medicine. It is, thus, a proof that the longstanding relationship of man with nature has again led him to explore the field to natural medicines. Production of plant-based pharmaceuticals is also a matter of worldwide interest. Plant "molecular farming" (PMF) is the practice of producing human therapeutic proteins in plants; ZMapp, a drug to combat Ebola virus, was produced in tobacco leaves by Mapp Biopharmaceutical Inc. of USA (Arntzen, 2015; Yao et al., 2015).

Major classes of secondary metabolites include flavonoids, terpenes, and terpenoids, saponins, etc. that are obtained from a diverse range of medicinal plants. However, the process of isolation of bioactive compounds is extremely cumbersome and leads to extraction of minimal amounts of crude compounds; loss of compounds also occurs during downstream processing and purification of these bioactive compounds. Hence, their commercialization is limited to a large extent (Misawa, 2011). Major efforts have been made in this direction, including costly techniques like total chemical synthesis and high-throughput screening. Also, non-resolution of the

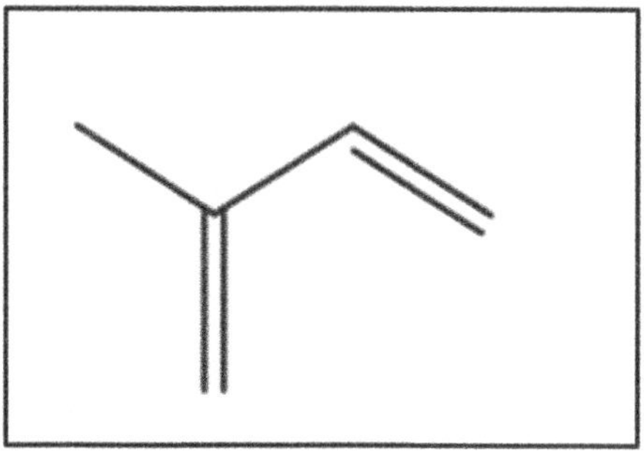

Fig. 18.3 Isoprene unit, backbone all major and economically important terpene derivatives

complex biosynthetic pathways of these bioactive compounds is a major drawback. Consequently, researchers have resorted to metabolic engineering for the production of pharmaceutical metabolites from plants. Terpenes are the largest class of secondary metabolites present in plants that possess defined physiological functions. The basic chemical unit of any plant secondary metabolite is a hydrocarbon chain known as isoprene unit (Fig. 18.3).

Terpene and terpene derivatives, obtained from plant sources, have a wide range of biological activities like effects against pathogenic bacteria, diabetes, soreness, rheumatism, etc. (Daniewski et al., 1989; Landau et al., 1994; Fortuna et al., 2001; Cabral et al., 2008a; Steinbach et al., 2008). Sawamura et al. have reported the antiviral potential of terpenoids isolated from *Alpinia officinarum* (Sawamura et al., 2010) against influenza virus. Diterpenes constitute a large family of secondary metabolites having four isoprene units (C20). They are synthesized via geranylgeranyl pyrophosphate (GGPP) through mevalonate pathway. Protonation of GGPP yields coplyl PP owing to a cyclization process and subsequently produces labdadienyl PP by an alternative process (Dewick, 2002). The most well-known member is taxane or taxol, commercially available under names such as paclitaxel, docetaxel, and cabazitaxel and used for chemotherapy. Other standard diterpenes include clerodane, kaurane, gibberellane, and so on (Morita & Itokawa, 1988; Gonzalez-Burgos & Gomez-Serranillos, 2012). Similarly, other groups of terpenes include monoterpenoids (C10), sesquiterpenoids (C15), triterpenoids (C30), and tetraterpenoids (C40). Artemisinin, the breakthrough anti-malarial drug, is a sesquiterpene lactone, and its biosynthetic pathway has been completely resolved—this can now be applied on an industrial scale, tissue culture application, etc. to fulfill its increased demand (Zhang et al., 2008; Teoh et al., 2009). Major pathways for synthesis of terpenes include the methylerythritol phosphate (MEP) pathway, which is localized in the plastids and the classical mevalonic acid (MVA) or melvonate pathway, comprising of six enzymatic steps, localized to the cytosol, endoplasmic reticulum, and peroxisomes (Banerjee & Sharkey, 2014; Gutensohn et al., 2014; Lu et al., 2016). Following the substrates, enzymes, inhibitors, and all the involved reagents in these pathways can make it theoretically possible to produce pharmaceutically important terpenoids in appropriate host organisms (Fig. 18.4).

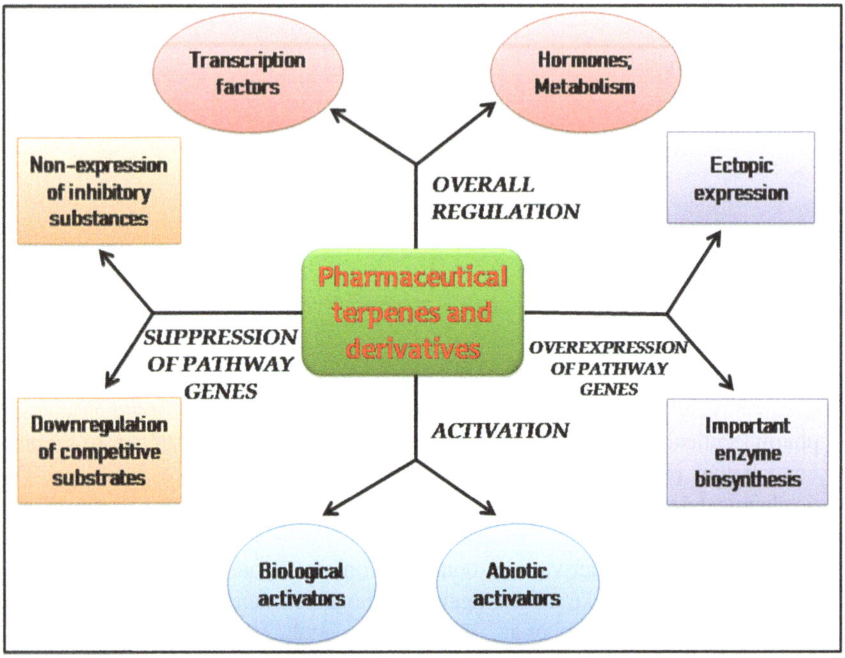

Fig. 18.4 Metabolic engineering to produce pharmaceutically important terpenes and terpenoids

Industrially Important Terpenes

Terpenes, along with fatty acids, are the major constituents of plant essential oils, which are economically important as they have wide applications as pharmaceutics, cosmetics, flavoring agents, fragrances, etc. The prices of these plant-based metabolites are highly prone to fluctuations due to geographical locations, climatic conditions, seasons, and political causes. Also, they are cultivated in a limited quantity, and the essential oils are further derived in low amounts as the entire process is highly labor intensive (Sangwan et al., 2001; Daviet & Schalk, 2010; Kanjilal et al., 2010). Metabolic engineering of illustrated MVA and MEP pathways in suitable hosts like microorganisms and other plants may lead to the production of valuable essential oils in a cost-effective manner. The production of these complex biomolecules in microbes like *Escherichia coli*, *Saccharomyces cerevisiae*, and others are very advantageous in terms of bioresource consumption, scaling up, and most importantly sustainability; large-scale fermentors can be used to produce terpenes on an industrial scale using low-cost substrate like molasses, sugar, etc. Genetic alterations have been made and successfully expressed in yeast cells to produce sesquiterpenes; patchoulol, an industrially important fragrant terpene molecule (also used as a flavoring agent), has been engineered in recombinant yeast strains (Takahashi et al., 2007; Asadollahi et al., 2008). The MVA pathway for terpene biosynthesis is considered to be most promising in *E. coli*—something that has been utilized for microbial synthesis of artemisinin (Tsuruta et al., 2009). Engineering or "tweeking" of the catalytic properties of bacterial P450 enzymes helps to achieve high turnover of terpene production in bacterial systems, in comparison to plant systems (Aharoni et al., 2006; Dietrich et al., 2009; Daviet & Schalk, 2010).

Metabolic engineering of pathways concerned with the biosynthesis of different terpene compounds (with desired traits) has been achieved in many plants. This process is much complex in plants than in microorganisms as plant has cellular compartmentalization in terms of synthesis. In spite of this, successful production of several monoterpenes, diterpenes, and sesquiterpenes has been achieved in many plants like tobacco; the key player in all these syntheses pathways has been the enzyme terpene synthase (Hohn & Ohlrogge, 1991; Wallaart et al., 2001; Aharoni et al., 2003). The compartmentalization of the terpene biosynthesis in plants, which is a challenging aspect, was overcome by changing the subcellular localization of the terpene synthases—this resulted in a much increased production of terpenes (Wu et al., 2006; Daviet & Schalk, 2010; Blanch et al., 2015).

Limitations and Challenges

Plants are one of the most interesting platforms for metabolic engineering—a scientific technique that has been adopted for the sustainable generation of value-added products; secondary metabolites or phytochemicals constitute a major chunk of such products. Hence, engineering of the synthesis pathways of these valuable metabolites may lead to their enhanced production. However, the path is not as easy as it seems. Even if all the metabolic pathways involved in phytochemical production is full deciphered and known, it may not account to its successful production by metabolic engineering (Fig. 18.5). This may be attributed to the fact that a comprehensive knowledge of the entire metabolism is lacking; even the interconnections between different hormonal, enzymatic, metabolic, kinetic and pathways are not well understood. Thus, the lack of metabolic crosstalk as well as substrate availability in the desired tissues is some major constraints in the metabolic engineering strategies

It has been suggested by a lot of investigators that the engineered metabolic pathways need to organized into complexes so that there is efficient utilization of the available substrates and limited accumulation of the intermediates—this would help in the linking of biosynthetic pathways towards generation of same product in an effective manner, than those operating independently (Wu et al., 2006; Wu & Chappell, 2008); linking of metabolic pathways have led to some success in this field. In addition, expression of plant metabolic pathways in microbial hosts may lead to distortions in the desired protein–protein interactions as the plant enzymes may not undergo the appropriate posttranslational modifications in the prokaryotes. A possible solution to this can be the expression of metabolic pathways, involved in the synthesis of valuable phytochemicals, in other plant systems that can be grown in controlled environment like algae or pteridophytes—such systems may allow suitable environment for gene recombination as well as appropriate pathway

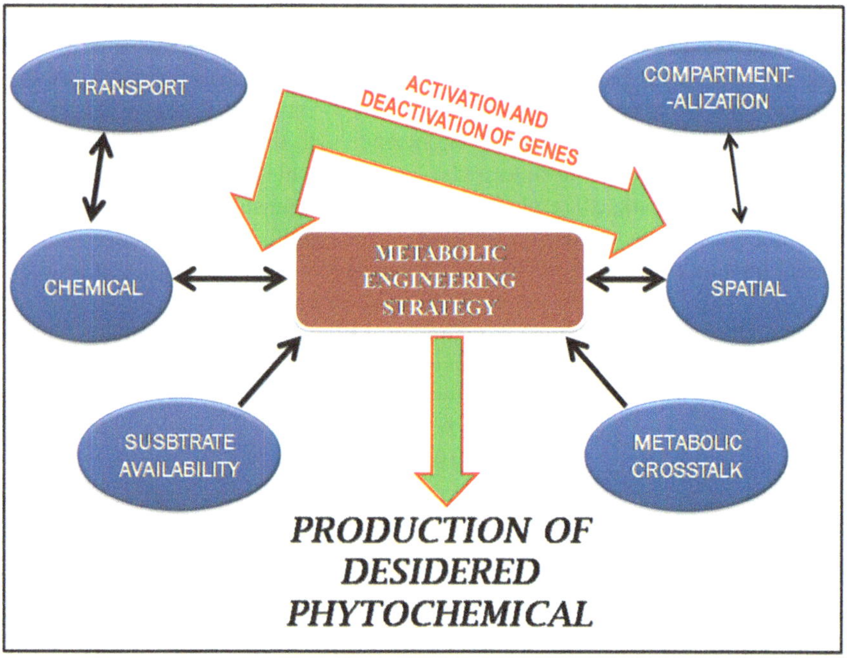

Fig. 18.5 Schematic representation of the challenges of metabolic engineering in plants

compartmentalization, thus alleviating the production of valuable plant compounds (Lau et al., 2014; Lynch et al., 2021; Renault et al., 2014).

Bioinformatics tools not only involve computational techniques but also advanced databases that can store and retrieve huge amount of genetic information. Hence, development of this sector is inevitable. In addition, chemicals obtained from synthetic pathways tend to have higher oxygen to carbon ratio in the biological molecules in comparison to natural synthesis, and extensive knowledge of such platforms is lacking. Most importantly, even when all the players and pathways involved in the biosynthesis of phytochemicals are well known, fact remains that the desired results of plant metabolic engineering are not obtained because not all results are predictable in biology (Dudareva & DellaPenna, 2013). As such, predicted high yields of biological chemicals are not achieved after metabolic engineering. Thus, further research is required for the success plant metabolic engineering as the challenges and limitations continue to exist.

Conclusion: A Shift Towards a Biobased Economy

Over the last 2 decades, plant metabolic engineering has progressed at a tremendous pace, owing to the developments in synthetic biology, omics platforms, MCA, etc. In silico plant models have helped researchers understand complex biosynthetic pathways to a large extent. Plants have always been attractive platforms for scientists, and terpenes are one of the most commercially important plant-derived phytochemical; gene manipulations and associated technology have aided in the in-depth understanding of terpene biosynthetic pathways. Metabolic engineering has paved the way for extraction of valuable essential oils on an industrial scale with limited optimization and investment—something that was not possible with the traditional extraction procedures. Engineering of terpene biosynthesis in microbial hosts, both for pharmaceutical and essential oil production, is largely advantageous due to easier gene manipulation, small genome size, flexibility for large scale production, etc. Thus, engineering of terpene syntheses in microbes provides an effective route for the production of economically important phytochemicals.

Understanding the fundamentals of plant metabolic engineering is extremely important for next-generation researchers. Engineering of pathways that are dedicated to the production of natural products may be able to generate sustainable platforms for all renewable resources in plants—production of medicines from plant sources with high yield and low cost will be able to boost the health sector tremendously, particularly in underdeveloped and developing countries. The technique of metabolic engineering can not only be used for secondary metabolite but can also be extended for the generation of primary production for the sustainable generation of major resources like food and fuel (Ralley et al., 2004; Kinney, 2006; Davies & Espley, 2013).

A lot has been achieved since the previous decade in plant metabolic engineering, especially after the developments of tools like SAMDI, yet challenges and limitations continue to exist. Shifting to more different forms of biobased platform phytochemicals, use of a single metabolic pathway for the generation of many phytochemicals and primary metabolites, and extensive knowledge and information about metabolic regulation and expression pattern of genes and development of many advanced tools and strategies for genetic modifications are the keys to overcoming the limitations of the proposed biobased platform chemicals from existing traditional biomass. Such a biobased platform has tremendous boost on the economic upliftment of a nation and help promote sustainable utilization of the existing bioresources.

References

Aharoni, A., Giri, A. P., Deuerlein, S., et al. (2003). Terpenoid metabolism in wild-type and transgenic *Arabidopsis* plants. *Plant Cell, 15*, 2866–2884. https://doi.org/10.1105/tpc.016253

Aharoni, A., Jongsma, M. A., Kim, T. Y., et al. (2006). Metabolic engineering of terpenoid biosynthesis in plants. *Phytochemistry Reviews, 5*, 49–58. https://doi.org/10.1007/s11101-005-3747-3

Alper, H. S., Paul, D., Meek, B. R., & Liddell, F. A., Jr. (2018). Metabolic pathway engineering. *Synthetic and Systems Biotechnology, 3*, 1–2. https://doi.org/10.1016/j.synbio.2018.01.002

Arntzen, C. (2015). Plant-made pharmaceuticals: From "Edible Vaccines" to Ebola therapeutics. *Plant Biotechnology Journal, 13*, 1013–1016. https://doi.org/10.1111/pbi.12460

Asadollahi, M. A., Maury, J., Møller, K., et al. (2008). Production of plant sesquiterpenes in *Saccharomyces cerevisiae*: Effect of ERG9 repression on sesquiterpene biosynthesis. *Biotechnology and Bioengineering, 99*, 666–677. https://doi.org/10.1002/bit

Banerjee, A., & Sharkey, T. (2014). Methyl erythritol 4-phosphate (MEP) pathway metabolic regulation. *Natural Product Reports, 00*, 1–13. https://doi.org/10.1039/x0xx00000x

Basak, S., Kesari, V., Ramesh, A. M., et al. (2017). Assessment of genetic variation among nineteen turmeric cultivars of Northeast India: Nuclear DNA content and molecular marker approach. *Acta Physiologiae Plantarum, 39*, 45. https://doi.org/10.1007/s11738-016-2341-1

Blanch, J. S., Peñuelas, J., Llusià, J., et al. (2015). Differences in photosynthesis and terpene content in leaves and roots of wild-type and transgenic *Arabidopsis thaliana* plants. *Russian Journal of Plant Physiology, 62*, 823–829. https://doi.org/10.1134/S1021443715060035

Bowsher, C. (2019). *Plant metabolism* (p. 1). McGraw Hill.

Cabral, C., Goncalves, M. J., Cavaleiro, C., et al. (2008a). *Vitex ferruginea* Schumach. Et. Thonn. subsp. *amboniensis* (Gurke) Verdc.: Glandular trichomes micromorphology, composition and antifungal activity of the essential oils. *Journal of Essential Oil Research, 20*, 86–90. https://doi.org/10.1080/10412905.2008.9699430

Cabral, M. E. S., Fortuna, A. M., De Riscala, E. C., et al. (2008b). Allelopathic activity of *Centaurea diffusa* and *Centaurea tweediei*: Effects of cnicin and onopordopicrin on seed germination, phytopathogenic bacteria and soil. *Allelopathy Journal, 21*, 183–190.

Chakrabartty, I., Vijayasekhar, A., & Rangan, L. (2019). Therapeutic potential of labdane diterpene isolated from *Alpinia nigra*: Detailed hemato-compatibility and antimicrobial studies. *Natural Product Research*. https://doi.org/10.1080/14786419.2019.1610756

Chamberlain, R. (2015). From bark to bacteria: The natural sources of modern medicine. *The Bichemist*, 42–44.

Colegate, S. M., & Molyneux, R. J. (2008). *Bioactive natural products. Detection, isolation, and structural determination* (2nd ed.). CRC Press.

Daniewski, W., Gumulka, M., Drozdz, B., et al. (1989). Sesquiterpene lactones. XXXVIII. Constituents of *Picris echoides* L. and their antifeedant activity. *Acta Societatis Botanicorum Poloniae, 58*, 351–354.

Dasgupta, A., Chowdhury, N., & De, R. K. (2020). Metabolic pathway engineering: Perspectives and applications. *Computer Methods and Programs in Biomedicine, 192*, 105,436. https://doi.org/10.1016/j.cmpb.2020.105436

Davies, K., & Espley, R. (2013). Opportunities and challenges for metabolic engineering of secondary metabolite pathways for improved human health characters in fruit and vegetable crops. *New Zealand Journal of Crop and Horticultural Science, 41*, 154–177. https://doi.org/10.1080/01140671.2013.793730

Daviet, L., & Schalk, M. (2010). Biotechnology in plant essential oil production: Progress and perspective in metabolic engineering of the terpene pathway. *Flavour and Fragrance Journal, 25*, 123–127. https://doi.org/10.1002/ffj

DellaPenna, D. (2001). Plant metabolic engineering. *Plant Physiology, 125*, 160–163. https://doi.org/10.1104/pp.125.1.160

Dewick, P. M. (2002). *Medicinal natural products: A biosynthetic approach* (2nd ed.). John Wiley & Sons Ltd.

Dietrich, J. A., Yoshikuni, Y., Fisher, K. J., et al. (2009). A novel semi-biosynthetic route for artemisinin production using engineered substrate-promiscuous P450BM3. *ACS Chemical Biology, 4*, 29.

Dudareva, N., & DellaPenna, D. (2013). Plant metabolic engineering: Future prospects and challenges. *Current Opinion in Biotechnology, 24*, 226–228. https://doi.org/10.1016/j.copbio.2013.02.002

Farnsworth, N. R., Akerele, O., Bingel, A. S., et al. (1985). Medicinal plants in therapy. *Bulletin of the World Health Organization, 63*, 965–981. https://doi.org/10.1016/0378-8741(87)90016-X

Fortuna, A. M., de Riscala, E. C., Catalan, C. A. N., et al. (2001). Sesquiterpene lactones from *Centaurea tweediei*. *Biochemical Systematics and Ecology, 29*, 967–971. https://doi.org/10.1016/S0305-1978(01)00042-4

Fu, R., Martin, C., & Zhang, Y. (2018). Next-generation plant metabolic engineering, inspired by an ancient Chinese irrigation system. *Molecular Plant, 11*, 47–57. https://doi.org/10.1016/j.molp.2017.09.002

Ghosh, S., & Rangan, L. (2013). *Alpinia*: The gold mine of future therapeutics. *3 Biotech, 3*, 173–185. https://doi.org/10.1007/s13205-012-0089-x

Gonzalez-Burgos, E., & Gomez-Serranillos, M. (2012). Terpene compounds in nature: A review of their potential antioxidant activity. *Current Medicinal Chemistry, 19*, 5319–5341.

Gutensohn, M., Nguyen, T. T. H., McMahon, R. D., et al. (2014). Metabolic engineering of monoterpene biosynthesis in tomato fruits *via* introduction of the non-canonical substrate neryl diphosphate. *Metabolic Engineering, 24*, 107–116. https://doi.org/10.1016/j.ymben.2014.05.008

Harvey, A. L., Edrada-Ebel, R., & Quinn, R. J. (2015). The re-emergence of natural products for drug discovery in the genomics era. *Nature Reviews. Drug Discovery, 14*, 111–129. https://doi.org/10.1038/nrd4510

Hohn, T. M., & Ohlrogge, J. B. (1991). Expression of a fungal sesquiterpene cyclase gene in transgenic tobacco. *Plant Physiology, 97*, 460–462. https://doi.org/10.1104/pp.97.1.460

Jöhrer, K., Obkircher, M., Neureiter, D., et al. (2012). Antimyeloma activity of the sesquiterpene lactone cnicin: Impact on Pim-2 kinase as a novel therapeutic target. *Journal of Molecular Medicine, 90*, 681–693. https://doi.org/10.1007/s00109-011-0848-x

Kanjilal, P. B., Kotoky, R., & Couladis, M. (2010). Essential oil composition of leaf and rhizome oil of *Alpinia nigra* (Gaertner) B.L.Burtt. from northeast India. *Journal of Essential Oil Research, 22*, 358–359. https://doi.org/10.1080/10412905.2010.9700345

Kinney, A. J. (2006). Metabolic engineering in plants for human health and nutrition. *Current Opinion in Biotechnology, 17*, 130–138. https://doi.org/10.1016/j.copbio.2006.02.006

Koffas, M., Roberge, C., Lee, K., & Stephanopoulos, G. (1999). Metabolic engineering. *Annual Review of Biomedical Engineering, 01*, 535–537. https://doi.org/10.1201/b16050-18

Kulkarni, R. (2016). Metabolic engineering: Biological art of producing useful chemicals. *Resonance, 21*, 233–237. https://doi.org/10.1007/s12045-016-0318-4

Lahlou, M. (2007). Screening of natural products for drug discovery. *Expert Opinion on Drug Discovery, 2*, 697–705. https://doi.org/10.1517/17460441.2.5.697

Landau, I., Müller-Schärer, H., & Ward, P. I. (1994). Influence of cnicin, a sesquiterpene lactone of *Centaurea maculosa* (Asteraceae), on specialist and generalist insect herbivores. *Journal of Chemical Ecology, 20*, 929–942. https://doi.org/10.1007/BF02059588

Lau, W., Fischbach, M. A., Osbourn, A., & Sattely, E. S. (2014). Key applications of plant metabolic engineering. *PLoS Biology, 12*, 1–5. https://doi.org/10.1371/journal.pbio.1001879

Lu, X., Tang, K., & Li, P. (2016). Plant metabolic engineering strategies for the production of pharmaceutical terpenoids. *Frontiers in Plant Science, 7*, 1–11. https://doi.org/10.3389/fpls.2016.01647

Lynch, J. H., Huang, X.-Q., & Dudareva, N. (2021). Silent constraints: The hidden challenges faced in plant metabolic engineering. *Current Opinion in Biotechnology, 69*, 112–117.

Misawa, N. (2011). Pathway engineering for functional isoprenoids. *Current Opinion in Biotechnology, 22*, 627–633. https://doi.org/10.1016/j.copbio.2011.01.002

Morita, H., & Itokawa, H. (1988). Cytotoxic and antifungal diterpenes from seeds of *Alpinia galanga*. *Planta Medica, 54*, 117–120.

Northwestern University. (2019). New method for engineering metabolic pathways: Two approaches provide faster way to create complex molecules. In: *ScienceDaily*

O'Kane, P. T., Dudley, Q. M., McMillan, A. K., et al. (2019). High-throughput mapping of CoA metabolites by SAMDI-MS to optimize the cell-free biosynthesis of HMG-CoA. *Science Advances, 5*. https://doi.org/10.1126/sciadv.aaw9180

Patwardhan, B., Vaidya, A. D. B., & Chorghade, M. (2004). Ayurveda and natural products drug discovery. *Current Science, 86*, 789–799.

Ralley, L., Enfissi, E. M. A., Misawa, N., et al. (2004). Metabolic engineering of ketocarotenoid formation in higher plants. *The Plant Journal, 39*, 477–486. https://doi.org/10.1111/j.1365-313X.2004.02151.x

Renault, H., Bassard, J. E., Hamberger, B., & Werck-Reichhart, D. (2014). Cytochrome P450-mediated metabolic engineering: Current progress and future challenges. *Current Opinion in Plant Biology, 19*, 27–34. https://doi.org/10.1016/j.pbi.2014.03.004

Sangwan, N. S., Farooqi, A. H. A., Shabih, F., & Sangwan, R. S. (2001). Regulation of essential oil production in plants. *Plant Growth Regulation, 34*, 3–21. https://doi.org/10.1023/A:1013386921596

Sawamura, R., Shimizu, T., Sun, Y., et al. (2010). *In vitro* and *in vivo* anti-influenza virus activity of diarylheptanoids isolated from *Alpinia officinarum*. *Antiviral Chemistry & Chemotherapy, 21*, 33–41. https://doi.org/10.3851/IMP1676

Singh, A., Jahan, I., Sharma, M., et al. (2016). Structural characterization, *in silico* studies and *in vitro* antibacterial evaluation of a furanoflavonoid from Karanj. *Planta Medica Letters, 3*, e91–e95. https://doi.org/10.1055/s-0042-105159

Skretas, G., & Kolisis, F. N. (2012). Combinatorial approaches for inverse metabolic engineering applications. *Computational and Structural Biotechnology Journal, 3*, e201210021. https://doi.org/10.5936/csbj.201210021

Steinbach, A., Scheidig, A. J., & Klein, C. D. (2008). The unusual binding mode of cnicin to the antibacterial target enzyme MurA revealed by X-ray crystallography. *Journal of Medicinal Chemistry, 51*, 5143–5147. https://doi.org/10.1021/jm800609p

Takahashi, S., Yeo, Y., Greenhagen, B. T., et al. (2007). Metabolic engineering of sesquiterpene metabolism in yeast. *Biotechnology and Bioengineering, 97*, 170–181. https://doi.org/10.1002/bit.21216

Teoh, K. H., Polichuk, D. R., Reed, D. W., & Covello, P. S. (2009). Molecular cloning of an aldehyde dehydrogenase implicated in artemisinin biosynthesis in *Artemisia annua*. *Botany, 87*, 635–642. https://doi.org/10.1139/B09-032

Tsuruta, H., Paddon, C. J., Eng, D., et al. (2009). High-level production of amorpha-4, 11-diene, a precursor of the antimalarial agent artemisinin, in *Escherichia coli*. *PLoS One, 4*, e4489. https://doi.org/10.1371/journal.pone.0004489

Wallaart, T. E., Bouwmeester, H. J., Hille, J., et al. (2001). Amorpha-4,11-diene synthase: Cloning and functional expression of a key enzyme in the biosynthetic pathway of the novel antimalarial drug artemisinin. *Planta, 212*, 460–465.

Wu, S., & Chappell, J. (2008). Metabolic engineering of natural products in plants; tools of the trade and challenges for the future. *Current Opinion in Biotechnology, 19*, 145–152. https://doi.org/10.1016/j.copbio.2008.02.007

Wu, S., Schalk, M., Clark, A., et al. (2006). Redirection of cytosolic or plastidic isoprenoid precursors elevates terpene production in plants. *Nature Biotechnology, 24*, 1441–1447. https://doi.org/10.1038/nbt1251

Yao, J., Weng, Y., Dickey, A., & Wang, K. Y. (2015). Plants as factories for human pharmaceuticals: Applications and challenges. *International Journal of Molecular Sciences, 16*, 28,549–28,565. https://doi.org/10.3390/ijms161226122

Yoon, J. M., Zhao, L., & Shanks, J. V. (2013). Metabolic engineering with plants for a sustainable biobased economy. *Annual Review of Chemical and Biomolecular Engineering, 4*, 211–237. https://doi.org/10.1146/annurev-chembioeng-061312-103320

Zhang, Y., Teoh, K. H., Reed, D. W., et al. (2008). The molecular cloning of artemisinic aldehyde Δ11(13) reductase and its role in glandular trichome-dependent biosynthesis of artemisinin in *Artemisia annua*. *The Journal of Biological Chemistry, 283*, 21,501–21,508. https://doi.org/10.1074/jbc.M803090200

Chapter 19
Ethical Perspectives and Limitations of Metabolic Engineering Technologies in Plants

Swarnavo Chakraborty and Aryadeep Roychoudhury

Abstract The prime goal of metabolic engineering technology in plants is to modulate the production of certain important compounds within naturally procured plant species. This technique takes into consideration both the cellular and metabolic system as a whole and promotes system manipulation, yet allows proper maintenance of the efficiency of the entire system. Basically, metabolic engineering aims to redirect some endogenous enzymatic steps in a metabolic process and promotes newer compound production within an organism or mediate improvement in production or degradation of preexisting chemicals within the system. This review deals with the limitations and ethical concerns raised as a part of this engineering technology. Several ethical issues have been raised when any decision, activity, or scenario associated with the metabolic engineering process has created conflicts with the existing moral principles of the society. Most of these issues have put forward challenging situations, as no so-called precedents and guidelines have been laid before in this context, making the situation really cumbersome to deal with. Therefore, development of superior analytical platforms should be the researcher's top priority in the coming years, in order to provide better solutions to the limitations and ethical perspectives of metabolic engineering.

Keywords Metabolic engineering · Cellular and metabolic system · Ethical concerns · Moral principles · Superior analytical platforms

Introduction

Every living organism constitute a series of biochemical pathways operating within their system. These biochemical reactions are involved in the survival of the particular organism, by the synthesis of important biological macromolecules having specialized role in metabolism, generation of fundamental building blocks

S. Chakraborty · A. Roychoudhury (✉)
Department of Biotechnology, St. Xavier's College (Autonomous), Kolkata, West Bengal, India

T. Aftab, K. R. Hakeem (eds.), *Metabolic Engineering in Plants*,
https://doi.org/10.1007/978-981-16-7262-0_19

associated with structural organization, and production of energy. The outcome of the abovementioned metabolic pathways results in the production of certain value-added chemical compounds, which have widespread applications to mankind. Such chemical compounds include certain bioactive secondary metabolites including anti-malarial drugs like artemisinin, raw materials which are required for the synthesis of important molecules like lactic acid, flavor imparting chemical compounds like terpenes, different biofuels, and their associated chemicals, for instance, butanol, ethanol, etc. (Kulkarni, 2016). Metabolic engineering is referred to as the process of modulating the metabolic pathways of an organisms, in order to generate sufficient quantities of a desired metabolite via genetic manipulations. This field of biotechnology mainly revolutionizes the ways through which commodity chemicals are generated, by virtue of the series of advantages of metabolic engineering over conventional chemical synthesis circuits.

Greater than 50% of the world population inhabiting in developing nations are under the effect of micronutrient deficiency, due to poor levels of essential minerals and vitamins, which cannot be generated by autotrophs in human diet (Mayer et al., 2008). The process of genetically improving these food crops to suffice for the shortages of vitamins and minerals is termed as biofortification. Another important drive is to carry out the development of "functional foods" for developed countries, in order to improve human health by decreasing the risks associated with obese conditions and resulting in improvement of athletic performance. This drive also aims to serve as a solution for a range of health problems in humans, including degenerative disorders and cancer. These functional food products exhibit increased quantities of certain metabolites, which might not be essential vitamins or do not have marked nutritive properties, yet display properties to render betterment of human health (Davies & Espley, 2013). Moreover, metabolic engineering has also been extensively exploited for the modulation of the profile of different secondary metabolites in crops and promote unwanted compound removal, in order to assure safe consumption by the consumers. For instance, reduction in the levels of toxic metabolites like cyanogenic glycosides in the roots of *Manihot esculenta* (cassava) was carried out via metabolic engineering, for making the product safe for human consumption (Sayre et al., 2011).

The most traditional way to utilize these chemical compounds involves the cultivation of the host organism responsible for the production of the particular biochemical, followed by harvesting of the chemical produced. This process generally involves the redirection of a series of enzymatic cascades, in order to generate newer compounds utilizing a particular organism as a model. The basic aim of metabolic engineering process is the improvement of the generation of certain existing compounds or degradation of these compounds as a whole (DellaPenna, 2001). It is usually found that simpler host systems like bacteria and yeast have been widely chosen as the pioneering hosts in metabolic engineering techniques, but recently other eukaryotic systems including fungi, plants, and animals are exploited for such technologies. The most common strategies for metabolic engineering revolves around genetic engineering techniques. The preliminary requirements for this technology involve the knowledge about the following:

1. The biosynthetic pathway associated with the chemical compound to be produced
2. The genes that encode for the associated enzymes in that pathway
3. Regulating the mode of action of these enzymes
4. Transfer of the concerned genes, followed by their expression or suppression in the host organism
5. Alteration of the properties of these enzymes, by mutating the respective genes either in vitro or in vivo
6. Generating an assembly of an array of important genes and expressing them within the host system

With technological advancements since the last few decades, there has been huge progress in plant metabolic engineering process, due to identifications of a series of novel mutations associated with important and informative metabolic pathways, thereby allowing the generation and testing of genetic models (Browse & Sommerville, 1991). For instance, the advancements of molecular genetic approaches helped in the proper dissection of a number of metabolic cascades associated with biosynthesis of different classes of important plant compounds like anthocyanins, amino acids, ascorbic acid, plant waxes, etc. (Radwanski & Last, 1995; Holton & Cornish, 1995; Post-Beittenmiller, 1996; Conklin et al., 1999; Conklin et al., 2000). One of the most striking example indicating the impact of the advancing technology on gene discovery for the purpose of engineering metabolic pathways in plants was the work on the carotenoid biosynthesis pathway and plant lipid biosynthesis mechanisms (Browse & Sommerville, 1991; Cunningham & Gantt, 1998). Metabolic engineering can be considered as a well-recognized biotechnological field, displaying immense commercial potential, and also serve as a biological tool for the economic and large-scale generation of important chemical compounds. Therefore, metabolic engineering technologies can pose immense benefits to mankind as a whole (Kulkarni, 2016).

Limitations and Ethical Perspectives of Metabolic Engineering

It is important to note that the technology of metabolic engineering is a rather young science and is a potentially growing field. The knowledge about plant metabolic pathways and the associated substrate-product relationships will not be sufficient to carry out the entire process; instead, detailed knowledge about the basic molecular biological techniques in this regard is mandatory. The role of molecular biology technologies including cloning, transformation in plants, analysis of promoter, targeting of proteins, and other methods associated with plant genetics is of immense importance in the engineering of metabolic cascades in plants (DellaPenna, 2001). It is evident that this engineering process exploits the power of living organisms to produce certain important biochemicals, by the copying of the associated biosynthesis pathways within the living system to allow their semi-natural production. This semi-natural mode of production can serve as an extraordinary alternative to the chemical synthesis or the isolation of such chemicals from natural resources. However, there are a series of disadvantages associated with such production, including the generation of certain detrimental by-products and even racemic mixtures, which constitute equal proportions of isomers of a particular optically active compound, which are mirror images of each other, commonly referred to as enantiomers (Kulkarni, 2016).

Since the last two decades, there has been a marked progress in our ability to decipher genes involved in many key metabolic pathways and their subsequent manipulation for the sake of improved or altered gene expression in transgenic organisms. However, the use of these sophisticated tools for engineering plant metabolic cascades has displayed a rather restricted success. This is because maximum attempts at engineering pathways involve the positive or negative modification of the expression of a single gene and not all or most of the genes associated with that pathway. Attempts of increasing flux of a particular compound through a metabolic pathway fail to provide proper predictions of the experimental outcomes. On the other hand, to obtain desired experimental outcomes, it is better to modify or convert an existing compound into the desired form. For instance, modifications of primary and intermediate steps of metabolic pathways result in limited success, as compared to manipulations to storage products in pathways or alterations in the steps of secondary metabolic pathways that provide more flexible outcomes and higher success rates in plants (Kinney, 1998; Stitt & Sonnewald, 1995).

With the advancement of technology, technological systems like power grids, Internet facility, digital computer systems, etc. have encompassed greater parts of our lives, and it has been our common practice that we take these systems for granted. The modular design strategy is responsible for the success of these systems and modulates the users to deal with varying levels of conflicting and complex situations. The "top-down" approach is usually employed for designing large systems, where each problem is subdivided into a set of sub-problems based on hierarchy. This process makes it easier to solve, implement, and design smaller sub-problems, with the information from the existing and previously characterized modules. After its successful verification and testing, a system can be guaranteed to be performing according to the specifications of its designing. However, in case of a non- modular or non-hierarchical design, the verification process is no longer tractable (Anderson et al., 2012). However, the ways in which modules are interconnected or the type of constituents of a module in biological system are still not clearly understood. Natural genetic circuits constitute modules which can be clearly explained by "network motifs," and these motifs can range from localized bimolecular interacting partners to entire pathways, for example, linear circuits or feed-forward loops (both coherent and non-coherent in nature). Different types of algorithms have been developed for the identification of these motifs in protein-protein interactions and genetic networks in various organisms, but the proper mode of function of the entire network is still not conceptually well defined (Milo et al., 2002). Interconnection between modules is a commonly encountered fact; however, this property of modularity is still a debatable topic (Mitchell, 2006; Purnick &

Weiss, 2009). It is still unclear whether modularity is an abstraction, which has been implemented by engineers for the simplification of the designs of complex systems or it is a natural property. In addition, it is mandatory to ensure that a particular design is robust to perturbations and noise. In order to achieve such robustness, apart for the use of feedback loops (Zhou et al., 1996), huge cost is incurred. Also in order to generate a robust system, the system often becomes quite fragile to perturbations, leading to catastrophic consequences. Therefore, it is necessary to maintain a proper balance between fragility and robustness in a particular system.

Another fundamental challenge faced by this field is the uncertain environment in which the host organism will inhabit. Such uncertainty can be imposed due to a range of factors, including huge levels of noise, cross-talk among different modules which are yet to be characterized, competition due to huge population size, evolution, and adaptability of that organism with the variable environment. These factors are not accounted for during the engineering process under laboratory conditions. The information about exact input and output module properties acts as the main obstacle in the way of designing robust and predictable biological circuits, in isolation, as well as when interconnected. Moreover, due to limited knowledge regarding the environmental variables like pressure, temperature, sources of energy, etc. and its effects on biological behavior, the process of metabolic engineering in certain systems is rather poorly characterized (Anderson et al., 2012).

Although a certain system in isolation might be fully characterized, there are lots of challenges that will govern the engineers' ability of predicting the behavior of interconnected modules in an ensemble. The most common obstacle is the feedback from downstream elements and events, which results in propagation of information in an opposite direction as anticipated previously. This obstacle is commonly referred to as "retroactivity" (Del Vecchio et al., 2008). For instance, an example of retroactivity in cell biology is the imposition of feedback upon an upstream promoter, due to certain downstream promoter activation (Ventura et al., 2010). Interestingly, long cascades of dephosphorylation and phosphorylation in biological systems often serve as an important attenuator of retroactivity (Ossareh et al., 2011). Moreover, different connected modules in an interconnected system tend to show variable sensitivity to inter- and intracellular environment (Cases & de Lorenzo, 2005), and such uncertain environment can even push the engineered organism to the limit of its robustness. It is often considered that engineered organisms are "evolutionary losers," as it is commonly found that these engineered organisms are competitively inferior, unlike their natural counterparts, and the engineered ones ultimately die off. This is due to the fact that the predictability of the potential of adaptation and evolution of an engineered organism is rather limited and not well understood. Due to low copy numbers of DNA and mitochondrial RNA, most biological systems experience higher loads of intrinsic stochasticity and extrinsic noise (Paulsson et al., 2000).

It is really a challenging task to engineer evolution within a biological system. This is because of the fact an engineered system might fail to exist in its future generations or might even undergo mutation to develop into something entirely different, thereby raising practical and ethical issues. The existing theories associated with modeling or robustness are not in accordance with the theories of evolution and mutation as a whole, yet lots of theoretical and experimental endeavors have been put by the scientific community (Ellis et al., 2009; Soyer & Goldstein, 2011). Therefore, it has been admitted that adaptation, evolution, and mutation lead to discrete uncertainty, having immense control over the long-term dynamics associated with an engineered system. Thus, it is mandatory for the future designs to keep evolution under consideration, to ward off the issues raised from a biosafety point of view, while at the same time to ensure that engineered organisms function the way it has been predicted previously (Anderson et al., 2012). So, in order to project the applications of metabolic engineering from the boundaries of the laboratories to the outside world, the entire process needs a much better solution.

Although we have been able to develop engineered systems, it still cannot be guaranteed that we can have full control over these systems, thereby indicating the need for consideration of the ethical implications of the engineering process. The most common ethical issue raised is about "playing God," i.e., this ethical concern focuses on the potential of biologists to create a new life via these engineering techniques (Coady, 2009; Douglas & Savulescu, 2010). Another important concern is the power of these engineering procedures to remove the clear distinction that must exist between an organism and a machine. This leads to the hampering of the moral status with respect to the engineered biological system, in relation to the non-living and the living (Deplazes & Huppenbauer, 2009). Moreover, one can never speculate with full accuracy how that engineered organism will behave in the future, upon their release in the natural environment, thereby raising another ethical concern. It is often believed that engineered organisms can have significant consequences upon the entire ecosystem and lead to irreversible environmental effects. Therefore, it is a challenging task to carry out a perfect prediction process and also to administer an absolute control over the entire system as a whole. So, the chances of potentially risking the environment upon release of such engineered products still remain as a lurking question.

In order to provide an answer for every ethical question raised, it is a paramount task to prevent misunderstandings regarding research by maintaining a dialogue between the public, industries, and scientists. People often consider food products with the slightest trace of recombinant DNA as highly obnoxious. However, at the same time, the common masses in the United Kingdom are ready to take drugs associated with recombinant DNA like interferons and insulin (Church, 2005). Although not scientifically correct, still the safety of genetically modified (GM) products is a very common debatable topic. Therefore, the concerns of the public about the product safety and the potential to self-regulate in a responsible manner need to be addressed, in order to prevent the huge impact imposed on industry and medicine by this biological field (Lentzos, 2009). Basically, it is a matter of faith of the public on the scientific community, i.e., for the acceptance of an engineered product by the public, a dialogue between the researchers and the public must be maintained from the inception of the engineering process. Moreover, industries should also play an important role in this regard, by framing a code of practice and setting up biosecurity and biosafety standards for the particular product, to ensure safe and proper use of the engineered biological system by the public.

Conclusion and Future Perspectives

Engineering of important plant metabolic circuits forms major focus of many plant biotechnological platforms since the last few years, with an aim to improve the health of the entire ecosystem, with humans being the central theme of focus. Genetic modification and breeding techniques like SDM (Site directed mutagenesis) and MAS (Marker assisted selection) have been exploited for the generation of targeted changes in biosynthesis pathways in plants, in order to improve the various parameters of plants associated with the betterment of the environment and society. With the improvement of tools in bioinformatics essential for the mining, analyzing and sorting of biological data, enhancement in the sophistication and sensitivity of various tools for biological investigation, and reduction in the costs for important technologies, for instance, de novo gene synthesis, the process of engineering biological systems has turned out to be rather simpler and easier. However, the applications of the abovementioned tools need to be propagated and well explained among the entire world for their better use and acceptance. This improvement of analytical and experimental procedures must promote the application of multivariate optimization and modularization by different research groups, in order to allow overproduction of a range of value-added compounds. It is evident that these demonstrations will definitely add to the information pool and will also allow metabolic engineers and biologists to carry out an objective-oriented designing of modules and promote the intelligent application of different technologies in various biological processes like scaffolding of enzymes and chromosomal integration.

It is not wrong to expect that the coming years will be filled with huge excitement, yet challenging for the metabolic engineers and biologists. It can be expected that a series of newer technologies, modulating and controlling cellular phenomenon, will crop up and ultimately add to the pool of different biological engineered products. However, these analytical techniques will always render themselves as a hurdle in the process of metabolic engineering, i.e., appear to be rate limiting. Moreover, the poor development of high-throughput screening techniques prevents the application of combinatorial strategies for the betterment of metabolic engineering techniques. Thus, development of newer, broader, better, and faster analytical platforms should become the top priority of the scientists in this new decade.

Acknowledgments Financial assistance from Science and Engineering Research Board, Government of India through the grant [EMR/2016/004799] and Department of Higher Education, Science and Technology and Biotechnology, Government of West Bengal, through the grant [264(Sanc.)/ST/P/S&T/1G-80/2017] to Dr. Aryadeep Roychoudhury is gratefully acknowledged.

References

Anderson, J., Strelkowa, N., Stan, G. B., Douglas, T., Savulescu, J., Barahona, M., & Papachristodoulou, A. (2012). Engineering and ethical perspectives in synthetic biology. *European Molecular Biology Organization Reports, 13*, 584–590.

Browse, J., & Sommerville, C. (1991). Glycerolipid synthesis: Biochemistry & regulation. *Annual Review of Physiology and Plant Molecular Biology, 42*, 467–506.

Cases, I., & de Lorenzo, V. (2005). Genetically modified organisms for the environment: stories of success and failure and what we have learned from them. *International Microbiology, 8*, 213–222.

Church, G. (2005). Let us go forth and safely multiply. *Nature, 438*, 423.

Coady, C. A. J. (2009). Playing God. In J. Savulescu & N. Bostrom (Eds.), *Human enhancement* (pp. 155–180). Oxford University Press.

Conklin, P. L., Norris, S. R., Wheeler, G. L., Williams, E. H., Smirnoff, N., & Last, R. L. (1999). Genetic evidence for the role of GDP-mannose in plant ascorbic acid (vitamin C) biosynthesis. *Proceedings of the National Academy of Sciences United States of America, 96*, 4198–4203.

Conklin, P. L., Saracco, S. A., Norris, S. R., & Last, R. L. (2000). Identification of ascorbic acid-deficient Arabidopsis thaliana mutants. *Genetics, 154*, 847–856.

Cunningham, F. X., & Gantt, E. (1998). Genes and enzymes of carotenoid biosynthesis in plants. *Annual Review of Physiology and Plant Molecular Biology, 49*, 557–583.

Davies, K. M., & Espley, R. V. (2013). Opportunities and challenges for metabolic engineering of secondary metabolite pathways for improved human health characters in fruit and vegetable crops. *New Zealand Journal of Crop and Horticultural Science, 41*, 154–177.

Del Vecchio, D., Ninfa, A. J., & Sontag, E. D. (2008). Modular cell biology: retroactivity and insulation. *Molecular Systems Biology, 4*, 161.

DellaPenna, D. (2001). Plant metabolic engineering. *Journal of Plant Physiology, 125*, 160–163.

Deplazes, A., & Huppenbauer, M. (2009). Synthetic organisms and living machines: Positioning the products of synthetic biology at the borderline between living and non-living matter. *Systems and Synthetic Biology, 2*, 55–63.

Douglas, T., & Savulescu, J. (2010). Synthetic biology and the ethics of knowledge. *Journal of Medical Ethics, 36*, 687–693.

Ellis, T., Wang, X., & Collins, J. J. (2009). Diversity-based, model-guided construction of synthetic gene networks with predicted functions. *Nature Biotechnology, 27*, 465–471.

Holton, T. A., & Cornish, E. C. (1995). Genetics and biochemistry of anthocyanin biosynthesis. *Plant Cell, 7*, 1071–1083.

Kinney, A. J. (1998). Manipulating flux through plant metabolic pathways. *Current Opinion in Plant Biology, 1*, 173–178.

Kulkarni, R. (2016). Metabolic engineering: Biological art of producing useful chemicals. *Resonance*, 233–237.

Lentzos, F. (2009). Synthetic biology in the social context: The UK debate to date. *BioSocieties, 10*, 305–315.

Mayer, J. E., Pfeiffer, W. H., & Beyer, P. (2008). Biofortified crops to alleviate micronutrient malnutrition. *Current Opinion in Plant Biology, 11*, 166–170.

Milo, R., Shen-Orr, S., Itzkovitz, S., Kashtan, N., Chklovskii, D., & Alon, U. (2002). Network motifs simple building blocks of complex networks. *Science, 298*, 824–827.

Mitchell, S. D. (2006). Modularity: More than a buzzword? Essay Review. *Biological Theory, 1*, 98–101.

Ossareh, H. R., Ventura, A. C., Merajver, S. D., & Del Vecchio, D. (2011). Long signaling cascades tend to attenuate retroactivity. *Biophysics Journal, 100*, 1617–1626.

Paulsson, J., Berg, O. G., & Ehrenberg, M. (2000). Stochastic focusing: Fluctuation-enhanced sensitivity of intracellular regulation. *Proceedings of the National Academy of Sciences United States of America, 97*, 7148–7153.

Post-Beittenmiller, D. (1996). Biochemistry and molecular biology of wax production in plants. *Annual Review of Physiology and Plant Molecular Biology, 47*, 405–430.

Purnick, P. E., & Weiss, R. (2009). The second wave of synthetic biology: From modules to systems. *Nature Reviews Molecular Cell Biology, 10*, 410–422.

Radwanski, E. R., & Last, R. L. (1995). Tryptophan biosynthesis and metabolism: biochemical and molecular genetics. *Plant Cell, 7*, 921–934.

Sayre, R., Beeching, J. R., Cahoon, E. B., Egesi, C., Fauquet, C., Fellman, J., et al. (2011). The BioCassava Plus program: Biofortification of cassava for Sub- Saharan Africa. *Annual Review of Plant Biology, 62*, 251–272.

Soyer, O. S., & Goldstein, R. A. (2011). Evolution of response dynamics underlying bacterial chemotaxis. *BMC Evolutionary Biology, 11*, 240.

Stitt, M., & Sonnewald, U. (1995). Regulation of metabolism in transgenic plants. *Annual Review of Physiology and Plant Molecular Biology, 46*, 341–368.

Ventura, A. C., Jiang, P., Van Wassenhove, L., Del Vecchio, M. S. D., & Ninfa, A. J. (2010). Signaling properties of a covalent modification cycle are altered by a downstream target. *Proceedings of the National Academy of Sciences United States of America, 107*, 10,032–10,037.

Zhou, K., Doyle, J. C., & Glover, K. (1996). *Robust and optimal control*. Prentice Hall.

The manufacturer's authorised representative in the EU is Springer Nature Customer Service Centre GmbH, Europaplatz 3, 69115 Heidelberg, Germany. If you have any concerns regarding our products, please contact ProductSafety@springernature.com

Printed and bound by CPI Group (UK) Ltd, Croydon, CR0 4YY
30/09/2026
02225239-0017